P9-APS-798

C^*-ALGEBRAS AND THEIR AUTOMORPHISM GROUPS

L.M.S. MONOGRAPHS

Editors: D. EDWARDS *and* H. HALBERSTAM

1. Surgery on Compact Manifolds *by* C. T. C. Wall, F.R.S.
2. Free Rings and Their Relations *by* P. M. Cohn
3. Abelian Categories with Applications to Rings and Modules *by* N. Popescu
4. Sieve Methods *by* H. Halberstam and H.-E. Richert
5. Maximal Orders *by* I. Reiner
6. On Numbers and Games *by* J. H. Conway
7. An Introduction to Semigroup Theory *by* J. M. Howie
8. Matroid Theory by D. J. A. Welsh
9. Subharmonic Functions, Volume 1 *by* W. K. Hayman, F.R.S. and P. B. Kennedy
10. Topos Theory *by* P. T. Johnstone
11. Extremal Graph Theory *by* Béla Bollobás
12. Spectral Theory of Linear Operators *by* H. R. Dowson
13. Rational Quadratic Forms *by* J. W. S. Cassels, F.R.S.
14. C^*-Algebras and Their Automorphism Groups *by* Gert K. Pedersen

Published for the London Mathematical Society
by Academic Press Inc. (London) Ltd.

C^*-ALGEBRAS AND THEIR AUTOMORPHISM GROUPS

GERT K. PEDERSEN

Mathematics Institute
University of Copenhagen
Copenhagen, Denmark

1979

ACADEMIC PRESS

LONDON NEW YORK SAN FRANCISCO

A Subsidiary of Harcourt Brace Jovanovich, Publishers

ACADEMIC PRESS INC. (LONDON) LTD.
24/28 Oval Road
London NW1

United States Edition published by
ACADEMIC PRESS INC.
111 Fifth Avenue
New York, New York 10003

British Library Cataloguing in Publication Data

Pedersen, Gert K
C*-algebras and their automorphism groups.—(London Mathematical Society. Monographs; 14 ISSN 0076–0560).
1. C*-algebras
I. Title II. Series
512'.55 QA326
ISBN 0-12-549450-5

LCCCN 78-18028

PRINTED IN GREAT BRITAIN BY
PAGE BROS (NORWICH) LTD, MILE CROSS LANE, NORWICH

Preface

The theory of C^*-algebras is the study of operators on Hilbert space with algebraic methods. The motivating example is the spectral theorem for a normal operator (which, in effect, is nothing but Gelfand transformation applied to the algebra generated by the operator). The applications of the theory range from group representations to model quantum field theory and quantum statistical mechanics.

Already the C^*-algebra theory has grown to a size where any comprehensive treatment would result in a series of volumes more suited as a source of references than as a textbook. The material presented here has been limited by the author's knowledge and prejudice to form a somewhat manageable version. Thus the aspects of the theory concerning partially ordered vector spaces are treated in great detail. Also, since C^*-algebra theory has benefited tremendously from impulses from mathematical physics, it seemed proper to give an account which would please the C^*-physicists. Therefore the problems connected with groups of automorphisms have received special attention in this treatise. In the converse direction, the theory of von Neumann algebras, often so dominantly exposed, has here been reduced to its proper place as ancilla C^*-algebrae.

At the end of each section a few remarks are inserted, with references to the bibliography. The intention is to give the reader a rough idea of the development of the subject. Such personal comments are bound to contain errors, and the author humbly asks forgiveness from the mathematicians who have undeservedly not been mentioned.

Many people were important for the completion of this book: Richard Kadison whose work has been a constant source of inspiration for me; Daniel Kastler who provided shelter and a two months raincurtain when the work was begun in 1974; colleagues who shouldered my teaching load while I was writing; and students at the University of Copenhagen who were exposed to the first wildly incorrect drafts. It is a pleasure to record my thanks to all of them.

Copenhagen
August 1978

Gert Kjærgård Pedersen

發行人：曾　蘭　英
地　址：新竹市東美路45之24
電　話：(035)234636
發行者：凡　異　出　版　社
局版台業字1596號
郵政劃撥108256號
(曾蘭英帳戶)

Contents

Chapter 1

Abstract C^*-algebras

1.1 Spectral theory

1.1.1. A *C^*-algebra* is a complex Banach algebra A with an involution, $*$, satisfying $\|x^*x\| = \|x\|^2$ for all x in A. Since $\|x^*x\| \leqslant \|x^*\| \|x\|$ we have $\|x\| \leqslant \|x^*\|$ for each x in A, whence $\|x\| = \|x^*\|$, so that the involution is isometric. An element x in A is *normal* if it commutes with its *adjoint* x^*; and it is *self-adjoint* if $x = x^*$. The self-adjoint part of a subset B of A is denoted by B_{sa}. For each x in A the elements $\frac{1}{2}(x + x^*)$ and $-\frac{1}{2}i(x - x^*)$ (the real resp. imaginary part of x) belong to A_{sa}. It follows that A_{sa} is a closed real subspace of A and that each element x in A has a unique decomposition $x = y + iz$ with y and z in A_{sa}.

1.1.2. In general a C^*-algebra need not have a unit. If however, the C^*-algebra A has a unit (denoted by 1_A, or just 1 when no confusion may arise) and $A \neq 0$, then $1_A^* = 1_A$ and $\|1_A\| = 1$.

If $1 \in A$ we say that an element u in A is *unitary* if $u^*u = uu^* = 1$. Note that each unitary is normal and has norm 1.

1.1.3. PROPOSITION. *For each C^*-algebra A there is a C^*-algebra $\tilde{A}$ with unit containing A as a closed ideal. If A has no unit then $\tilde{A}/A = \mathbf{C}$.*

Proof. Let π denote the left regular representation of A as operators on itself, i.e. $\pi(x)y = xy$ for all x and y in A. It is clear that π is a homomorphism and that $\|\pi(x)\| \leqslant \|x\|$. But since

$$\|x\|^2 = \|xx^*\| = \|\pi(x)x^*\| \leqslant \|\pi(x)\| \|x^*\|$$

we see that π is an isometry. Let 1 denote the identity operator on A and let $\tilde{A}$ be the algebra of operators on A of the form $\pi(x) + \alpha 1$ with x in A and α in $\mathbf{C}$. Since $\pi(A)$ is complete and $\tilde{A}/\pi(A) = \mathbf{C}$, $\tilde{A}$ is also complete. With the involution defined by $(\pi(x) + \alpha 1)^* = \pi(x^*) + \bar{\alpha}1$, $\tilde{A}$ becomes a C^*-algebra since for each $\varepsilon > 0$ there is a y in A with $\|y\| = 1$ such that

$$\|\pi(x) + \alpha 1\|^2 \leqslant \varepsilon + \|(x + \alpha)y\|^2$$
$$= \varepsilon + \|y^*(x^* + \bar{\alpha})(x + \alpha)y\| \leqslant \varepsilon + \|(x^* + \bar{\alpha})(x + \alpha)y\|$$
$$\leqslant \varepsilon + \|(\pi(x) + \alpha 1)^*(\pi(x) + \alpha 1)\|.$$

1.1.4. For each x in a C^*-algebra A we define the *spectrum* of x in A (written $\mathrm{Sp}_A(x)$) as the set of complex numbers λ such that $\lambda 1 - x$ is not invertible in $\tilde{A}$. Note that $0 \in \mathrm{Sp}_A(x)$ whenever $A \neq \tilde{A}$. By straightforward computations it follows that if $\lambda \neq 0$, then $\lambda \notin \mathrm{Sp}_A(x)$ if and only if there is a y in A such that $xy = yx = \lambda^{-1}x + \lambda y$ (corresponding to the fact that $\lambda^{-1} - y = (\lambda 1 - x)^{-1}$ in $\tilde{A}$).

If $x \in A_{\mathrm{sa}}$ and $\nu(x)$ is the spectral radius of x then by repeated use of the equality $\|x^2\| = \|x\|^2$ we obtain

$$\nu(x) = \mathrm{Lim}\,\|x^{2^n}\|^{2^{-n}} = \|x\|.$$

If x is just normal, then from the preceding,

$$\nu(x)^2 \leqslant \|x\|^2 = \|x^*x\| = \mathrm{Lim}\,\|(x^*x)^n\|^{n^{-1}}$$
$$\leqslant \mathrm{Lim}(\|(x^*)^n\|\,\|x^n\|)^{n^{-1}} = \nu(x)^2,$$

whence again $\nu(x) = \|x\|$.

1.1.5. LEMMA. *If $x \in A_{\mathrm{sa}}$ then $\mathrm{Sp}(x) \subset \boldsymbol{R}$. If $1 \in A$ and u is unitary, $\mathrm{Sp}(u)$ is contained in the unit circle.*

Proof. If $\lambda \in \mathrm{Sp}(u)$ then $\lambda^{-1} \in \mathrm{Sp}(u^{-1})$. Since $u^{-1} = u^*$ we have $|\lambda| \leqslant 1$ and $|\lambda^{-1}| \leqslant 1$, whence $|\lambda| = 1$ which proves the second assertion in the lemma.

Take now x in A_{sa}. The power series $\Sigma(n!)^{-1}(\mathrm{i}x)^n$ converges in $\tilde{A}$ to an element $\exp(\mathrm{i}x)$ which is unitary (in $\tilde{A}$) since

$$\exp(\mathrm{i}x)^* = \exp(-\mathrm{i}x) = \exp(\mathrm{i}x)^{-1}.$$

If $\lambda \in \mathrm{Sp}(x)$ then $\exp(\mathrm{i}\lambda) \in \mathrm{Sp}(\exp(\mathrm{i}x))$ by computation, whence $|\exp(\mathrm{i}\lambda)| = 1$ by the first part of the proof. It follows that $\mathrm{Sp}(x) \subset \boldsymbol{R}$ as desired.

1.1.6. Let A be a commutative Banach algebra. The *spectrum* $\hat{A}$ of A is the set of non-zero homomorphisms of A onto $\boldsymbol{C}$. Each element in $\hat{A}$ belongs to the unit ball of the dual A^* of A, and since $\hat{A} \cup \{0\}$ is the weak* closed subset of A^* consisting of functionals t such that $t(xy) = t(x)t(y)$ for all x, y in A, we see that $\hat{A}$ is a locally compact Hausdorff space in the weak* topology. The *Gelfand transform* on A is the homomorphism $x \to \hat{x}$ of A into $C_0(\hat{A})$ given by $\hat{x}(t) = t(x)$ for all x in A and t in $\hat{A}$.

1.1.7. THEOREM. *If A is a commutative C^*-algebra then the Gelfand transform is a $*$-preserving isometry of A onto $C_0(\hat{A})$.*

Proof. If $t \in \hat{A}$ and $x \in A$ then $\ker t$ is a maximal ideal of A, whence $t(x) \in \mathrm{Sp}(x)$ (and conversely, if $\lambda \in \mathrm{Sp}(x)\backslash\{0\}$ then $\lambda = t(x)$ for some t in $\hat{A}$). If therefore $x = x^*$ then $t(x) \in \boldsymbol{R}$ by 1.1.5. It follows that $t(x^*) = \overline{t(x)}$ for each x in A which shows that the map $x \to \hat{x}$ is $*$-preserving (using complex conjugation of functions as involution in $C_0(\hat{A})$). Moreover, $\|\hat{x}\|$ is the spectral radius of x, whence $\|\hat{x}\| = \|x\|$ by 1.1.4 as each x in A is normal. Thus $x \to \hat{x}$ is a $*$-preserving isometry of A into $C_0(\hat{A})$ and since the set of functions $\{\hat{x} \mid x \in A\}$ separates points in $\hat{A}$ and does not all vanish at any point we conclude from the Stone–Weierstrass theorem that the image of A is all of $C_0(\hat{A})$.

1.1.8. PROPOSITION. *Let x be a normal element of a C^*-algebra A, and let B denote the smallest C^*-subalgebra of A containing x. Then $B = C_0(\mathrm{Sp}_A(x)\backslash\{0\})$ and $\mathrm{Sp}_A(x)\backslash\{0\} = \mathrm{Sp}_B(x)\backslash\{0\}$.*

Proof. Since B is a singly generated commutative C^*-algebra it follows that $\hat{B} = \mathrm{Sp}_B(x)\backslash\{0\}$, whence $B = C_0(\mathrm{Sp}_B(x)\backslash\{0\})$ by 1.1.7 (suppressing the isomorphism). If therefore $\lambda \in \mathrm{Sp}_B(x)\backslash\{0\}$ there is for each $\varepsilon > 0$ an element b in B with $\|b\| = 1$ such that $\|\lambda b - xb\| < \varepsilon$. This shows that $\lambda 1 - x$ is not invertible in $\tilde{A}$, thus $\lambda \in \mathrm{Sp}_A(x)$. It is immediate from 1.1.4 that if $\lambda \notin \mathrm{Sp}_B(x)$ and $\lambda \neq 0$ then $\lambda \notin \mathrm{Sp}_A(x)$, and the proposition follows.

1.1.9. If x is a normal element of A and $f \in C_0(\mathrm{Sp}(x)\backslash\{0\})$ we denote by $f(x)$ the element of A corresponding to f via the embedding of $C_0(\mathrm{Sp}(x)\backslash\{0\})$ into A given by 1.1.8.

If f is a continuous function on $\boldsymbol{C}$ vanishing at 0 then it can be approximated uniformly by polynomials on any given compact set. It follows that if $\{x_n\}$ is a sequence of normal elements converging to x, then $\{f(x_n)\}$ converges to $f(x)$.

1.1.10. Let x be a self-adjoint element of A. By 1.1.5 $\mathrm{Sp}(x) \subset \boldsymbol{R}$. We write x_+ for $f_1(x)$, where $f_1(t) = t \vee 0$; x_- for $f_2(x)$, where $f_2(t) = -(t \wedge 0)$ and $|x|$ for $f_3(x)$, where $f_3(t) = |t|$. Then $x = x_+ - x_-$, $|x| = x_+ + x_-$ and $x_+x_- = 0$. We say that x_+ and x_- are the *positive* and *negative part* of x and that $|x|$ is the *absolute value* of x. If $\mathrm{Sp}(x) \subset \boldsymbol{R}_+$ we write $x^{1/2}$ for $f_4(t)$, where $f_4(t) = t^{1/2}$. It will be shown in 1.3.3 that $\mathrm{Sp}(x^*x) \subset \boldsymbol{R}_+$ for any x in A. We will then define $|x| = (x^*x)^{1/2}$ to be the absolute value of x.

1.1.11. If $x \in A_{sa}$ and $\|x\| \leqslant 1$ then $u = x + i(1 - x^2)^{1/2}$ is a normal element with $u^* = x - i(1 - x^2)^{1/2}$. Since $u^*u = 1$, u is unitary. But $x = \frac{1}{2}(u + u^*)$ which shows that each element in A can be written as a linear combination of (four) unitary elements.

An elementary calculation shows that if x and y are invertible in A and $x^*x = y^*y$ then the element xy^{-1} is unitary. This is used in the proof of 1.1.12.

1.1.12. PROPOSITION. *If A is a C^*-algebra with unit then the unit ball in A is the closed convex hull of the unitary elements in A.*

Proof. If $x \in A$ and $\|x\| < 1$ the spectrum of $1 - xx^*$ is strictly positive so that the element

$$f(x, \lambda) = (1 - xx^*)^{-1/2}(1 + \lambda x)$$

exists in A and is invertible for each λ in $\mathbf{C}$ with $|\lambda| = 1$. Using the power series expansion $(1 - xx^*)^{-1} = \Sigma(xx^*)^n$ we see that $x^*(1 - xx^*)^{-1} = (1 - x^*x)^{-1}x^*$, whence

$$\begin{aligned} f(x, \lambda)^* f(x, \lambda) + 1 &= (1 + \bar{\lambda}x^*)(1 - xx^*)^{-1}(1 + \lambda x) + 1 \\ &= (1 - xx^*)^{-1} + (1 - x^*x)^{-1}\bar{\lambda}x^* + (1 - xx^*)^{-1}\lambda x \\ &\quad + (1 - x^*x)^{-1}. \end{aligned}$$

This expression is unchanged when exchanging x by x^* and λ by $\bar{\lambda}$ and we conclude that

$$f(x, \lambda)^* f(x, \lambda) = f(x^*, \bar{\lambda})^* f(x^*, \bar{\lambda}).$$

It follows that for each λ in $\mathbf{C}$ with $|\lambda| = 1$ the element $u_\lambda = f(x, \lambda) f(x^*, \bar{\lambda})^{-1}$ is unitary (cf. 1.1.11).

The function

$$u(\lambda) = (1 - xx^*)^{-1/2}(\lambda + x)(1 + \lambda x^*)^{-1}(1 - x^*x)^{1/2}$$

is holomorphic in a neighbourhood of the closed unit disc and $u(\lambda) = \lambda u_{\bar{\lambda}}$ when $|\lambda| = 1$. Moreover,

$$u(0) = (1 - xx^*)^{-1/2}x(1 - x^*x)^{1/2} = (1 - xx^*)^{-1/2}(1 - xx^*)^{1/2}x = x.$$

It follows from Cauchy's integral formula (A4, Appendix) that

$$x = (2\pi)^{-1} \int_0^{2\pi} u(e^{it})\, dt.$$

Since the measure $(2\pi)^{-1} dt$ on $[0, 2\pi]$ can be approximated by convex combinations of point measures, and since the elements $u(e^{it})$ are unitary in A, the open unit ball of A is contained in the closed convex hull of the unitary elements in A, from which the proposition follows.

1.1.13. If A is a C^*-algebra with unit then 1 is an extreme point in the unit ball of A. For if $1 = \frac{1}{2}(x + y)$ with x and y in A_{sa} then x commutes with y and by spectral theory $x = y = 1$. In the general case we have $1 = \frac{1}{2}[\frac{1}{2}(x + x^*) + \frac{1}{2}(y + y^*)]$ whence $\frac{1}{2}(x + x^*) = \frac{1}{2}(y + y^*) = 1$. Thus x and y are normal elements and again $x = y = 1$ from spectral theory.

Since multiplication by a unitary element is a linear isometry of A it follows from the above that every unitary in A is an extreme point in the unit ball of A. From 1.1.12 we see that the unit ball of a C^*-algebra with unit is the closed convex hull of its extreme points. This is remarkable since the unit ball is not in general compact either in the norm topology or in any other vector space topology on A.

1.1.14. Notes and remarks. The axioms for an abstract C^*-algebra were formulated in 1943 by Gelfand and Naimark [92]. With the aid of an extra axiom (namely that the spectrum of x^*x is positive for every x) they showed that any C^*-algebra is isomorphic to an algebra of operators on a Hilbert space. The theory of operator algebras had been developed during the thirties by Murray and von Neumann in a series of papers [175, 169, 170, 176, 171], dealing mainly with the weakly closed algebras (= von Neumann algebras). The name C^*-algebra was coined by Segal in [235] where the foundations for representation theory were laid. Presumably the C is meant to indicate that a C^*-algebra is a non-commutative analogue of $C(T)$, whereas the * recalls the importance of the involution.

The result in 1.1.11 is an early discovery; that in 1.1.12 is more recent [189].

1.2. Examples

1.2.1. As mentioned in 1.1.5 there is a bijective correspondence between commutative C^*-algebras and locally compact Hausdorff spaces. Non-commutative examples of C^*-algebras arise by considering the set $B(H)$ of bounded linear operators on a (complex) Hilbert space H. With the operator sum, product and norm and with the adjoint operation as involution, $B(H)$ becomes a C^*-algebra which is non-commutative when $\dim(H) > 1$. We shall study $B(H)$ and its subalgebras in some detail in the next chapter. When $\dim(H) = n < \infty$ we may identify $B(H)$ with the algebra M_n of (complex) $n \times n$ matrices.

1.2.2. Given two C^*-algebras A and B there are in general several ways of completing the algebraic tensor product $A \otimes B$ (which is an algebra with involution in a natural way) to obtain a C^*-algebra. We shall content ourselves here with the case where one of the factors is commutative so that this unpleasantness does not occur.

Let T be a locally compact Hausdorff space and A a C^*-algebra. By $C^b(T, A)$ we understand the set of bounded continuous functions x from T to A and by $C_0(T, A)$ the subset of functions x vanishing at infinity, i.e. the function $t \to \|x(t)\|$ belongs to $C_0(T)$. With pointwise sum, product and involution, and with $\|x\| = \sup \|x(t)\|$ for each x in $C_0(T, A)$ we obtain a C^*-algebra such that $C_0(T) \otimes A$ form a dense subset.

1.2.3. The simplest example of a non-commutative, infinite-dimensional C^*-algebra is probably $C_0(\mathbf{N}, \mathbf{M}_2)$. If one prefers an algebra with unit then $C(\mathbf{N} \cup \{\infty\}, \mathbf{M}_2)$ is a good example. This last algebra also has some C^*-subalgebras which are useful when trying to find counter-examples. For instance the set of sequences in $C(\mathbf{N} \cup \{\infty\}, \mathbf{M}_2)$ which tend to a diagonal matrix; the set of sequences which tend to a multiple of the identity matrix (nothing but $C_0(\mathbf{N}, \mathbf{M}_2)^{\sim}$ as defined in 1.1.3); or the set of sequences x such that $(x(\infty))_{ij} = 0$ unless $i = j = 1$.

1.2.4. Let $\{A_i \mid i \in I\}$ be a family of C^*-algebras. The set of functions x from I into $\cup A_i$ such that $x_i \in A_i$ for each i in I and such that the function $i \to \|x_i\|$ is bounded, is a C^*-algebra with pointwise sum, product and involution. We shall denote this C^*-algebra by ΠA_i and call it the *direct product* of the A_i's. Considering instead the elements in ΠA_i such that $\|x_i\| \to 0$ as $i \to \infty$ (with I as a discrete space) we obtain the *direct sum* of the A_i's, which we denote by $\oplus A_i$. When I is a finite set we may of course write $A_1 \oplus A_2 \ldots \oplus A_n$ instead of $\oplus A_i$ $(= \Pi A_i)$.

If $A_i = A$ for all i in I then

$$\prod A_i = C^b(I, A) \quad \text{and} \quad \bigoplus A_i = C_0(I, A).$$

1.2.5. The most important non-commutative infinite-dimensional C^*-algebra is the C^*-subalgebra $C(H)$ of $B(H)$ consisting of the *compact operators* on H. Since $C(H)$ is a minimal closed ideal of $B(H)$, being the closure of the finite-dimensional operators, $C(H)$ is *simple*, i.e. contains no non-zero closed ideals. This means that $C(H)$ cannot be decomposed into smaller algebras, which explains its role as building block in more complicated C^*-algebras. Note that $1 \notin C(H)$ if $\dim H = \infty$ and that $C(H)$ is separable if H is separable.

1.2.6. Notes and remarks. The general theory of tensor products of C^*-algebras can be found in Sakai's book [231]. We return in Chapter 6 to tensor products of matrix algebras as a means to generate new algebras by an inductive limit procedure (infinite tensor products). See also 8.15.15.

1.3. Positive elements and order

1.3.1. LEMMA. *The following four conditions on an element x in A are equivalent:*

(i) *x is normal and* $\mathrm{Sp}(x) \subset \mathbf{R}_+$;

(ii) *$x = y^2$ with y in A_{sa};*

(iii) *$x = x^*$ and $\|t1 - x\| \leqslant t$ for any $t \geqslant \|x\|$;*

(iv) *$x = x^*$ and $\|t1 - x\| \leqslant t$ for some $t \geqslant \|x\|$.*

Proof. (i) ⇒ (ii). Using 1.1.8 we set $y = x^{1/2}$ and have $y^2 = x$. (ii) ⇒ (i). Embedding x and y in a commutative C^*-subalgebra we see that $x = x^*$ and $\mathrm{Sp}(x) \subset \boldsymbol{R}_+$. (i) ⇒ (iii). From 1.1.6 we have $\|z\| = \sup\{|\lambda| \mid \lambda \in \mathrm{Sp}(z)\}$ for each normal element z of $\tilde{A}$. Applying this to $t1 - x$ with $t \geqslant \|x\|$ we have

$$\|t1 - x\| = \sup\{|t - \lambda| \mid \lambda \in \mathrm{Sp}(x)\} \leqslant t.$$

(iii) ⇒ (iv) is immediate. (iv) ⇒ (i). If $\lambda \in \mathrm{Sp}(x)$ then $t - \lambda \in \mathrm{Sp}(t1 - x)$, whence

$$|t - \lambda| \leqslant \|t1 - x\| \leqslant t.$$

Therefore $\lambda > 0$ since $\lambda \leqslant t$.

1.3.2. The elements x of a C^*-algebra A satisfying the conditions in 1.3.1 are called *positive* (in symbols $x \geqslant 0$), and the positive part of a subset B of A is denoted by B_+.

1.3.3. THEOREM. *The set A_+ is a closed real cone in A_{sa}, and $x \in A_+$ if and only if $x = y^*y$ for some y in A.*

Proof. From 1.3.1 (iii) it is clear that A_+ is a closed subset of A_{sa} stable under multiplication with positive scalars. To prove that A_+ is a cone take x and y in A_+. By 1.3.1 (iii) we have

$$\|(\|x\| + \|y\|)1 - (x + y)\| = \|(\|x\|1 - x) + (\|y\|1 - y)\|$$

$$\leqslant \|\|x\|1 - x\| + \|\|y\|1 - y\| \leqslant \|x\| + \|y\|,$$

whence $x + y \in A_+$ by 1.3.1 (iv) since $\|x\| + \|y\| \geqslant \|x + y\|$.

Assume now that $x = y^*y$. Then $x = x^*$ so that $x = x_+ - x_-$ by 1.1.8. Moreover

$$(yx_-^{1/2})^*(yx_-^{1/2}) = x_-^{1/2}y^*yx_-^{1/2} = x_-^{1/2}(x_+ - x_-)x_-^{1/2} = -x_-^2 \in -A_+.$$

Put $yx_-^{1/2} = a + ib$ with a and b in A_{sa}. Then

$$(yx_-^{1/2})(yx_-^{1/2})^* = 2(a^2 + b^2) - (yx_-^{1/2})^*(yx_-^{1/2}) \in A_+$$

since A_+ was a cone. But, zero apart, the spectrum of a product does not depend on the order of the factors (A1, Appendix), whence

$$\mathrm{Sp}(x_-^2) \subset \boldsymbol{R}_+ \cap -\boldsymbol{R}_+ = 0,$$

so that $x_- = 0$ and $x \geqslant 0$.

1.3.4. Since $A_+ - A_+ = A_{sa}$ and $A_+ \cap (-A_+) = 0$, A_{sa} becomes a partially ordered real vectorspace by defining $x \leqslant y$ whenever $y - x \in A_+$. When A is non-commutative A_{sa} is not a vector lattice.

1.3.5. PROPOSITION. *If $0 \leqslant x \leqslant y$ then $a^*xa \leqslant a^*ya$ for each a in A and $\|x\| \leqslant \|y\|$.*

Proof. Since $y - x = b^*b$ from 1.3.3 we have $a^*ya - a^*xa = (ba)^*(ba) \in A_+$.
Adjoining a unit to A we have $y \leqslant \|y\| 1$ from spectral theory. Then $x \leqslant \|y\| 1$, whence $\|x\| \leqslant \|y\|$.

1.3.6. PROPOSITION. *If* $1 \in A$ *and* x *and* y *are invertible elements in* A_+ *with* $x \leqslant y$ *then* $y^{-1} \leqslant x^{-1}$.

Proof. From 1.3.5 we have $y^{-1/2}xy^{-1/2} \leqslant 1$, whence $\|x^{1/2}y^{-1/2}\| \leqslant 1$ and thus $\|x^{1/2}y^{-1}x^{1/2}\| \leqslant 1$ which implies that $x^{1/2}y^{-1}x^{1/2} \leqslant 1$. By 1.3.5 $y^{-1} \leqslant x^{-1/2}1x^{-1/2} = x^{-1}$.

1.3.7. We say that a continuous real function f on an interval in $\boldsymbol{R}$ is *operator monotone* (increasing) if $x \leqslant y$ implies $f(x) \leqslant f(y)$ whenever the spectra of x and y belong to the interval of definition for f.

For each $\alpha > 0$ define f_α on $]-1/\alpha, \infty[$ by

$$f_\alpha(t) = (1 + \alpha t)^{-1}t = [1 - (1 + \alpha t)^{-1}]/\alpha.$$

Since the process of taking inverses is operator monotone decreasing by 1.3.6, it is easy to see that f_α is operator monotone increasing on $]-1/\alpha, \infty[$. The family of functions $\{f_\alpha\}$ will be used repeatedly in the sequel. Note that $f_\alpha(t) < \mathrm{Min}\{t, 1/\alpha\}$ and that $\mathrm{Lim} f_\alpha(t) = t$ uniformly on compact subsets of $\boldsymbol{R}$ when $\alpha \to 0$. Moreover, $f_\alpha \geqslant f_\beta$ when $\alpha \leqslant \beta$ and $f_\alpha \circ f_\beta = f_{\alpha+\beta}$ on $]-(\alpha+\beta)^{-1}, \infty[$. Finally, if $t > 0$ then $\mathrm{Lim}\, \alpha f_\alpha(t) = 1$ uniformly on compact subsets of $]0, \infty[$ when $\alpha \to \infty$.

1.3.8. PROPOSITION. *If* $0 < \beta \leqslant 1$ *the function* $t \to t^\beta$ *is operator monotone on* $\boldsymbol{R}_+$.

Proof. If $0 \leqslant x \leqslant y$ then $f_\alpha(x) \leqslant f_\alpha(y)$ with f_α as in 1.3.7. Now

$$\int_0^\infty f_\alpha(t)\alpha^{-\beta}\, d\alpha = \int_0^\infty (1 + \alpha t)^{-1} t\alpha^{-\beta}\, d\alpha$$

$$= \int_0^\infty (1 + \alpha)^{-1} t\alpha^{-\beta} t^\beta t^{-1}\, d\alpha = t^\beta \int_0^\infty (1 + \alpha)^{-1}\alpha^{-\beta}\, d\alpha = \gamma t^\beta$$

with γ in $\boldsymbol{R}_+$. For all t in $[0, \|y\|]$ and $\varepsilon > 0$ there is therefore a large n and an equidistant division $0 = \alpha_0 < \alpha_1 < \cdots < \alpha_m = n$ of the interval $[0, n]$ such that

$$|t^\beta - (\gamma m)^{-1} n \sum_{k=1}^{m} f_{\alpha_k}(t)\alpha_k^{-\beta}| < \varepsilon.$$

It follows that $y^\beta - x^\beta \geqslant -2\varepsilon$, and since ε is arbitrary $x^\beta \leqslant y^\beta$.

1.3.9. PROPOSITION. *If* $0 \leqslant x \leqslant y$ *implies* $x^\beta \leqslant y^\beta$ *for some* $\beta > 1$ *and all* x, y *in a* C^**-algebra* A, *then* A *is commutative.*

Proof. By iteration we see that if the exponent β preserves order then so does β^n for every n in $\boldsymbol{N}$. Using 1.3.8 we see that the exponents which preserve order form a segment of $\boldsymbol{R}_+$. It suffices therefore to prove the proposition with $\beta = 2$.

Take x, y in A_+ and $\varepsilon > 0$. Then $x \leqslant x + \varepsilon y$ whence

$$x^2 \leqslant (x + \varepsilon y)^2 = x^2 + \varepsilon(xy + yx) + \varepsilon^2 y^2.$$

This gives $0 \leqslant xy + yx + \varepsilon y^2$ for any $\varepsilon > 0$, thus

$$(*) \qquad xy + yx \geqslant 0.$$

Set $xy = a + ib$ with a and b in A_{sa}. Clearly $a \geqslant 0$. But $(*)$ is valid for any product of positive elements and

$$(**) \qquad x(yxy) = a^2 - b^2 + i(ab + ba)$$

from which we conclude that $a^2 - b^2 \geqslant 0$.

The set of numbers $\alpha \geqslant 1$ such that $\alpha b^2 \leqslant a^2$ for all x and y in A_+ with $xy = a + ib$ is therefore non-empty. The set E is also closed, so if it was bounded it would have a largest element, say λ. Thus if x, y belongs to A_+ and $xy = a + ib$ then $a^2 - \lambda b^2 \geqslant 0$ and therefore by $(*)$

$$(***) \qquad 0 \leqslant b^2(a^2 - \lambda b^2) + (a^2 - \lambda b^2)b^2 = b^2a^2 + a^2b^2 - 2\lambda b^4.$$

From $(**)$ we now have

$$\lambda(ab + ba)^2 \leqslant (a^2 - b^2)^2,$$

that is

$$\lambda[ab^2a + ba^2b + a(bab) + (bab)a] \leqslant a^4 + b^4 - a^2b^2 - b^2a^2.$$

On the left-hand side we have $a(bab) + (bab)a \geqslant 0$ by $(*)$ and $ba^2b \geqslant \lambda b^4$ by assumption and finally, $ab^2a \geqslant 0$. Using this, and inserting $(***)$ on the right-hand side we get

$$\lambda^2 b^4 \leqslant a^4 + (1 - 2\lambda)b^4,$$

that is

$$(\lambda^2 + 2\lambda - 1)b^4 \leqslant a^4.$$

By 1.3.8 this implies that $(\lambda^2 + 2\lambda - 1)^{1/2} b^2 \leqslant a^2$ for all a and b with $a + ib = xy$ and x, y in A_+. But then $(\lambda^2 + 2\lambda - 1)^{1/2} \in E$ in contradiction with our choice of λ as the largest element. It follows that E is unbounded, whence $\alpha b^2 \leqslant a^2$ for all $\alpha \geqslant 1$, i.e. $b = 0$, and A is commutative.

1.3.10. We say that a continuous real function f on an interval in $\boldsymbol{R}$ is *operator convex* if for any two operators x, y with spectrum in this interval and any λ in $[0, 1]$

$$f(\lambda x + (1 - \lambda)y) \leqslant \lambda f(x) + (1 - \lambda)f(y).$$

We say that f is *operator concave* if $-f$ is operator convex.

1.3.11. PROPOSITION. *The functions* f_α, $\alpha \geqslant 0$, *the functions* $t \to t^\beta$, $0 < \beta \leqslant 1$, *and the functions* $t \to \log(\varepsilon + t)$, $\varepsilon > 0$, *are all operator concave on* $\boldsymbol{R}_+$.

Proof. If a is a positive invertible operator then from spectral theory we have for each λ in $[0, 1]$

$$[\lambda + (1 - \lambda)a]^{-1} \leqslant \lambda + (1 - \lambda)a^{-1}.$$

If x and y are positive invertible operators put $a = x^{-1/2}yx^{-1/2}$. Multiplying the inequality above with $x^{-1/2}$ from both sides, we get by 1.3.5

$$[\lambda x + (1 - \lambda)y]^{-1} \leqslant \lambda x^{-1} + (1 - \lambda)y^{-1},$$

which shows that the function $t \to t^{-1}$ is operator convex on $]0, \infty[$. It follows immediately from the formula in 1.3.7 that the functions f_α, $\alpha \geqslant 0$, are operator concave on $\boldsymbol{R}_+$.

Since operator concavity like operator monotonicity is preserved under limits (uniformly on compact subsets) and under convex combinations we see exactly as in the proof of 1.3.8 that the functions $t \to t^\beta$, $0 < \beta \leqslant 1$, are operator concave. Finally, for each $\varepsilon > 0$ put

$$g_\alpha(t) = (\alpha + 1)^{-1}(\alpha + t + \varepsilon)^{-1}(t + \varepsilon - 1) = (\alpha + 1)^{-1} - (\alpha + t + \varepsilon)^{-1}.$$

The last expression shows that g_α is operator concave for $\alpha \geqslant 0$. The first expression shows that for each $t \geqslant 0$ the function $\alpha \to g_\alpha(t)$ is integrable. An elementary calculation yields

$$\int_0^\infty g_\alpha(t)\, d\alpha = \log(\varepsilon + t)$$

and consequently the functions $t \to \log(\varepsilon + t)$, $\varepsilon > 0$ are operator concave. Incidentally, the argument also shows that the functions are operator monotone.

1.3.12. Notes and remarks. The result in 1.3.3, due to Kelley and Vaught [148], shows that the extra axiom $(x^*x \geqslant 0)$ in the original definition of a C^*-algebra (see 1.1.14) was redundant, as Gelfand and Naimark also suspected.

Operator monotone functions were characterized by Löwner [163] as being those continuous functions $f: I \to \boldsymbol{R}$ which admit a holomorphic extension $\tilde{f}$ to the upper half plane $\boldsymbol{C}_+ = \{\operatorname{Im} z > 0\}$ such that $\tilde{f}(\boldsymbol{C}_+) \subset \boldsymbol{C}_+$. It follows by a slight variation of Herglotz's formula that each function f which is operator monotone on $\boldsymbol{R}_+$ has a unique representation $f(t) = \int_0^\infty f_\alpha(t)\, d\mu(\alpha)$ for some positive measure μ on $\boldsymbol{R}_+$. The result in 1.3.9 is found in [179]. Operator convexity and monotonicity is treated in [16].

1.4. Approximate units and factorization theorems

1.4.1. Let A be a C^*-algebra. A net $\{u_\lambda \mid \lambda \in \Lambda\}$ in A_+ with $\|u_\lambda\| \leqslant 1$ for all λ is called an *approximate unit* for A if $\lambda < \mu$ implies $u_\lambda \leqslant u_\mu$ and if $\operatorname{Lim} \|x(1 - u_\lambda)\| = 0$ for each x in A. Then, of course, $\operatorname{Lim} \|(1 - u_\lambda)x\| = 0$ as well.

1.4.2. THEOREM. *Each C^*-algebra contains an approximate unit.*

Proof. Consider the set Λ of elements u in A_+ such that $\|u\| < 1$. We claim that Λ is an increasing net in the partial ordering on A_{sa}. To see this take u and v in Λ. Then the elements $a = (1 - u)^{-1}u$ and $b = (1 - v)^{-1}v$ belong to A_+ by 1.1.9. Set $w = (1 + a + b)^{-1}(a + b)$. Then $w \in \Lambda$. By 1.3.7 we have

$$w = f_1(a + b) \geqslant f_1(a) = [1 + (1 - u)^{-1}u]^{-1}[(1 - u)^{-1}u] = u,$$

and similarly $w \geqslant v$, which proves that Λ is an increasing net in A_+.

To show that Λ is an approximate unit it suffices to prove that the net $\{\|x(1 - u)x\| \mid u \in \Lambda\}$, which is decreasing by 1.3.5, converges to zero for each x in A_+, since

$$\|(1 - u)x\|^2 = \|x(1 - u)^2x\| \leqslant \|x(1 - u)x\|,$$

and A is linearly spanned by A_+.

With f_α as in 1.3.7 note that $\alpha f_\alpha(x) \in \Lambda$ and that

$$x(1 - \alpha f_\alpha(x))x = (1 + \alpha x)^{-1}x^2 \leqslant \alpha^{-1}x,$$

whence $\|x(1 - \alpha f_\alpha(x))x\| \leqslant \alpha^{-1}\|x\|$ which tends to zero as $\alpha \to \infty$.

1.4.3. The common definition of an approximate unit for a C^*-algebra (or just a Banach algebra) A does not specify that the approximate unit has to belong to A_+, nor does it have to be increasing. The more restrictive definition given here facilitates some of the computations later.

It is clear that there is nothing unique about an approximate unit; in fact every subnet of the one constructed in 1.4.2 will work equally well. However, the approximate unit constructed in 1.4.2 contains all other approximate units (if they are scaled down a little so as not to touch the unit sphere of A) and we shall refer to it as the *canonical approximate unit* for A.

If A is separable then it may be convenient to be able to work with a countable approximate unit for A. Take a dense sequence $\{x_n\}$ in A and choose an increasing sequence $\{u_n\}$ in the canonical approximate unit for A such that $\|x_k(1 - u_n)\| < 1/n$ for $k \leqslant n$. Then $\{u_n\}$ is an approximate unit for A.

1.4.4. LEMMA. *Let x, y and a be elements of a C^*-algebra A such that $a \geqslant 0$ and $x^*x \leqslant a^\alpha$, $yy^* \leqslant a^\beta$ with $\alpha + \beta > 1$. Then the sequence with elements*

$u_n = x[(1/n) + a]^{-1/2}y$ *is norm convergent to an element* u *in* A *with* $\|u\| \leq \|a^{(\alpha+\beta-1)/2}\|$.

Proof. Put $d_{nm} = [(1/n) + a]^{-1/2} - [(1/m) + a]^{-1/2}$. Then

$$\begin{aligned}\|u_n - u_m\|^2 &= \|xd_{nm}y\|^2 = \|y^*d_{nm}x^*xd_{nm}y\| \\ &\leq \|y^*d_{nm}a^\alpha d_{nm}y\| = \|a^{\alpha/2}d_{nm}y\|^2 \\ &= \|a^{\alpha/2}d_{nm}yy^*d_{nm}a^{\alpha/2}\| \\ &\leq \|a^{\alpha/2}d_{nm}a^\beta d_{nm}a^{\alpha/2}\| = \|d_{nm}a^{(\alpha+\beta)/2}\|^2.\end{aligned}$$

From spectral theory we see that the sequence $\{[(1/n) + a]^{-1/2}a^{(\alpha+\beta)/2}\}$ is increasing and thus by Dini's theorem uniformly convergent to $a^{(\alpha+\beta-1)/2}$. Consequently $\|d_{nm}a^{(\alpha+\beta)/2}\| \to 0$ so that $\{u_n\}$ is norm convergent to an element u in A. We have

$$\|u_n\| = \|x[(1/n) + a]^{-1/2}y\| \leq \|a^{\alpha/2}[(1/n) + a]^{-1/2}a^{\beta/2}\| \leq \|a^{(\alpha+\beta-1)/2}\|,$$

reasoning as above; which shows that $\|u\| \leq \|a^{(\alpha+\beta-1)/2}\|$.

1.4.5. PROPOSITION. *Let* x *and* a *be elements in a* C^**-algebra* A *such that* $a \geq 0$ *and* $x^*x \leq a$. *If* $0 < \alpha < \frac{1}{2}$ *there is an element* u *in* A *with* $\|u\| \leq \|a^{\frac{1}{2}-\alpha}\|$ *such that* $x = ua^\alpha$.

Proof. Define $u_n = x[(1/n) + a]^{-\frac{1}{2}}a^{\frac{1}{2}-\alpha}$. From 1.4.4 we see that $\{u_n\}$ is convergent to an element u in A with

$$\|u\| \leq \|a^{\frac{1}{2}(1+1-2\alpha-1)}\| = \|a^{\frac{1}{2}-\alpha}\|.$$

Furthermore,

$$\begin{aligned}\|x - u_na^\alpha\|^2 &= \|x(1 - [(1/n) + a]^{-1/2}a^{1/2})\|^2 \\ &\leq \|a^{1/2}(1 - [(1/n) + a]^{-1/2}a^{1/2})\|^2 \to 0\end{aligned}$$

as $n \to \infty$ by spectral theory (Dini's theorem). It follows that $x = ua^\alpha$.

1.4.6. Since by definition $x^*x = |x|^2$ (see 1.1.8) we have by 1.4.5 for each x in A a factorization $x = u|x|^\alpha$ with $0 < \alpha < 1$. A factorization with $\alpha = 1$ (polar decomposition) is not possible in a general C^*-algebra but can be performed in algebras for which spectral theory admits the use of Borel functions (see 2.2.9).

1.4.7. PROPOSITION. *The extreme points in the unit ball of a* C^**-algebra* A *are precisely those elements* x *in* A *such that* $(1 - xx^*)A(1 - x^*x) = 0$. *In particular*, x^*x *and* xx^* *are idempotents (projections) and* A *has a unit* $1 = xx^* + x^*x - xx^{*2}x$.

Proof. If x^*x is not an idempotent, there is a λ in $\mathrm{Sp}(x^*x)$ with $\lambda^2 \neq \lambda$. By 1.4.5 and 1.4.6 we may write $x = ua$ with $a = (x^*x)^{1/3}$ and $\|u\| \leq 1$ (assuming that $\|a\| \leqslant 1$) and obtain a decomposition

$$x = \tfrac{1}{2}[u(2a - a^2) + ua^2] = \tfrac{1}{2}[x(2 - a) + xa],$$

with both $u(2a - a^2)$ and ua^2 in the unit ball of A. We have $x \neq xa$, since $x^*x \neq x^*x^{5/3}$; and $x \neq x(2 - a)$ which shows that x is not an extreme point.

If x^*x (and therefore also xx^*) is an idempotent, put $p = x^*x$ and $q = xx^*$. Then $x = xp = qx$. If y is a non-zero element of the form $(1 - q)z(1 - p)$ with $\|z\| \leqslant 1$ then

$$\begin{aligned}\|x + y\|^2 &= \|qxp + (1 - q)z(1 - p)\|^2 \\ &= \|px^*qxp + (1 - p)z^*(1 - q)z(1 - p)\| = \max\{\|x\|^2, \|y\|^2\} = 1\end{aligned}$$

and thus the decomposition $x = \frac{1}{2}[(x + y) + (x - y)]$ is non-trivial so that x is not extreme.

Assume now that $x \in A$ such that $(1 - xx^*)A(1 - x^*x) = 0$. Then in particular

$$0 = x^*(1 - xx^*)x(1 - x^*x) = x^*x(1 - x^*x)^2.$$

It follows from spectral theory that x^*x is an idempotent. Since $\mathrm{Sp}(x^*x)\backslash\{0\} = \mathrm{Sp}(xx^*)\backslash\{0\}$ (see A1, Appendix) this implies that xx^* is an idempotent too. Let $p = x^*x$ and $q = xx^*$, and assume that $x = \frac{1}{2}(y + z)$ with y and z in the unit ball of A. Then $x = xp = \frac{1}{2}(yp + zp)$, whence

$$p = x^*x \leqslant \tfrac{1}{2}(py^*yp + pz^*zp) \leqslant p.$$

Since py^*yp and pz^*zp belong to the unit ball of the C^*-algebra pAp for which p is the unit, it follows from spectral theory that $p = py^*yp = pz^*zp$. But

$$p = x^*x = \tfrac{1}{4}(py^*yp + py^*zp + pz^*yp + pz^*zp),$$

so that

$$p = \tfrac{1}{2}(py^*zp + pz^*yp).$$

Since p is extreme in the unit ball of pAp by 1.1.11, we conclude that $p = py^*zp = pz^*yp$, whence by computation $p(y - z)^*(y - z)p = 0$, so that $yp = zp$. Similarly $qy = qz$ and thus by the assumption on p and q we have

$$y - z = (1 - q)(y - z)(1 - p) = 0,$$

so that x is an extreme point.

If $\{u_\lambda\}$ is an approximate unit for A then $(1 - xx^*)u_\lambda(1 - x^*x)y = 0$ for each y in A. In the limit $(1 - xx^*)(1 - x^*x)y = 0$ which shows that $1 = xx^* + x^*x - xx^{*2}x \in A$.

1.4.8. PROPOSITION. *Let p_1 and p_2 be idempotents in A_+. The extreme points in the unit ball of the subspace p_1Ap_2 are precisely those elements x in p_1Ap_2 such that $(p_1 - xx^*)A(p_2 - x^*x) = 0$.*

Proof. Replace in the proof of 1.4.7 the symbol 1 by p_1 if it stands to the left of x and by p_2 if it stands to the right. The arguments carry over verbatim.

1.4.9. Recall that a partially ordered Banach space E over the reals satisfies the Riesz decomposition property if for any three positive elements a, b and c in E with $a \leqslant b + c$ there is a decomposition $a = d + e$ in E with $d \leqslant b$ and $e \leqslant c$. If E is a vector lattice it has the Riesz decomposition, and if E satisfies the Riesz decomposition property E^* is a vector lattice. It will later become quite apparent that if A is a C^*-algebra then A_{sa} satisfies the Riesz decomposition property if and only if A is commutative.

The next result can be viewed as the non-commutative version of the Riesz decomposition property and will for general C^*-algebras replace the ordinary one.

1.4.10. PROPOSITION. *Let x, y and z be elements in a C^*-algebra A. If $x^*x \leqslant yy^* + zz^*$ then there are elements u and v in A with $u^*u \leqslant y^*y$ and $v^*v \leqslant z^*z$ such that $xx^* = uu^* + vv^*$.*

Proof. Let $a = yy^* + zz^*$. By 1.4.5 the sequences with elements

$$u_n = x[(1/n) + a]^{-1/2}y, \qquad v_n = x[(1/n) + a]^{-1/2}z$$

are norm convergent in A with limits u and v, respectively. We have

$$\begin{aligned} u_n^*u_n &= y^*[(1/n) + a]^{-1/2}x^*x[(1/n) + a]^{-1/2}y \\ &\leqslant y^*[(1/n) + a]^{-1/2}a[(1/n) + a]^{-1/2}y \leqslant y^*y, \end{aligned}$$

and since A_+ is closed this implies that $u^*u \leqslant y^*y$. Similarly $v^*v \leqslant z^*z$. Finally,

$$\begin{aligned} \|xx^* - u_nu_n^* - v_nv_n^*\| &= \|x(1 - [(1/n) + a]^{-1/2}(yy^* + zz^*)[(1/n) + a]^{-1/2})x^*\| \\ &= \|x[(1/n) + a]^{-1}(1/n)x^*\| \\ &\leqslant \|a^{1/2}[(1/n) + a]^{-1}(1/n)a^{1/2}\| \leqslant 1/n, \end{aligned}$$

and thus $xx^* = uu^* + vv^*$.

1.4.11. Notes and remarks. The existence of approximate units for C^*-algebras was shown by Segal [235]. The canonical approximate unit was found by Dixmier around 1968. The extreme points were characterized by Kadison [124]; the generalized Riesz decomposition appears in [193].

1.5 Hereditary algebras, ideals and quotients

1.5.1. A cone M in the positive part of a C^*-algebra A is called *hereditary* if $0 \leqslant x \leqslant y$, $y \in M$, implies $x \in M$ for each x in A. A $*$-subalgebra B of A is *hereditary* if B_+ is hereditary in A_+. Given a hereditary cone M in A_+ define

$$L(M) = \{x \in A \mid x^*x \in M\}.$$

1.5.2. THEOREM. *For each C^*-algebra A the mappings $B \to B_+$, $M \to L(M)$ and $L \to L \cap L^*$ define bijective, order preserving correspondences between the sets of hereditary C^*-subalgebras of A, closed hereditary cones of A_+ and closed left ideals of A.*

Proof. If B is a hereditary C^*-subalgebra of A then by definition B_+ is a closed hereditary cone in A_+.

If M is a closed hereditary cone in A_+ then $L(M)$ is closed in A. Moreover, if $x \in L(M)$, $y \in A$, then by 1.3.5

$$(yx)^*(yx) = x^*(y^*y)x \leqslant \|y\|^2 x^*x \in M,$$

whence $yx \in L(M)$. If $x, y \in L(M)$ then

$$(x + y)^*(x + y) \leqslant (x + y)^*(x + y) + (x - y)^*(x - y) = 2(x^*x + y^*y) \in M,$$

so that $x + y \in L(M)$. It follows that $L(M)$ is a closed left ideal in A.

If L is a closed left ideal in A then $L \cap L^*$ is a C^*-subalgebra of A whose positive part coincides with L_+. Assume that $x^*x \leqslant y$ for some x in A and y in L. Then $y^{1/3} \in L_+$ and by 1.4.5 $x = uy^{1/3}$ for some u in A. Since L is a left ideal $x \in L$. It follows that $x^*x \in L_+$ so that L_+ is a closed hereditary cone in A_+, but also that $x \in L$ if and only if $x^*x \in L_+$. This shows that $L(L_+) = L$ and that $L(M)_+ = M$ for any closed hereditary cone M in A_+. The correspondences $L \to L_+$ and $M \to L(M)$ are thus the inverses of each other.

Finally, since for any closed hereditary cone M in A_+ the set $L(M) \cap L(M)^*$ is a C^*-algebra whose positive part is equal to M, we must have $L(M) \cap L(M)^* = \mathrm{Span}(M)$ so that the correspondences $B \to B_+$ and $M \to \mathrm{Span}(M)$ are the inverses of each other.

It is clear that the correspondences constructed above are order preserving.

1.5.3. COROLLARY. *If I is a closed ideal of a C^*-algebra then $I = I^*$.*

Proof. If $x^* \in I$ then $x^*x \in I_+$, and since I is also a closed left ideal this implies that $x \in I$.

1.5.4. LEMMA. *Let I be a closed ideal of A and $\{u_\lambda\}$ an approximate identity for I. Then for each x in A*

$$\operatorname*{Inf}_{y \in I} \|x + y\| = \operatorname{Lim} \|x(1 - u_\lambda)\|.$$

Proof. Let α denote the left-hand side of the equation and let β be the limit of the decreasing net $\{\|x(1 - u_\lambda)x^*\|\}$. Since

$$\alpha^2 \leqslant \|x(1 - u_\lambda)\|^2 = \|x(1 - u_\lambda)^2 x^*\| \leqslant \|x(1 - u_\lambda)x^*\|,$$

it suffices to prove that $\alpha^2 \geqslant \beta$. For $\varepsilon > 0$ take y in I such that $\alpha + \varepsilon \geqslant \|x + y\|$. Then

$$(\alpha + \varepsilon)^2 \geqslant \|x + y\| \, \|1 - u_\lambda\| \, \|x^* + y^*\| \geqslant \|(x + y)(1 - u_\lambda)(x^* + y^*)\| \to \beta,$$

since both $\|y(1 - u_\lambda)\|$ and $\|(1 - u_\lambda)y^*\|$ tend to zero. Thus $(\alpha + \varepsilon)^2 \geqslant \beta$, whence $\alpha^2 \geqslant \beta$.

1.5.5. COROLLARY. *If I is a closed ideal of A, then A/I equipped with its natural operations is a C^*-algebra.*

Proof. It is clear that A/I is a Banach algebra with involution. Let $\{u_\lambda\}$ be an approximate unit for I. Then, with $\dot{x}$ the image of x in A/I,

$$\begin{aligned} \|\dot{x}^*\dot{x}\| &= \operatorname{Lim} \|x^*x(1 - u_\lambda)\| \geqslant \operatorname{Lim} \|(1 - u_\lambda)x^*x(1 - u_\lambda)\| \\ &= \operatorname{Lim} \|x(1 - u_\lambda)\|^2 = \|\dot{x}\|^2. \end{aligned}$$

1.5.6. By a *morphism* ρ between C^*-algebras A and B we mean a *-preserving homomorphism of A into B.

1.5.7. THEOREM. *Each morphism ρ between C^*-algebras A and B is norm decreasing, and $\rho(A)$ is a C^*-subalgebra of B. If ρ is injective then it is isometric.*

Proof. If $x \in A_{sa}$ then $\mathrm{Sp}_B(\rho(x))\backslash\{0\} \subset \mathrm{Sp}_A(x)\backslash\{0\}$. Since $\|x\|$ is equal to the spectral radius of x by 1.1.5 we obtain for each y in A

$$\|\rho(y)\|^2 = \|\rho(y^*y)\| \leqslant \|y^*y\| = \|y\|^2,$$

which proves that ρ is norm decreasing.

Assume now that ρ is injective. As seen above it suffices to show that ρ is isometric on positive elements. Restricting to the subalgebra generated by a positive element we may therefore assume that A is commutative. Exchanging B with the closure of $\rho(A)$, we may assume that B is also commutative and that $\rho(A)$ is dense in B. We may also adjoin units to A and B and assume that $\rho(1_A) = 1_B$. Let S and T be the compact Hausdorff spaces such that $A = C(S)$ and $B = C(T)$. Let ρ^* denote the continuous map of T onto a closed subset of S obtained by transposition of ρ. Since ρ is an isomorphism, $\rho^*(T) = S$. Since $\rho(A)$ is dense in B, ρ^* is injective, and consequently a homeomorphism. It follows that ρ is an isometry of A onto B.

Returning to the case of a general morphism ρ of A into B, we observe that the kernel I of ρ is a closed ideal of A and that ρ induces an isomorphism from the C^*-algebra A/I into B, with image $\rho(A)$. Since an isomorphism is isometric, $\rho(A)$ is closed in B, and the theorem is proved.

1.5.8. COROLLARY. *Let I be a closed ideal of a C^*-algebra A and let B be a C^*-subalgebra of A. Then $I + B$ is equal to the C^*-algebra generated by I and B.*

Proof. Evidently $I + B$ is a *-subalgebra of A containing I and B. Let ρ be the morphism of A onto A/I. By 1.5.7 $\rho(B)$ is closed in A/I, whence $I + B = \rho^{-1}(\rho(B))$ is closed in A.

1.5.9. PROPOSITION. *Let I and J be closed ideals of a C^*-algebra. Then $(I + J)_+ = I_+ + J_+$.*

Proof. From 1.5.8 we know that $I + J$ is a C^*-algebra, and clearly $(I + J)_{sa} = I_{sa} + J_{sa}$. Suppose $x \in (I + J)_+$. Then $x = y + z$ with y in I_{sa} and z in J_{sa}. Consequently $x \leqslant |y| + |z|$, whence $x = uu^* + vv^*$ with $u^*u \leqslant |y|$ and $v^*v \leqslant |z|$ by 1.4.10. Since I is hereditary by 1.5.2 and 1.5.3, $u \in I$ and $v \in J$. Thus $uu^* \in I_+$ and $vv^* \in J_+$ so that $x \in I_+ + J_+$ as desired.

1.5.10. PROPOSITION. *Let ρ be a morphism of a C^*-algebra A onto a C^*-algebra B. If $a \in A_+$ and $x \in B$ with $x^*x \leqslant \rho(a)$ then $x = \rho(y)$ for some y in A with $y^*y \leqslant a$.*

Proof. Take z in A with $\rho(z) = x$. If $b = (z^*z - a)_+$ then $b \in A_+$ and $\rho(b) = 0$. Moreover, $z^*z \leqslant a + b$. Define $y_n = z[(1/n) + a + b]^{-1/2}a^{1/2}$ and let y be the limit of the sequence $\{y_n\}$ which is norm convergent by 1.4.4. Since

$$\begin{aligned} y_n^*y_n &= a^{1/2}[(1/n) + a + b]^{-1/2}z^*z[(1/n) + a + b]^{-1/2}a^{1/2} \\ &\leqslant a^{1/2}[(1/n) + a + b]^{-1/2}(a + b)[(1/n) + a + b]^{-1/2}a^{1/2} \leqslant a \end{aligned}$$

we have $y^*y \leqslant a$. However, ρ preserves spectral functions whence $\rho(y_n) = x[(1/n) + \rho(a)]^{-1/2}\rho(a)^{1/2}$. Since $x^*x \leqslant \rho(a)$ we have

$$\rho(y) = \operatorname{Lim} \rho(y_n) = x.$$

1.5.11. COROLLARY. *The image of a hereditary *-algebra (respectively a left ideal) under a surjective morphism between C^*-algebras is a hereditary *-subalgebra (respectively a left ideal).*

1.5.11. Notes and remarks. The correspondences in 1.5.2. goes back to Effros [74]. The important structure theorems 1.5.3, 1.5.5 and 1.5.7 were established by Segal [236]. They show that the collection of C^*-algebras is a category with the morphisms as morphisms. The results in 1.5.9 and 1.5.10 can be found in [243] and [35], respectively.

Chapter 2

Concrete C^*-algebras

Most of the non-commutative C^*-algebras that naturally arise in the theory and its applications are given concretely as algebras of operators on a Hilbert space. If H is a (complex) Hilbert space and $\boldsymbol{B}(H)$ denotes the C^*-algebra of bounded linear operators on H (with the adjoint operation as involution) then each norm closed subalgebra of $\boldsymbol{B}(H)$ which is closed under the adjoint operation is a C^*-algebra. Conversely, we shall show in Chapter 3 that each C^*-algebra can be realized as a C^*-subalgebra of $\boldsymbol{B}(H)$ for some H. This Hilbert space H is, however, not unique which is the reason why we have developed the theory of abstract C^*-algebras in Chapter 1.

2.1. Topologies on $\boldsymbol{B}(H)$

2.1.1. There are eight important topologies on $\boldsymbol{B}(H)$. We shall content ourselves here with only three of these: The norm topology (or uniform topology), the strong topology and the weak topology.

The *strong* topology on $\boldsymbol{B}(H)$ is the locally convex vector space topology associated with the family of semi-norms of the form $x \to \|x\xi\|$, $x \in \boldsymbol{B}(H)$, $\xi \in H$.

The *weak* topology on $\boldsymbol{B}(H)$ is the locally convex vector space topology associated with the family of semi-norms of the form $x \to |(x\xi \mid \eta)|$, $x \in \boldsymbol{B}(H)$, $\xi, \eta \in H$.

2.1.2. The expression

$$(*) \qquad \|(xy - x_0y_0)\xi\| \leqslant \|x\| \, \|(y - y_0)\xi\| + \|(x - x_0)y_0\xi\|$$

shows that the map $x, y \to xy$ is strongly continuous if the first factor remains in a bounded subset of $\boldsymbol{B}(H)$. For the weak topology we only have that the maps $x \to xy$, $x \in \boldsymbol{B}(H)$ and $x \to yx$, $x \in \boldsymbol{B}(H)$ are weakly continuous for each fixed y in $\boldsymbol{B}(H)$. On the other hand, the involution $x \to x^*$ is weakly continuous but not strongly continuous if H is infinite dimensional.

Since the unit ball of H is weakly compact the unit ball of $B(H)$ is weakly compact. If H is separable and $\{\xi_n\}$ is a dense sequence in the unit sphere of H then the norms $x \to \Sigma 2^{-n} \|x\xi_n\|$ and $x \to \Sigma 2^{-n-m} |(x\xi_n \mid \xi_m)|$ determine the strong and weak topologies, respectively, on the unit ball of $B(H)$. It follows that the unit ball of $B(H)$ is metrizable and separable, hence second countable, in both the strong and the weak topology.

Note that the weak topology is weaker than the strong topology, which in turn is weaker than the norm topology.

If $\{x_i\}$ is a bounded net in $B(H)_+$ which is weakly convergent to zero then $\{x_i\}$ is strongly convergent to zero, since $\{x_i^{1/2}\}$ is strongly convergent to zero and $x_i \leqslant \|x_i^{1/2}\| x_i^{1/2}$. A less obvious result which we shall use in Chapter 4 is contained in the following:

2.1.3. LEMMA. *If $\{x_i\}$ is a weakly convergent net in $B(H)_+$ with limit x, such that $x_i \geqslant x \geqslant \varepsilon 1$ for all i and some $\varepsilon > 0$ then $\{x_i^{-1}\}$ is a bounded net in $B(H)_+$ converging strongly to x^{-1}.*

Proof. Clearly $0 \leqslant x_i^{-1} \leqslant \varepsilon^{-1} 1$. Moreover, by 1.3.5

$$(x^{-1} - x_i^{-1})^2 = |x_i^{-1}(x_i - x)x^{-1}|^2$$
$$= x^{-1}(x_i - x)x_i^{-2}(x_i - x)x^{-1} \leqslant \|(x_i - x)^{1/2} x_i^{-2}(x_i - x)^{1/2}\| x^{-1}(x_i - x)x^{-1}$$
$$= \|x_i^{-1}(x_i - x)x_i^{-1}\| x^{-1}(x_i - x)x^{-1} \leqslant \varepsilon^{-1} x^{-1}(x_i - x)x^{-1}.$$

Thus $\{(x^{-1} - x_i^{-1})^2\}$ converges weakly to zero, which means that $\{x^{-1} - x_i^{-1}\}$ converges strongly to zero.

2.1.4. Let H be a Hilbert space and n a cardinal number. Let K be the orthogonal sum of n copies of H, and for each $i \leqslant n$ let p_i be the projection of K onto the ith copy of H. Each element x in $B(K)$ has a representation as a matrix (x_{ij}), $1 \leqslant i \leqslant n$, $1 \leqslant j \leqslant n$ with x_{ij} in $B(H)$, which is obtained simply by identifying the operator $p_i x p_j$ with an element x_{ij} in $B(H)$. When n is finite then, conversely, each such matrix corresponds to an element in $B(K)$. When n is infinite this is no longer true, as one can easily see by taking $\dim H = 1$.

We define an injective morphism ρ of $B(H)$ into $B(K)$ by taking $(\rho(x))_{ij} = 0$ if $i \neq j$ and $(\rho(x))_{ii} = x$ otherwise, i.e. by repeating the element x in $B(H)$ along the diagonal. The image algebra $\rho(B(H))$ is called an *amplification* of $B(H)$ of multiplicity n and is sometimes denoted by $1_n \otimes B(H)$. If $n < \infty$ and $\dim H < \infty$ then we know from algebra that $K = C^n \otimes H$ and $B(K) = M_n \otimes B(H)$. Denoting by 1_n the identity matrix in M_n we may then identify $\rho(x)$ with $1_n \otimes x$. In the infinite dimensional case the same formulas are valid if we define suitable topological tensor products.

2.1.5. THEOREM. *The following conditions on a linear functional ϕ on $B(H)$ are equivalent:*

(i) $\phi(x) = \Sigma_{k=1}^{n}(x\xi_k \mid \eta_k)$ *for some* ξ_k *and* η_k *in* H *and all* x *in* $B(H)$;

(ii) ϕ *is weakly continuous;*

(iii) ϕ *is strongly continuous.*

Proof. (i) $\Rightarrow$ (ii) $\Rightarrow$ (iii): Evident. (iii) $\Rightarrow$ (i). By assumption there exists vectors $\xi_1 \ldots \xi_n$ such that $\text{Max}\,\|x\xi_k\| \leq 1$ implies $|\phi(x)| \leq 1$ for all x in $B(H)$. It follows that

$$|\phi(x)| \leq \left(\sum_{k=1}^{n} \|x\xi_k\|^2\right)^{1/2}.$$

With the notation as in 2.1.4 (with $n < \infty$) we define $\xi = \xi_1 \oplus \cdots \oplus \xi_n$ in K and observe that the definition

$$\psi(\rho(x)\xi) = \phi(x)$$

gives a linear functional on the closed subspace of K spanned by the vectors $\rho(x)\xi$, $x \in B(H)$, such that $|\psi(\rho(x)\xi)| \leq \|\rho(x)\xi\|$. It follows from the Riesz–Frechet theorem that there is a vector $\eta = \eta_1 \oplus \cdots \oplus \eta_n$ in K such that

$$\phi(x) = (\rho(x)\xi \mid \eta) = \sum_{k=1}^{n} (x\xi_k \mid \eta_k).$$

2.1.6. COROLLARY. *Each strongly closed convex set in $B(H)$ is weakly closed.*

2.1.7. Notes and remarks. The eight vector space topologies on $B(H)$ are: The norm topology, the strong topology, the strong* topology, the σ-strong (or ultra-strong) topology, the σ-strong* topology, the Mackey topology, the weak topology, the σ-weak topology. On bounded subsets of $B(H)$ weak = σ-weak and strong = σ-strong. Thus we can avoid the σ-topologies as long as we work on bounded sets only. The *-topologies are simply obtained from strengthening the non-starred by requiring that the involution be continuous. Thus we can avoid them by working only in the self-adjoint part of $B(H)$. Nevertheless the reader may enjoy Kadison's observation, that the involution is strongly continuous when restricted to normal operators [132].

2.2. Von Neumann's bicommutant theorem

2.2.1. For each subset M of $B(H)$ let M' denote the *commutant* of M, i.e.

$$M' = \{x \in B(H) \mid \forall y \in M : xy = yx\}.$$

Clearly M' is a weakly closed algebra and if $M = M^*$ then M' is a C^*-algebra. We shall write M'', M''' etc. instead of $(M')'$, $((M')')'$ etc.

If $M_1 \subset M_2$ then $M_1' \supset M_2'$. Since $M \subset M''$ this implies that $M' = M'''$ and $M'' = M''''$ for every subset of $B(H)$.

2.2.2. THEOREM. *Let $\mathcal{M}$ be a C^*-subalgebra of $B(H)$ containing the identity operator. The following conditions are equivalent:*

(i) $\mathcal{M} = \mathcal{M}''$.

(ii) *$\mathcal{M}$ is weakly closed.*

(iii) *$\mathcal{M}$ is strongly closed.*

Proof. The implications (i) $\Rightarrow$ (ii) $\Leftrightarrow$ (iii) are clear from 2.2.1 and 2.1.6. (iii) $\Rightarrow$ (i): For a fixed ξ in H let p be the projection on the closure of the subspace of vectors $x\xi$, $x \in \mathcal{M}$. Note that $p\xi = \xi$ since $1 \in \mathcal{M}$. Since $pxp = xp$ for each x in $\mathcal{M}$ we have $p \in \mathcal{M}'$. Let y be a fixed element of $\mathcal{M}''$. Then $py = yp$, whence $y\xi \in pH$. Thus for each $\varepsilon > 0$ there is an x in $\mathcal{M}$ with $\|(y - x)\xi\| < \varepsilon$.

Take $\xi_1, \xi_2, \ldots, \xi_n$ in H. With notation as in 2.1.4 put $\xi = \xi_1 \oplus \cdots \oplus \xi_n$. An immediate calculation shows that

$$\rho(\mathcal{M})' = \{x \in B(K) \mid x_{ij} \in \mathcal{M}'\}.$$

Therefore $\rho(y) \in \rho(\mathcal{M})''$. We can then apply the first part of the proof with $\rho(\mathcal{M})$, $\rho(y)$ and K in place of $\mathcal{M}$, y and H and obtain an x in $\mathcal{M}$ such that

$$\sum_{k=1}^{n} \|(y - x)\xi_k\|^2 = \|(\rho(y) - \rho(x))\xi\|^2 < \varepsilon^2.$$

It follows that we can approximate y strongly from $\mathcal{M}$, whence $y \in \mathcal{M}$ and the theorem is proved.

2.2.3. LEMMA. *Let $\{x_\lambda\}$ be an increasing net of self-adjoint operators in $B(H)$, i.e. $\lambda < \mu$ implies $x_\lambda \leq x_\mu$. If $\|x_\lambda\| \leq \gamma$ for some γ in $\mathbf{R}$ and all λ then $\{x_\lambda\}$ is strongly convergent to an element x in $B(H)_{sa}$ with $\|x\| \leq \gamma$.*

Proof. The proof is left to the reader.

2.2.4. We say that a C^*-subalgebra A of $B(H)$ acts *non-degenerately* on H if for each non-zero vector ξ there is an element x in A with $x\xi \neq 0$. If $\{u_\lambda\}$ is an approximate unit for A then $\{u_\lambda\}$ is a bounded increasing net in $B(H)_{sa}$, hence strongly convergent to an element e by 2.2.3. For each x in A, $\{u_\lambda x\}$ is norm convergent to x and strongly convergent to ex. It follows that $x = ex$ so that e is a unit for A. But then $x = ex$ for each x in the weak closure of A as well. In particular $e = e^2$ so that e is a projection in $B(H)$. It is clear that A acts non-degenerately on eH and, since $A(1 - e) = 0$, it will act degenerately on any larger space. Since $A + \mathbf{C}e$ is a C^*-algebra with the same strong closure as A, we have the following corollary to 2.2.2.

2.2.5. COROLLARY. *Let A be a C^*-subalgebra of $B(H)$ with strong closure $\mathcal{M}$. Then $\mathcal{M}$ is a weakly closed C^*-algebra with unit and if A acts non-degenerately on H then $\mathcal{M} = A''$.*

2.2.6. A *von Neumann algebra* is a strongly (= weakly) closed C^*-subalgebra of $B(H)$. Apart from the case where the unit of a von Neumann algebra $\mathcal{M}$ is not the identity operator on H, a case we can always avoid by working on a smaller Hilbert space, we see from 2.2.2 that von Neumann algebras are characterized by the condition $\mathcal{M} = \mathcal{M}''$. Since the spectral projection of a normal operator x (indeed every spectral Borel function of x) commutes with all operators that commutes with x we see that $\mathcal{M}$ contains with each normal element all its spectral projections. Thus $\mathcal{M}$ is the norm closure of the linear span of its projections.

2.2.7. For each operator x in $B(H)$ we define the *range projection* of x (denoted by $[x]$) as the projection on the closure of xH. If $x \geqslant 0$ then the sequence $\{[(1/n) + x]^{-1}x\}$ is monotone increasing to $[x]$. If p and q are projections then $p \vee q = [p + q]$ and thus $p \wedge q = 1 - [2 - (p + q)]$. Since $[x]H$ is the orthogonal complement of the null space of x^* we have $[x] = [xx^*]$. If therefore $\mathcal{M}$ is a von Neumann algebra in $B(H)$ then $[x] \in \mathcal{M}$ for each x in $\mathcal{M}$.

2.2.8. An operator u on H is called a *partial isometry* if u^*u is a projection (cf. 1.4.7). Since

$$(uu^*)^3 = u(u^*u)(u^*u)u^* = (uu^*)^2$$

this implies that also uu^* is a projection. Note that u maps the space u^*uH isometrically onto uu^*H and maps $(1 - u^*u)H$ to zero; hence the name partial isometry. We next prove the existence of a *polar decomposition.*

2.2.9. PROPOSITION. *For each element x in a von Neumann algebra $\mathcal{M}$ there is a unique partial isometry u in $\mathcal{M}$ with $u^*u = [|x|]$ and $x = u|x|$.*

Proof. Consider the sequence $u_n = x[(1/n) + |x|]^{-1}$. Since $x = x[|x|]$ we have $u_n = u_n[|x|]$. A short computation shows that

$$(u_n - u_m)^*(u_n - u_m) = ([(1/n) + |x|]^{-1} - [(1/m) + |x|]^{-1})^2 |x|^2,$$

and this tends strongly, hence weakly, to zero by spectral theory. It follows that $\{u_n\}$ is strongly convergent to an element u in $\mathcal{M}$ with $u[|x|] = u$. Since $\{u_n|x|\}$ is norm convergent to x we have $x = u|x|$. Then $x^*x = |x|u^*u|x|$ which implies that $u^*u \geqslant [|x|]$. Hence $u^*u = [|x|]$, in particular u is a partial isometry.

If $x = v|x|$ and $v^*v = [|x|]$ then from $v|x| = u|x|$ we get $v = v[|x|] = u$, so that u is unique.

2.2.10. It is perhaps of interest to note that in a von Neumann algebra $\mathcal{M}$ each element x can be written $x = v|x|$ with v an extreme point in the unit ball of $\mathcal{M}$, but then no longer in a unique way. To see this note first that by 2.2.9 the set V of elements v in the unit ball of $\mathcal{M}$ for which $x = v|x|$ is non-empty, convex and weakly compact. If $v \in V$ and v^*v is not a projection then as in the proof of 1.4.7 we have a non-trivial decomposition

$$v = \tfrac{1}{2}[v(2 - a) + va]$$

in the unit ball of $\mathcal{M}$, with $a = (v^*v)^{1/3}$. But since $x = v|x|$ we have $|x|^2 = |x|v^*v|x|$, whence $v^*v|x| = |x|$ so that also $a|x| = |x|$. It follows that $va|x| = x$ and $v(2 - a)|x| = x$. If therefore v is extreme in V then v must be a partial isometry. Put $p = v^*v$ and $q = vv^*$. If $y = (1 - q)z(1 - p)$ with $\|z\| \leq 1$ then $v \pm y \in V$ since $(1 - p)|x| = 0$. As v is extreme this implies that $y = 0$, whence $(1 - q)\mathcal{M}(1 - p) = 0$; so that v is extreme in the unit ball of $\mathcal{M}$ by 1.4.7.

2.2.11. Notes and remarks. The bicommutant theorem appears in [173]. It is probably the most important single theorem in von Neumann algebra theory. The polar decomposition goes back to [174].

2.3. Kaplansky's density theorem

2.3.1. We say that a continuous function f on $\boldsymbol{R}$ is *strongly continuous* if for each strongly continuous net $\{x_i\}$ in $\boldsymbol{B}(H)_{\text{sa}}$ with limit x the net $\{f(x_i)\}$ is strongly convergent to $f(x)$.

2.3.2. PROPOSITION. *Each continuous function f on $\boldsymbol{R}$ such that $f(0) = 0$ and $|f(t)| \leq \alpha|t| + \beta$ for some positive α and β is strongly continuous.*

Proof. Let S denote the set of strongly continuous functions on $\boldsymbol{R}$ and let S^b denote the bounded elements of S. Then S is a uniformly closed self-adjoint vector space of functions and the equation $(*)$ in 2.1.2 shows that $S^bS \subset S$. In particular S^b is a C^*-algebra.

Set $e(t) = (1 + t^2)^{-1}t$. If $x, y \in B(H)_{\text{sa}}$ then

$$e(y) - e(x) = (1 + y^2)^{-1}[y(1 + x^2) - (1 + y^2)x](1 + x^2)^{-1}$$

$$= (1 + y^2)^{-1}(y - x)(1 + x^2)^{-1} + (1 + y^2)^{-1}y(x - y)x(1 + x^2)^{-1}.$$

Since $\|(1 + y^2)^{-1}\| \leq 1$ and $\|(1 + y^2)^{-1}y\| \leq 1$ we see that $e(y)$ tend strongly to $e(x)$ when y tend strongly to x. Therefore $e \in S^b$ and thus $t \to [1 + (\varepsilon t)^2]^{-1}\varepsilon t$ belongs to S^b for each $\varepsilon > 0$. But these functions separate points of $\boldsymbol{R}\setminus\{0\}$ so that by the Stone–Weierstrass theorem $C_0(\boldsymbol{R}\setminus\{0\}) \subset S^b$.

Let f be as specified in the proposition. Then $t \to f(t)(1 + t^2)^{-1}$ belongs to $C_0(\boldsymbol{R}\setminus\{0\})$. Since the function $t \to t$ belongs to S this implies that $t \to f(t)(1 + t^2)^{-1}t$ belongs to S. But this function is bounded, so that also $t \to f(t)(1 + t^2)^{-1}t^2$ belongs to S. It follows that

$$f = f \cdot (1 + t^2)^{-1}t^2 + f \cdot (1 + t^2)^{-1} \in S.$$

2.3.3. THEOREM. *Let A be a C^*-subalgebra of $B(H)$ with strong closure $\mathcal{M}$. Then the unit ball A^1 of A is strongly dense in the unit ball $\mathcal{M}^1$ of $\mathcal{M}$. Furthermore, A^1_{sa} (respectively A^1_+) is strongly dense in $\mathcal{M}^1_{sa}$ (respectively $\mathcal{M}^1_+$). Finally, if $1 \in A$ the unitary group of A is strongly dense in the unitary group of $\mathcal{M}$.*

Proof. Since A_{sa} is convex its strong and weak closures coincide by 2.1.6. As the involution is weakly continuous this closure is $\mathcal{M}_{sa}$. If therefore $x \in \mathcal{M}^1_{sa}$ there is a strongly convergent net $\{x_i\}$ in A_{sa} with limit x. Let $f(t) = (t \wedge 1) \vee (-1)$. Then f is strongly continuous by 2.3.2 so that $f(x_i) \to f(x)$. Since $f(x) = x$ and $\{f(x_i)\} \subset A^1_{sa}$ we see that A^1_{sa} is strongly dense in $\mathcal{M}^1_{sa}$. If $x \in \mathcal{M}^1_+$ we take instead $f(t) = (t \wedge 1) \vee 0$ and have $f(x_i) \to f(x) = x$ with $\{f(x_i)\} \subset A^1_+$, which prove that A^1_+ is strongly dense in $\mathcal{M}^1_+$.

If u is a unitary element in $\mathcal{M}$ and log is a discontinuous but Borel measurable branch of the logarithm function defined on a neighbourhood of the unit circle, then the element $x = i \log u$ is an element of $\mathcal{M}_{sa}$. From the above there is a strongly convergent net $\{x_i\}$ in A_{sa} with limit x. If $1 \in A$ we can form $u_i = \exp(ix_i)$ in A, and by 2.3.2 the net $\{u_i\}$ converges strongly to $\exp(ix) = u$.

To prove the general case note first that by 2.2.4 the unit belongs to the strong closure of A^1, so that A^1 and $\tilde{A}^1$ have the same strong closure. There is therefore no lack of generality in assuming that $1 \in A$. By 1.1.10 each element x in the unit ball of $\mathcal{M}$ can be approximated in norm by a finite convex combination of unitaries. Each unitary in $\mathcal{M}$ can be approximated strongly by unitaries in A from what we proved above; and then the convex combination of unitaries from A, which belongs to A^1, will approximate x strongly. This completes the proof.

2.3.4. *Notes and remarks*. The density theorem is Kaplansky's great gift to mankind [139]. It can be used every day, and twice on Sundays. The converse of 2.3.2 is proved in [132].

2.4. The up–down theorem

2.4.1. For each subset M of $B(H)_{sa}$ let M_σ (resp. M_δ) denote the set of operators in $B(H)_{sa}$ which can be obtained as strong limits of monotone increasing (resp.

decreasing) sequences from M. Let M^m and M_m denote the sets obtained by using instead increasing and decreasing nets from M. Clearly $M \subset M_\sigma \subset M^m$ and $M_\delta = -(-M)_\sigma$. If M is strongly closed then $M^m = M_m = M$. We shall investigate the converse of this statement.

2.4.2. LEMMA. *Let A be a C^*-subalgebra of $B(H)$ with strong closure $\mathscr{M}$. If p is a projection in $\mathscr{M}$ then for each sequence $\{\xi_i\}$ of unit vectors in H there is an element y in $((A_+^1)_\sigma)_\delta$ such that $y(1-p)\xi_i = 0$ and $(1-y)p\xi_i = 0$ for all i.*

Proof. We shall approximate p strongly on vectors of the form $p\xi_i$ and $(1-p)\xi_i$. By 2.3.3 we can find a sequence $\{x_n\}$ in A_+^1 such that $\|p\xi_i - x_n p\xi_i\| \leqslant n^{-1}$ and $\|x_n(1-p)\xi_i\| < n^{-1}2^{-n}$ for all $i \leqslant n$.

For $n < m$ define

$$y_{nm} = \left(1 + \sum_{k=n}^{m} kx_k\right)^{-1} \sum_{k=n}^{m} kx_k.$$

By spectral theory $y_{nm} \in A_+^1$ and $y_{nm} \leqslant \Sigma_{k=n}^m kx_k$. Thus for $i \leqslant n$

$$(*) \qquad (y_{nm}(1-p)\xi_i \mid (1-p)\xi_i) \leqslant \sum_{k=n}^{m} 2^{-k} < 2^{-n+1}.$$

Since $\Sigma_{k=n}^m kx_k \geqslant mx_m$ we have $y_{nm} \geqslant (1 + mx_m)^{-1} mx_m$ by 1.3.7, hence

$$1 - y_{nm} \leqslant (1 + mx_m)^{-1} \leqslant (1+m)^{-1}(1 + m(1 - x_m)).$$

It follows that for $i \leqslant m$

$$(**) \qquad ((1 - y_{nm})p\xi_i \mid p\xi_i) \leqslant 2(1+m)^{-1}.$$

For fixed n the sequence $\{y_{nm}\}$ is monotone increasing, hence strongly convergent to an element y_n in $(A_+^1)_\sigma$. Since $y_{n+1,m} \leqslant y_{nm}$, we see that $y_{n+1} \leqslant y_n$, and thus the sequence $\{y_n\}$ is monotone decreasing to an element y in $((A_+^1)_\sigma)_\delta$. From $(*)$ and $(**)$ we have

$$(y_n(1-p)\xi_i \mid (1-p)\xi_i) \leqslant 2^{-n+1} \quad \text{and} \quad ((1-y_n)p\xi_i \mid p\xi_i) \leqslant 0,$$

and therefore, since $0 \leqslant y \leqslant 1$,

$$y(1-p)\xi_i = 0 \quad \text{and} \quad (1-y)p\xi_i = 0$$

for all i.

2.4.3. THEOREM. *Let A be a C^*-subalgebra of $B(H)$ with strong closure $\mathscr{M}$. If H is separable then $\mathscr{M}_+^1 = ((A_+^1)_\sigma)_\delta$ and $\mathscr{M}_{sa} = ((A_{sa})_\sigma)_\delta$.*

Proof. With $\{\xi_i\}$ as a dense sequence in the unit ball of H we see from 2.4.2 that each projection in $\mathscr{M}$ belongs to $((A_+^1)_\sigma)_\delta$. Assuming that A acts nondegenerately on H we see that 1 is the largest element in $\mathscr{M}_+^1$, whence $1 \in (A_+^1)_\sigma$.

For each x in $\mathcal{M}_+^1$ there is a sequence of spectral projections $\{p_k\}$ such that x is the norm limit of $\Sigma_{k=1}^n 2^{-k}p_k$. To see this take p_1 corresponding to the spectrum in $]\frac{1}{2},1]$, p_2 corresponding to $]\frac{1}{4},\frac{1}{2}] \cup]\frac{3}{4},1]$, p_3 corresponding to $]\frac{1}{8},\frac{1}{4}] \cup]\frac{3}{8},\frac{1}{2}] \cup]\frac{5}{8},\frac{3}{4}] \cup]\frac{7}{8},1]$ etc. Let $\{z_{km}\}$ be a sequence in $(A_+^1)_\sigma$ which decreases to p_k and define

$$x_n = \sum_{k=1}^{n} 2^{-k} z_{kn} + 2^{-n}.$$

Since $(A_+^1)_\sigma$ is convex, $x_n \in (A_+^1)_\sigma$. Furthermore,

$$x_n - x_{n+1} = \sum_{k=1}^{n} 2^{-k}(z_{kn} - z_{k,n+1}) + 2^{-n} - (2^{-n-1} z_{n+1,n+1} + 2^{-n-1}) \geqslant 0$$

so that $\{x_n\}$ is decreasing. Since

$$x_n - x \leqslant \sum_{k=1}^{n} 2^{-k}(z_{kn} - p_k) + 2^{-m}$$

when $n > m$, we have $\operatorname{Lim}(x_n - x) \leqslant 2^{-m}$ for all m, and thus $x \in ((A_+^1)_\sigma)_\delta$.

To show that $\mathcal{M}_{\text{sa}} = ((A_{\text{sa}})_\sigma)_\delta$ note that any x in $\mathcal{M}_{\text{sa}}$ can be written in the form $\alpha y - \beta$ with α and β positive and y in $\mathcal{M}_+^1$. Then $\alpha y \in ((A_{\text{sa}})_\sigma)_\delta$ from the above and

$$-\beta \in -(A_{\text{sa}})_\sigma = (A_{\text{sa}})_\delta \subset ((A_{\text{sa}})_\sigma)_\delta.$$

Since $((A_{\text{sa}})_\sigma)_\delta$ is closed under addition $x \in ((A_{\text{sa}})_\sigma)_\delta$.

2.4.4. THEOREM. *A C^*-subalgebra $\mathcal{M}$ of $B(H)$ is a von Neumann algebra if and only if $(\mathcal{M}_{\text{sa}})^m = \mathcal{M}_{\text{sa}}$.*

Proof. The condition is obviously necessary. Assume now that $\mathcal{M}_{\text{sa}}$ is monotone closed. By 2.2.4 we may then assume that $1 \in \mathcal{M}$. To prove that $\mathcal{M}$ is a von Neumann algebra it suffices to show that each projection p in the strong closure of $\mathcal{M}$ belongs to $\mathcal{M}$. If $\xi \in pH$ and $\eta \in (1-p)H$ there is by 2.4.2 an element y in $\mathcal{M}_+$ such that $y\xi = \xi$ and $y\eta = 0$. The range projection $p_{\xi\eta}$ of y belongs to $\mathcal{M}$ and $p_{\xi\eta}\xi = \xi$, $p_{\xi\eta}\eta = 0$. The projections $p_{\xi\eta_1} \wedge p_{\xi\eta_2} \wedge \cdots \wedge p_{\xi\eta_n}$ form a decreasing net in $\mathcal{M}_+$ when $\{\eta_1 \ldots \eta_n\}$ runs through the finite subsets of $(1-p)H$. Thus the limit projection $p_\xi \leqslant p$. Clearly p is the limit of the increasing net of projections $p_{\xi_1} \vee p_{\xi_2} \cdots p_{\xi_n}$ where $\{\xi_1, \xi_2 \ldots \xi_n\}$ runs through the finite subsets of pH. Thus $p \in \mathcal{M}$ and the theorem is proved.

2.4.5. Notes and remarks. The up–down theorem (2.4.3) is due to the author [196, 197], strongly influenced by the paper of Kadison [127] where 2.4.4 is proved.

2.5. Normal morphisms and ideals

2.5.1. Let $\mathcal{M}$ and $\mathcal{N}$ be von Neumann algebras in $\boldsymbol{B}(H)$ and $\boldsymbol{B}(K)$, respectively. A positive linear map ρ of $\mathcal{M}$ into $\mathcal{N}$ is said to be *normal* if for each bounded, monotone increasing net $\{x_i\}$ in $\mathcal{M}_{sa}$ with limit x, the net $\{\rho(x_i)\}$ increases to $\rho(x)$ in $\mathcal{N}_{sa}$. Since the internal structure that distinguishes C^*-subalgebras of $\boldsymbol{B}(H)$ and von Neumann algebras in $\boldsymbol{B}(H)$ is the monotone completeness of von Neumann algebras (cf. 2.4.4), the normalcy of maps between von Neumann algebras is a natural condition.

It is evident that if a positive map $\rho:\mathcal{M}\to\mathcal{N}$ is strong–strong continuous then ρ is normal. The converse is false. To see this take $\mathcal{M}=\boldsymbol{B}(H)$ with $\dim H=\infty$ and $\mathcal{N}=\boldsymbol{B}(K)$ and let ρ be the amplification of $\boldsymbol{B}(H)$ into $\boldsymbol{B}(K)$ described in 2.1.4, where K is the orthogonal sum of an infinite number of copies of H. Let $\{\xi_n\}$ be an orthonormal basis for H and define $\eta=\oplus n^{-1}\xi_n$ in K. The functional

$$x\to(\rho(x)\eta\,|\,\eta)=\sum_{n=1}^{\infty}n^{-2}(x\xi_n\,|\,\xi_n)$$

is not strongly continuous on $\boldsymbol{B}(H)$ by 2.1.5. Therefore the morphism ρ, which is evidently normal, cannot be strong–strong continuous.

2.5.2. PROPOSITION. *Any isomorphism between von Neumann algebras is normal.*

Proof. If $\rho:\mathcal{M}\to\mathcal{N}$ is an isomorphism and $\{x_i\}$ is a bounded, monotone increasing net in $\mathcal{M}_{sa}$ with limit x then $\{\rho(x_i)\}$ is increasing with a limit $y\leqslant\rho(x)$. However, $\{\rho^{-1}(\rho(x_i))\}$ is then increasing to the limit $x\leqslant\rho^{-1}(y)$, whence $y=\rho(x)$.

2.5.3 THEOREM. *Let ρ be a normal morphism between von Neumann algebras $\mathcal{M}$ and $\mathcal{N}$. Then* $\ker\rho$ *is strongly closed in $\mathcal{M}$ and $\rho(\mathcal{M})$ is strongly closed in $\mathcal{N}$.*

Proof. We know that $\ker\rho$ is a norm closed ideal of $\mathcal{M}$ and that $\rho(\mathcal{M})$ is a C^*-subalgebra of $\mathcal{N}$. Since $\rho(1)$ is a unit for the strong closure of $\rho(\mathcal{M})$ we may assume that $\rho(1)=1$.

Let $\{x_i\}$ be a bounded increasing net in $(\ker\rho)_{sa}$ with limit x. Then $\rho(x)=\operatorname{Lim}\rho(x_i)=0$ whence $x\in\ker\rho$ and so $\ker\rho$ is a von Neumann algebra by 2.4.4.

Let $\{x_i\}$ be a bounded increasing net in $\rho(\mathcal{M})$ with limit x in $\mathcal{N}$. We may assume that $0\leqslant x_i\leqslant 1$ for all i. Let Λ denote the set of elements y in $\mathcal{M}_+$ such that $\rho(y)=x_i$ for some i. We claim that Λ is an increasing net in $\mathcal{M}_+$. For if $\rho(y_i)=x_i$ and $\rho(y_j)=x_j$ choose x_k such that $x_i\leqslant x_k$ and $x_j\leqslant x_k$. Then choose z_i and z_j in $\mathcal{M}_+$ such that

$$\rho(z_i)=\rho(z_j)=x_k,\qquad z_i\geqslant y_i,\qquad z_j\geqslant y_j.$$

Put $y_k = z_i + |z_j - z_i|$ and note that $y_k \in \Lambda$ with $y_k \geq y_i$ and $y_k \geq y_j$. For each $\varepsilon > 0$ the net

$$\{(1 + \varepsilon y)^{-1} y \,|\, y \in \Lambda\}$$

is bounded (by ε^{-1}) and increasing in $\mathcal{M}_+$ with a limit y_0. Since

$$\rho((1 + \varepsilon y)^{-1} y) = (1 + \varepsilon x_i)^{-1} x_i \nearrow (1 + \varepsilon x)^{-1} x,$$

we conclude from the normalcy of ρ that $(1 + \varepsilon x)^{-1} x = \rho(y_0)$. Since $\rho(\mathcal{M})$ is norm closed and $\|x - (1 + \varepsilon x)^{-1} x\| \leq \varepsilon$ we conclude that $x \in \rho(\mathcal{M})$. Thus $\rho(\mathcal{M})$ is monotone closed and therefore a von Neumann algebra.

2.5.4. PROPOSITION. *Let $\mathcal{M}$ be a von Neumann algebra. For each strongly closed hereditary C*-subalgebra $\mathcal{N}$ of $\mathcal{M}$ there is a unique projection p in $\mathcal{M}$ such that $\mathcal{N} = p\mathcal{M}p$. For each strongly closed left ideal $\mathcal{L}$ of $\mathcal{M}$ there is a unique projection p in $\mathcal{M}$ such that $\mathcal{L} = \mathcal{M}p$. If $\mathcal{L}$ is a two-sided ideal then $p \in \mathcal{M} \cap \mathcal{M}'$.*

Proof. If $\mathcal{N}$ is a von Neumann subalgebra of $\mathcal{M}$ it has a unit p. If $\mathcal{N}_+$ is hereditary then $pxp \in \mathcal{N}_+$ for each x in $\mathcal{M}_+$ so that $p\mathcal{M}p \subset \mathcal{N}$. Since p is a unit for $\mathcal{N}$ we have $\mathcal{N} \subset p\mathcal{M}p$, whence $\mathcal{N} = p\mathcal{M}p$.

If $\mathcal{L}$ is a strongly closed left ideal of $\mathcal{M}$ then $\mathcal{L}^*$ is strongly closed by 2.1.6 and thus $\mathcal{L} \cap \mathcal{L}^* = p\mathcal{M}p$ from the above, using 1.5.2. Since $x \in \mathcal{L}$ if and only if $x^*x = \mathcal{L}_+$ we have $\mathcal{L} = \mathcal{M}p$.

If, furthermore, $\mathcal{L}$ is a two-sided ideal then $u^*pu \leq p$ for every unitary element u. Since the unitary elements in $\mathcal{M}$ form a group this implies that $u^*pu = p$, i.e. $pu = up$. By 1.1.9, $\mathcal{M}$ is linearly spanned by its unitary elements, whence $p \in \mathcal{M}'$ as desired.

2.5.5. COROLLARY. *The image of a von Neumann algebra $\mathcal{M}$ under a normal morphism is isomorphic to $\mathcal{M}q$ for some projection q in $\mathcal{M} \cap \mathcal{M}'$.*

Proof. The kernel of a normal morphism ρ is $\mathcal{M}p$ for some projection p in $\mathcal{M} \cap \mathcal{M}'$ by 2.5.3 and 2.5.4. Put $q = 1 - p$. Then $\mathcal{M} = \mathcal{M}p \oplus \mathcal{M}q$ and ρ is an isomorphism of $\mathcal{M}q$ onto $\rho(\mathcal{M})$.

2.5.6. Notes and remarks. The results in this section belong to the stock in trade of von Neumann algebras. Theorem 2.5.3 was established by Dixmier [57].

2.6. The central cover

2.6.1. If $\mathcal{M}$ is a von Neumann algebra its *centre* $\mathcal{Z} = \mathcal{M} \cap \mathcal{M}'$ is also a von Neumann algebra. Since $\mathcal{Z}$ is commutative $\mathcal{Z}_{sa}$ is a real vector lattice, and since $\mathcal{Z}_{sa}$ is monotone closed by 2.4.4, this vector lattice is complete. The spectrum of $\mathscr{Z}$ (see 1.1.6) is usually too large to be of much help. Since $\mathscr{Z}$ is generated by its projections we see that the topology on $\hat{\mathscr{Z}}$ must have a basis consisting of sets which are both open and closed!

We say that $\mathscr{M}$ is a *factor* if $\mathscr{Z}$ consists only of the scalar multiples of 1. We shall see later that any von Neumann algebra can be 'decomposed' into factors.

2.6.2. The following concept is useful when dealing with a von Neumann algebra $\mathscr{M}$ with a large centre.

For each x in $\mathscr{M}_{sa}$ we define the *central cover* of x (denoted by $c(x)$) as the infimum of all z in $\mathscr{Z}_{sa}$ with $z \geqslant x$. This infimum exists since $\mathscr{Z}_{sa}$ is a complete lattice. If $x \geqslant 0$ then since $x \leqslant \|x\| 1$ we have $x \leqslant c(x) \leqslant \|x\| 1$, whence $\|x\| = \|c(x)\|$. If p is a projection in $\mathscr{M}$ then

$$p = p^2 \leqslant (c(p))^2 \leqslant c(p),$$

whence $c(p)^2 = c(p)$ so that $c(p)$ is a central projection.

2.6.3. LEMMA. *For each projection p in $\mathscr{M}$ we have $c(p) = \vee u^*pu$, where u runs through the unitary group of $\mathscr{M}$ or one of its weakly dense subgroups.*

Proof. Let G be a weakly dense subgroup of the unitary group of $\mathscr{M}$. If $z \in \mathscr{Z}_{sa}$ and $z \geqslant p$ then $z \geqslant u^*pu$. Thus $c(p)$ is a projection majorizing all projections u^*pu, $u \in G$, therefore also majorizing their supremum $q = \vee u^*pu$. Since G is a group $u^*qu = q$ for all u in G, whence $qu = uq$. The elements which commute with q form a weakly closed algebra, and since $\mathscr{M}$ is linearly spanned by its unitaries by 1.1.9, we have $q \in \mathscr{M}'$, whence $q = c(p)$.

The next three lemmas belong naturally to this section although they will not be used before Chapter 4.

2.6.4. LEMMA. *If $x \in \mathscr{M}_{sa}$ and $z \in \mathscr{Z}_{sa}$ then $c(x + z) = c(x) + z$.* If, moreover, *$z \geqslant 0$, then $c(xz) = c(x)z$.*

Proof. Since $x + z \leqslant c(x) + z$ we have $c(x + z) \leqslant c(x) + z$. Replacing z with $-z$ and x with $x + z$ gives $c(x) \leqslant c(x + z) - z$ and the equality follows.

If $z \geqslant 0$ then $xz \leqslant c(x)z$ whence $c(xz) \leqslant c(x)z$. Replacing z with $(z + \varepsilon 1)^{-1}$ and x with $x(z + \varepsilon 1)$ gives $c(x) \leqslant c(x(z + \varepsilon 1))(z + \varepsilon 1)^{-1}$. But $x(z + \varepsilon 1) \leqslant xz + \varepsilon \|x\| 1$ whence $c(x(z + \varepsilon 1)) \leqslant c(xz) + \varepsilon \|x\| 1$. With $\varepsilon \to 0$ we obtain $zc(x) \leqslant c(xz)$, hence $zc(x) = c(xz)$.

2.6.5. LEMMA. *If $\{x_i\}$ is an increasing net in $\mathscr{M}_{sa}$ with limit x, then $\{c(x_i)\}$ increases to $c(x)$.*

Proof. Since $\{c(x_i)\}$ form an increasing net in $\mathscr{Z}_{sa}$ bounded above by $c(x)$, it has a limit z in $\mathscr{Z}_{sa}$ with $z \leqslant c(x)$. But $x_i \leqslant z$ for all i so that $x \leqslant z$ and consequently $c(x) \leqslant z$. Thus $c(x) = z$.

2.6.6. LEMMA. *If* $p_1 \leqslant p_2 \leqslant \cdots \leqslant p_n$ *are projections in* $\mathcal{M}$ *then* $c(\Sigma p_k) = \Sigma c(p_k)$.

Proof. Suppose $z \in \mathcal{Z}_{sa}$ with $z \geqslant \Sigma p_k$. Then in particular $z \geqslant p_n$ and thus $z \geqslant c(p_n)$. Since $c(p_n) - p_n$ is a projection in $\mathcal{M}$ orthogonal to all p_k, and since the three elements z, $c(p_n) - p_n$ and Σp_k commute mutually we conclude from spectral theory that since z majorizes each of the other two it must majorize their sum; i.e.

$$z \geqslant c(p_n) - p_n + \sum_{k=1}^{n} p_k = c(p_n) + \sum_{k=1}^{n-1} p_k.$$

It follows that

$$c\left(\sum_{k=1}^{n} p_k\right) = c\left(c(p_n) + \sum_{k=1}^{n} p_k\right) = c(p_n) + c\left(\sum_{k=1}^{n-1} p_k\right)$$

using 2.6.4. The proof now proceeds by induction.

2.6.7. PROPOSITION. *Let* $\mathcal{M}$ *be a von Neumann algebra and* p *a projection in* $\mathcal{M}'$. *Then the map* $\rho: x \to xp$ *is a normal morphism of* $\mathcal{M}$ *and* $\mathcal{M}p$ *is isomorphic to* $\mathcal{M}c(p)$.

Proof. It is clear that ρ is a strong–strong continuous morphism of $\mathcal{M}$ into $B(H)$. Thus $\mathcal{M}p$ is a von Neumann algebra by 2.5.3 and ker ρ is equal to $\mathcal{M}q$ for some central projection q by 2.5.4. Since $qp = 0$ we have $p \leqslant 1 - q$, whence $c(p) \leqslant 1 - q$ ($\mathcal{M}$ and $\mathcal{M}'$ have the same centre). But $(1 - c(p))p = 0$, whence $1 - c(p) \leqslant q$. Thus $c(p) = 1 - q$ and $\mathcal{M}c(p)$ is isomorphic to $\mathcal{M}p$.

2.6.8. COROLLARY. *If* $\mathcal{M}$ *is a factor then* $\mathcal{M}$ *is isomorphic to* $\mathcal{M}p$ *for any non-zero projection* p *in* $\mathcal{M}'$.

2.6.9. Notes and remarks. The central cover (or central support) of a projection is an ancient notion. The (not very deep) idea to extend the notion to arbitrary positive elements is presented in [198].

2.7. A generalization of Lusin's theorem

2.7.1. A well-known result of Lusin states that if μ is a Radon measure on a locally compact Hausdorff space T then for each f in $L_\mu^\infty(T)$ and each $\varepsilon > 0$ there is a Borel set $E \subset T$ with $\mu(T \setminus E) < \varepsilon$ and a g in $C_0(T)$ such that $f = g$ on E. We shall see in 3.4 that $L_\mu^\infty(T)$ is the prototype of commutative von Neumann algebras and that a bounded measure μ can be represented as a strongly continuous linear functional on $L_\mu^2(T)$. It will then become clear that the result in this section is indeed a generalization of Lusin's theorem, and also that the 'up–down' theorem (2.4.3) is a generalization of a familiar result from measure theory.

2.7.2. LEMMA. *Let A be a C^*-subalgebra of $B(H)$ with strong closure $\mathcal{M}$. For each x in $\mathcal{M}$, each projection p_0 in $\mathcal{M}$, each $\varepsilon > 0$ and each set $\{\xi_1, \ldots, \xi_n\}$ in H there is a projection p_1 in $\mathcal{M}$ with $p_1 \leqslant p_0$ and $\|(p_0 - p_1)\xi_k\| \leqslant \varepsilon$ for all k, and an element y in A with $\|y\| \leqslant \|xp_0\|$ such that $\|(x - y)p_1\| \leqslant \varepsilon$. If $x \in \mathcal{M}_{sa}$ we can find y in A_{sa} with $\|(x - y)p_1\| \leqslant \varepsilon$, but then only with $\|y\| \leqslant \text{Min}\{2\|xp_0\|, \|x\|\}$.*

Proof. For each b in $B(H)$ define $\phi(b) = \Sigma(b\xi_k \mid \xi_k)$. By 2.1.4, ϕ is weakly continuous. By 2.3.3 there is a net $\{y_i\}$ in A converging strongly to xp_0 with $\|y_i\| \leqslant \|xp_0\|$. Then the net $\{p_0(x - y_i)^*(x - y_i)p_0\}$ converges weakly to zero. We can therefore find y in A with $\|y\| \leqslant \|xp_0\|$ such that $\phi(p_0(x - y)^*(x - y)p_0) \leqslant \varepsilon^4$. Let p_1 be the spectral projection of $p_0(x - y)^*(x - y)p_0$ corresponding to the interval $[0, \varepsilon^2[$ multiplied with p_0. Then $p_1 \leqslant p_0$ and

$$p_0 - p_1 \leqslant \varepsilon^{-2} p_0(x - y)^*(x - y)p_0.$$

Since ϕ is order preserving and linear, this implies that $\phi(p_0 - p_1) \leqslant \varepsilon^2$, so that

$$\|(p_0 - p_1)\xi_k\|^2 = ((p_0 - p_1)\xi_k \mid \xi_k) \leqslant \varepsilon^2$$

for all k. Moreover,

$$\|(x - y)p_1\|^2 = \|p_1 p_0(x - y)^*(x - y)p_0 p_1\| \leqslant \varepsilon^2.$$

If $x = x^*$ we must distinguish two cases:

If $\|x\| \leqslant 2\|xp_0\|$ we approximate x and choose y in A_{sa} with $\|y\| \leqslant \|x\|(\leqslant 2\|xp_0\|)$ such that $\phi(p_0(x - y)^*(x - y)p_0) \leqslant \varepsilon^4$. Then we complete the argument as above.

If $\|x\| \geqslant 2\|xp_0\|$ then, since

$$\begin{aligned}\|x - (1 - p_0)x(1 - p_0)\| &= \|xp_0 + p_0x(1 - p_0)\| \\ &\leqslant \|xp_0\| + \|p_0x\| = 2\|xp_0\|,\end{aligned}$$

we can find y in A_{sa} approximating $x - (1 - p_0)x(1 - p_0)$ with $\|y\| \leqslant 2\|xp_0\|(\leqslant \|x\|)$ such that

$$\begin{aligned}&\phi(p_0(x - y)^*(x - y)p_0) \\ &\quad = \phi(p_0(x - (1 - p_0)x(1 - p_0) - y)^*(x - (1 - p_0)x(1 - p_0) - y)p_0) \leqslant \varepsilon^4.\end{aligned}$$

From there on we proceed as before and complete the proof.

2.7.3. THEOREM. *Let A be a C^*-subalgebra of $B(H)$ with strong closure $\mathcal{M}$. For each x in $\mathcal{M}$, each projection p_0 in $\mathcal{M}$, each $\varepsilon > 0$ and each set $\{\xi_1, \ldots, \xi_n\}$ in H*

there is a projection p in $\mathcal{M}$ with $p \leq p_0$ and $\|(p_0 - p)\xi_k\| < \varepsilon$ for all k, and an element y in A with $\|y\| \leq \|xp_0\| + \varepsilon$ such that $xp = yp$. If $x \in \mathcal{M}_{sa}$ we can find y in A_{sa} with $xp = yp$ but then only with $\|y\| \leq \mathrm{Min}\{2\|xp_0\|, \|x\|\} + 2\varepsilon$.

Proof. By 2.7.2 we find a projection p_1 in $\mathcal{M}$ and y_1 in A such that $p_1 \leq p_0$, $\|(p_0 - p_1)\xi_k\| \leq 2^{-1}\varepsilon$ for all k and $\|(x - y_1)p_1\| < 2^{-1}\varepsilon$. In general $\|y_1\| \leq \|xp_0\|$ but if $x = x^*$ then $y_1 = y_1^*$ and $\|y_1\| \leq \mathrm{Min}\{2\|xp_0\|, \|x\|\}$. We now repeat the argument with $x - y_1$, p_1 and $2^{-2}\varepsilon$ instead of x, p_0 and $2^{-1}\varepsilon$, and obtain p_2 and y_2 with $p_2 \leq p_1$ and $\|(p_1 - p_2)\xi_k\| \leq 2^{-2}\varepsilon$ and $\|(x - y_1 - y_2)p_2\| \leq 2^{-2}\varepsilon$. In general $\|y_2\| \leq 2^{-1}\varepsilon$ but if $x = x^*$ then $y_2 = y_2^*$ and

$$\|y_2\| \leq \mathrm{Min}\{2\|(x - y_1)p_1\|, \|x - y_1\|\} \leq 2 \cdot 2^{-1}\varepsilon.$$

Continuing in this fashion we obtain a decreasing sequence $\{p_n\}$ of projections in $\mathcal{M}$ and a sequence $\{y_n\}$ in A such that

$$\|(p_{n-1} - p_n)\xi_k\| < 2^{-n}\varepsilon \qquad \text{and} \qquad \left\|\left(x - \sum_{k=1}^{n} y_k\right)p_n\right\| < 2^{-n}\varepsilon.$$

In general $\|y_{n+1}\| \leq 2^{-n}\varepsilon$, but if $x = x^*$ then $y = y^*$ and $\|y_{n+1}\| < 2^{-n}2\varepsilon$. Let p be the limit of the p_n's and let $y = \Sigma\, y_n$. Then for all k

$$\|(p_0 - p)\xi_k\|^2 = ((p_0 - p)\xi_k \mid \xi_k) = \sum_{k=1}^{\infty} ((p_{n-1} - p_n)\xi_k \mid \xi_k)$$

$$\leq \sum_{n=1}^{\infty} (2^{-n}\varepsilon)^2 < \varepsilon^2$$

as desired. Clearly $xp = yp$. In general

$$\|y\| \leq \|y_1\| + \sum_{n=2}^{\infty} \|y_n\| \leq \|xp_0\| + \sum_{n=2}^{\infty} 2^{-n+1}\varepsilon = \|xp_0\| + \varepsilon.$$

If $x = x^*$ then $y = y^*$ and

$$\|y\| \leq \|y_1\| + \sum_{n=2}^{\infty} \|y_n\| \leq \mathrm{Min}\{2\|xp_0\|, \|x\|\} + 2\varepsilon.$$

2.7.4. COROLLARY. *If p_0 is a finite-dimensional projection in $\mathcal{M}$ then for each x in $\mathcal{M}$ and $\varepsilon > 0$ there is a y in A with $xp_0 = yp_0$ and $\|y\| \leq \|xp_0\| + \varepsilon$. If $x = x^*$ we can choose $y = y^*$ but then $\|y\| \leq \mathrm{Min}\{2\|xp_0\|, \|x\|\} + 2\varepsilon$.*

Proof. If $\{\xi_1, \ldots, \xi_n\}$ form a basis for p_0H, then $p \leq p_0$ and $\|(p_0 - p)\xi_k\|^2 < 1/n$ for all k implies that $p = p_0$. The statement now follows immediately from 2.7.3.

2.7.5. THEOREM. *Let A be a C^*-subalgebra of $B(H)$ with strong closure $\mathcal{M}$. For any finite-dimensional projection p_0 in $\mathcal{M}$ and any x in $p_0\mathcal{M}p_0$ there is a y in A with $\|y\| = \|x\|$ and $yp_0 = x$. If $x = x^*$ we can choose $y = y^*$. If $x \geqslant 0$ we can choose $y \geqslant 0$. If x is unitary on p_0H and $1 \in A$ we can choose y unitary on H.*

Proof. Assume first that $x = x^*$. By 2.7.4 we can choose y in A_{sa} such that $yp_0 = x$. Since

$$y^2p_0 = yx = yp_0x = x^2,$$

it follows from the Stone–Weierstrass theorem that $f(y)p_0 = f(x)p_0$ for each continuous function f on $\boldsymbol{R}$. Let $f(t) = (t \wedge \|x\|) \vee (-\|x\|)$. Then $f(y) \in A_{sa}$, $\|f(y)\| \leqslant \|x\|$ and $f(y)p_0 = x$. If $x \geqslant 0$ we choose instead $f(t) = (t \wedge \|x\|) \vee 0$ and have $f(y) \geqslant 0$ and $f(y)p_0 = x$ (cf. the proof of 2.3.3).

If x is unitary on p_0H we can write $x = \exp(ia)p_0$ with a self-adjoint in $p_0\mathcal{M}p_0$. From the first part of the proof $a = zp_0$ for some z in A_{sa}, and if $1 \in A$ we can form the unitary operator $y = \exp(iz)$ in A with $yp_0 = \exp(ia)p_0 = x$.

In the general case we write $x = u|x|$ with u a unitary operator on the finite dimensional space p_0H. Then from the above, $|x| = zp_0$ and $u = vp_0$ with $z \in A_+$ and v unitary in $\tilde{A}$. Put $y = vz$. Then $y \in A$, $\|y\| = \|z\| = \|x\|$ and

$$yp_0 = vzp_0 = vp_0|x| = u|x| = x.$$

2.7.6. Notes and remarks. The generalized Lusin theorem (2.7.3) is due to Tomita [265], see also Saitô [220]. The idea of using it to obtain 2.7.5 (from which Kadison's celebrated transitivity theorem (3.13.2 (iv)) is an immediate corollary) is borrowed from Takesaki [261].

2.8. Maximal commutative subalgebras

2.8.1. A commutative C^*-subalgebra A of a C^*-algebra B is said to be *maximal commutative* if it is not contained in any larger commutative C^*-subalgebra of B. It is clear then that A contains the centre of B. If A is a commutative C^*-subalgebra of a von Neumann algebra $\mathcal{M}$ then $A \subset A' \cap \mathcal{M}$. Now it is clear that A together with any normal element from $A' \cap \mathcal{M}$ will generate a commutative subalgebra of $\mathcal{M}$. Therefore A is a maximal commutative in $\mathcal{M}$ if and only if $A = A' \cap \mathcal{M}$. In particular A is maximal commutative in $B(H)$ if and only if $A = A'$.

For simplicity assume that H is separable and let $\{\xi_n\}$ be an orthonormal basis for H. If p_n denotes the one-dimensional projection on $\boldsymbol{C}\xi_n$ then the von Neumann algebra $\mathfrak{A}$ generated by the set $\{p_n\}$ is commutative and isomorphic to $l^\infty(\boldsymbol{N})$. Moreover $\mathfrak{A}$ is maximal commutative, a fact which can be proved directly (or inferred from 2.8.3 since $\Sigma 2^{-n}\xi_n$ is cyclic for $\mathfrak{A}$). This indicates the rôle of maximal commutative subalgebras of $\boldsymbol{B}(H)$ as 'generalized bases'.

2.8.2. Let A be a C^*-subalgebra of $\boldsymbol{B}(H)$. For each ξ_0 in H we let $[A\xi_0]$ denote the projection on the closure of the subspace $A\xi_0$, and we call $[A\xi_0]$ the *cyclic projection* generated by ξ_0 and A. As we saw in the proof of 2.2.2, $[A\xi_0] \in A'$. We say that ξ_0 is *cyclic* for A if $[A\xi_0] = 1$.

If $x\xi_0 = 0$ implies $x = 0$ for each x in A we say that ξ_0 is *separating* for A. Since $xx'\xi_0 = x'x\xi_0$ it is immediate that ξ_0 is separating for A' if it is cyclic for A. But the converse is also true; for $(1 - [A\xi_0])\xi_0 = 0$ and $[A\xi_0] \in A'$.

2.8.3. PROPOSITION. *Let $\mathfrak{A}$ be a commutative von Neumann algebra in $\boldsymbol{B}(H)$. If there is a cyclic vector for $\mathfrak{A}$ then $\mathfrak{A}$ is maximal commutative. The converse is true if H is separable.*

Proof. Let ξ_0 be a cyclic vector for $\mathfrak{A}$, and take x' in $\mathfrak{A}'$. There is a sequence $\{x_n\}$ in $\mathfrak{A}$ such that $x_n\xi_0 \to x'\xi_0$. Since all elements in $\mathfrak{A}$ are normal we have

$$\|x_n\xi_0 - x_m\xi_0\| = \|x_n^*\xi_0 - x_m^*\xi_0\|,$$

which shows that $x_n^*\xi_0 \to \eta$ for some η in H. However, for each x in $\mathfrak{A}$

$$\begin{aligned}(\eta \mid x\xi_0) &= \operatorname{Lim}(x_n^*\xi_0 \mid x\xi_0) = \operatorname{Lim}(x^*\xi_0 \mid x_n\xi_0)\\ &= (x^*\xi_0 | x'\xi_0) = (x'^*\xi_0 | x\xi_0),\end{aligned}$$

whence $\eta = x'^*\xi_0$. Since for each x in $\mathfrak{A}$ we have $\|x_n(x\xi_0)\| = \|x_n^*(x\xi_0)\|$ we obtain in the limit

$$\|x'(x\xi_0)\| = \operatorname{Lim}\|xx_n\xi_0\| = \operatorname{Lim}\|xx_n^*\xi_0\| = \|x'^*(x\xi_0)\|$$

which implies that x' is normal. We have shown that $\mathfrak{A}'$ consists entirely of normal elements, whence $\mathfrak{A}'$ is commutative, i.e. $\mathfrak{A}' \subset \mathfrak{A}'' = \mathfrak{A}$ and thus $\mathfrak{A}' = \mathfrak{A}$.

To prove the converse take by Zorn's lemma a maximal set of unit vectors $\{\xi_n\}$ such that the corresponding cyclic projections $\{[\mathfrak{A}\xi_n]\}$ are pairwise orthogonal. Since H is separable the set $\{\xi_n\}$ is countable, and since $\mathfrak{A}$ is maximal commutative each $[\mathfrak{A}\xi_n]$ belongs to $\mathfrak{A}$. If $\Sigma[\mathfrak{A}\xi_n] \neq 1$ then we could find a unit vector ξ orthogonal to all subspaces $\{\mathfrak{A}\xi_n\}$ in contradiction with the maximality of the family $\{\xi_n\}$. Thus $\Sigma[\mathfrak{A}\xi_n] = 1$. Let $\xi_0 = \Sigma 2^{-n}\xi_n$. Then

$$\mathfrak{A}\xi_0 \supset \mathfrak{A}[\mathfrak{A}\xi_n]\xi_0 = \mathfrak{A}\xi_n.$$

It follows that $\mathfrak{A}\xi_0$ is dense in H and the proof is complete.

2.8.4. THEOREM. *Let $\mathcal{M}$ be a C^*-subalgebra of $B(H)$. Then the following conditions are equivalent.*

(i) *$\mathcal{M}$ is a von Neumann algebra.*

(ii) *$\mathcal{M}$ contains the limit of each bounded increasing net of pairwise commuting elements from $\mathcal{M}_{sa}$.*

(iii) *Each maximal commutative C^*-subalgebra of $\mathcal{M}$ is a von Neumann algebra.*

(iv) *$\mathcal{M}$ contains all spectral projections of each element in $\mathcal{M}_{sa}$ and $\mathcal{M}$ contains the sum of any set of pairwise orthogonal projections from $\mathcal{M}$.*

Proof. (i) $\Rightarrow$ (ii) is obvious. (ii) $\Rightarrow$ (iii) follows from 2.4.4 since condition (ii) implies that each maximal commutative C^*-subalgebra of $\mathcal{M}$ is monotone closed. (iii) $\Rightarrow$ (iv) is also obvious. In order to prove (iv) $\Rightarrow$ (i) we shall need two lemmas. We say that a C^*-subalgebra of $B(H)$ is a *concrete AW^*-algebra* if it satisfies condition (iv).

2.8.5. LEMMA. *If $\mathcal{M}$ is a concrete AW^*-algebra it contains the supremum and infimum of any set of projections from $\mathcal{M}$.*

Proof. Let $\{p_i\}$ be a family of projections in $\mathcal{M}$ and put $p = \vee p_i$. Let $\{q_j\}$ be a maximal family of non-zero pairwise orthogonal projections from $\mathcal{M}$ majorized by p. Then $q = \Sigma q_j \in \mathcal{M}$ and $q \leqslant p$. If $p - q \neq 0$ then $p_i(p - q) \neq 0$ for some p_i in $\{p_i\}$. The range projection

$$q_0 = [(1 - q)p_i(1 - q)] = [(p - q)p_i(p - q)]$$

belongs to $\mathcal{M}\backslash\{0\}$ and yet $q_0 \leqslant 1 - q$ and $q_0 \leqslant p$. This contradicts the maximality of $\{q_j\}$ and thus $p = q \in \mathcal{M}$.

From the above we see that $\mathcal{M}$ has a unit, which we may assume to be 1. Then the equation

$$\wedge p_i = 1 - \vee (1 - p_i)$$

shows that $\wedge p_i \in \mathcal{M}$.

2.8.6. LEMMA. *If $\mathcal{M}$ is a concrete AW^*-algebra and p is a projection in the strong closure of $\mathcal{M}$ then for each pair of vectors ξ, η in pH and $(1 - p)H$, respectively, there is a projection $q_{\xi\eta}$ in $\mathcal{M}$ such that $q_{\xi\eta}\xi = \xi$ and $q_{\xi\eta}\eta = 0$.*

Proof. Take $\varepsilon > 0$ and put $\varepsilon_n = \frac{1}{2}4^{-n}\varepsilon$. Let $p_0 = 1$ and $x_0 = 0$. Choose $p_1 = 1$ and choose x_1 in $\mathcal{M}_+^1$ such that

$$\|(x_1 - p)\xi\|^2 < \varepsilon_1, \qquad \|x_1\eta\| < 1, \qquad \|(x_1 - p)\xi\| < 1.$$

Suppose that for all k with $1 \leqslant k \leqslant n$ we have found x_k in $\mathcal{M}_+^1$ and projections p_k in $\mathcal{M}$ such that

(1) $\|(x_k - p)p_k\xi\|^2 < \varepsilon_k, \qquad \|x_k\eta\| < 1/k, \qquad \|(x_k - p)\xi\| < 1/k$

(2) $p_k(x_k - x_{k-1})^2 p_k \leqslant 2^{-k+1}$

(3) $p_k \leqslant p_{k-1}, \qquad \|(p_{k-1} - p_k)\xi\|^2 \leqslant 2^{-k+1}\varepsilon.$

The map $a \to (a\xi \mid \xi)$ restricted to the spectral algebra generated by $p_n(x_n - p)^2 p_n$ is a bounded Radon measure on the spectrum of this element. Since the latter is a closed subset of $[0, 1]$ we can therefore find a continuous real function g on $[0, 1]$ such that $g = 1$ on an open interval $I \subset [2^{-n-1}, 2^{-n}]$ so small that

$$(g(p_n(x_n - p)^2 p_n)\xi \mid \xi) < \tfrac{1}{2}\varepsilon_{n+1}.$$

Let f denote the characteristic function for an interval $]-\infty, t_0[$ with t_0 in I and choose a function h with $0 \leqslant h \leqslant 1$ and support in I such that the function $k = f + h$ is continuous. For each x in $\mathcal{M}_+^1$ let $y = p_n(x - x_n)^2 p_n$ and $z = p_n(x - p)^2 p_n$. Then

$$\begin{aligned} f(y)zf(y) &= (k(y) - h(y))z(k(y) - h(y)) \\ &\leqslant 2k(y)zk(y) + 2h(y)zh(y) \\ &\leqslant 2k(y)zk(y) + 2h(y)^2 \leqslant 2k(y)zk(y) + 2g(y), \end{aligned}$$

using the inequality $(a + b)^*(a + b) \leqslant 2a^*a + 2b^*b$ plus the facts that $\|z\| \leqslant 1$ and $h^2 \leqslant g$. By 2.3.3 we can let x approximate p strongly. Then $k(y)zk(y)$ tends strongly to zero since each factor is bounded and convergent (k is bounded and continuous) by 2.3.2. Since also g is continuous, $g(y)$ tends strongly to $g(p_n(x_n - p)^2 p_n)$. It follows that we can find x_{n+1} in $\mathcal{M}_+^1$ such that, with $p_{n+1} = f(p_n(x_{n+1} - x_n)^2 p_n)p_n$ we have

$$(p_{n+1}p_n(x_{n+1} - p)^2 p_n p_{n+1}\xi \mid \xi) < \varepsilon_{n+1},$$

and moreover such that

$$(p_n(x_{n+1} - x_n)^2 p_n\xi \mid \xi) < \varepsilon_n,$$

$$\|x_{n+1}\eta\| < 1/(n+1), \qquad \|(x_{n+1} - p)\xi\| < 1/(n+1).$$

Since

$$p_{n+1} \leqslant p_n,$$

we see that the pair x_{n+1}, p_{n+1} satisfies (1). Also

$$p_{n+1}(x_{n+1} - x_n)^2 p_{n+1} \leqslant 2^{-n}$$

since $f(t) = 0$ for $t \geq 2^{-n}$, so that (2) is satisfied. Finally, since $f(t) = 1$ for $t < 2^{-n-1}$,

$$p_n - p_{n+1} \leq 2^{n+1} p_n(x_{n+1} - x_n)^2 p_n$$

which shows that

$$((p_n - p_{n+1})\xi \mid \xi) \leq 2^{n+1}\varepsilon_n = 2^{-n}\varepsilon,$$

so that p_{n+1} satisfies (3).

By induction we can thus find a sequence $\{x_n\}$ in $\mathcal{M}_+^1$ and a sequence of projections $\{p_n\}$ in $\mathcal{M}$ satisfying (1), (2) and (3). Let p_0 be the infimum of the decreasing sequence $\{p_n\}$. Then $p_0 \in \mathcal{M}$ by 2.8.5 and

$$((1 - p_0)\xi \mid \xi) = \sum_{n=1}^{\infty} ((p_n - p_{n+1})\xi \mid \xi) \leq \sum_{n=1}^{\infty} 2^{-n}\varepsilon = \varepsilon.$$

For all n we have by (2)

$$\|(x_{n+1} - x_n)p_0\|^2 \leq \|(x_{n+1} - x_n)p_{n+1}\|^2 \leq 2^{-n},$$

which shows that the sequence $\{x_n p_0\}$ is norm convergent. Let y denote the limit of the sequence $\{x_n p_0 x_n\}$. From (1) we have $\|y\eta\| = 0$, and moreover

$$(y\xi \mid \xi) = \operatorname{Lim}(p_0 x_n \xi \mid x_n \xi) = (p_0 \xi \mid \xi) \geq 1 - \varepsilon.$$

Thus $[y]\eta = 0$ and $\|[y]\xi\|^2 \geq 1 - \varepsilon$.

We can construct the projection $[y]$ in $\mathcal{M}$ for each $\varepsilon > 0$. Then letting $q_{\xi\eta}$ be the supremum of all the $[y]$'s we have $q_{\xi\eta} \in \mathcal{M}$ by 2.8.5, $q_{\xi\eta}\eta = 0$ and $q_{\xi\eta}\xi = \xi$.

2.8.7. The proof of the implication (iv) $\Rightarrow$ (i) in 2.8.4 is now easy. For each projection p in the strong closure of $\mathcal{M}$ we fix ξ in pH and let q_ξ denote the infimum of all projections $q_{\xi\eta}$, $\eta \in (1 - p)H$ with $q_{\xi\eta}$ as in 2.8.6. Then $q_\xi \in \mathcal{M}$ by 2.8.5, $q_\xi \xi = \xi$ and $q_\xi \leq p$. Clearly p is the supremum of all projections q_ξ, $\xi \in pH$, whence $p \in \mathcal{M}$ and the theorem is proved.

2.8.8. COROLLARY. *Let $\mathcal{M}$ be a C^*-subalgebra of $\boldsymbol{B}(H)$, with H a separable Hilbert space. Then $\mathcal{M}$ is a von Neumann algebra if it contains the spectral projections of each element in $\mathcal{M}_{sa}$.*

Proof. If $\{p_i\}$ is a family of pairwise orthogonal projections in $\mathcal{M}$ then since H is separable only countably many p_i's are non-zero. By deleting some of the zero projections we obtain a countable family $\{p_n\}$ with the same sum in $\boldsymbol{B}(H)$. The element $x = \Sigma 2^{-n} p_n$ belongs to $\mathcal{M}_{sa}$ since $\mathcal{M}$ is a C^*-algebra and the range projection of x, which belongs to $\mathcal{M}$ by assumption, is equal to $\Sigma\, p_n$. By 2.8.4 $\mathcal{M}$ is a von Neumann algebra.

2.8.9. Notes and remarks. The idea of characterizing a von Neumann algebra as an algebra of operators each of whose maximal commutative subalgebras is weakly closed, goes back to Kaplansky [140]. The result in 2.8.4 is due to the author [199]. Its corollary (2.8.8) characterizes von Neumann algebras (on separable Hilbert spaces) as the only C^*-algebras in which the spectral theorem can be used in its full force.

For any C^*-subalgebra A of $B(H)$ define $a(A)$ as the smallest C^*-subalgebra of $B(H)$ containing all spectral projections of each self-adjoint element in A. It is easy to verify that $A \subset a(A) \subset A''$. If H is separable is then $A'' = a(A)$? This failing, is $A'' = a(a \ldots a(A) \ldots)$ (finitely many steps)? Note that by 2.8.8 a transfinite (but countable) application of the operation a will produce A''.

Chapter 3

Functionals and Representations

In this chapter the abstract C^*-algebra theory is linked with the concrete theory of operators on Hilbert spaces. The main tool is the Gelfand–Naimark–Segal construction which to every positive functional associates a representation on a Hilbert space, in close analogy with the L^2-space associated with a measure.

A global version of this construction yields the universal (i.e. all-containing) representation of any C^*-algebra A, and the bi-commutant A'' of A in this representation, the so-called enveloping von Neumann algebra, then serves as a convenient reference frame for the representation theory. For example, the (equivalence classes of) representations of A correspond bijectively to the central projections in A''.

Another global representation of A as continuous affine functions on its state space stresses the important connections between C^*-algebra theory and the theory of convex compact sets. The main result is the bijective correspondence between indecomposable (irreducible) representations and extreme points in the state space (pure states).

3.1. Positive functionals

3.1.1. Let A^* denote the dual space of a C^*-algebra A (this notation should not lead to confusion with the set of adjoints as long as it is applied only to self-adjoint sets). For each ϕ in A^* define ϕ^* by $\phi^*(x) = \overline{\phi(x^*)}$. We say that ϕ is *self-adjoint* if $\phi = \phi^*$. This is clearly equivalent to the condition that $\phi(A_{sa}) \subset \boldsymbol{R}$. The expression

$$\phi = \tfrac{1}{2}(\phi + \phi^*) + \tfrac{1}{2}(\phi - \phi^*)$$

shows that each ϕ in A^* has a unique decomposition in a self-adjoint and a skew-adjoint part. With $(A^*)_{sa}$ as the real Banach space of self-adjoint elements of A^* we note that $(A^*)_{sa}$ is isometrically isomorphic with the dual of the real Banach space A_{sa}, i.e. $(A^*)_{sa} = (A_{sa})^*$.

3.1.2. A linear functional ϕ on A is called *positive* (in symbols $\phi \geqslant 0$) if $\phi(A_+) \subset \boldsymbol{R}_+$. A positive functional is self-adjoint. If ϕ is positive then for any sequence $\{x_n\}$ in A_+^1 we have

$$\sum_{k=1}^{n} 2^{-k}\phi(x_k) \leqslant \phi\left(\sum_{k=1}^{\infty} 2^{-k}x_k\right),$$

which shows that a positive functional is necessarily bounded on A_+^1 and thus belongs to A^*. The following version of the *Cauchy–Schwarz inequality* provides the link between abstract C^*-algebras and concrete algebras of operators on Hilbert spaces.

3.1.3. THEOREM. *If ϕ is a positive functional on a C^*-algebra A then for all x, y in A*

$$|\phi(y^*x)|^2 \leqslant \phi(y^*y)\phi(x^*x).$$

Proof. For each complex λ we have $\phi((\lambda x + y)^*(\lambda x + y)) \geqslant 0$. With $\lambda = t\phi(x^*y)|\phi(y^*x)|^{-1}$ (note that $\overline{\phi(y^*x)} = \phi(x^*y)$) and t in $\boldsymbol{R}$ this gives

$$t^2\phi(x^*x) + 2t|\phi(y^*x)| + \phi(y^*y) \geqslant 0,$$

from which the theorem follows.

3.1.4. PROPOSITION. *An element ϕ in A^* is positive if and only if $\lim \phi(u_\lambda) = \|\phi\|$ for some approximate unit $\{u_\lambda\}$ in A.*

Proof. If $\phi \geqslant 0$ then $\{\phi(u_\lambda)\}$ is an increasing net in $\boldsymbol{R}_+$ with limit $\alpha \leqslant \|\phi\|$. For each x in A^1 we have, by 3.1.3,

$$|\phi(u_\lambda x)|^2 \leqslant \phi(u_\lambda^2)\phi(x^*x) \leqslant \phi(u_\lambda)\ \|\phi\| \leqslant \alpha\,\|\phi\|.$$

Since $\{u_\lambda\}$ is an approximate unit for A and ϕ is continuous this implies that $|\phi(x)|^2 \leqslant \alpha\,\|\phi\|$, whence $\|\phi\|^2 \leqslant \alpha\,\|\phi\|$ and $\alpha = \|\phi\|$.

To prove the converse suppose that $\{\phi(u_\lambda)\}$ is convergent to $\|\phi\|$. Take x in A_{sa}^1 and write $\phi(x) = \alpha + i\beta$ with α, β in $\boldsymbol{R}$. Adjusting the sign of x we may assume that $\beta \geqslant 0$. Take u_λ such that $\|xu_\lambda - u_\lambda x\| < 1/n$. Then

$$\|nu_\lambda - ix\|^2 = \|n^2u_\lambda^2 + x^2 - in(xu_\lambda - u_\lambda x)\| \leqslant n^2 + 2.$$

However,

$$\operatorname{Lim}|\phi(nu_\lambda - ix)|^2 = (n\,\|\phi\| + \beta)^2 + \alpha^2.$$

This implies that

$$(n\,\|\phi\| + \beta)^2 + \alpha^2 \leqslant (n^2 + 2)\|\phi\|^2$$

for all n, whence $\beta = 0$. Therefore $\phi = \phi^*$.

If $x \in A_+^1$ then $u_\lambda - x \in A_{sa}^1$ so that $\phi(u_\lambda - x) \leqslant \|\phi\|$. Passing to the limit this implies that $\phi(x) \geqslant 0$, whence $\phi \geqslant 0$ and the proof is complete.

3.1.5. LEMMA. *Let $\tilde{A}$ be the C^*-algebra obtained by adjoining a unit to the C^*-algebra A. For each positive functional ϕ on A define an extension $\tilde{\phi}$ on $\tilde{A}$ by setting $\tilde{\phi}(1) = \|\phi\|$. Then $\tilde{\phi}$ is positive on $\tilde{A}$ and $\|\tilde{\phi}\| = \|\phi\|$.*

Proof. If $\{u_\lambda\}$ is an approximate unit for A then for each x in A and each complex λ,

$$\begin{aligned}\operatorname{Lim\,sup} \|\lambda u_\lambda + x\|^2 &= \operatorname{Lim\,sup} \| \, |\lambda|^2 u_\lambda^2 + \bar{\lambda} u_\lambda x + \lambda x^* u_\lambda + x^* x\| \\ &\leqslant \operatorname{Lim\,sup} \| \, |\lambda|^2 1 + \bar{\lambda} u_\lambda x + \lambda x^* u_\lambda + x^* x\| = \|\lambda 1 + x\|^2,\end{aligned}$$

using 1.3.5. It follows from 3.1.4 that

$$|\tilde{\phi}(\lambda 1 + x)| = \operatorname{Lim} |\phi(\lambda u_\lambda + x)| \leqslant \|\lambda 1 + x\| \, \|\phi\|,$$

so that $\|\tilde{\phi}\| = \|\phi\|$. Since $\tilde{\phi}(1) = \|\tilde{\phi}\|$, we have $\tilde{\phi} \geqslant 0$ by 3.1.4.

3.1.6. PROPOSITION. *Let B be a C^*-subalgebra of A. For each positive functional ϕ on B there is a norm preserving extension of ϕ to a positive functional on A. If B is hereditary this extension is unique.*

Proof. Adjoin a unit 1 to A and let $\tilde{A}$ (resp. $\tilde{B}$) be the C^*-algebra generated by 1 and A (resp. B). Even though B already had a unit the argument in 3.1.5 shows that the definition $\tilde{\phi}(1) = \|\phi\|$ gives a positive norm preserving extension of ϕ to $\tilde{B}$. By the Hahn–Banach theorem we may then extend $\tilde{\phi}$ to a functional ψ on $\tilde{A}$ such that $\|\psi\| = \|\tilde{\phi}\|$. Since $\|\psi\| = \psi(1)$, $\psi \geqslant 0$ by 3.1.4. The restriction of ψ to A gives the desired extension of ϕ.

Assume now that B is hereditary and let $\{u_\lambda\}$ be an approximate unit for B. Then $\|\psi\| = \operatorname{Lim} \phi(u_\lambda)$ by 3.1.4 so that $\psi(1 - u_\lambda) \to 0$. Since $u_\lambda \tilde{A} u_\lambda \subset B$ it follows from 3.1.3 that

$$\psi(x) = \operatorname{Lim} \psi(u_\lambda x u_\mu) = \operatorname{Lim} \phi(u_\lambda x u_\lambda)$$

for every x in $\tilde{A}$.

3.1.7. Notes and remarks. If A is commutative, so that $A = C_0(T)$ for some locally compact Hausdorff space T, then the positive functionals on A correspond precisely to the positive bounded measures on T. Thus all the ideas, techniques and results from ordinary measure theory should be tested against the more general setting of positive functionals on a (non-commutative) C^*-algebra. The Cauchy–Schwarz inequality (3.1.3) is the first step along the road.

3.2. The Jordan decomposition

3.2.1. A *state* of a C^*-algebra A is a positive functional of norm one. The set of states of A is denoted by S_A (or just S if no confusion can arise). From 3.1.4 we see that the norm is an additive function on the set of positive functionals, which implies that S is a convex set.

The self-adjoint part of the unit ball of A^* can be identified with the unit ball of $(A_{sa})^*$ and is therefore convex and weak* compact. The positive part of the unit ball of $(A_{sa})^*$, which we shall denote by Q is convex and weak* closed, hence weak* compact. It is easy to see that Q is the convex span of S and the zero functional.

If $1 \in A$ then by 3.1.4

$$S = \{\phi \in A^* \mid \|\phi\| \leqslant 1, \phi(1) = 1\},$$

which shows that S is weak* closed in Q and thus weak* compact. In general this is not true and that is the reason why we have to use the space Q (the quasi-states) although our real interest lies with the state space S.

3.2.2. LEMMA. *The unit ball of $(A_{sa})^*$ is the convex span of S and $-S$.*

Proof. Let K denote the convex span of Q and $-Q$. Then K is also the convex span of S and $-S$ and is a weak* compact subset of the unit ball of $(A_{sa})^*$ (since Q and $-Q$ are both weak* compact).

If $x \in A_{sa}$ and $\lambda \in \mathrm{Sp}(x)\backslash\{0\}$ then $\lambda = \omega(x)$ for some non-zero homomorphism ω on the C^*-subalgebra B of A generated by x. We may regard ω as a state of B and by 3.1.6 there is then a state ϕ of A which extends ω. This shows that

$$\mathrm{Sp}(x)\backslash\{0\} \subset \{\phi(x) \mid \phi \in S\},$$

thus in particular

$$(*) \qquad \|x\| = \mathrm{Sup}\{|\phi(x)| \mid \phi \in S\}.$$

Suppose now that ψ was an element in the unit ball of $(A_{sa})^*$ not belonging to K. By the Hahn–Banach theorem there is an element x in A_{sa} and α in $\boldsymbol{R}$ such that $\psi(x) > \alpha$ but $\phi(x) \leqslant \alpha$ for all ϕ in K. Since K is symmetric this implies that $|\phi(x)| \leqslant \alpha$ for all ϕ in K, whence $\|x\| \leqslant \alpha$ by $(*)$. This contradicts $\psi(x) > \alpha$ and thus K equals the unit ball of $(A_{sa})^*$.

3.2.3. LEMMA. *Let ϕ and ψ be positive functionals on a C^*-algebra A with unit. The following conditions are equivalent:*

(i) $\|\phi - \psi\| = \|\phi\| + \|\psi\|$.

(ii) *For every $\varepsilon > 0$ there is a z in A_+^1 such that $\phi(1 - z) < \varepsilon$ and $\psi(z) < \varepsilon$.*

Proof. (i) $\Rightarrow$ (ii). Since $\phi - \psi \in (A_{sa})^*$ there is an element x in A^1_{sa} such that

$$\phi(x) - \psi(x) + \varepsilon \geqslant \|\phi - \psi\|.$$

But then

$$\phi(x) - \psi(x) + \varepsilon \geqslant \|\phi\| + \|\psi\| = \phi(1) + \psi(1),$$

whence $\phi(1 - x) + \psi(1 + x) < \varepsilon$. Since $0 \leqslant 1 - x \leqslant 2$ and $0 \leqslant 1 + x \leqslant 2$ we can choose $z = \frac{1}{2}(1 + x)$ so that $1 - z = \frac{1}{2}(1 - x)$.

(ii) $\Rightarrow$ (i). Clearly $\|\phi - \psi\| \leqslant \|\phi\| + \|\psi\|$. But if $\phi(1 - z) < \varepsilon$ and $\psi(z) < \varepsilon$ then

$$\begin{aligned}\|\phi\| + \|\psi\| &= \phi(1) + \psi(1) \leqslant \phi(2z - 1) + \psi(1 - 2z) + 4\varepsilon \\ &= (\phi - \psi)(2z - 1) + 4\varepsilon \leqslant \|\phi - \psi\| + 4\varepsilon,\end{aligned}$$

since $\|2z - 1\| \leqslant 1$. As ε is arbitrary, $\|\phi\| + \|\psi\| \leqslant \|\phi - \psi\|$.

3.2.4. We say that two positive functionals ϕ and ψ satisfying the conditions in 3.2.3 are *orthogonal* and write $\phi \perp \psi$. If A has no unit one must replace condition (ii) in 3.2.3 with the existence of x and y in A_+ with $\|x + y\| \leqslant 1$ such that $\phi(x) > \|\phi\| - \varepsilon$ and $\psi(y) > \|\psi\| - \varepsilon$. We shall not need this version.

3.2.5. THEOREM. *For each self-adjoint functional ϕ on a C^*-algebra A there is a unique pair ϕ_+ and ϕ_- of positive functionals such that $\phi = \phi_+ - \phi_-$ and $\phi_+ \perp \phi_-$.*

Proof. By 3.2.2 we have $\phi = \alpha\phi_1 - (1 - \alpha)\phi_2$ with ϕ_1 and ϕ_2 positive and of norm less than or equal to $\|\phi\|$. Put $\phi_+ = \alpha\phi_1$ and $\phi_- = (1 - \alpha)\phi_2$ and observe that

$$\|\phi_+\| + \|\phi_-\| = \alpha\|\phi_1\| + (1 - \alpha)\|\phi_2\| \leqslant \|\phi\|,$$

so that $\phi_+ \perp \phi_-$.

To prove uniqueness of the decomposition assume that $\phi_1, \phi_2, \psi_1, \psi_2$ are positive functionals on A with $\phi_1 \perp \psi_1$, $\phi_2 \perp \psi_2$ and $\phi_1 - \psi_1 = \phi_2 - \psi_2$. Since $\phi_1 + \psi_2 = \psi_1 + \phi_2$ we may adjoin a unit to A and extend the functionals as in 3.1.5 without destroying the relations $\phi_1 - \psi_1 = \phi_2 - \psi_2$ and $\phi_1 \perp \psi_1$, $\phi_2 \perp \psi_2$. Thus we may assume that A has a unit. Using 3.2.3 we have for each $\varepsilon > 0$ that $\phi_1(1 - z) < \varepsilon$ and $\psi_1(z) < \varepsilon$ with $0 \leqslant z \leqslant 1$. Then

$$\phi_2(z) \geqslant \phi_2(z) - \psi_2(z) = \phi_1(z) - \psi_1(z) > \phi_1(1) - 2\varepsilon.$$

Likewise, $\psi_2(1 - z) > \psi_1(1) - 2\varepsilon$ so that

$$\phi_2(z) + \psi_2(1 - z) > \|\phi_1\| + \|\psi_1\| - 4\varepsilon = \|\phi_2\| + \|\psi_2\| - 4\varepsilon.$$

It follows that $\phi_2(1-z)+\psi_2(z)<4\varepsilon$. Since $\phi_1-\phi_2=\psi_1-\psi_2$ we have for each x in A,

$$\begin{aligned}\phi_1(x)-\phi_2(x)&=\phi_1(xz)-\phi_2(xz)+\phi_1(x(1-z))-\phi_2(x(1-z))\\&=\psi_1(xz)-\psi_2(xz)+\phi_1(x(1-z))-\phi_2(x(1-z)).\end{aligned}$$

We have

$$\begin{aligned}|\psi_1(xz)|^2&\leqslant\psi_1(xx^*)\psi_1(z^2)\leqslant\|x\|^2\,\|\psi_1\|\,\psi_1(z)\\&\leqslant\|x\|^2\,\|\phi_1-\psi_1\|\,\psi_1(z),\end{aligned}$$

and similarly for the other summands, whence

$$\begin{aligned}|\phi_1(x)-\phi_2(x)|&\leqslant\|x\|\,\|\phi_1-\psi_1\|^{1/2}(\varepsilon^{1/2}+(4\varepsilon)^{1/2}+\varepsilon^{1/2}+(4\varepsilon)^{1/2})\\&\leqslant\|x\|\,\|\phi_1-\psi_1\|^{1/2}\,6\varepsilon^{1/2}.\end{aligned}$$

Since ε is arbitrary, $\phi_1=\phi_2$ and thus $\psi_1=\psi_2$.

3.2.6. COROLLARY. *Each element in the unit sphere of* $(A_{sa})^*$ *has a unique representation in* $\mathrm{Conv}(S\cup(-S))$.

3.2.7. Notes and remarks. The result in 3.2.5 is due to Grothendieck [104], and generalizes the ordinary Jordan decomposition of a signed measure. The proof of the uniqueness of the decomposition is taken from [194].

3.3. The Gelfand–Naimark–Segal construction

3.3.1. A *representation* of a C^*-algebra A is a pair (π,H) consisting of a Hilbert space H and a morphism π of A into $B(H)$. If $\{(\pi_i,H_i)\}$ is a family of representations we define a representation $\oplus\,\pi_i$ on the Hilbert space $\oplus\,H_i$ by

$$(\oplus\pi_i)(x)\oplus\xi_i=\oplus\pi_i(x)\xi_i$$

for all x in A. In the converse direction, if (π,H) is a representation and p is a projection in $\pi(A)'$ then we define *subrepresentations* π_1 and π_2 on pH and $(1-p)H$ simply by restricting the operators $\pi(x)$ to these subspaces. For the new representations we may write $\pi_1\oplus\pi_2=\pi$.

3.3.2. We say that a representation (π,H) is *non-degenerate* if $\pi(A)$ is non-degenerate on H (see 2.2.4). We say that (π,H) is a *cyclic representation* if there is a cyclic vector in H for $\pi(A)$ (see 2.8.2). A cyclic representation is clearly non-degenerate. Using Zorn's lemma we can for a given non-degenerate representation (π,H) write 1 as a sum of cyclic projections p_i in $\pi(A)'$. Then with π_i as the restriction of π to p_iH we have $\pi=\oplus\,\pi_i$ and each π_i is a cyclic representation. We may therefore concentrate our attention on the cyclic representations of the C^*-algebra.

3.3.3. THEOREM. *For each positive functional ϕ on a C^*-algebra A there is a cyclic representation (π_ϕ, H_ϕ) of A with a cyclic vector ξ_ϕ such that $(\pi_\phi(x)\xi_\phi \mid \xi_\phi) = \phi(x)$ for all x in A.*

Proof. Define the *left kernel* of ϕ as the set

$$L_\phi = \{x \in A \mid \phi(x^*x) = 0\}.$$

Since the set $\{x \in A_+ \mid \phi(x) = 0\}$ is clearly a closed hereditary cone in A_+ we see from 1.5.2 that L_ϕ is a left ideal. Let $x \to \xi_x$ denote the map from A onto the quotient space $A - L_\phi$. Using 3.1.3 we see that the sesquilinear form on $A - L_\phi$ given by $(\xi_x \mid \xi_y) = \phi(y^*x)$ is well-defined and determines a pre-Hilbert space structure on $A - L_\phi$. Let H_ϕ denote the completed Hilbert space. For each x, y in A define $\pi_\phi(x)\xi_y = \xi_{xy}$. Since L_ϕ is a left ideal this gives a homomorphism π_ϕ of A as linear operators on $A - L_\phi$. Furthermore,

$$\|\pi_\phi(x)\xi_y\|^2 = \|\xi_{xy}\|^2 = \phi(y^*x^*xy) \leqslant \|x\|^2\, \phi(y^*y) = \|x\|^2\, \|\xi_y\|^2,$$

which shows that we may extend each $\pi_\phi(x)$ to H_ϕ and obtain a norm-decreasing homomorphism of A into $\boldsymbol{B}(H_\phi)$. Finally,

$$(\pi_\phi(x)\xi_y \mid \xi_z) = \phi(z^*xy) = (\xi_y \mid \pi_\phi(x^*)\xi_z)$$

which shows that $\pi_\phi(x)^* = \pi_\phi(x^*)$ so that (π_ϕ, H_ϕ) is a representation.

If $\{u_\lambda\}$ is an approximate unit for A then for $\lambda < \mu$

$$\|\xi_{u_\mu} - \xi_{u_\lambda}\|^2 = \phi((u_\mu - u_\lambda)^2) \leqslant \phi(u_\mu - u_\lambda).$$

Since $\phi(u_\lambda) \to \|\phi\|$, the net $\{\xi_{u_\lambda}\}$ is convergent with a limit ξ_ϕ in H_ϕ. For each x in A we have

$$\pi_\phi(x)\xi_\phi = \operatorname{Lim} \pi_\phi(x)\xi_{u_\lambda} = \operatorname{Lim} \xi_{xu_\lambda} = \xi_x,$$

since the map $x \to \xi_x$ is continuous (ϕ is continuous). Thus ξ_ϕ is a cyclic vector, and since

$$(\pi_\phi(x^*x)\xi_\phi \mid \xi_\phi) = (\xi_x \mid \xi_x) = \phi(x^*x),$$

we have $(\pi_\phi(y)\xi_\phi \mid \xi_\phi) = \phi(y)$ for all y in A by linearity.

3.3.4. For any positive functional ϕ on A we say that the representation (π_ϕ, H_ϕ) constructed in 3.3.3 is the cyclic representation *associated with* ϕ. It is sometimes convenient to write $(\pi_\phi, H_\phi, \xi_\phi)$ for the representation π_ϕ of A on H with cyclic vector ξ_ϕ. Note that $\|\xi_\phi\|^2 = \|\phi\|$.

In the converse direction, if (π, H) is a non-degenerate representation of A then for each unit vector ξ in H we define a state of A, the *vector state* in (π, H) determined by ξ, by the map

$$x \to (\pi(x)\xi \mid \xi).$$

3.3.5. PROPOSITION. *Let ϕ be a positive functional on a C^*-algebra A and let $(\pi_\phi, H_\phi, \xi_\phi)$ be its associated representation. For each positive functional $\psi \leqslant \phi$, there is a unique element a in $\pi_\phi(A)'$ with $0 \leqslant a \leqslant 1$ such that $\psi(x) = (\pi_\phi(x)a\xi_\phi \mid \xi_\phi)$ for all x in A.*

Proof. If $a \in \pi_\phi(A)'$ and $0 \leqslant a \leqslant 1$ then for each x in A_+

$$0 \leqslant a^{1/2}\pi_\phi(x)a^{1/2} = \pi_\phi(x)a = \pi_\phi(x)^{1/2}a\pi_\phi(x)^{1/2} \leqslant \pi_\phi(x)$$

which shows that $0 \leqslant \psi \leqslant \phi$.

Conversely, if $0 \leqslant \psi \leqslant \phi$, define a sesquilinear form on $A - L_\phi$ (notations as in 3.3.3) by $\langle \xi_x \mid \xi_y \rangle = \psi(y^*x)$. This sesquilinear form is well defined, positive definite and bounded by ϕ, and so there is a unique operator a on H_ϕ such that

$$(a\xi_x \mid \xi_y) = \psi(y^*x) \qquad \text{and} \qquad 0 \leqslant a \leqslant 1.$$

Since

$$(\pi_\phi(x)a\xi_y \mid \xi_z) = (a\xi_y \mid \xi_{x^*z}) = \psi(z^*xy) = (a\xi_{xy} \mid \xi_z) = (a\pi_\phi(x)\xi_y \mid \xi_z)$$

we have $a \in \pi_\phi(A)'$.

3.3.6. We say that two representations (π_1, H_1) and (π_2, H_2) of a C^*-algebra A are *spatially equivalent* (or *unitarily equivalent*) if there is an isometry u of H_1 onto H_2 such that $u\pi_1(x)u^* = \pi_2(x)$ for all x in A. We say that the representations are *equivalent* (or *quasi-equivalent*) if there is an isomorphism ρ of $\pi_1(A)''$ onto $\pi_2(A)''$ such that $\rho(\pi_1(x)) = \pi_2(x)$ for all x in A. Note that spatial equivalence implies equivalence since the map $\rho: a \to uau^*$ of $\boldsymbol{B}(H_1)$ onto $\boldsymbol{B}(H_2)$ is strongly continuous and takes $\pi_1(A)$ onto $\pi_2(A)$, wherefore ρ gives an isomorphism of $\pi_1(A)''$ onto $\pi_2(A)''$.

3.3.7. PROPOSITION. *Two cyclic representations (π_1, H_1) and (π_2, H_2) of a C^*-algebra A with cyclic vectors ξ_1 and ξ_2 are spatially equivalent with an isometry u such that $u\xi_1 = \xi_2$, if and only if $(\pi_1(x)\xi_1 \mid \xi_1) = (\pi_2(x)\xi_2 \mid \xi_2)$ for all x in A.*

Proof. If $u\xi_1 = \xi_2$ then for each x in A

$$(\pi_1(x)\xi_1 \mid \xi_1) = (u^*\pi_2(x)u\xi_1 \mid \xi_1) = (\pi_2(x)\xi_2 \mid \xi_2).$$

Conversely, if we have this equality, define a linear map u from $\pi_1(A)\xi_1$ onto $\pi_2(A)\xi_2$ by $u\pi_1(x)\xi_1 = \pi_2(x)\xi_2$. Since

$$\|u\pi_1(x)\xi_1\|^2 = (\pi_2(x^*x)\xi_2 \mid \xi_2) = (\pi_1(x^*x)\xi_1 \mid \xi_1) = \|\pi_1(x)\xi_1\|^2,$$

we see that u extends to an isometry of $[\pi_1(A)\xi_1]$ onto $[\pi_2(A)\xi_2]$. But ξ_1 and ξ_2 are cyclic vectors, wherefore u is an isometry of H_1 onto H_2. Finally,

$$u\pi_1(x)\pi_1(y)\xi_1 = \pi_2(xy)\xi_2 = \pi_2(x)u\pi_1(y)\xi_1$$

for all x and y in A which shows that $u\pi_1(x) = \pi_2(x)u$ since they coincide on a dense set of vectors. It follows that (π_1, H_1) and (π_2, H_2) are spatially equivalent.

3.3.8. COROLLARY. *If ϕ and ψ are positive functionals on a C^*-algebra A and ψ is dominated by a multiple of ϕ then the representation (π_ψ, H_ψ) is spatially equivalent to a subrepresentation of (π_ϕ, H_ϕ).*

Proof. By 3.3.5 $\psi(x) = (\pi_\phi(x)a^{1/2}\xi_\phi \mid a^{1/2}\xi_\phi)$ with a in $\pi_\phi(A)'$. Thus by 3.3.7 (π_ψ, H_ψ) is spatially equivalent to the subrepresentation of (π_ϕ, H_ϕ) determined by the projection $[\pi_\phi(A)a^{1/2}\xi_\phi]$ in $\pi_\phi(A)'$. (This projection is actually the range projection of a.)

3.3.9. PROPOSITION. *Let ϕ be a normal positive functional on a von Neumann algebra $\mathcal{M}$. Then its associated representation $(\pi_\phi, H_\phi, \xi_\phi)$ is a normal morphism of $\mathcal{M}$.*

Proof. Let $\{x_i\}$ be a bounded monotone increasing net in $\mathcal{M}_{sa}$ with limit x. Then $\{\pi_\phi(x_i)\}$ increases to a limit y in $B(H_\phi)$ and $y \leqslant \pi_\phi(x)$. However, for each z in $\mathcal{M}$

$$(\pi_\phi(x)\xi_z \mid \xi_z) = \phi(z^*xz) = \operatorname{Lim} \phi(z^*x_iz) = \operatorname{Lim}(\pi_\phi(x_i)\xi_z \mid \xi_z) = (y\xi_z \mid \xi_z).$$

It follows that $(\pi_\phi(x)-y)\xi_z = 0$, so that $\pi_\phi(x) = y$ on a dense subspace of H_ϕ, and thus everywhere.

3.3.10. Notes and remarks. The GNS construction (3.3.3) appears in [92] and, in perfected form, in [235], where also 3.3.5 is proved. Note how the Hilbert space H_ϕ, associated with a positive functional ϕ, generalizes the L^2-space associated with a (bounded) measure on a locally compact Hausdorff space (cf. 3.1.7). Thus 3.3.5 may be regarded as a version of the Radon–Nikodyn theorem.

3.4. Commutative von Neumann algebras

3.4.1. Let T be a locally compact Hausdorff space. A positive functional on $C_0(T)$ can be identified with a bounded Radon measure μ on T by the Riesz representation theorem. The construction carried out in 3.3.3 will in the commutative case give $H_\mu = L^2_\mu(T)$ and π_μ will be the representation of $C_0(T)$ as multiplication operators on $L^2_\mu(T)$. As shown in 3.3.7 these are essentially the only cyclic representations of $C_0(T)$.

Let $L^\infty_\mu(T)$ denote the set of equivalence classes of essentially bounded Borel functions on T. We may and shall identify $L^\infty_\mu(T)$ with a C^*-algebra of multiplication operators on $L^2_\mu(T)$.

3.4.2. PROPOSITION. *For each bounded Radon measure μ on a locally compact Hausdorff space T, the strong closure of $\pi_\mu(C_0(T))$ on $L^2_\mu(T)$* is $L^\infty_\mu(T)$.

Proof. Since $\pi_\mu(C_0(T))''$ is commutative with a cyclic vector it is maximal commutative by 2.8.3. As $L^\infty_\mu(T)$ commutes with $\pi_\mu(C_0(T))$ it commutes with its strong closure and consequently

$$L^\infty_\mu(T) \subset \pi_\mu(C_0(T))' = \pi_\mu(C_0(T))''.$$

Since $\pi_\mu(C_0(T)) \subset L^\infty_\mu(T)$ it now remains to prove that $L^\infty_\mu(T)$ is strongly closed. Let $\{x_i\}$ be a bounded increasing net in $L^\infty_\mu(T)_{sa}$. There is then an increasing sequence $\{x_n\}$ in $\{x_i\}$ such that $\operatorname{Lim} \mu(x_n) = \operatorname{Lim} \mu(x_i)$. By Lebesgue's monotone convergence theorem there is an element x in $L^\infty_\mu(T)$ such that $\operatorname{Lim} \mu(x_n y) = \mu(xy)$ for every y in $L^1_\mu(T)$. But this means precisely that $\{x_n\}$ converges strongly to x. For each i there is an increasing sequence $\{x'_n\}$ in $\{x_i\}$ such that $x_i \leqslant x'_n$ and $x_n \leqslant x'_n$. Thus $x'_n \nearrow x'$ where $x_i \leqslant x'$ and $x \leqslant x'$. Since $\mu(x) = \mu(x')$ it follows that $x = x'$. Consequently $x_i \leqslant x$, whence $x_i \to x$ strongly. We have shown that $L^\infty_\mu(T)$ is monotone closed, and therefore strongly closed by 2.4.4.

3.4.3. PROPOSITION. *Each maximal commutative von Neumann algebra on a separable Hilbert space is spatially isomorphic to an algebra $L^\infty_\mu(T)$ on $L^2_\mu(T)$ for some locally compact, second countable Hausdorff space T.*

Proof. If $\mathfrak{A}$ is maximal commutative in $B(H)$ and H is separable there is a cyclic vector ξ_0 for $\mathfrak{A}$ by 2.8.3. Since $B(H)^1$ is a second countable in the strong topology (see 2.4.2), $\mathfrak{A}^1$ is second countable and thus $\mathfrak{A}$ is separable in the strong topology. There is therefore a separable C^*-subalgebra A of $\mathfrak{A}$ such that $A'' = \mathfrak{A}$.

Since A is separable, $A = C_0(T)$ for some second countable, locally compact Hausdorff space T. The map $x \to (x\xi_0 \mid \xi_0)$ on A determines a bounded Radon measure μ on T. We then have the identical representation of A on H with strong closure $\mathfrak{A}$ and the representation on $L^2_\mu(T)$ with strong closure $L^\infty_\mu(T)$. Since these two cyclic representations are associated with the same positive functional we see from 3.3.7 that there is an isometry u of $L^2_\mu(T)$ onto H such that $uL^\infty_\mu(T)u^* = \mathfrak{A}$.

3.4.4. THEOREM. *Each commutative von Neumann algebra on a separable Hilbert space is isomorphic to an algebra $L^\infty_\mu(T)$ for some locally compact, second countable Hausdorff space T and some probability measure μ on T.*

Proof. If $\mathfrak{A}$ is a commutative von Neumann algebra in $B(H)$ and H is separable, let $\{\xi_n\}$ be a maximal family of unit vectors in H such that the cyclic projections $[\mathfrak{A}\xi_n]$ in $\mathfrak{A}'$ are pairwise orthogonal. Necessarily then $\Sigma[\mathfrak{A}\xi_n] = 1$, and since H is separable the family $\{\xi_n\}$ is countable. Consider the vector $\xi_0 = \Sigma 2^{-n}\xi_n$ and the cyclic projection $p = [\mathfrak{A}\xi_0]$. If $x \in \mathfrak{A}$ and $x\xi_0 = 0$ then since $\{x\xi_n\}$ is an orthogonal sequence we have $x\xi_n = 0$, whence $x[\mathfrak{A}\xi_n] = 0$ for all n so that $x = 0$. It follows that $\mathfrak{A}$ is isomorphic to $\mathfrak{A}p$ (cf. 2.6.7) and since ξ_0 is cyclic for $\mathfrak{A}p$ the theorem follows from 2.8.3 and 3.4.3.

3.4.5. Notes and remarks. A characterization of arbitrary commutative von Neumann algebras is also available. The underlying space T need no longer be second countable, and the measure μ is no longer finite (or even σ-finite); but $\mathfrak{A} = L^{\infty}_{\mu}(T)$ as before. See Segal [237, 238, 239].

3.5. The σ-weak topology on $B(H)$

3.5.1. For each positive operator x on some Hilbert space H we define the *trace* of x as

$$(*) \qquad \operatorname{Tr}(x) = \sum (x\xi_i \mid \xi_i) \in [0, \infty],$$

where $\{\xi_i\}$ is an orthonormal basis for H. It is well-known that the trace is independent of the choice of basis, i.e. $\operatorname{Tr}(u^*xu) = \operatorname{Tr}(x)$ for every unitary u on H, from which it follows that $\operatorname{Tr}(x^*x) = \operatorname{Tr}(xx^*)$ for every x in $B(H)$.

We say that x in $B(H)$ is of *trace class* if $\operatorname{Tr}(|x|) < \infty$, and we denote by $T(H)$ the elements in $B(H)$ of trace class. We say that x is a *Hilbert–Schmidt* operator if $\operatorname{Tr}(x^*x) < \infty$, and we denote by $HS(H)$ the set of Hilbert–Schmidt operators in $B(H)$. It is well-known that $T(H)$ consists of those x for which the sum $\Sigma|(x\xi_i \mid \eta_i)|$ is convergent for every two orthonormal bases $\{\xi_i\}$ and $\{\eta_i\}$ in H; and that $HS(H)$ is a Hilbert space with the inner product $(x \mid y) = \operatorname{Tr}(y^*x)$. It follows that both $HS(H)$ and $T(H)$ are norm dense ideals of $C(H)$ and that $T(H) \subset HS(H) \subset C(H)$. Using the definition $(*)$ we can now extend Tr to a linear functional on $T(H)$.

3.5.2. LEMMA. *If* $x \in B(H)$ *and* $y \in T(H)$, *then* $|\operatorname{Tr}(xy)| \leqslant \|x\| \operatorname{Tr}(|y|)$.

Proof. Let $y = u|y|$ be the polar decomposition of y (see 2.2.9). Using the Cauchy–Schwarz inequality on the product $(xu|y|^{1/2})|y|^{1/2}$ we get

$$|\operatorname{Tr}(xy)|^2 \leqslant \operatorname{Tr}(xu|y|u^*x^*)\operatorname{Tr}(|y|)$$
$$= \operatorname{Tr}(|y|^{1/2}u^*x^*xu|y|^{1/2})\operatorname{Tr}(|y|) \leqslant \|x\|^2 \operatorname{Tr}(|y|)^2.$$

3.5.3. If $x, y \in T(H)$ and $x + y = u|x + y|$ is the polar decomposition then by 3.5.2,

$$\operatorname{Tr}(|x + y|) = \operatorname{Tr}(u^*(x + y)) \leqslant \operatorname{Tr}(|x|) + \operatorname{Tr}(|y|).$$

It follows that $x \to \mathrm{Tr}(|x|)$ is a norm on $T(H)$, and since $\mathrm{Tr}(|x|) \geqslant \|x\|$ for all x we see that $(T(H), \mathrm{Tr}(|\cdot|))$ is complete and thus a Banach space.

3.5.4. THEOREM. *The dual of* $\boldsymbol{C}(H)$ *is* $(\boldsymbol{T}(H), \mathrm{Tr}(|\cdot|))$. *The dual of* $(\boldsymbol{T}(H), \mathrm{Tr}(|\cdot|))$ *is* $\boldsymbol{B}(H)$. *The dualities are implemented by the form* $(x, y) \to \mathrm{Tr}(xy)$.

Proof. If ϕ is a bounded linear functional on $\boldsymbol{C}(H)$ then for each x in $\boldsymbol{HS}(H)$ we have

$$|\phi(x)|^2 \leqslant \|\phi\|^2 \|x^*x\| \leqslant \|\phi\|^2 \mathrm{Tr}(x^*x).$$

It follows that ϕ is a continuous functional on the Hilbert space $\boldsymbol{HS}(H)$, whence $\phi(x) = \mathrm{Tr}(xy)$ for some y in $\boldsymbol{HS}(H)$ and all x in $\boldsymbol{HS}(H)$. Let $y = u|y|$ be the polar decomposition of y and let p be a finite-dimensional projection. Then

$$\mathrm{Tr}(p|y|) = \mathrm{Tr}(pu^*y) = \phi(pu^*) \leqslant \|\phi\|.$$

Since this holds for all p, we have $\mathrm{Tr}(|y|) \leqslant \|\phi\|$ and $y \in \boldsymbol{T}(H)$. Conversely by 3.5.2, $\|\phi\| \leqslant \mathrm{Tr}(|y|)$ so that the map $\phi \to y$ is an isometry. It is clear from 3.5.2 that each y in $\boldsymbol{T}(H)$ gives a bounded functional on $\boldsymbol{C}(H)$ and thus

$$\boldsymbol{C}(H)^* = (\boldsymbol{T}(H), \mathrm{Tr}(|\cdot|)).$$

Suppose now that ψ is a bounded functional on $(\boldsymbol{T}(H), \mathrm{Tr}(|\cdot|))$. For each finite dimensional projection p we have $p\boldsymbol{C}(H) = p\boldsymbol{T}(H)$ and thus from the first part of the proof there is an x_p in $\boldsymbol{T}(H)$ such that $\psi(py) = \mathrm{Tr}(x_p y)$ for all y in $\boldsymbol{T}(H)$. Furthermore,

$$\begin{aligned}\|x_p\| &= \mathrm{Sup}\{|\mathrm{Tr}(x_p y)| \mid \dim(yH) = 1, \|y\| = 1\} \\ &\leqslant \mathrm{Sup}\{|\mathrm{Tr}(x_p y)| \mid y \in \boldsymbol{T}(H), \mathrm{Tr}(|y|) = 1\} \leqslant \|\psi\|.\end{aligned}$$

Since the unit ball in $\boldsymbol{B}(H)$ is weakly compact we may assume that the net $\{x_p\}$ is weakly convergent to an x in $\boldsymbol{B}(H)$, with $\|x\| \leqslant \|\psi\|$, when p runs through the monotone increasing net of finite dimensional projections, which form an approximate unit for $\boldsymbol{T}(H)$ under the norm $\mathrm{Tr}(|\cdot|)$. It follows that

$$\psi(y) = \mathrm{Lim}\, \psi(py) = \mathrm{Lim}\, \mathrm{Tr}(x_p y) = \mathrm{Tr}(xy),$$

since for each y in $\boldsymbol{T}(H)$ the functional $x \to \mathrm{Tr}(xy)$ is weakly continuous on the unit ball of $\boldsymbol{B}(H)$. By 3.5.2 $\|\psi\| \leqslant \|x\|$, whence $\|\psi\| = \|x\|$, so that the map $\psi \to x$ is an isometry. It is clear from 3.5.2 that each x from $\boldsymbol{B}(H)$ gives a bounded functional on $(\boldsymbol{T}(H), \mathrm{Tr}(|\cdot|))$ and thus

$$(\boldsymbol{T}(H), \mathrm{Tr}(|\cdot|))^* = \boldsymbol{B}(H).$$

3.5.5. One remarkable fact which we learn from 3.5.4 is that $\boldsymbol{B}(H)$, regarded as a Banach space, is the dual of a Banach space (it is even the bi-dual of a Banach space). We define the *σ-weak topology* on $B(H)$ (also known as the *ultra-weak* topology) as the locally convex vector space topology associated with the family of semi-norms of the form $y \to |\mathrm{Tr}(xy)|$, $y \in B(H)$, $x \in T(H)$. Writing x in $T(H)$ as the product of two Hilbert–Schmidt operators and selecting an orthonormal basis for H it is easy to verify that each functional $y \to \mathrm{Tr}(xy)$ has the form

$$y \to \sum_n (y\xi_n \mid \eta_n), \qquad \xi_n \in H, \qquad \eta_n \in H$$

where $\Sigma_n \|\xi_n\|^2 < \infty$ and $\Sigma_n \|\eta_n\|^2 < \infty$, and that conversely, each functional of this form is σ-weakly continuous. It is then clear that the σ-weak topology is stronger than the weak topology; but as we have already noticed the two topologies coincide on the unit ball of $B(H)$, and therefore coincide on every bounded subset of $B(H)$. It follows from 2.3.3 that a C^*-subalgebra of $B(H)$ is a von Neumann algebra if and only if it is σ-weakly closed.

3.5.6. COROLLARY. *Each von Neumann algebra is isomorphic as a Banach space to the dual of a Banach space.*

Proof. If $\mathcal{M}$ is a von Neumann algebra in $B(H)$ let N denote the set of operators x in $T(H)$ for which the functional $\mathrm{Tr}(x\,\cdot)$ vanishes on $\mathcal{M}$. Since $\mathcal{M}$ is a σ-weakly closed subspace of $B(H)$ it is isomorphic to the dual of the quotient space $T(H) - N$.

3.5.7. Notes and remarks. See Dixmier [54] and [57].

3.6. Normal functionals

3.6.1. Expanding the terminology from 2.5.1 we say that a bounded functional ϕ on a von Neumann algebra $\mathcal{M}$ is *normal* if for each bounded, monotone increasing net $\{x_i\}$ in $\mathcal{M}_{sa}$ with limit x the net $\{\phi(x_i)\}$ converges to $\phi(x)$. From the linearity of ϕ we see that this is equivalent to the condition that $\{\phi(x_i)\}$ converges to zero for each bounded, monotone decreasing net $\{x_i\}$ in $\mathcal{M}_+$ with limit zero. We denote by $\mathcal{M}_*$ the set of normal functionals on $\mathcal{M}$.

3.6.2. PROPOSITION. *The set* $\mathcal{M}_*$ *is a *-invariant norm closed subspace of* $\mathcal{M}^*$. *If* $\phi \in \mathcal{M}_*$ *and* $x \in \mathcal{M}$ *the functionals* $\phi(\cdot\, x)$ *and* $\phi(x\,\cdot)$ *are normal. If, moreover,* $\phi = \phi^*$ *then* ϕ_+ *and* ϕ_- *are both normal.*

Proof. It is clear that $\mathcal{M}_*$ is a *-invariant subspace of $\mathcal{M}^*$. If $\{\phi_n\}$ is a norm convergent sequence in $\mathcal{M}_*$ with limit ϕ then

$$|\phi(x_i)| \leqslant \|x_i\|\,\|\phi - \phi_n\| + |\phi_n(x_i)|.$$

If therefore $\{x_i\}$ is a bounded monotone decreasing net in $\mathcal{M}_+$ with limit zero then $\{\phi(x_i)\}$ converges to zero; whence $\phi \in \mathcal{M}_*$.

If $\phi \in \mathcal{M}_*$ then $\phi(x^* \cdot x) \in \mathcal{M}_*$ for every x in $\mathcal{M}$.

The formula

$$4yx = \sum_{k=0}^{3} i^k(x + i^k)^* y(x + i^k)$$

shows that the functional $y \to \phi(yx)$ is a linear combination of normal functionals, whence $\phi(\cdot x) \in \mathcal{M}_*$; and $\phi(x \cdot) \in \mathcal{M}_*$ since $\mathcal{M}_*$ is *-invariant.

If $\phi = \mathcal{M}_*$ and $\phi = \phi^*$ then for $\varepsilon > 0$ we can, by 3.2.6, find z in $\mathcal{M}_+^1$ such that $\phi_+(1 - z) < \varepsilon$ and $\phi_-(z) < \varepsilon$. Then for each x in $\mathcal{M}$

$$|\phi_+(x) - \phi(zx)| \leqslant |\phi_+((1 - z)x)| + |\phi_-(zx)|$$

$$\leqslant \phi_+(1 - z)^{1/2} \|\phi_+\|^{1/2} \|x\| + \phi_-(z)^{1/2} \|\phi_-\|^{1/2} \|x\| < 2\varepsilon^{1/2} \|\phi\|^{1/2} \|x\|.$$

Since $\phi(z \cdot) \in \mathcal{M}_*$ and $\mathcal{M}_*$ is norm closed it follows that ϕ_+ and therefore also ϕ_- belong to $\mathcal{M}_*$.

3.6.3. LEMMA. *If ϕ is a normal state of $\mathcal{M}$ there is a set $\{p_i\}$ of pairwise orthogonal projections in $\mathcal{M}$ with $\Sigma p_i = 1$ such that each functional $\phi(\cdot p_i)$ is weakly continuous.*

Proof. Let $\{p_i\}$ be a maximal family of pairwise orthogonal projections in $\mathcal{M}$ such that $\phi(\cdot p_i)$ is weakly continuous for each i. If $p_0 = \Sigma p_i \neq 1$ take a unit vector ξ in $(1 - p_0)H$ and put $\psi(x) = 2(x\xi \mid \xi)$ for each x in $\mathcal{M}$. Let $\{q_j\}$ be a maximal family of pairwise orthogonal projections in $(1 - p_0)\mathcal{M}(1 - p_0)$ such that $\phi(q_i) \geqslant \psi(q_i)$ for all i: and put $q_0 = \Sigma q_j$. Since ϕ and ψ are normal, $\phi(q_0) \geqslant \psi(q_0)$. Therefore $q_0 \neq 1 - p_0$. Put $p_1 = 1 - p_0 - q_0$. Then $p_1 \neq 0$ and for each projection $p \leqslant p_1$ we have $\phi(p) < \psi(p)$ by the maximality of q_0. Since each element in $\mathcal{M}_+$ can be approximated in norm by positive linear combinations of projections this shows that $\phi \leqslant \psi$ on $p_1\mathcal{M}p_1$. But then

$$|\phi(xp_1)|^2 \leqslant \phi(p_1x^*xp_1) \leqslant \psi(p_1x^*xp_1) = 2\,\|xp_1\xi\|^2,$$

which implies that the functional $\phi(\cdot p_1)$ is strongly continuous and therefore also weakly continuous by 2.1.5. This contradicts the maximality of the family $\{p_i\}$ and consequently $\Sigma p_i = 1$.

3.6.4. THEOREM. *Let ϕ be a bounded functional on a von Neumann algebra $\mathcal{M}$ in* $\boldsymbol{B}(H)$. *The following conditions are equivalent:*

(i) ϕ *is normal*;

(ii) ϕ *is weakly continuous on the unit ball of* $\mathscr{M}$;

(iii) ϕ *is* σ*-weakly continuous*;

(iv) *There is an operator* x *of trace class on* H *such that* $\phi(y) = \mathrm{Tr}(xy)$ *for all* y *in* $\mathscr{M}$.

Proof. (i) $\Rightarrow$ (ii). By 3.6.2 we may assume that ϕ is positive. Using 3.6.3 we can then (by adding a sufficient number of the p_i's) find a projection p in $\mathscr{M}$ such that $\phi(\cdot p)$ is weakly continuous and $\phi(1 - p) < \varepsilon$. If $\{x_i\}$ is a bounded net converging weakly to zero then

$$|\phi(x_i)| \leqslant |\phi(x_i p)| + \|x_i\| \, \|\phi\|^{1/2} \varepsilon^{1/2},$$

which shows that $\{\phi(x_i)\}$ converges to zero.

(ii) $\Rightarrow$ (iii). We see that ϕ is σ-weakly continuous on the unit ball of $\mathscr{M}$ and thus σ-weakly continuous on any ball around the origin. Since the σ-weak topology on $\mathscr{M}$ is the weak* topology associated with $T(H) - N$ (notations as in 3.5.6), the Krein–Smulian theorem (A2, Appendix) tells us that ϕ is σ-weakly continuous. Moreover, we see that ϕ is associated with an element of $T(H) - N$, which proves (iii) $\Rightarrow$ (iv). We have already noticed that (iv) $\Rightarrow$ (ii) and evidently (ii) $\Rightarrow$ (i) which completes the proof.

3.6.5. From 3.6.4 and 3.5.6 we see that for each von Neumann algebra $\mathscr{M}$ the Banach space $\mathscr{M}_*$ satisfies $(\mathscr{M}_*)^* = \mathscr{M}$. We say that $\mathscr{M}_*$ is the *pre-dual* of $\mathscr{M}$.

3.6.6. PROPOSITION. *For each positive normal functional* ϕ *on a von Neumann algebra* $\mathscr{M}$ *in* $B(H)$ *there is a positive element* x *in* $T(H)$ *such that* $\phi(y) = \mathrm{Tr}(xy)$ *for all* y *in* $\mathscr{M}$. *In particular* $\|\phi\| = \mathrm{Tr}(x)$.

Proof. We know from 3.6.4 that $\phi = \mathrm{Tr}(z\,\cdot)$ for some z in $T(H)$ so that there exists sequences $\{\xi_n\}$ and $\{\eta_n\}$ in H with $\Sigma\|\xi_n\|^2 < \infty$ and $\Sigma\|\eta_n\|^2 < \infty$ such that $\phi(y) = \Sigma(y\xi_n \,|\, \eta_n)$ for all y in $\mathscr{M}$.

With K as the orthogonal sum of an infinite number of copies of H we consider the amplification ρ of $B(H)$ into $B(K)$ described in 2.1.4. Let $\xi = \oplus\, \xi_n$ and $\eta = \oplus\, \eta_n$ so that $\phi(y) = (\rho(y)\xi|\eta)$. If $y \geqslant 0$ we have

$$\begin{aligned}(\rho(y)(\xi + \eta)\,|\,(\xi + \eta)) &\geqslant (\rho(y)(\xi + \eta)\,|\,(\xi + \eta)) - (\rho(y)(\xi - \eta)\,|\,(\xi - \eta)) \\ &= 2((\rho(y)\xi\,|\,\eta) + (\rho(y)\eta\,|\,\xi)) \\ &= 2(\phi(y) + \phi^*(y)) = 4\phi(y).\end{aligned}$$

By 3.3.5 this implies that

$$\phi(y) = (\rho(y)a^{1/2}(\xi + \eta)\,|\,a^{1/2}(\xi + \eta))$$

where a is a positive element in the commutant of $\rho(\mathcal{M})[\rho(\mathcal{M})(\xi + \eta)]$. It follows that

$$\phi(y) = \sum (y\zeta_n \mid \zeta_n)$$

for all y in $\mathcal{M}$, with $\oplus \zeta_n = a^{1/2}(\xi + \eta)$. Choose an orthonormal basis $\{\zeta_n'\}$ for the subspace of H spanned by the ζ_n and let x_1 be the Hilbert–Schmidt operator on H such that $x_1\zeta_n' = \zeta_n$. Then with $x = x_1 x_1^*$ we get

$$\phi(y) = \mathrm{Tr}(x_1^* y x_1) = \mathrm{Tr}(xy)$$

as desired.

3.6.7. PROPOSITION. *To each normal functional ϕ on a von Neumann algebra $\mathcal{M}$ there is a unique positive normal functional $|\phi|$ such that $\| |\phi| \| = \|\phi\|$ and $|\phi(x)|^2 \leqslant \|\phi\| \, |\phi|(x^*x)$ for every x in $\mathcal{M}$. Moreover, there is a partial isometry u $\mathcal{M}$ such that $\phi = |\phi|(u\,\cdot)$ and $|\phi| = \phi(u^*\,\cdot)$.*

Proof. We may assume that $\|\phi\| = 1$. Since ϕ is weakly continuous on $\mathcal{M}^1$ by 3.6.4, the set

$$\{x \in \mathcal{M}^1 \mid \phi(x) = 1\}$$

is a non-empty weakly closed face of $\mathcal{M}^1$, and as $\mathcal{M}^1$ is weakly compact this face contains an extreme point u^* of $\mathcal{M}^1$. By 1.4.7 u (and u^*) is a partial isometry.

The functional $|\phi| = \phi(u^*\,\cdot)$ is evidently normal and since

$$\| |\phi| \| \geqslant |\phi|(1) = \phi(u^*) = 1 \geqslant \| |\phi| \|,$$

we see from 3.1.4 that $|\phi| \geqslant 0$ and $\| |\phi| \| = 1$.

Put $p = u^*u$ and take x in $\mathcal{M}$. Then

$$\begin{aligned}\|nu^* + (1-p)x\|^2 &= \|(nu + x^*(1-p))(nu^* + (1-p)x)\| \\ &= \|n^2uu^* + x^*(1-p)x\| \leqslant n^2 + \|x\|^2.\end{aligned}$$

It follows that

$$|n + \phi((1-p)x)|^2 \leqslant n^2 + \|x\|^2$$

which implies that $\mathrm{Re}\,\phi((1-p)x) \leqslant 0$. Since this holds for any x we conclude that $\phi((1-p)x) = 0$, whence

$$\phi(x) = \phi(px) = \phi(u^*ux) = |\phi|(ux).$$

It follows from the Cauchy–Schwarz inequality that $|\phi(x)|^2 \leqslant |\phi|(x^*x)$ for all x in $\mathcal{M}$.

If ψ were another positive functional on $\mathcal{M}$ with $\|\psi\| = 1$ and $|\phi(x)|^2 \leqslant \psi(x^*x)$ then

$$||\phi|(x)|^2 = |\phi(u^*x)|^2 \leqslant \psi(x^*uu^*x) \leqslant \psi(x^*x).$$

With $x = x^*$ and $\varepsilon > 0$ this implies that

$$(|\phi|(1 + \varepsilon x))^2 \leqslant \psi((1 + \varepsilon x)^2)$$

and since $|\phi|(1) = \psi(1)$ we obtain

$$2|\phi|(x) + \varepsilon(|\phi|(x))^2 \leqslant 2\psi(x) + \varepsilon\psi(x^2),$$

whence $|\phi|(x) \leqslant \psi(x)$ for all x in $\mathcal{M}_{sa}$, i.e. $|\phi| = \psi$.

3.6.8. LEMMA. *If* $\phi \in \mathcal{M}_*$ *and* $x \in \mathcal{M}$ *then* $|\phi(x\cdot)| \leqslant \|x\|\,|\phi|$.

Proof. From 3.6.7 we have partial isometries u and v in $\mathcal{M}$ such that $\phi = |\phi|(u\cdot)$ and $|\phi(x\cdot)| = \phi(xv^*\cdot)$. Put $\psi = |\phi(x\cdot)|$, $\omega = |\phi|$ and $y = uxv^*$ so that $\psi = \omega(y\cdot)$. Then

$$\omega(y\cdot) = \psi = \psi^* = \omega^*(\cdot y^*) = \omega(\cdot y^*).$$

If therefore $z \in \mathcal{M}_+$ then

$$\omega(yz)^2 = \omega(yz^{1/2}z^{1/2})^2 \leqslant \omega(yzy^*)\omega(z)$$
$$= \omega(y^2z)\omega(z).$$

Similarly $\omega(y^2z)^2 \leqslant \omega(y^4z)\omega(z)$ and by induction

$$\omega(yz)^{2^n} \leqslant \omega(y^{2^n}z)\omega(z)^{2^n-1} \leqslant \|y\|^{2^n}\|z\|\,\|\omega\|\,\omega(z)^{2^n-1}.$$

Passing to the limit this yields $\omega(yz) \leqslant \|y\|\,\omega(z)$ and thus

$$|\phi(x\cdot)| = \psi = \omega(y\cdot) \leqslant \|y\|\,\omega \leqslant \|x\|\,|\phi|.$$

3.6.9. LEMMA. *If* $\{\phi_n\}$ *is a norm convergent sequence in* $\mathcal{M}_*$ *with limit* ϕ *then* $\{|\phi_n|\}$ *converges to* $|\phi|$.

Proof. Since $\operatorname{Lim}\|\phi_n\| = \|\phi\|$ we may assume that $\|\phi_n\| = 1$ for all n. By 3.6.7 there are partial isometries u_n and u in $\mathcal{M}$ such that $\phi_n = |\phi_n|(u_n\cdot)$ and $|\phi| = \phi(u^*\cdot)$. Then $\operatorname{Lim}|\phi_n|(u_nu^*) = 1$ and it follows from the Cauchy–Schwarz inequality that $\{|\phi_n|((1 - u_nu^*)\cdot)\}$ converges to zero in norm. But

$$|\phi_n|(u_nu^*\cdot) = \phi_n(u^*\cdot) \to \phi(u^*\cdot) = |\phi|,$$

and thus $\{|\phi_n|\}$ converges to $|\phi|$.

3.6.10. LEMMA. *If* ϕ, ψ *belongs to* $(\mathcal{M}_*)_+$ *with* $\psi \leqslant \phi$ *there is a sequence* $\{x_n\}$ *in* $\mathcal{M}$ *such that* $\{\phi(x_n\cdot)\}$ *converges to* ψ *in norm.*

Proof. By 3.3.5 we have $\psi(x) = (\pi_\phi(x)a\xi_\phi \mid \xi_\phi)$ with a in $\pi_\phi(\mathscr{M})'$. Since ξ_ϕ is cyclic for π_ϕ there is a sequence $\{x_n\}$ in $\mathscr{M}$ such that $a\xi_\phi = \operatorname{Lim} \pi_\phi(x_n^*)\xi_\phi$, whence

$$|\psi(x) - \phi(x_n x)| = |(\pi_\phi(x)\xi_\phi \mid a\xi_\phi - \pi_\phi(x_n^*)\xi_\phi)|$$
$$\leqslant \|x\| \, \|\xi_\phi\| \, \|a\xi_\phi - \pi_\phi(x_n^*)\xi_\phi\| \to 0.$$

3.6.11. THEOREM. *Let $\mathscr{M}$ be a von Neumann algebra with pre-dual $\mathscr{M}_*$. The mappings $p \to (\mathscr{M}p)^\perp$ and $E \to E_+$ define bijective correspondences between the classes of:*

(i) *Projections in $\mathscr{M}$;*

(ii) *Norm closed subspaces of $\mathscr{M}_*$ which are invariant under multiplication from the left by elements of $\mathscr{M}$;*

(iii) *Norm closed hereditary cones in $(\mathscr{M}_*)_+$.*

Proof. If p is a projection in $\mathscr{M}$ the set

$$E = (\mathscr{M}p)^\perp = \{\phi \in \mathscr{M}_* \mid \phi(\mathscr{M}p) = 0\}$$

is a norm closed, left invariant subspace of $\mathscr{M}_*$. Moreover, from the Cauchy–Schwartz inequality it follows that

$$E_+ = \{\phi \in (\mathscr{M}_*)_+ \mid \phi(p) = 0\},$$

so that E_+ is a norm closed hereditary cone in $(\mathscr{M}_*)_+$.

If E is a norm closed, left invariant subspace of $\mathscr{M}_*$ the set $E^\perp$ is a σ-weakly closed left ideal in $\mathscr{M}$ and thus $E^\perp = \mathscr{M}q$ for some projection q in $\mathscr{M}$ by 2.5.4. Since $\mathscr{M}$ and $\mathscr{M}_*$ are in duality the bipolar theorem yields $E = E^{\perp\perp} = (\mathscr{M}q)^\perp$. If E was of the form $(\mathscr{M}p)^\perp$ then the bipolar theorem gives

$$\mathscr{M}p = (\mathscr{M}p)^{\perp\perp} = \mathscr{M}q,$$

whence $p = q$. We have thus established the bijective correspondence between (i) and (ii).

If E is a norm closed, left invariant subspace of $\mathscr{M}_*$ then, as we saw above, E_+ is a hereditary cone in $(\mathscr{M}_*)_+$. Conversely, if P is a norm closed hereditary cone in $(\mathscr{M}_*)_+$, define

$$E = \{\phi \in \mathscr{M}_* \mid |\phi| \in P\}.$$

Since P is hereditary it follows from 3.6.8 that E is left invariant. Moreover, by 3.6.9 E is norm closed. Take ϕ and ψ in E. By 3.6.7 there are partial isometries u, v and w in $\mathscr{M}$ such that

$$|\phi + \psi| = (\phi + \psi)(u^* \cdot), \qquad \phi = |\phi|(v \cdot), \qquad \psi = |\psi|(w \cdot).$$

Furthermore, if $\omega = |\phi| + |\psi|$ then $\omega \in P$, and by 3.6.10 there are sequences $\{x_n\}$ and $\{y_n\}$ in $\mathcal{M}$ such that $|\phi| = \operatorname{Lim} \omega(x_n \cdot)$ and $|\psi| = \operatorname{Lim} \omega(y_n \cdot)$. Consequently,

$$|\phi + \psi| + \omega_n = \omega((x_n vu^* + y_n wu^*) \cdot)$$

and $\operatorname{Lim} \|\omega_n\| = 0$. It follows from 3.6.8 that $||\phi + \psi| + \omega_n| \in P$ and therefore $|\phi + \psi| \in P$ by 3.6.9. This proves that E is a vector space and since evidently $E_+ = P$ we have established the correspondence between (ii) and (iii).

3.6.12. Notes and remarks. Theorem 3.6.4 is due to Dixmier ([57] and [58]). Since any isomorphism between von Neumann algebras is normal (by 2.5.2), the theorem implies that isomorphisms are homeomorphisms in the σ-weak topology. The polar decomposition of normal functionals (3.6.7) is a result of Sakai [223], while 3.6.11 was proved by Effros in [74].

3.7. The universal representation

3.7.1. Let A be a C^*-algebra with state space S. A subset F of S is said to be *separating* for A if $\phi(x) = 0$ for all ϕ in F implies $x = 0$ for each x in A_+. Note that S itself is separating for A by 3.2.2. Moreover, each subset of S whose convex hull is weak* dense in S is separating for A, again by 3.2.2.

3.7.2. If A is separable then the unit ball of A^* is weak* metrizable and compact, hence second countable. Therefore S is second countable and if $\{\phi_n\}$ is a dense sequence in S then the state $\Sigma 2^{-n} \phi_n$ will be a separating set. However, for positive linear maps between C^*-algebras, in particular for morphisms and positive functionals it is customary to say that ϕ is *faithful* if $\phi(x) = 0$ implies $x = 0$ for every positive element, i.e. if $\{\phi\}$ is separating. Since the kernel of a morphism of a C^*-algebra is generated by its positive elements, a morphism is faithful if and only if it is an injection.

3.7.3. For each ϕ in S let $(\pi_\phi, H_\phi, \xi_\phi)$ denote the cyclic representation associated with ϕ. For each subset F of S we form the Hilbert space $H_F = \oplus_{\phi \in F} H_\phi$ and the representation $\pi_F = \oplus_{\phi \in F} \pi_\phi$ on H_F as in 3.3.1.

3.7.4 PROPOSITION. *If F is a separating family of states of a C^*-algebra A then π_F is a faithful representation of A into $\boldsymbol{B}(H_F)$.*

Proof. If x is a positive element in the kernel of π_F then

$$\phi(x) = (\pi_\phi(x)\xi_\phi \mid \xi_\phi) = 0$$

for each ϕ in F, whence $x = 0$. Therefore the kernel of π_F is zero.

3.7.5. COROLLARY. *Every C^*-algebra A has a faithful representation as an algebra of operators on a Hilbert space H. If A is separable H can be chosen separable.*

Proof. Since S is separating for A, π_S is faithful. If A is separable we can choose a faithful state ϕ on A by 3.7.2. Then π_ϕ is faithful by 3.7.4. Since the map $x \to \xi_x$ of A into H_ϕ is continuous, H_ϕ contains a countable dense set. Consequently H_ϕ is separable.

3.7.6. We say that the space H_S is the *universal Hilbert space* for A and that π_S is the *universal representation*. The *enveloping von Neumann algebra* of A is the strong closure of $\pi_S(A)$. It will henceforth be denoted by A''. Since π_S is faithful we may and shall from now on regard A as a C^*-subalgebra of A'' and delete the symbol π_S. The universal representation has, as befits its name, a universal property.

3.7.7. THEOREM. *For each non-degenerate representation (π, H) of a C^*-algebra A there is a unique normal morphism π'' of A'' onto $\pi(A)''$ which extends π.*

Proof. Assume first that (π, H) has a cyclic vector ξ with $\|\xi\| = 1$. Then the vector functional ϕ defined by ξ is a state and (π, H) is spatially equivalent to (π_ϕ, H_ϕ) by 3.3.7. Let p_ϕ be the projection of H_S onto H_ϕ. Then $p_\phi x = x p_\phi = \pi_\phi(x)$ for all x in A which shows that p_ϕ belongs to the commutant of A''. Consequently we can construct a normal morphism π'' of A'' into $\boldsymbol{B}(H)$ by composing the map $x \to x p_\phi$ with the spatial isomorphism of $\boldsymbol{B}(H_\phi)$ onto $\boldsymbol{B}(H)$. Since $\pi''(x) = \pi(x)$ for each x in A, π'' is an extension of π and by normality $\pi''(A'') = \pi(A)''$.

Since each non-degenerate representation of A is the direct sum of cyclic representations (see 3.3.2) and since the direct sum of normal morphisms of A'' to a direct sum of von Neumann algebras is a normal morphism, the existence of π'' is assured. The uniqueness of π'' is clear since two normal morphisms of A'' which coincide on A must coincide on the smallest C^*-subalgebra B containing A for which $B_{sa}^m = B_{sa}$. But $B = A''$ by 2.4.4 and the proof is complete.

3.7.8. PROPOSITION. *The enveloping von Neumann algebra A'' of a C^*-algebra A is isomorphic, as a Banach space, to the second dual of A.*

Proof. Each state of A is a vector state in (π_S, H_S) and therefore a normal state on A''. Since by 3.2.2 each element of A^* is a linear combination of elements from S we can therefore define a map from A^* into A''_*.

As A is σ-weakly dense in A'' this map will be a linear isometry and each element ϕ of A''_* will be the image of the element $\phi \mid A$ in A^*. Thus $A^* = A''_*$ and consequently $A^{**} = A''$ by 3.6.5.

3.7.9. COROLLARY. *For each C^*-subalgebra B of A the strong closure of B in A'' is isomorphic to B''.*

Proof. The bi-transpose of the inclusion map of B into A is a linear isometry of B'' onto the weak closure of B in A''. Since this map is $*$-preserving and multiplicative on B, and B is weakly dense in B'', the map is an isomorphism of the von Neumann algebra B'' onto the weak closure of B in A''.

3.7.10. Notes and remarks. Corollary 3.7.5 is the celebrated result of Gelfand and Naimark [92], which gave the abstract characterization of algebras of operators on a Hilbert space. The universal representation and the enveloping von Neumann algebra are essentially only trivial extensions of the GNS construction. However, the systematic use of A'' as a universe in which all relevant information about the C^*-algebra A can be stored, has been one of the most fruitful ideas in the theory. If there is a generation gap in the C^*-family, it goes between those members for which the statement $A'' = A^{**}$ is a theorem, and the youngsters, for whom it is a first principle.

3.8. The enveloping von Neumann algebra

3.8.1. With each (non-degenerate) representation (π, H) of a C^*-algebra A we associate the projection $c(\pi)$ in the centre of A'' for which $A''c(\pi)$ is isomorphic to $\pi(A)''\,(=\pi''(A''))$. We say that $c(\pi)$ is the *central cover* of (π, H). The following theorem in principle reduces the study of representations of A—up to equivalence—to the study of central projections of A''.

3.8.2. THEOREM. *Two representations (π_1, H_1) and (π_2, H_2) of a C^*-algebra A are equivalent if and only if $c(\pi_1) = c(\pi_2)$, and the map $(\pi, H) \to c(\pi)$ gives a bijective correspondence between equivalence classes of representations of A and non-zero central projections in A''.*

Proof. For each central projection $p \neq 0$ in A'' the map $x \to xp$, $x \in A$ is a representation of A on pH_S with central cover p, since its normal extension is $x \to xp, x \in A''$. If (π, H) is a representation of A then (π, H) is equivalent to the representation $\tilde{\pi}: x \to xc(\pi)$ on $c(\pi)H_S$, with the restriction of π'' to $A''c(\pi)$ as the intertwining isomorphism between $\tilde{\pi}(A)''$ and $\pi(A)''$. From this the theorem follows.

3.8.3. We are primarily interested in separable C^*-algebras and representations on separable Hilbert spaces (which we just call *separable representations*). However, it is convenient to be able to phrase our results in terms of the enveloping von Neumann algebra which in general acts on a highly non-separable Hilbert space.

We say that a von Neumann algebra $\mathcal{M}$ is *σ-finite* (or countably decomposable) if each set of pairwise orthogonal non-zero projections in $\mathcal{M}$ is countable. Likewise, a projection p in $\mathcal{M}$ is called *σ-finite* if $p\mathcal{M}p$ is σ-finite. If $\mathcal{M}$ acts on a separable Hilbert space then it is clearly σ-finite. The converse is not true in general.

3.8.4. PROPOSITION. *A von Neumann algebra $\mathcal{M}$ has a faithful normal representation on a separable Hilbert space if and only if $\mathcal{M}$ is σ-finite and contains a strongly dense sequence (is countably generated).*

Proof. If $\mathcal{M} \subset B(H)$ and H is separable then $\mathcal{M}$ is σ-finite and since $B(H)^1$ is second countable in the strong topology (see 2.1.2), $\mathcal{M}^1$ is second countable and so $\mathcal{M}$ is separable in the strong topology.

Conversely, if $\mathcal{M} \subset B(H)$ we can for each unit vector ξ in H define the cyclic projection $[\mathcal{M}'\xi]$ in $\mathcal{M}$ (see 2.8.2). Let $\{[\mathcal{M}'\xi_n]\}$ be a maximal family of such projections which are pairwise orthogonal. Then $\Sigma[\mathcal{M}'\xi_n] = 1$, and if $\mathcal{M}$ is σ-finite the family $\{\xi_n\}$ is countable. Define a normal state ϕ on $\mathcal{M}$ by $\phi(x) = \Sigma 2^{-n}(x\xi_n | \xi_n)$. If $x \in \mathcal{M}_+$ and $\phi(x) = 0$, then $x\xi_n = 0$ for all n, whence $\mathcal{M}'x\xi_n = 0$, i.e. $x[\mathcal{M}'\xi_n] = 0$. Since $\Sigma[\mathcal{M}'\xi_n] = 1$, this implies that $x = 0$; and so ϕ is faithful. It follows from 3.3.9 and 3.7.4 that (π_ϕ, H_ϕ) is a faithful normal representation of $\mathcal{M}$. If $\mathcal{M}$ is countably generated then there is a separable C^*-algebra A which is strongly dense in $\mathcal{M}$. For each x in $\mathcal{M}$ we can therefore find a net $\{x_i\}$ in A converging strongly to x, with $\|x_i\| \leqslant \|x\|$ by 2.3.3. Then

$$\|\xi_x - \xi_{x_i}\|^2 = \phi((x - x_i)^*(x - x_i)) = \sum 2^{-n}\|(x - x_i)\xi_n\|^2$$

which tends to zero as $x_i \to x$. Since A is separable, H_ϕ contains a dense separable subspace and consequently H_ϕ is separable.

3.8.5. COROLLARY. *A representation (π, H) of a separable C^*-algebra A is equivalent to a separable representation if and only if $c(\pi)$ is σ-finite in A''.*

3.8.6. COROLLARY. *Each separable representation of a separable C^*-algebra is equivalent to a cyclic representation.*

3.8.7. PROPOSITION. *Let $\rho_0 : \mathcal{M}_1 \to \mathcal{M}_2$ be an isomorphism between von Neumann algebras $\mathcal{M}_1$ and $\mathcal{M}_2$ on Hilbert spaces H_1 and H_2, respectively. There is then a von Neumann algebra $\mathcal{M}$ (isomorphic to $\mathcal{M}_1$ and $\mathcal{M}_2$) on a Hilbert space H, projections p_1 and p_2 in $\mathcal{M}'$ with $c(p_1) = c(p_2) = 1$ and partial isometries $u_1 : H_1 \to p_1H$ and $u_2 : H_2 \to p_2H$ such that $u_1^*p_1\mathcal{M}u_1 = \mathcal{M}_1$, $u_2^*p_2\mathcal{M}u_2 = \mathcal{M}_2$ and the diagram below is commutative.*

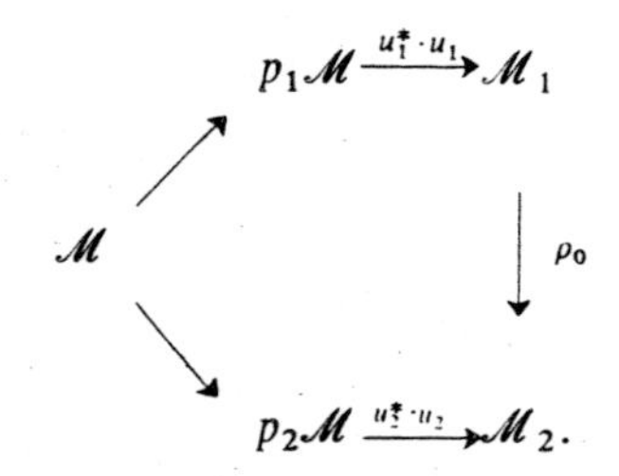

Proof. Choose by Zorn's lemma a maximal family of unit vectors $\{\xi_i\}$ in H_2 such that the cyclic projections $[\mathscr{M}_2\xi_i]$ in $\mathscr{M}_2'$ are pairwise orthogonal. Necessarily then $\Sigma[\mathscr{M}_2\xi_i] = 1$. For each i consider the normal state ϕ_i on $\mathscr{M}_1$ given by $\phi_i(x) = (\rho_0(x)\xi_i \mid \xi_i)$, where ρ_0 is the isomorphism from $\mathscr{M}_1$ onto $\mathscr{M}_2$. By 3.6.6 there is a sequence $\{\eta_{in}\}$ in H_1 such that $\phi_i(x) = \Sigma_n(x\eta_{in} \mid \eta_{in})$. For each i and n let H_{in} be a copy of H_1 and define $H = \oplus H_{in}$. Let ρ be the amplification of $B(H_1)$ into $B(H)$ described in 2.1.4 and put $\mathscr{M} = \rho(\mathscr{M}_1)$. Each of the subspaces H_{in} are invariant for $\mathscr{M}$. We choose one of them and let p_1 be the projection of H on this H_{in} and u_1 the partial isometry of H_1 into H which is the identity map of H_1 onto H_{in}. Then $p_1 \in \mathscr{M}'$, $c(p_1) = 1$ and $u_1^*\mathscr{M}p_1u_1 = \mathscr{M}_1$.

For each i define $\eta_i = \oplus\, \eta_{in}$ in H and form $[\mathscr{M}\eta_i]$ in $\mathscr{M}'$. The projections $[\mathscr{M}\eta_i]$ is an orthogonal family since $[\mathscr{M}\eta_i]H \subset \oplus_n H_{in}$. We define $p_2 = \Sigma[\mathscr{M}\eta_i]$. For each x in $\mathscr{M}_1$

$$(\rho(x)\eta_i \mid \eta_i) = \sum_n (x\eta_{in} \mid \eta_{in}) = (\rho_0(x)\xi_i \mid \xi_i).$$

By 3.3.7 there is an isometry u_i of $[\mathscr{M}_2\xi_i]H_2$ onto $[\mathscr{M}\eta_i]H$ such that

$$u_i^*\mathscr{M}[\mathscr{M}\eta_i]u_i = \mathscr{M}_2[\mathscr{M}_2\xi_i].$$

Since the initial and final projections of the u_i's are pairwise orthogonal the operator $u_2 = \Sigma\, u_i$ is an isometry of H_2 onto p_2H, such that

$$u_2^*\mathscr{M}p_2u_2 = \mathscr{M}_2.$$

Since $u_2^*p_2\rho(\mathscr{M}_1)u_2 = \rho_0(\mathscr{M}_1)$, it is clear that $c(p_2) = 1$, and the proof is complete.

3.8.8. COROLLARY. *If (π_1, H_1) and (π_2, H_2) are equivalent representations of a C^*-algebra A there is a representation (π, H) of A, equivalent to (π_1, H_1) and (π_2, H_2) such that both (π_1, H_1) and (π_2, H_2) are spatially equivalent to subrepresentations of (π, H).*

3.8.9. PROPOSITION. *Let $\rho_0 : \mathscr{M}_1 \to \mathscr{M}_2$ be an isomorphism between von Neumann algebras $\mathscr{M}_1$ and $\mathscr{M}_2$. There are then projections q_1 and q_2 in $\mathscr{M}_1'$ and $\mathscr{M}_2'$, respectively, with $c(q_1) = 1$ and $c(q_2) = 1$, and a spatial isomorphism*

$\rho: q_1\mathcal{M}_1 \to q_2\mathcal{M}_2$ such that the diagram below is commutative.

$$\begin{array}{ccc} \mathcal{M}_1 & \xrightarrow{\rho_0} & \mathcal{M}_2 \\ \downarrow & & \downarrow \\ q_1\mathcal{M}_1 & \xrightarrow{\rho} & q_2\mathcal{M}_2. \end{array}$$

Proof. From 3.8.7 we see that it is enough to prove the theorem under the assumption that $\mathcal{M}_1 = \mathcal{M}p_1$ and $\mathcal{M}_2 = \mathcal{M}p_2$, where $\mathcal{M}$ is a von Neumann algebra in $\boldsymbol{B}(H)$ and p_1, p_2 belongs to $\mathcal{M}'$ with $c(p_1) = c(p_2) = 1$.

Since the unit ball of the subspace $p_2\mathcal{M}'p_1$ is convex and σ-weakly compact it contains by Krein–Milman's theorem an extreme point u. Put $q_1 = u^*u$ and $q_2 = uu^*$. Then $(p_2 - q_2)\mathcal{M}'(p_1 - q_1) = 0$ by 1.4.8, so that for each unitary w in $\mathcal{M}'$ we have $w^*(p_2 - q_2)w \perp (p_1 - q_1)$. By 2.6.3 this implies that $c(p_2 - q_2) \perp (p_1 - q_1)$, whence $c(p_2 - q_2) \perp c(p_1 - q_1)$. Now,

$$1 = c(p_1) \leqslant c(p_1 - q_1) + c(q_1),$$

and similarly $1 \leqslant c(p_2 - q_2) + c(q_2)$. It follows that

$$1 \leqslant c(p_2 - q_2) + c(q_2) \leqslant 1 - c(p_1 - q_1) + c(q_2) \leqslant c(q_1) + c(q_2) = 2c(q_1),$$

whence $c(q_1) = c(q_2) = 1$.

3.8.10. COROLLARY. *If (π_1, H_1) and (π_2, H_2) are equivalent representations of a C^*-algebra A then they have spatially equivalent subrepresentations which are equivalent to the original ones.*

3.8.11. THEOREM. *Let (π_1, H_1) and (π_2, H_2) be non-degenerate representations of a C^*-algebra A. The following conditions are equivalent:*

(i) $c(\pi_1) \perp c(\pi_2)$.

(ii) $((\pi_1 \oplus \pi_2)A)'' = \pi_1(A)'' \oplus \pi_2(A)''$.

(iii) $((\pi_1 \oplus \pi_2)A)' = \pi_1(A)' \oplus \pi_2(A)'$.

(iv) *There are no non-zero spatially equivalent subrepresentations of (π_1, H_1) and (π_2, H_2).*

Proof. (i) $\Rightarrow$ (ii). We extend π_1, π_2 and $\pi_1 \oplus \pi_2$ to A'' as in 3.7.7. Since $(\pi_1 \oplus \pi_2)''(x) = 0$ if and only if $\pi_1''(x) = 0$ and $\pi_2''(x) = 0$ and since $\ker \pi'' = A''(1 - c(\pi))$ for any representation π of A, we have

$$\begin{aligned} \ker(\pi_1 \oplus \pi_2)'' &= \ker \pi_1'' \cap \ker \pi_2'' = A''(1 - c(\pi_1))(1 - c(\pi_2)) \\ &= A''(1 - (c(\pi_1) + c(\pi_2))). \end{aligned}$$

It follows that $(\pi_1 \oplus \pi_2)''(A'')$ is isomorphic to $A''(c(\pi_1) + c(\pi_2))$ which equals $A''c(\pi_1) \oplus A''c(\pi_2)$, whence

$$((\pi_1 \oplus \pi_2)A)'' = \pi_1(A)'' \oplus \pi_2(A)''.$$

(ii) $\Rightarrow$ (iii) follows from the bicommutant theorem (2.2.2).

(iii) $\Rightarrow$ (iv). If there is a pair of spatially equivalent subrepresentations of (π_1, H_1) and (π_2, H_2) then there is a partial isometry u from H_1 into H_2 such that $u^*u \in \pi_1(A)'$, $uu^* \in \pi_2(A)'$ and $u^*(\pi_2(x)uu^*)u = \pi_1(x)u^*u$ for all x in A. Regarding u as an element of $B(H_1 \oplus H_2)$ we have

$$(\pi_1 \oplus \pi_2)(x)u = (\pi_1(x) + \pi_2(x))u = \pi_2(x)u$$
$$= u\pi_1(x) = u(\pi_1(x) + \pi_2(x))$$
$$= u(\pi_1 \oplus \pi_2)(x),$$

which shows that $u \in (\pi_1 \oplus \pi_2)(A)'$. By assumption

$$((\pi_1 \oplus \pi_2)(A))' \subset B(H_1) \oplus B(H_2),$$

whence $u = 0$.

(iv) $\Rightarrow$ (i). The subrepresentations of (π_1, H_1) and (π_2, H_2) determined by the projections $\pi_1''(c(\pi_2))$ and $\pi_2''(c(\pi_1))$ in $\pi_1(A)'$ and $\pi_2(A)'$, respectively, are equivalent by 3.8.2, with central cover $c(\pi_1)c(\pi_2)$. If $c(\pi_1)c(\pi_2) \neq 0$ these subrepresentations have non-zero spatially equivalent subrepresentations by 3.8.10.

3.8.12. Two representations satisfying the conditions in 3.8.11 are said to be *disjoint*. Simple computations with central projections in A'' show that for any two representations (π_1, H_1) and (π_2, H_2) of a C^*-algebra A there are central projections p_1 and p_2 in $\pi_1(A)'$ and $\pi_2(A)'$, respectively, such that the subrepresentations on p_1H_1 and p_2H_2 are equivalent and those on $(1 - p_1)H_1$ and $(1 - p_2)H_2$ are disjoint. Moreover, since p_1 and p_2 are central projections, the subrepresentations on p_1H_1 and $(1 - p_1)H_1$, respectively p_2H_2 and $(1 - p_2)H_2$, are also disjoint.

3.8.13. A non-degenerate representation (π, H) of a C^*-algebra A is called a *factor representation* if $\pi(A)''$ is a factor. Since the centre of $\pi(A)''$ is the image of the centre of A'' we immediately conclude that (π, H) is a factor representation if and only if $c(\pi)$ is a minimal projection in the centre of A''. From 3.8.12 we see that two factor representations are either equivalent or disjoint.

3.8.14. Notes and remarks. The von Neumann algebra results in 3.8.7 and 3.8.9 on the structure of normal morphisms are due to Dixmier [58], who also drew the consequences 3.8.8, 3.8.10 and 3.8.11 concerning representations of C^*-algebras.

3.9. Abstract von Neumann algebras

3.9.1. From 3.7.5 we know that any C^*-algebra A can be represented as a C^*-subalgebra of some $B(H)$. One may ask what conditions should be imposed on

A in order for it to be representable as a von Neumann subalgebra of some $B(H)$. We shall examine the three best known conditions of this kind.

3.9.2. A C^*-algebra A is said to be *monotone complete* if each bounded increasing net in A_{sa} has a least upper bound in A_{sa}. We say that A is an AW^*-*algebra* if each maximal commutative C^*-subalgebra of A is monotone complete.

A positive linear map ϕ between monotone complete C^*-algebras A and B is *normal* if $\text{L.U.B.}\,\phi(x_i) = \phi(\text{L.U.B.}\,x_i)$ wherever $\{x_i\}$ is a bounded increasing net in A_{sa}. If A and B are AW^*-algebras we say that ϕ is *completely additive* if it is normal on each maximal commutative C^*-subalgebra of A.

3.9.3. THEOREM. *If A is a monotone complete C^*-algebra with a separating family of normal states then there is a normal isomorphism of A onto a von Neumann algebra.*

Proof. The result is contained in 3.9.4, but can be derived independently by combining 2.4.4, 2.5.3, 3.3.9 and 3.7.4.

3.9.4. THEOREM. *If A is an AW^*-algebra with a separating family of completely additive states then there is a completely additive isomorphism of A onto a von Neumann algebra.*

Proof. Let B be a maximal commutative C^*-subalgebra of A. If e is the least upper bound for an approximate unit in B_+ then e is a unit for B. But then $(1-e)x(1-e)$ commutes with B for any x in A. Since B is maximal commutative $(1-e)x(1-e)\in B$ whence

$$(1-e)x(1-e) = (1-e)x(1-e)e = 0.$$

It follows that $e = 1$ i.e. A has a unit.

Let F denote the set of completely additive states of A and consider the representation $\pi_F = \bigoplus_{\phi\in F}\pi_\phi$ on $H_F = \bigoplus_{\phi\in F} H_\phi$. By 3.7.4 (π_F, H_F) is faithful. If now $\{x_i\}$ is a bounded monotone increasing net in B_{sa} then $\{x_i\}$ has a least upper bound x in B_{sa} and $\{\pi_F(x_i)\}$ has a least upper bound y in $B(H_F)$ with $y \leqslant \pi_F(x)$. However, if $\phi\in F$ and $(\pi_\phi, H_\phi, \xi_\phi)$ is the cyclic representation associated with ϕ then for each unitary u in A

$$\begin{aligned}(\pi_\phi(x)\pi_\phi(u)\xi_\phi \mid \pi_\phi(u)\xi_\phi) &= \phi(u^*xu)\\ &= \text{L.U.B.}\,\phi(u^*x_iu) = \text{L.U.B.}(\pi_\phi(x_i)\pi_\phi(u)\xi_\phi \mid \pi_\phi(u)\xi_\phi)\\ &= (y\pi_\phi(u)\xi_\phi \mid \pi_\phi(u)\xi_\phi).\end{aligned}$$

This implies that $(\pi_\phi(x) - y)\pi_\phi(u)\xi_\phi = 0$, and since by 1.1.11, A is linearly spanned by its unitaries we have $(\pi_\phi(x) - y)[\pi_\phi(A)\xi_\phi] = 0$. As $H_F = \bigoplus_{\phi\in F} H_\phi$ we conclude that $\pi_\phi(x) = y$. This shows that $\pi_F(B)$ is monotone closed and therefore strongly closed by 2.4.4, and since this is true for any maximal commutative C^*-subalgebra B of A we see from 2.8.4 that $\pi_F(A)$ is a von Neumann algebra.

3.9.5. The usual definition of a completely additive state ϕ on a von Neumann algebra $\mathcal{M}$ requires only that $\phi(\Sigma p_i) = \Sigma \phi(p_i)$ for any set $\{p_i\}$ of pairwise orthogonal projections from $\mathcal{M}$. This definition is the formal equivalent of the additivity of a measure on a σ-algebra of sets (= projections), just as the normality condition is patterned after Lebesgue's monotone convergence theorem. In measure theory one derives the second condition from the first. This can also be done here.

3.9.6. PROPOSITION. *Let ϕ be a state of a von Neumann algebra $\mathcal{M}$ which is completely additive on projections. Then ϕ is normal.*

Proof. Let $\{x_i\}$ be a monotone increasing net in $\mathcal{M}_+$ with limit x, consisting of pairwise commuting elements. For each monotone increasing function f in $C_0(]0, \infty])$ the net $\{f(x_i)\}$ increases to $f(x)$. By spectral theory the same is true for each monotone increasing lower semi-continuous function vanishing at zero, since each such function is the increasing limit of a sequence of monotone increasing elements from $C_0(]0, \infty])$. In particular $\{f(x_i)\}$ increases to $f(x)$ for each characteristic function f corresponding to an interval $]\alpha, \infty[$. Since ϕ is completely additive on projections it follows that $\operatorname{Lim} \phi(f(x_i)) = \phi(f(x))$. We can approximate the function $t \to t$ uniformly on the interval $[0, \|x\|]$ by a convex combination of characteristic functions corresponding to intervals $]\alpha, \infty[$ which proves that $\operatorname{Lim} \phi(x_i) = \phi(x)$. Thus ϕ is completely additive in the sense used in 3.9.4. Taking F to be a separating family of normal states of $\mathcal{M}$ together with ϕ we see from 3.9.4 that (π_F, H_F) is an isomorphism of $\mathcal{M}$ onto a von Neumann algebra. Then π_F is a normal map by 2.5.2 and since ϕ is a vector state in (π_F, H_F) it is also normal.

3.9.7. Dixmier [55] showed that the C^*-algebra A of bounded Borel functions on $\boldsymbol{R}$ modulo the ideal of functions vanishing outside a set of first category is a commutative AW^*-algebra (hence a monotone complete C^*-algebra) with no non-zero completely additive (= normal) functionals. If A had a faithful representation as a von Neumann algebra then this representation would be normal, since any isomorphism between monotone complete C^*-algebras is normal (cf. the proof of 2.5.2). But this is not possible by 3.9.3.

The last characterization (by Sakai) of abstract von Neumann algebras is without doubt the most elegant. It is simply the converse of 3.5.6, and helps to understand the profound rôle of the pre-dual of von Neumann algebras (3.6.5).

3.9.8. THEOREM. *Let A be a C^*-algebra and V a Banach space such that A is isomorphic (as a Banach space) to the dual of V. Then A has a faithful representation as a von Neumann algebra with $A_* = V$.*

Proof. Consider the weak* topology on A arising from V, and identify V with the weak* continuous elements of A^*. Since A^1 is weak* compact it has an extremal point by Krein–Milman's theorem, whence $1 \in A$ by 1.4.7.

We claim that A_{sa} is weak* closed. To prove this it is enough, by the Krein–Smulian theorem (A2, Appendix) to show that A^1_{sa} is weak* closed. Let $\{x_i\}$ be a weak* convergent net in A^1_{sa} and write the limit as $x + iy$ with x and y in A_{sa}. Then $\{x_i + in\}$ is weak* convergent to $x + i(n + y)$ for every n. Since $\|x_i + in\| \leqslant (1 + n^2)^{1/2}$ and the norm is weak* lower semi-continuous this implies that

$$(1 + n^2)^{1/2} \geqslant \|x + i(n + y)\| \geqslant \|n + y\|.$$

If $y \neq 0$ we may assume that Sp(y) contains a number $\lambda > 0$ (passing if necessary to $\{-x_i\}$). But then

$$\lambda + n \leqslant \|n + y\| \leqslant (n^2 + 1)^{1/2}$$

for all n, a contradiction. Thus $y = 0$.

We next claim that A_+ is weak* closed. Again it suffices to show that A^1_+ is weak* closed. But $A^1_+ = \frac{1}{2}(A^1_{sa} + 1)$.

It now follows that V_+ is separating for A. For if $x \in A_{sa}$ and $-x \notin A_+$ then since A_+ is a weak* closed cone in A_{sa} there is by Hahn–Banach's theorem a ϕ in V_{sa} such that $\phi(A_+) \geqslant 0$ and $\phi(x) > 0$.

We claim that A is monotone complete. To see this let $\{x_i\}$ be a bounded, monotone increasing net in A_{sa}. Since A^1_{sa} is weak* compact there is a subnet $\{x_j\}$ of $\{x_i\}$ which is weak* convergent to an element x in A_{sa}. For each x_i we have eventually $x_j \geqslant x_i$ and thus $x \geqslant x_i$ since A_+ is weak* closed. Therefore x is a majorant for $\{x_i\}$. If $y \in A_{sa}$ and $y \geqslant x_i$ for all i then $y \geqslant x_j$ for all j and thus $y \geqslant x$ since A_+ is weak* closed. It follows that x is the least upper bound of $\{x_i\}$ in A_{sa}, so that A is monotone complete. At the same time we see that each ϕ in V_+ is normal because

$$\operatorname{Lim} \phi(x_i) \leqslant \phi(x) = \operatorname{Lim} \phi(x_j) \leqslant \operatorname{Lim} \phi(x_i).$$

We have proved that A is a monotone complete C^*-algebra with a separating family (namely V^1_+) of normal states. Therefore A has a faithful representation as a von Neumann algebra by 3.9.3. From the construction of this representation it follows that $V_+ \subset A_*$. We showed above that if $x \in A_{sa}$ and $x \neq 0$ then $\phi(x) \neq 0$ for some ϕ in V_+. Thus the linear span of V_+ is norm dense in V, whence $V \subset A_*$. Since the compact topology on A^1 is unique, the weak* and the σ-weak topology coincide, and thus $A_* = V$.

3.9.9. Notes and remarks. The notion of AW^*-algebra was introduced by Kaplansky [140] (see also Rickart [214] for a similar idea) in order to obtain an algebraic, i.e. Hilbert space-free theory for von Neumann algebras. It was soon realized that there are (even commutative) AW^*-algebras which are not von Neumann algebras [55], and that the missing condition was the existence of normal functionals. Working instead with monotone complete C^*-algebras Kadison [127] established 3.9.3. The corresponding result for AW^*-algebras (3.9.4) was proved by the author [199]. Incidentally, the theory of monotone complete C^*-algebras is fully as interesting (especially from an order-theoretic point of view) as the AW^*-algebra theory. See Kadison and Pedersen [135] for the comparison and trace theory, and Maitland Wright [276] for some elegant constructions of monotone complete factors which are not von Neumann algebras. However, the AW^*-theory (and its ring-theoretic analogues) is far better developed, see Berberian [17], and even today draws the attention of young experts. May we propose a voluntary limit of one AW^*-algebra paper per customer?

Sakai's characterization of von Neumann algebras (3.9.8) appeared in [221]. Note that a von Neumann algebra is a dual space with a *unique* pre-dual, in contrast with many other dual Banach spaces. It is fairly easy to show (using 3.5.4 and 3.8.4) that a von Neumann algebra $\mathcal{M}$ can be represented faithfully on a separable Hilbert space if and only if $\mathcal{M}_*$ is (norm) separable. Readers who prefer a space-free description of von Neumann algebras may substitute 'with separable pre-dual' for the more earthy formulation 'on a separable Hilbert space' throughout the book.

3.10 Kadison's function representation

3.10.1. The Gelfand representation of a commutative C^*-algebra A represents the elements as continuous functions on the spectrum $\hat{A}$ of A. The elements of $\hat{A}$ are the non-zero homomorphisms of A into $\mathbf{C}$ and, as is well known, a positive linear functional ϕ on A with $\|\phi\| \leq 1$ is precisely a homomorphism if it cannot be written as a non-trivial convex combination of positive elements from the unit ball of A^*.

For a non-commutative C^*-algebra A we have already seen the importance of the state space S and the quasi-state space Q introduced in 3.2.1. Since Q is convex and weak* compact it is well supplied with extremal points by the Krein–Milman theorem. Clearly 0 is extremal in Q whereas all other extremal points of Q belongs to S. We say that the non-zero extremal points of Q are the *pure states* of A. Their importance in representation theory will be explained in 3.13. It is no longer true that pure states are homomorphisms and, in general, A will not have very many non-zero homomorphisms, if any. Instead of representing A as functions on the pure state space we shall therefore find it more convenient to represent A as affine functions on all of Q.

3.10.2. If Q is a convex set in a real topological vector space we denote by $A(Q)$ the Banach space of all continuous bounded functions a on Q which are affine in the sense that

$$a(\lambda\phi + (1-\lambda)\psi) = \lambda a(\phi) + (1-\lambda)a(\psi)$$

for every convex combination of points ϕ and ψ in Q. We denote by $B(Q)$ the Banach space of all bounded affine functions on Q. If $0 \in Q$ then $A_0(Q)$, respectively $B_0(Q)$, denotes the set of elements in $A(Q)$, respectively $B(Q)$, that vanish at zero.

3.10.3. THEOREM. *Let A be a C^*-algebra with enveloping von Neumann algebra A'' and let Q be the quasi-state space of A. There is then an order preserving linear isometry* $\hat{}$ *of A''_{sa} onto $B_0(Q)$ such that $(A_{sa})\hat{} = A_0(Q)$.*

Proof. By 3.7.8 we may identify A'' and A^{**} so that A''_{sa} can be identified with the second dual of A_{sa} (as a real Banach space). For each x in A''_{sa} and ϕ in Q define $\hat{x}(\phi) = \phi(x)$. Since by 3.2.2 $\mathrm{Conv}(Q \cup (-Q)) = (A^*)^1_{sa}$ we see that $\hat{}$ is an order preserving linear isometry of A''_{sa} into $B_0(Q)$. That $(A_{sa})\hat{} \subset A_0(Q)$ follows from the definition of the weak* topology on A^*.

If $b \in B_0(Q)$ there is a unique extension of b to a bounded, symmetric, affine function on $(A^*)^1_{sa}$. By homogeneity b can then be extended to a bounded linear functional x on $(A^*)_{sa}$, i.e., to an element of A^{**}_{sa}. Since $\hat{x} = b$ we have shown that $\hat{}$ is surjective. Finally, if $x \in A''_{sa}$ and $\hat{x} \in A_0(Q)$ then x will be weak* continuous on $(A^*)^1_{sa}$ and therefore on all of $(A^*)_{sa}$. But this means that $x \in A_{sa}$, and so $(A_{sa})\hat{} = A_0(Q)$.

3.10.4. We say that an element h in a C^*-algebra A is *strictly positive* if $\phi(h) > 0$ for every non-zero positive linear functional ϕ on A. If A has a unit then clearly 1 is strictly positive. There are, however, many C^*-algebras without units that contain strictly positive elements.

3.10.5. PROPOSITION. *Let A be a C^*-algebra. The following conditions are equivalent:*

(i) *There is a strictly positive element h in A_+.*

(ii) *There is an element h in A_+ such that $[h] = 1$ in A''.*

(iii) *There is a countable approximate unit for A.*

Proof. (i) ⇒ (ii). If h is strictly positive then $(h\xi \mid \xi) > 0$ for each non-zero vector ξ in the universal Hilbert space for A (see 3.7.6), whence $[h] = 1$ in A''.

(ii) ⇒ (iii). Take an increasing sequence $\{f_n\}$ of positive continuous functions on $\mathbf{R}_+$ such that $f_n(0) = 0$ and $\mathrm{Lim}\, f_n(t) = 1$ whenever $t > 0$. Put $u_n = f_n(h)$. Then $u_n \nearrow 1$ in A'' by spectral theory, so that for each x in A and each ϕ in Q we have

$$\phi(x^*(1-u_n)^2 x) = (x^*(1-u_n)^2 x)\hat{}(\phi) \searrow 0.$$

Since $\hat{}$ is an isometry by 3.10.3, we conclude from Dini's theorem that $\|(1 - u_n)x\|^2 \to 0$ and thus $\{u_n\}$ is a countable approximate unit.

(iii) $\Rightarrow$ (i). If $\{u_n\}$ is a countable approximate unit for A put $h = \Sigma 2^{-n} u_n$. If ϕ is positive linear functional with $\phi(h) = 0$ then $\phi(u_n) = 0$ for all n whence $\phi = 0$ by 3.1.4. Thus h is strictly positive.

3.10.6. From 1.4.3 we see that if A is a separable C^*-algebra then it contains strictly positive elements. This is, however, not a necessary condition. If $A = C_0(T)$ with T a locally compact Hausdorff space then A contains strictly positive elements precisely when T is a countable union of compact sets (σ-compact).

Note that the proof of 3.10.5 shows that if A contains a strictly positive element then it has a countable approximate unit contained in a commutative C^*-subalgebra (viz. the C^*-algebra generated by h).

3.10.7 THEOREM. *Let A be a C^*-algebra. The mappings $L \to L^\perp$ and $E \to E \cap Q$ define bijective correspondences between the classes of*

(i) *Closed left ideals of A;*

(ii) *Weak* closed, left invariant subspaces of A^*;*

(iii) *Weak* closed faces of Q containing zero.*

Proof. The bi-polar theorem shows that the mappings $L \to L^\perp$ and $E \to E^\perp$ are each other's inverses, and give a bijective correspondence between class (i) and class (ii).

If L is a closed left ideal in A it follows from the Cauchy–Schwarz inequality that

$$(L^\perp)_+ = \{\phi \in A^*_+ \mid \phi(L_+) = 0\},$$

so that $(L^\perp)_+$ is a hereditary cone in A^*_+. But then $L^\perp \cap Q$ is a face of Q containing zero. Conversely, if F is a weak* closed face of Q containing zero then $\boldsymbol{R}_+ F$ is a weak* closed cone in A^*_+ by the Krein–Smulian theorem. Moreover, $\boldsymbol{R}_+ F$ is hereditary. Indeed, if $0 \leqslant \phi \leqslant \lambda\psi$ with $\lambda > 0$ and $\psi \in F$ then

$$\psi = \lambda^{-1}\|\phi\|\,\phi_1 + \|\psi - \lambda^{-1}\phi\|\,\psi_1 + (1 - \|\psi\|)0,$$

where ϕ_1, ψ_1 and 0 belong to Q. Since

$$\lambda^{-1}\|\phi\| + \|\psi - \lambda^{-1}\phi\| + 1 - \|\psi\| = \lambda^{-1}\phi(1) + \psi(1) - \lambda^{-1}\phi(1) + 1 - \psi(1) = 1,$$

and F is a face of Q it follows that $\|\phi\|^{-1}\phi = \phi_1 \in F$, whence $\phi \in \boldsymbol{R}F$. Consequently,

$$\boldsymbol{R}F = \{\phi \in A^*_+ \mid \phi(p) = 0\}$$

for some projection p in A'' by 3.6.11. We claim that the corresponding left invariant subspace

$$E = \{\phi \in A^* \mid \phi(A''p) = 0\} = \{\phi \in A^* \mid |\phi| \in \mathbf{R}F\}$$

is weak* closed. It suffices to show that the unit ball of E is weak* closed; so let $\{\phi_i\}$ be a weak* convergent net in E^1 with limit ϕ. Then $\{|\phi|\} \subset F$, and since F is weak* compact we may assume that $|\phi_i| \to \psi$ in F. For each x in A

$$|\phi(x)|^2 = \operatorname{Lim} |\phi_i(x)|^2 \leqslant \operatorname{Lim\,sup} |\phi_i|(x^*x) = \psi(x^*x)$$

by 3.6.7. Thus if $x \in A$ and $y \in A_+^1$ then

$$|\phi(xy)|^2 \leqslant \psi(y^*x^*xy) \leqslant \|x\|^2 \psi(y).$$

Since p belongs to the weak closure of A_+ and ϕ and ψ are weakly continuous on A'' it follows that

$$|\phi(xp)|^2 \leqslant \|x\|^2 \psi(p) = 0,$$

for all x in A, thus also for all x in A'', whence $\phi \in E$. We have shown that E is a left invariant, weak* closed subspace of A^* and that $E \cap Q = F$, which completes the proof.

3.10.8. COROLLARY. *The mappings $I \to I^\perp$ and $E \to E \cap Q$ define bijective correspondences between the classes of*

(i) *Closed ideals of A;*

(ii) *Invariant (left and right) weak* closed subspaces of A^*;*

(iii) *Weak* closed faces of Q containing zero which are invariant under all transformations $\phi \to \phi(u^* \cdot u)$ with u unitary in $\tilde{A}$.*

Proof. The only point which requires proof is the statement that if F is a face in class (iii) then the left invariant subspace

$$E = \{\phi \in A^* \mid |\phi| \in \mathbf{R}F\}$$

is right invariant. Note however that if $\phi \in E$ then $\phi(u^* \cdot u) \in E$ for every unitary u in $\tilde{A}$ since

$$|\phi(u^* \cdot u)| = |\phi|(u^* \cdot u).$$

Since E is left invariant this implies that $\phi(\cdot u) \in E$ for every unitary u in $\tilde{A}$, so that E is right invariant (cf. 1.1.11).

3.10.9. Notes and remarks. Today we all know that a Banach space should be represented as affine functions on the unit ball of its dual space, or on some suitable subset of it. In 1950 when Kadison proved 3.10.3 (see [123]) it was not obvious. The representation is still a rich source of inspiration (see sections 3.11 and 3.12). The notion of strictly positive elements (which appears here because it gives us the opportunity for an instant application of Kadison's function representation) appears in Aarnes and Kadison [1]. The far deeper results in 3.10.7 and 3.10.8 are due to Effros [74].

3.11. Semi-continuity

3.11.1. We shall use the function representation given in 3.10.3 to characterize certain classes of elements in the enveloping von Neumann algebra by their topological or measure theoretic properties. The applications in this chapter will be to the classes of multipliers which are uninteresting if the algebra has a unit.

Assume therefore that A is a C^*-algebra without unit and consider $\tilde{A}$ as the C^*-subalgebra of A'' generated by A and the unit 1 of A''. If $\{u_\lambda\}$ is an approximate unit for A then $\operatorname{Lim} \phi(u_\lambda) = \|\phi\|$ for every positive functional ϕ on A. On the other hand, $\{u_\lambda\}$ is strongly convergent to 1 in A''. It follows that $\hat{1}$ is the unique affine function on Q which vanishes at zero and is constant equal to one on S. Moreover, $\hat{1}$ is the pointwise limit of the continuous functions $\hat{u}_\lambda$ and is therefore lower semi-continuous.

It will be clear from the context that the two next results are valid in a much more general situation. However, we shall stick to the C^*-algebra formalism.

3.11.2. LEMMA. *For each lower semi-continuous element b in $B(Q)$ there is an increasing net $\{a_i\}$ in $A(Q)$ which converges pointwise to b.*

Proof. Let $G(b)$ be the supergraph of b in $A_{sa}^* \times \boldsymbol{R}$, i.e.

$$G(b) = \{(\phi, \alpha) \in Q \times \boldsymbol{R} \mid b(\phi) \leqslant \alpha\}.$$

By assumption $G(b)$ is a closed convex set in $A_{sa}^* \times \boldsymbol{R}$. If therefore $\psi \in Q$ and $\beta < b(\psi)$, so that $(\psi, \beta) \notin G(b)$, then by the Hahn–Banach theorem we can separate (ψ, β) and $G(b)$ strictly with a closed hyperplane in $A_{sa}^* \times \boldsymbol{R}$. Since $(\psi, b(\psi)) \in G(b)$ this hyperplane is not 'vertical' to $\boldsymbol{R}$ and is therefore the graph of a continuous affine function on A_{sa}^*, whose restriction to Q is an element a in $A(Q)$ with $a(\psi) > \beta$ and $a(\phi) < b(\phi)$ for all ϕ in Q. This already shows that b is the upper envelope of elements from $A(Q)$, which are strictly smaller than b.

Assume now that a_1 and a_2 are elements in $A(Q)$ such that $a_1 < b$ and $a_2 < b$. For $i = 1, 2$ let $G(a_i)$ be the cutoff subgraph of a_i, i.e.

$$G(a_i) = \{(\phi, \alpha) \in Q \times \boldsymbol{R} \mid a_i(\phi) \geqslant \alpha \geqslant -\|a_i\|\}.$$

Then $G(a_i)$ is convex and compact, and $G(a_i) \cap G(b) = \varnothing$. Since a_1, a_2 and b are affine this implies that

$$\operatorname{Conv}(G(a_1) \cup G(a_2)) \cap G(b) = \varnothing,$$

and since $\text{Conv}(G(a_1) \cup G(a_2))$ is convex and compact it can be separated strictly from $G(b)$ with a closed hyperplane in $A_{sa}^* \times \boldsymbol{R}$. As before this gives an element a in $A(Q)$ with $a > a_1, a > a_2$ and $a < b$. Thus the minorants of b in $A(Q)$ form an increasing net which by the first part of the proof converges pointwise to b.

3.11.3. LEMMA. *If $b \in B_0(Q)$ and $b|S$ is lower semi-continuous there is a net $\{a_i\}$ in $A_0(Q)$ and a net $\{\alpha_i\}$ in $\boldsymbol{R}$ such that $a_i + \alpha_i \hat{1} \leqslant b$ for each i, and the net $\{a_i + \alpha_i \hat{1}\}$ is pointwise convergent to b.*

Proof. Let $G(b)$ be the weak* closure in $A_{sa}^* \times \boldsymbol{R}$ of the convex set

$$\{(\phi, \alpha) \in Q \times \boldsymbol{R} \mid b(\phi) \leqslant \alpha\}.$$

Given $\varepsilon > 0$ and a finite set $\{\psi_k\}$ in S we shall find an element a in $A(Q)$ such that $a < b$ and $a(\psi_k) > b(\psi_k) - \varepsilon$ for all k. Toward this end let K be the convex hull (compact) in $Q \times \boldsymbol{R}$ of the set $\{(\psi_k, b(\psi_k) - \varepsilon)\}$. If $(\phi, \alpha) \in K \cap G(b)$ then $\phi = \Sigma \lambda_k \psi_k$,

$$\alpha = \sum \lambda_k (b(\psi_k) - \varepsilon) = b(\phi) - \varepsilon.$$

Moreover, there is a net $\{(\phi_i, \alpha_i\}$ converging to (ϕ, α) with $b(\phi_i) \leqslant \alpha_i$. Since $\phi \in S$ and $\hat{1}$ is lower semi-continuous we have

$$1 = \hat{1}(\phi) \leqslant \text{Lim Inf}\, \hat{1}(\phi_i) \leqslant 1,$$

so that $\{\|\phi_i\|^{-1} \phi_i\}$ converges to ϕ in S. But as $b|S$ is lower semi-continuous this implies that

$$b(\phi) \leqslant \text{Lim Inf} \|\phi_i\|^{-1} b(\phi_i) = \text{Lim Inf}\, b(\phi_i)$$

$$\leqslant \text{Lim Inf}\, \alpha_i = \alpha = b(\phi) - \varepsilon,$$

a contradiction. Thus $G(b) \cap K = \varnothing$, and by the Hahn–Banach theorem the two sets can be separated strictly by a closed hyperplane in $A_{sa}^* \times \boldsymbol{R}$. As in 3.11.2 this produces an element a in $A(Q)$ with $a < b$, $a(\psi_k) > b(\psi_k) - \varepsilon$ for all k.

From the first part of the proof there is thus a net $\{b_i\}$ in $A(Q)$ with $b_i < b$ such that $\{b_i | S\}$ converges pointwise to $b|S$. Put $\alpha_i = b_i(0)$ and $a_i = b_i - \alpha_i$. Then $a_i \in A_0(Q)$ and since

$$(a_i + \alpha_i \hat{1}) | S = b_i | S \leqslant b | S$$

and $(a_i + \alpha_i \hat{1})(0) = 0$ we conclude that $a_i + \alpha_i \hat{1} \leqslant b$ and that $\{a_i + \alpha_i \hat{1}\}$ is pointwise convergent to b.

3.11.4. Recall from 2.4.1 that if M is a subset of $\boldsymbol{B}(H)_{sa}$ then M^m denotes the set of elements in $\boldsymbol{B}(H)_{sa}$ which can be obtained as strong limits of bounded, monotone increasing nets from M. Also M^- denotes the norm closure of M.

We shall relate $(A_{sa})^m$ and $(\tilde{A}_{sa})^m$ to the sets of elements in $B_0(Q)$ which are lower semi-continuous on Q and S, respectively.

3.11.5. PROPOSITION. *Let A be a C^*-algebra and x an element in A''_{sa}. The following conditions are equivalent:*

(i) $x + \varepsilon 1 \in (A_{sa})^m$ *for each* $\varepsilon > 0$;

(ii) $x \in ((A_{sa})^m)^-$;

(iii) $\hat{x}$ *is lower semi-continuous on* Q;

(iv) *There is a bounded, monotone increasing net* $\{x_i + \alpha_i 1\}$ *in* $\tilde{A}_{sa}$ *with limit* x *such that* $x_i \in A_{sa}$, $\alpha_i \in \mathbf{R}$ *and* $\alpha_i \nearrow 0$.

Proof. (i) $\Rightarrow$ (ii). Evident.

(ii) $\Rightarrow$ (iii). The set of lower semi-continuous elements in $B_0(Q)$ is norm closed and closed under monotone increasing limits. Since it contains $(A_{sa})^\wedge$ by 3.10.3 and since $^\wedge$ is an isometry, it contains $((A_{sa})^m)^-)^\wedge$.

(iii) $\Rightarrow$ (iv). By 3.11.2 there is an increasing net $\{a_i\}$ in $A(Q)$ which converges pointwise to $\hat{x}$. Put $\alpha_i = a_i(0)$ and let x_i be the unique element in A_{sa} for which $\hat{x}_i = a_i - \alpha_i$. Note that $\alpha_i \nearrow 0$. For each ϕ in S we have

$$\phi(x_i + \alpha_i 1) = \phi(x_i) + \alpha_i = a_i(\phi).$$

This implies that $\{x_i + \alpha_i 1\}$ is a monotone increasing net in A''_{sa} which converges to x.

(iv) $\Rightarrow$ (i). Fix $\delta > 0$ such that $3\delta < \varepsilon$ and assume, as we may, that $\alpha_i + \delta > 0$ for all i. Let $\{u_\lambda\}$ be the canonical approximate unit for A (see 1.4.3), and consider elements of the form

$$y_{ir\lambda} = x_i + (\alpha_i + r)u_\lambda$$

with $\varepsilon > r > 3\delta$. We claim that $\{y_{ir\lambda}\}$ form a monotone increasing net in A_{sa}. To see this take $y_{ir\lambda}$ and $y_{js\mu}$. Choose k such that $x_k + \alpha_k 1 \geqslant x_i + \alpha_i 1$ and $x_k + \alpha_k 1 \geqslant x_j + \alpha_j 1$. Furthermore take t strictly larger than r and s but less than ε. Finally choose u_ν such that $u_\nu \geqslant u_\lambda$, $u_\nu \geqslant u_\mu$ and also such that

$$u_\nu \geqslant (t - r + |x_k - x_i|)^{-1}|x_k - x_i|,$$

$$u_\nu \geqslant (t - s + |x_k - x_j|)^{-1}|x_k - x_j|.$$

Then with $x_k - x_i = z$ and $\alpha_k - \alpha_i = \gamma$ we have

$$\begin{aligned} y_{kt\nu} - y_{ir\lambda} &= z + (\alpha_k + t)u_\nu - (\alpha_i + r)u_\lambda \\ &\geqslant z + (\gamma + t - r)u_\nu \geqslant z + (\gamma + t - r)(t - r + |z|)^{-1}|z| \\ &= (t - r + |z|)^{-1}[(t - r)z + |z|\, z + (\gamma + t - r)|z|] \\ &= (t - r + |z|)^{-1}[(t - r)(z + |z|) + (z + \gamma)|z|] \geqslant 0, \end{aligned}$$

since $z + y \geqslant 0$. Similarly $y_{k\iota\nu} \geqslant y_{js\mu}$. It follows that the net $\{y_{ir\lambda}\}$ is monotone increasing in A_{sa}, and it is clear that it converges strongly to $x + \varepsilon 1$.

3.11.6. PROPOSITION. *If x is a positive element of $((A_{sa})^m)^-$ then $x + \varepsilon 1 \in (A_+)^m$ for each $\varepsilon > 0$.*

Proof. Returning to the proof of the implication (iv) $\Rightarrow$ (i) in 3.11.5 we note that the assumption $\alpha_i + \delta > 0$ implies that each element $(x_i + (\alpha_i + \delta)1)^\wedge$ is lower semi-continuous on Q (since $\hat{1}$ is lower semi-continuous by 3.11.1). If therefore $x \geqslant 0$ then eventually $(x_i + (\alpha_i + \delta)1)^\wedge \geqslant -\delta$ by a standard compactness argument. Thus $\phi(x_i + (\alpha_i + 2\delta)1) \geqslant 0$ for every ϕ in S, whence $x_i + (\alpha_i + 2\delta)1 \geqslant 0$. From the construction of the elements $y_{ir\lambda}$ we see that the subset consisting of those elements $y_{ir\lambda} = x_i + (\alpha_i + r)u_\lambda$ for which $u_\lambda \geqslant (\delta + |x_i|)^{-1}|x_i|$ is a subnet. However, for these elements we have

$$
\begin{aligned}
y_{ir\lambda} &\geqslant x_i + (\alpha_i + 3\delta)(\delta + |x_i|)^{-1}|x_i| \\
&= (\delta + |x_i|)^{-1}[\delta x_i + x_i|x_i| + (\alpha_i + 3\delta)|x_i|] \\
&= (\delta + |x_i|)^{-1}[(x_i + \alpha_i + 2\delta)|x_i| + \delta(x_i + |x_i|)] \geqslant 0.
\end{aligned}
$$

It follows that $x + \varepsilon 1 \in (A_+)^m$.

3.11.7. PROPOSITION. *Let A be a C*-algebra and x an element of A''_{sa}. The following conditions are equivalent:*

(i) $x \in (A_{sa})^m$;

(ii) $x \in ((A_{sa})^m)^- + \boldsymbol{R}1$;

(iii) *There is a lower semi-continuous element b in $B(Q)$ such that $\hat{x}|S = b|S$.*

Proof. (i) $\Rightarrow$ (ii). If $\{x_i + \alpha_i 1\}$ is a bounded, monotone increasing net in $\tilde{A}_{sa}$ with limit x then $\{\alpha_i\}$ is a bounded monotone increasing net in $\boldsymbol{R}$ $(= \tilde{A}_{sa}/A_{sa})$, hence convergent to an α in $\boldsymbol{R}$. It follows from condition (iv) in 3.11.5 that $x - \alpha 1 \in ((A_{sa})^m)^-$.

(ii) $\Rightarrow$ (iii). If $x = y + \alpha 1$ with y in $((A_{sa})^m)^-$ define $b = \hat{y} + \alpha$ on Q. Clearly $b \in B(Q)$ and $b|S = \hat{x}|S$. By 3.11.5, b is lower semi-continuous on Q.

(iii) $\Rightarrow$ (i). Define $a = b - b(0)$. Then a is lower semi-continuous and belongs to $B_0(Q)$, and by 3.11.5 $a = \hat{y}$ with y in $((A_{sa})^m)^-$. Again by 3.11.5 $y + 1 \in (A_{sa})^m$ and so $(y + 1) + (b(0)1 - 1) \in (\tilde{A}_{sa})^m$. For each ϕ in S

$$\phi(y + b(0)1) = a(\phi) + b(0) = b(\phi) = \phi(x),$$

whence $x = y + b(0)1 \in (\tilde{A}_{sa})^m$.

3.11.8. PROPOSITION. *Let A be a C*-algebra and x an element of A''_{sa}. The following conditions are equivalent:*

(i) $x \in ((\tilde{A}_{sa})^m)^-$;

(ii) *$\hat{x}$ is lower semi-continuous on S*;

(iii) $(1 - \varepsilon x)^{-1} \in ((A_{sa})^m)^-$ *when* $\varepsilon x < 1$ *and* $\varepsilon > 0$;

Proof. (i) ⇒ (ii). If $x \in \tilde{A}_{sa}$ then $\hat{x}$ is continuous on S. Consequently $\hat{x}$ is lower semi-continuous on S for each x in $(\tilde{A}_{sa})^m$ and, since ^ is an isometry, also for each x in $((\tilde{A}_{sa})^{m-})$.

(ii) ⇒ (iii). By 3.11.3 there is a net $\{x_i\}$ in $\tilde{A}_{sa}$ which converges weakly to x in A'' and satisfies $x_i \leqslant x$ for all i. If $\varepsilon x < 1$ then $1 - \varepsilon x_i \geqslant 1 - \varepsilon x > 0$ and thus by 2.1.3 the net $\{(1 - \varepsilon x_i)^{-1}\}$ is strongly convergent to $(1 - \varepsilon x)^{-1}$ with $(1 - \varepsilon x_i)^{-1} \leqslant (1 - \varepsilon x)^{-1}$ for all i. Since $(1 - \varepsilon x_i)^{-1} \in \tilde{A}_+$ it has the form $y_i + \alpha_i 1$ with y_i in A_{sa} and $\alpha_i \geqslant 0$, and thus $((1 - \varepsilon x_i)^{-1})^\wedge$ is lower semi-continuous on Q ($\hat{1}$ is lower semi-continuous). It follows that $((1 - \varepsilon x)^{-1})^\wedge$ is lower semi-continuous on Q, whence $(1 - \varepsilon x)^{-1} \in ((A_{sa})^m)^-$ by 3.11.5.

(iii) ⇒ (i). By 3.11.7 we have

$$\varepsilon^{-1}((1 - \varepsilon x)^{-1} - 1) = (1 - \varepsilon x)^{-1} x \in (\tilde{A}_{sa})^m,$$

and when $\varepsilon \to 0$ this implies that

$$\|(1 - \varepsilon x)^{-1} x - x\| = \|(1 - \varepsilon x)^{-1} \varepsilon x^2\| \to 0,$$

whence $x \in ((\tilde{A}_{sa})^m)^-$.

3.11.9. PROPOSITION. *Let A be a C*-algebra and p a projection in A''. The following conditions are equivalent:*

(i) $p \in (A_+)^m$;

(ii) $p \in ((\tilde{A}_{sa})^m)^-$;

(iii) *The face* $F = \{\phi \in Q \mid \phi(p) = 0\}$ *is weak* closed*;

(iv) *p belongs to the strong closure in A'' of the hereditary C*-subalgebra* $pA''p \cap A$ *of A.*

Proof. (i) ⇒ (ii) is obvious.

(ii) ⇒ (iii). If $p \in ((\tilde{A}_{sa})^m)^-$ then

$$(1 - \varepsilon)(1 - \varepsilon p)^{-1} = (1 - \varepsilon)(1 - p) + p \in ((A_{sa})^m)^-$$

by 3.11.8 whenever $0 < \varepsilon < 1$. With $\varepsilon \to 1$ we obtain $p \in ((A_{sa})^m)^-$ so that $\hat{p}$ is lower semi-continuous on Q by 3.11.5. Consequently F is weak* closed.

(iii) ⇒ (iv). By 3.10.7 the left invariant subspace

$$E = \{\phi \in A^* \mid \phi(A''p) = 0\}$$

is weak* closed. Its annihilator in A is the left ideal

$$L = \{x \in A \mid xp = x\} = A''p \cap A;$$

its annihilator in A'' is of course $A''p$. Since A'' is the bi-dual of A this implies that L is weakly dense in $A''p$. Thus p is in the strong closure of L_{sa} and $L_{sa} = (L \cap L^*)_{sa}$ which is the self-adjoint part of the hereditary C^*-subalgebra $pA''p \cap A$.

(iv)$\Rightarrow$(i). Put $B = pA''p \cap A$. By Kaplansky's density theorem each element in $pA''p$ can be approximated strongly with elements of the form xyx where $x \in B^1_+$ and $y \in A$. Since B is hereditary this means that the strong closure of B in A'' is precisely $pA''p$. If therefore $\{u_\lambda\}$ is an approximate unit for B then $\{u_\lambda\}$ is strongly convergent to an element b such that $bB = B$ and $b \leqslant p$. This implies that $b = p$ and consequently. $p \in (B_+)^m \subset (A_+)^m$.

3.11.10. A projection satisfying the conditions in 3.11.9 is called *open.* A projection whose complement is open is said to be *closed.* From 3.10.7 (and 3.6.11) we infer that open projections are in bijective correspondence with the hereditary C^*-subalgebras of A and with the weak * closed faces of Q containing zero. In particular the central open projections correspond bijectively to the closed ideals of A.

If A is commutative, viz. $A = C_0(\hat{A})$, then as is well known the open subsets of $\hat{A}$ correspond to the closed ideals of A. Thus the open projections in A'' correspond to the open subsets of $\hat{A}$. We shall investigate the generalization of this correspondence in the next chapter.

3.11.11. Notes and remarks. This section is borrowed from Akemann and Pedersen [4]. No examples are known for which $(A_{sa})^m \neq ((A_{sa})^m)^-$, but it is shown in [4] that one may have $(\tilde{A}_{sa})^m \neq ((\tilde{A}_{sa})^m)^-$. Thus it may well be that for non-commutative C^*-algebras without unit there are four different classes, each of which possesses some semi-continuity properties. If $1 \in A$ then $A = \tilde{A}$, whence $(A_{sa})^m = ((\tilde{A}_{sa})^m)^-$ by 3.11.7, so that all four classes coincide. Semi-continuity plays an unexpected rôle in the theory of derivations and one-parameter groups of automorphisms; see 8.5 and 8.6.

3.12. Multipliers

3.12.1. Let A be a non-degenerate C^*-subalgebra of $\boldsymbol{B}(H)$. We say that an operator x in $\boldsymbol{B}(H)$ is a *left*, respectively *right, multiplier* for A if $xA \subset A$, respectively $Ax \subset A$. We say that x is a *two-sided multiplier* (or just a *multiplier*) if x is both a left and a right multiplier. We say that x is a *quasi-multiplier* if $AxA \subset A$. If $1 \in A$ there are, of course, no (left, right or quasi-) multipliers outside of A. Since $xA \subset A$ implies $xA'' \subset A''$ and $1 \in A''$ (A is non-degenerate) this implies that all multipliers of A belong to A''.

In order to show that the multipliers of A do not depend on the particular representation in $\boldsymbol{B}(H)$ we introduce the notion of centralizers as follows.

A linear map $\rho: A \to A$ is called a *left*, respectively *right*, *centralizer* if $\rho(xy) = \rho(x)y$, respectively $\rho(xy) = x\rho(y)$, for all x and y in A. A *double centralizer* is a pair (ρ_1, ρ_2) consisting of a right and a left centralizer such that $\rho_1(x)y = x\rho_2(y)$ for all x and y in A. A *quasi-centralizer* is a bilinear map $\rho: A \times A \to A$ such that for each fixed x in A the map $\rho(x, \cdot)$ is a left centralizer and the map $\rho(\cdot, x)$ is a right centralizer.

3.12.2. LEMMA. *Each left centralizer or quasi-centralizer of a C^*-algebra A is bounded.*

Proof. Let ρ be a left centralizer of A. If ρ was not bounded we could find x_n in A with $\|x_n\| < 1/n$ but $\|\rho(x_n)\| > n$. Let $a = \Sigma x_n x_n^*$. Then by 1.4.5 there is for each n an element u_n in A with $\|u_n\| \leqslant \|a^{1/6}\|$ such that $x_n = a^{1/3}u_n$. But then

$$\|\rho(x_n)\| = \|\rho(a^{1/3})u_n\| \leqslant \|\rho(a^{1/3})\| \, \|a^{1/6}\|,$$

a contradiction. Thus ρ is bounded. If $\rho(\cdot, \cdot)$ is a quasi-centralizer of A then for each x in A^1 the left centralizer $\rho(x, \cdot)$ is a bounded map, say $\|\rho(x, \cdot)\| = \lambda_x$, from the above. However, for a fixed y in A the right centralizer $\rho(\cdot, y)$ is also bounded. It follows from the uniform boundedness theorem that the set $\{\lambda_x \mid x \in A^1\}$ is bounded. Consequently $\rho(\cdot, \cdot)$ is a bounded map on $A \times A$.

3.12.3. PROPOSITION. *Let A be a non-degenerate C^*-subalgebra of $B(H)$. There is then a bijective correspondence between left (respectively right, double and quasi-) centralizers of A and left (respectively right, two-sided and quasi-) multipliers of A in A''.*

Proof. If ρ is a left centralizer of A and $\{u_\lambda\}$ is an approximate unit for A then $\{\rho(u_\lambda)\}$ is a bounded net by 3.12.2 and has therefore a weak limit point x in A''. However, for each y in A the net $\{u_\lambda y\}$ is norm convergent to y and so, again by 3.12.2, $\{\rho(u_\lambda y)\}$ is norm convergent to $\rho(y)$. Since $\rho(u_\lambda y) = \rho(u_\lambda)y$ this implies that $xy = \rho(y)$. It follows that x is a left multiplier corresponding to ρ. If z was another element in A'' such that $zy = \rho(y)$ then $(x - z)y = 0$ for all y in A, hence also for all y in A'' and thus $x = z$. This shows that the correspondence between left centralizers and left multipliers is a bijection.

If (ρ_1, ρ_2) is a double centralizer of A then from the above there is a unique pair (x_1, x_2) corresponding to (ρ_1, ρ_2) consisting of a right and a left multiplier of A. However, for all y and z in A:

$$yx_1z = \rho_1(y)z = y\rho_2(z) = yx_2z,$$

whence $x_1 = x_2$.

If $\rho(\cdot, \cdot)$ is a quasi-centralizer let x be a weak limit point in A'' of the net $\{\rho(u_\lambda, u_\lambda)\}$, where $\{u_\lambda\}$ as before is an approximate unit for A. Then for every y and z in A we have

$$\rho(y, z) = \text{Lim}\, \rho(yu_\lambda, u_\lambda z) = \text{Lim}\, y\rho(u_\lambda, u_\lambda)z = yxz,$$

which prove that x is a quasi-multiplier for A in A'' corresponding to $\rho(\cdot, \cdot)$. The unicity of x is verified as above, and the proof is complete.

3.12.4. Given a C^*-algebra A we consider it as an algebra of operators on its universal Hilbert space (3.7.6) and denote by $LM(A)$, $RM(A)$, $M(A)$ and $QM(A)$ the sets of left, right, two-sided and quasi-multipliers of A in A'', respectively.

It is clear that

$$M(A) = LM(A) \cap RM(A)$$

and that

$$LM(A) + RM(A) \subset QM(A).$$

Moreover, $LM(A)$, $RM(A)$ and $QM(A)$ are norm closed subspaces in A''; $QM(A)$ is $*$-invariant, whereas $LM(A) = RM(A)$. The best behaved class is $M(A)$ which is a C^*-subalgebra of A''.

As an immediate consequence of 3.12.3 we have the following result.

3.12.5. COROLLARY. *If A is a C^*-algebra and (π, H) is a faithful representation of A then its normal extension (π'', H) to A'' maps $QM(A)$ isometrically onto the set of quasi-multipliers of $\pi(A)$ in $\pi''(A'')$.*

3.12.6. If A is a commutative C^*-algebra, and thus of the form $C_0(\hat{A})$, then we can consider the faithful representation of A as operators on $l^2\hat{A}$—the square summable functions on $\hat{A}$—where $(xf)(t) = x(t)f(t)$ for x in A, f in $l^2\hat{A}$ and t in $\hat{A}$. It is easy to determine the multipliers of A in this representation and we see from 3.12.5 that $M(A)$ is isomorphic to $C^b(\hat{A})$—the bounded, continuous functions on $\hat{A}$. If $\beta(\hat{A})$ denotes the Stone–Čech compactification of $\hat{A}$ then $C^b(\hat{A}) = C(\beta(\hat{A}))$ and thus $M(A)$ is the algebraic counterpart of the maximal compactification of $\hat{A}$, just as $\tilde{A}$ is the algebraic counterpart of the minimal (one-point) compactification of $\hat{A}$.

3.12.7. If A is an ideal in a C^*-algebra B then we say that A is *essential* in B if each non-zero closed ideal of B has a non-zero intersection with A.

3.12.8. PROPOSITION. *If a C^*-algebra A is an essential ideal in a C^*-algebra B then there is an injection of B into $M(A)$ which is the identity map on A.*

Proof. By 3.7.9 we may identify A'' with the strong closure of A in B'', and by 3.11.10 (or just 2.5.4) this is equal to $B''p$ for some (open) central projection p in B''. The morphism $\pi: B \to A''$ given by $\pi(x) = xp$ is the identity map on A and takes B into $M(A)$, since for each y in A and x in B

$$\pi(x)y = xpy = xy \in A.$$

The kernel of π is therefore a closed ideal of B which has zero intersection with A. Since A is essential in B this implies that π is injective.

3.12.9. THEOREM. *Let A be a C^*-algebra. Then*

$$M(A)_{sa} = (\tilde{A}_{sa})^m \cap (\tilde{A}_{sa})_m. \qquad QM(A)_{sa} = ((\tilde{A}_{sa})^m)^- \cap ((\tilde{A}_{sa})_m)^-.$$

Proof. Let $\{u_\lambda\}$ be an approximate unit for A and take x in $M(A)_+^1$. Then the nets $\{x^{1/2}u_\lambda x^{1/2}\}$ and $\{(1-x)^{1/2}u_\lambda(1-x)^{1/2}\}$ are monotone increasing in A with limits x and $1-x$, respectively, so that both x and $1-x$ belong to $(A_{sa})^m$. Consequently

$$x \in 1 - (A_{sa})^m = 1 + (A_{sa})_m \subset (\tilde{A}_{sa})_m,$$

whence $x \in (\tilde{A}_{sa})^m \cap (\tilde{A}_{sa})_m$. Since $M(A)_{sa}$ is linearly spanned by $M(A)_+^1$ and $(\tilde{A}_{sa})^m \cap (\tilde{A}_{sa})_m$ is a real vector space we have proved that it contains $M(A)_{sa}$.

Conversely, if $x \in (\tilde{A}_{sa})^m \cap (\tilde{A}_{sa})_m$ let $\{x_i\}$ be a net in $\tilde{A}_{sa}$ which is monotone increasing to x and let $\{y_j\}$ be a net in $\tilde{A}_{sa}$ which is monotone decreasing to x. Then for each a in A the net $\{a^*(y_j - x_i)a\}$ is monotone decreasing to zero in A_+, since A is an ideal of $\tilde{A}$. From the function representation of A (3.10.3) and Dini's theorem it follows that the net is norm convergent to zero, whence

$$\|xa - x_i a\|^2 = \|a^*(x - x_i)^2 a\|$$
$$\leqslant \|x - x_i\| \, \|a^*(x - x_i)a\| \leqslant \|x - x_i\| \, \|a^*(y_j - x_i)a\| \to 0.$$

Since $x_i a \in A$ it follows that $xa \in A$ so that $xA \subset A$, and as $x = x^*$, $x \in M(A)$.

Assume now that $x \in QM(A)_{sa}$. If $\phi \in S$ then for each $\varepsilon > 0$ we can find u_λ such that $\phi(u_\lambda) > 1 - \varepsilon^2$ by 3.1.4. It follows from the Cauchy–Schwarz inequality that

$$\|\phi - \phi(u_\lambda \cdot u_\lambda)\| \leqslant \|\phi((1 - u_\lambda)\cdot)\| + \|\phi(u_\lambda \cdot (1 - u_\lambda))\| < 2\varepsilon.$$

If now $\{\phi_i\}$ is a net in S which is weak* convergent to ϕ then eventually $\phi_i(u_\lambda) > 1 - \varepsilon^2$ whence $\|\phi_i - \phi_i(u_\lambda \cdot u_\lambda)\| < 2\varepsilon$. Since $x \in QM(A)$ we have $u_\lambda x u_\lambda \in A$, whence $\operatorname{Lim} \phi_i(u_\lambda x u_\lambda) = \phi(u_\lambda x u_\lambda)$. Since ε was arbitrary it follows that $\{\phi_i(x)\}$ converges to $\phi(x)$, which proves that $\hat{x}$ is continuous when restricted to S. Consequently $x \in ((\tilde{A}_{sa})^m)^- \cap ((\tilde{A}_{sa})_m)^-$ by 3.11.8.

Finally take x in $((\tilde{A}_{sa})^m)^- \cap ((\tilde{A}_{sa})_m)^-$. Given $\varepsilon > 0$ we can find nets $\{y_i\}$ and $\{z_j\}$ in $\tilde{A}_{sa}$ such that $\{y_i\}$ is monotone increasing to y, $\{z_j\}$ is monotone decreasing to z and $\|x - y\| < \varepsilon$, $\|x - z\| < \varepsilon$. Thus if $a \in A$ then $a^*ya \in (A_{sa})^m$, $a^*za \in (A_{sa})_m$ and $\|a^*xa - a^*ya\| < \varepsilon\|a\|^2$, $\|a^*xa - a^*za\| \leqslant \varepsilon\|a\|^2$. Consequently $a^*xa \in ((A_{sa})^m)^- \cap ((A_{sa})_m)^-$ which by 3.11.5 implies that $(a^*xa)^\wedge$ is continuous on Q and thus $a^*xa \in A$ by 3.10.3. The polarization identity

$$4b^*xc = \sum_{k=0}^{3} i^k(c + i^k b)^* x(c + i^k b)$$

shows that $b^*xc \in A$ for each b and c in A, whence $x \in QM(A)$.

3.12.10. PROPOSITION. *Let π be a surjective morphism between separable C^*-algebras A and B. Then π extends to a surjective morphism of $M(A)$ onto $M(B)$.*

Proof. By 3.7.7 there is a normal extension of π to a surjective morphism π'' of A'' onto B''. Clearly $\pi''(M(A)) \subset M(B)$. Since A and B are separable Q_A and Q_B are second countable. If therefore $b \in M(B)_{sa}$ there are sequences $\{x_n\}$ and $\{y_n\}$ in $\tilde{B}_{sa}$ such that $x_n \nearrow b$ and $y_n \searrow b$ (utilizing 3.12.9). We can also choose a countable approximate unit $\{u_n\}$ for the kernel of π in A.

Choose self-adjoint elements v_1 and w_1 in $\tilde{A}$ such that $\pi(v_1) = x_1$, $\pi(w_1) = y_1$ and $v_1 \leqslant w_1$. Put

$$v_1' = v_1 + (w_1 - v_1)^{1/2} u_1 (w_1 - v_1)^{1/2}.$$

Then $\pi(v_1') = x_1$ and $v_1 \leqslant v_1' \leqslant w_1$. Suppose that we have chosen $\{v_k\}$, $\{v_k'\}$ and $\{w_k\}$ in $\tilde{A}_{sa}$ for $1 \leqslant k \leqslant n-1$ satisfying the conditions:

(i) $v_{k-1}' \leqslant v_k \leqslant w_k \leqslant w_{k-1}$;

(ii) $\pi(v_k) = \pi(v_k') = x_k, \pi(w_k) = y_k$;

(iii) $v_k' = v_k + (w_k - v_k)^{1/2} u_k (w_k - v_k)^{1/2}$.

Using 1.5.10 we then choose v_n in $\tilde{A}_{sa}$ such that $\pi(v_n) = x_n$ and $v_{n-1}' \leqslant v_n \leqslant w_{n-1}$. With the same argument we choose w_n in $\tilde{A}_{sa}$ such that $\pi(w_n) = y_n$ and $v_n \leqslant w_n \leqslant w_{n-1}$. We then define

$$v_n' = v_n + (w_n - v_n)^{1/2} u_n (w_n - v_n)^{1/2}.$$

The elements v_n, v_n' and w_n satisfy conditions (i), (ii) and (iii) and we can thus by induction find sequences $\{v_n\}$, $\{v_n'\}$ and $\{w_n\}$ in $\tilde{A}_{sa}$ satisfying (i), (ii) and (iii). It follows that there exists a in $(\tilde{A}_{sa})^m$ and c in $(\tilde{A}_{sa})_m$ such that $v_n \nearrow a$, $v_n' \nearrow a$ and $w_n \searrow c$. Since π'' is normal we have $\pi''(a) = \pi''(c) = b$. Let p be the open central projection in A'' such that $u_n \nearrow p$ and note that by 3.11.9 the kernel of π'' in A'' is precisely $A''p$. Thus $(c - a)p = (c - a)$. On the other hand, since all the strongly convergent sequences involved are bounded and the square root function is strongly continuous we see from condition (iii) that in the limit

$$a = a + (c - a)^{1/2} p (c - a)^{1/2}.$$

It follows that $c = a$, whence

$$a \in (\tilde{A}_{sa})^m \cap (\tilde{A}_{sa})_m = M(A)_{sa}$$

by 3.12.9, and thus $\pi''(M(A)) \supset M(B)$.

3.12.11. If A is a commutative C^*-algebra, i.e. $A = C_0(\hat{A})$ then a morphism of A consists of restricting the functions on $\hat{A}$ to a closed subset $\hat{B}$ of $\hat{A}$. As mentioned in 3.12.6, $M(A) = C^b(\hat{A})$ and $M(B) = C^b(\hat{B})$. Thus 3.12.10 is in the commutative case nothing but Tietze's extension theorem for second countable, locally compact Hausdorff spaces. Since there are non-normal, locally compact Hausdorff spaces the separability assumption in 3.12.10 is necessary already in the commutative case.

3.12.12. Let B be a C^*-subalgebra of a C^*-algebra A. Then $B'' \subset A''$ by 3.7.9 so that $M(B) \subset A''$. Assume that $\{u_\lambda\}$ is an approximate unit for A contained in B. Then for each x in A and y in $M(B)$ we have

$$yx = \operatorname{Lim} yu_\lambda x \in BA \subset A,$$

and similarly $xy \in A$; so that $M(B) \subset M(A)$.

If A is separable it has a strictly positive element h by 3.10.6. Thus with $B = C^*(h)$ we have a commutative C^*-subalgebra of A containing an approximate unit for A. If $1 \notin A$ then 0 is not an isolated point in $\mathrm{Sp}(h)$ whence $C^b(\mathrm{Sp}(h)\backslash\{0\})$ is non-separable. Since B is isomorphic to $C_0(\mathrm{Sp}(h)\backslash\{0\})$ and $M(B)$ is isomorphic $C^b(\mathrm{Sp}(h)\backslash\{0\})$ (cf. 3.12.6) we conclude from the above that the multiplier algebra of a (separable) C^*-algebra without unit is never separable.

3.12.13. In 1.4.2 we showed the existence of approximate units in any C^*-algebra A and in 3.10.6 we discussed the possibility of finding a commutative approximate unit for A.

Assume now that A is contained as a closed ideal of a C^*-algebra B. We say that an approximate unit $\{u_\lambda \mid \lambda \in \Lambda\}$ for A is *quasi-central* for B if

$$\operatorname{Lim} \|u_\lambda x - xu_\lambda\| = 0 \qquad \forall x \in B.$$

3.12.14. THEOREM. *Let $\{v_i \mid i \in I\}$ be an approximate unit for a C^*-algebra A contained as a closed ideal of a C^*-algebra B. There is then an approximate unit for A contained in* $\mathrm{Conv}\{v_i \mid i \in I\}$, *which is quasi-central for B.*

Proof. Let Λ denote the collection of all (non-empty) finite subsets of B and for each λ in Λ let $|\lambda|$ denote the cardinality of λ. Given i and λ let $M_{i\lambda}$ denote the set of elements u in

$$\mathrm{Conv}\{v_j \mid j \succ i\}$$

such that $\|ux - xu\| < |\lambda|^{-1}$ for all x in λ. We claim that $M_{i\lambda} \neq \varnothing$.

To see this fix $x_1, x_2, \ldots, x_n$ in B and let $C = \oplus_{k=1}^n B$. Consider the net $\{c_j \mid j \succ i\}$ in C where $c_{jk} = v_j x_k - x_k v_j$, $1 \leqslant k \leqslant n$. Working in B'' we know from 3.11.10 that $v_j \nearrow p$ where p is the open central projection in B'' for which $A = pB'' \cap B$. It follows that $c_j \to 0$ σ-weakly in C''. But since C in the σ-weak topology and the norm topology has the same continuous functionals we conclude from Hahn–Banach's theorem that $\mathrm{Conv}\{c_j\}$ contains zero as a limit point in norm. Consequently, there is a convex combination $u = \Sigma_j \gamma_j v_j$, $j \succ i$, such that $\|ux_k - x_k u\| < 1/n$ for every $k \leqslant n$. With $\lambda = \{x_1, \ldots, x_n\}$ we see that $u \in M_{i\lambda}$.

Invoking the axiom of choice we select an element $u_{i\lambda}$ from each $M_{i\lambda}$, $i \in I$, $\lambda \in \Lambda$. We define a partial order in the set U of these elements by $u_{i\lambda} \prec u_{j\mu}$ if $i \prec j$, $\lambda \subset \mu$ and $u_{i\lambda} \leqslant u_{j\mu}$. To show that U is a directed set take $u_{i\lambda}$ and $u_{j\mu}$ in U; say $u_{i\lambda} = \Sigma \gamma_n v_n$ and $u_{j\mu} = \Sigma \gamma_m v_m$. Find k in I such that $k \succ i$, $k \succ j$ and $k \succ n$, $k \succ m$ for all n and m occurring in the expressions for $u_{i\lambda}$ and $u_{j\mu}$, respectively. Take $\nu \supset \mu \cup \lambda$ and consider the element $u_{k\nu}$ in U. We have $i \prec k$, $\lambda \subset \nu$ and also $u_{i\lambda} \leqslant u_{k\nu}$; because if $u_{k\nu} = \Sigma \gamma_l v_l$ then $l \succ k \succ n$ for all l and n, whence $v_l \geqslant v_n$ for all l and n, so that finally $u_{k\nu} \geqslant u_{i\lambda}$. Consequently $u_{k\nu} \succ u_{i\lambda}$ and similarly $u_{k\nu} \succ u_{j\mu}$, whence U is directed.

By construction the net U is contained in $\mathrm{Conv}\{v_i\}$ and $u_{i\lambda} \geqslant v_i$ for all i and λ. It follows that U is an approximate unit for A. Moreover, U is quasi-central, since $\|u_{i\lambda}x - xu_{i\lambda}\| \leqslant |\lambda|^{-1}$ for each x in λ.

3.12.15. COROLLARY. *If A is a separable C^*-algebra it contains a commutative approximate unit which is quasi-central for $M(A)$.*

Proof. From 3.10.6 we know that A contains a (countable) commutative approximate unit; viz. $u_n = h^{1/n}$, where h is a strictly positive element for A. Applying 3.13.2 to the commutative family $\mathrm{Conv}\{u_n\}$ we obtain the desired result.

3.12.16. COROLLARY. *If A is a closed ideal in a separable C^*-algebra B, then A contains a countable commutative approximate unit which is quasi-central for B.*

3.12.17. Notes and remarks. Centralizers of non-commutative algebras were studied by Johnson [121] and 3.12.3 was established by Busby [29]. The connection between multipliers and semi-continuity was recognized in [198]. Further developments appeared in [4] and in [7] where 3.12.10 was proved. The strict topology on $M(A)$ is the vector space topology generated by semi-norms of the form $x \to \|xa\| + \|ax\|$, where $a \in A$. It was proved in [29] that $M(A)$ is the strict completion of A. This fact can be used to compute $M(A)$ in a number of interesting cases (see Section 3 of [7]).

Quasi-central approximate units appear in [14] and [5]. Although the idea behind the proof is simple, it has been used with great success [5, 14, 84, 6].

3.13. Pure states and irreducible representations

3.13.1. If (π, H) is a representation of a C^*-algebra A then a linear subspace K of H is said to be *reducing* for π if $\pi(A)K \subset K$. This explains the term *irreducible representations* employed for the representations which satisfy the conditons in the next theorem.

3.13.2. THEOREM. *Let (π, H) be a non-zero representation of a C^*-algebra A. The following conditions are equivalent:*

(i) *There are no non-trivial reducing subspaces for π.*

(ii) *The commutant of $\pi(A)$ is the scalar multiples of 1.*

(iii) *$\pi(A)$ is strongly dense in $\boldsymbol{B}(H)$.*

(iv) *For any two vectors ξ, η in H with $\xi \neq 0$ there is a y in A such that $\pi(y)\xi = \eta$.*

(v) *Each non-zero vector in H is cyclic for $\pi(A)$.*

(vi) *(π, H) is spatially equivalent to a cyclic representation associated with a pure state of A.*

Proof. (i) ⇒ (ii). If p is a projection in $\pi(A)'$ then pH is a closed reducing subspace for π. Thus either $p = 0$ or $p = 1$, whence $\pi(A)' = \boldsymbol{C}1$.

(ii) ⇒ (iii) follows from the bicommutant theorem (2.2.5).

(iii) ⇒ (iv) follows from 2.7.4 by taking p_0 as the projection on the subspace spanned by ξ and η and x as any operator in $\boldsymbol{B}(H)$ with $x\xi = \eta$.

(iv) ⇒ (i). If K is a non-zero reducing subspace then it contains a non-zero vector ξ. But then $H = \pi(A)\xi \subset K$ so that $K = H$.

(iv) ⇒ (v) is immediate.

(v) ⇒ (ii). If p is a non-zero projection in $\pi(A)'$ then pH contains a non-zero vector ξ. Since pH is reducing for π we have $\pi(A)\xi \subset pH$, whence $p = 1$. Since $\pi(A)'$ is generated by its projections $\pi(A)' = \boldsymbol{C}1$.

(ii) ⇒ (vi). Let ξ_0 be a unit vector in H and ϕ the corresponding vector state. If $(\pi_\phi, H_\phi, \xi_\phi)$ is the cyclic representation associated with ϕ then by 3.3.7 $(\pi, H. \xi_0)$ and $(\pi_\phi, H_\phi, \xi_\phi)$ are spatially equivalent with an isometry u such that $u\xi_0 = \xi_\phi$ (since ξ_0 is cyclic by (v)).

Assume now that ψ_1 and ψ_2 are states of A with $\alpha\psi_1 + (1 - \alpha)\psi_2 = \phi$ and $0 \leqslant \alpha \leqslant 1$. Then $0 \leqslant \alpha\psi_1 \leqslant \phi$ and so by 3.3.5 there is a positive a in $\pi_\phi(A)'$ such that $\alpha\psi_1(x) = (\pi_\phi(x)a\xi_\phi \mid \xi_\phi)$. But then $u^*au \in \pi(A)'$, whence $u^*au = \beta 1$ by (ii) and thus $a = \beta 1$. Since

$$\alpha\psi_1(x) = (\pi_\phi(x)\beta\xi_\phi \mid \xi_\phi) = \beta\phi(x)$$

and $\|\psi_1\| = \|\phi\| = 1$ we have $\alpha = \beta$ and $\psi_1 = \phi$; $\psi_2 = \phi$. It follows that ϕ is extremal and thus a pure state.

(vi) ⇒ (ii). Let $(\pi_\phi, H_\phi, \xi_\phi)$ be the cyclic representation associated with a pure state ϕ. If $\pi_\phi(A)'$ is not the scalars then it contains a projection p different from both 0 and 1. Let $\psi_1(x) = (\pi_\phi(x)p\xi_\phi \mid \xi_\phi)$ and $\psi_2(x) = (\pi_\phi(x)(1 - p)\xi_\phi \mid \xi_\phi)$. Then ψ_1 and ψ_2 are positive functionals with $\psi_1 + \psi_2 = \phi$, and $\|\psi_1\| = \|p\xi_\phi\|^2$, $\|\psi_2\| = \|(1 - p)\xi_\phi\|^2$. If $p\xi_\phi = 0$ then $p\pi_\phi(A)\xi_\phi = 0$, whence $p = 0$ since ξ_ϕ is cyclic. This and $(1 - p)\xi_\phi = 0$ is therefore excluded so that $0 < \|\psi_1\| < 1$. Since ϕ is pure this implies that $\psi_1 = \|\psi_1\|\phi$ and $\psi_2 = \|\psi_2\|\phi$.

But then

$$(p\xi_x \mid \xi_y) = \psi_1(y^*x) = \|\psi_1\|(\xi_x \mid \xi_y)$$

for all x, y in A, whence $p = \|\psi_1\| 1$. This is impossible since $p^2 = p$ and consequently we must have $\pi_\phi(A)' = \mathbf{C}1$.

If (π, H) is spatially equivalent to a representation $(\pi_\phi, H_\phi, \xi_\phi)$ associated with a pure state then since spatially equivalent representations have isomorphic commutants we conclude that $\pi(A)' = \mathbf{C}1$, and the theorem is proved.

3.13.3. COROLLARY. *Two irreducible representations (π_1, H_1) and (π_2, H_2) of a C^*-algebra A are either disjoint (3.8.12) or spatially equivalent.*

Proof. If (π_1, H_1) and (π_2, H_2) are not disjoint they have non-zero spatially equivalent subrepresentations by 3.8.11. However, from condition (i) in 3.13.2 we see that irreducible representations have only the two trivial subrepresentations, zero and the representation itself.

3.13.4. PROPOSITION. *Let ϕ and ψ be pure states of a C^*-algebra A. If $\|\phi - \psi\| < 2$ then (π_ϕ, H_ϕ) and (π_ψ, H_ψ) are equivalent. If (π_ϕ, H_ϕ) and (π_ψ, H_ψ) are equivalent then $\psi = \phi(u^* \cdot u)$ for some unitary u in $\tilde{A}$.*

Proof. If (π_ϕ, H_ϕ) and (π_ψ, H_ψ) are not (spatially) equivalent then they are disjoint by 3.13.3. Thus $c(\pi_\phi) \perp c(\pi_\psi)$ by 3.8.11. Since $\phi(c(\pi_\phi)) = 1$ and $\psi(c(\pi\psi)) = 1$ (regarding ϕ and ψ as normal functionals on A'') this implies that $\phi(c(\pi_\psi)) = 0$ and $\psi(c(\pi_\phi)) = 0$, whence

$$\|\phi - \psi\| \geq (\phi - \psi)(c(\pi_\phi) - c(\pi_\psi)) = 2.$$

Assume now that (π_ϕ, H_ϕ) and (π_ψ, H_ψ) are (spatially) equivalent. Then $\psi(x) = (\pi_\phi(x)\xi \mid \xi)$ for some unit vector ξ in H_ϕ and all x in A. Clearly there is a unitary element on the subspace spanned by ξ and ξ_ϕ which takes ξ_ϕ to ξ. By 2.7.5 there is therefore a unitary u in $\tilde{A}$ such that $\pi_\phi(u)\xi_\phi = \xi$. Consequently

$$\begin{aligned}\psi(x) &= (\pi_\phi(x)\pi_\phi(u)\xi_\phi \mid \pi_\phi(u)\xi_\phi) \\ &= (\pi_\phi(u^*xu)\xi_\phi \mid \xi_\phi) = \phi(u^*xu)\end{aligned}$$

for all x in A, which completes the proof.

3.13.5. LEMMA. *Each closed left ideal L in a C^*-algebra A is the intersection of those left kernels of pure states of A which contains L.*

Proof. By 3.10.7 there is a unique weak* closed face F of Q containing zero such that

$$L = \{x \in A \mid \phi(x^*x) = 0, \phi \in F\} = \bigcap_{\phi \in F} L_\phi.$$

Since F is a closed face of Q the extremal points of F are pure states of A (together with zero). If ∂F denotes the pure states of A contained in F then from the Krein–Milman theorem we have

$$L = \bigcap_{\phi \in F} L_\phi = \bigcap_{\phi \in \partial F} L_\phi,$$

which completes the proof.

3.13.6. PROPOSITION. *Let ϕ be a state of a C^*-algebra A. The following conditions are equivalent:*

(i) *ϕ is a pure state.*

(ii) *The null-space for ϕ is $L_\phi + L_\phi^*$.*

(iii) *There is an open projection p in A'' such that $\phi(p) = 0$ and $1 - p$ is a minimal projection in A''.*

(iv) *L_ϕ is a regular maximal left ideal of A.*

Proof. (i) ⇒ (ii). Let N_ϕ denote the null-space for ϕ. From the Cauchy–Schwarz inequality it is evident that $L_\phi + L_\phi^* \subset N_\phi$. Assume now that ϕ is pure and take x in N_ϕ. This means that $(\xi_x \mid \xi_\phi) = 0$ in the cyclic representation $(\pi_\phi, H_\phi, \xi_\phi)$ associated with ϕ (3.3.3). Since this representation is irreducible by 3.13.2, there is an element y in A such that $\pi_\phi(y)\xi_\phi = 0$ and $\pi_\phi(y)\xi_x = \xi_x$ by 2.7.5, and we may further choose $y = y^*$. Thus $y \in (L_\phi)_{sa}$ and $yx - x \in L_\phi$. Consequently

$$x = x - yx + (x^*y)^* \in L_\phi + L_\phi^*,$$

so that $N_\phi \subset L_\phi + L_\phi^*$.

(ii) ⇒ (iii). Let p be the open projection in A'' such that $L_\phi = A''p \cap A$. Regarding ϕ as a normal functional on A'' we have $\phi(p) = 0$. Let q be a non-zero projection in A'' with $q \leqslant 1 - p$ and let ψ be a state of A such that $\psi(q) = 1$. Then $\psi(p) = 0$ so that $L_\phi \subset L_\psi$ and consequently

$$N_\phi = L_\phi + L_\phi^* \subset L_\psi + L_\psi^* \subset N_\psi.$$

Since N_ϕ and N_ψ have co-dimension one this implies that $N_\phi = N_\psi$ so that ψ is proportional to ϕ and thus equal to ϕ since $\|\phi\| = \|\psi\|$. Using this on any functional of the form $\psi(q)^{-1}\psi(q \cdot q)$ it follows that $\psi(q \cdot q) = \psi(q)\phi$ for each ψ in $(A^*)_+$. But then $\psi(q) = \psi(q)\phi(q)$ so that $\phi(q) = 1$ whenever q is non-zero. Since this holds for any $q \leqslant 1 - p$ we conclude that $1 - p$ is minimal in A''.

(iii) ⇒ (iv). The correspondence between left ideals and open projections given in 3.11.10 is order preserving, and since p by assumption is a maximal projection in A'' it is clear that $A''p \cap A$ is a maximal closed left ideal in A. However, since $\phi(p) = 0$, $A''p \cap A \subset L_\phi$, whence $A''p \cap A = L_\phi$. As will be seen from the last step in the proof each maximal closed left-ideal of A is regular.

(iv) ⇒ (i). Assume only that L_ϕ is maximal among all closed left ideals of A. By 3.13.5 there is then a pure state ψ of A such that $L_\phi = L_\psi$. From the implication (i) ⇒ (ii) it follows that

$$N_\psi = L_\psi + L_\psi^* = L_\phi + L_\phi^* \subset N_\phi$$

whence as before $N_\psi = N_\phi$ and $\psi = \phi$, so that ϕ is pure. By 3.13.2 there is an element e in A such that $\pi_\phi(e)\xi_\phi = \xi_\phi$ which means that $xe - x \in L_\phi$ for each x in A and thus L_ϕ is a regular left ideal.

3.13.7. An ideal I in a C^*-algebra A is *prime* if $xAy \subset I$ implies $x \in I$ or $y \in I$ for all x, y in A. Equivalently, I is prime if $I_1 I_2 \subset I$ implies $I_1 \subset I$ or $I_2 \subset I$ for any two (left, right, or two-sided) ideals I_1 and I_2 of A.

We say that I is a *primitive ideal* if $I = \ker \pi$ for some irreducible representation (π, H) of A. If zero is a primitive ideal, i.e. if A has a faithful irreducible representation on some Hilbert space, we say that A is a *primitive C^*-algebra*. If A is simple (1.2.5) then it is primitive by 3.13.2, since each non-zero C^*-algebra has pure states. As the example $A = \boldsymbol{B}(H)$ shows, the converse is false. If ϕ is a state of A then the kernel of $(\pi_\phi, H_\phi, \xi_\phi)$ consists precisely of those x in A for which

$$\phi(z^*xy) = (\pi_\phi(x)\xi_y \,|\, \xi_z) = 0$$

for all y and z in A. Thus

$$\ker \pi_\phi = \{x \in A \,|\, xy \in L_\phi, \forall\, y \in A\}.$$

It follows that $\ker \pi_\phi$ may be characterized as the largest (right) ideal of A contained in L_ϕ. As an immediate consequence of 3.13.5 we therefore have the following result.

3.13.8. COROLLARY. *Each closed ideal of a C^*-algebra is the intersection of those primitive ideals which contain it.*

3.13.9. LEMMA. *If ϕ is a pure state of a C^*-algebra A, and L_1 and L_2 are left ideals of A such that $L_1 L_2 \subset L_\phi$ then $L_1 \subset L_\phi$ or $L_2 \subset L_\phi$.*

Proof. If $L_2 \not\subset L_\phi$ then $\pi_\phi(L_2)\xi_\phi$ is a non-zero reducing subspace for π_ϕ in H_ϕ. Since (π_ϕ, H_ϕ) is irreducible this implies that $\pi_\phi(L_2)\xi_\phi = H_\phi$, in particular $\xi_\phi \in \pi_\phi(L_2)\xi_\phi$. By assumption

$$0 = \pi_\phi(L_1 L_2)\xi_\phi = \pi_\phi(L_1)\pi_\phi(L_2)\xi_\phi \supset \pi_\phi(L_1)\xi_\phi;$$

whence $L_1 \subset L_\phi$.

3.13.10. PROPOSITION. *Each primitive ideal of a C^*-algebra is prime.*

Proof. If I is a primitive ideal and $I_1 I_2 \subset I$ let ϕ be a pure state such that $I = \ker \pi_\phi$. Then $I_1 I_2 \subset L_\phi$, whence $I_1 \subset L_\phi$ or $I_2 \subset L_\phi$ by 3.13.9, and since by 3.13.7 I is the largest ideal contained in L_ϕ this implies that $I_1 \subset I$ or $I_2 \subset I$.

3.13.11. Notes and remarks. The correspondence between pure states and irreducible representations was established by Segal [235]. It is vital for the decomposition theory in Chapter 4. The equivalence between conditions (iv) and (v) in 3.13.2 is Kadison's transitivity theorem [128]. It says that for representations of C^*-algebras topological irreducibility is the same as algebraic irreducibility. The result in 3.13.4 appears in Glimm and Kadison [99]. From 3.13.6 (ii) it is easy to see that for a pure state ϕ the quotient space $A - L_\phi$ is a Hilbert space in the quotient norm, i.e. $\phi(x^*x) = \|x + L\|^2$ for all x in A, see [255]. The result in 3.13.10 has a (partial) converse in 4.3.6.

Chapter 4

Decomposition Theory

In this chapter we describe the various attempts at decomposing (separable) C^*-algebras, their functionals and their representations, into smaller and presumably more manageable portions. Central to these investigations is the notion of spectrum of a C^*-algebra.

It is natural to distinguish three stages of the theory: *topological* decomposition (sections 1–4), *measurable* decomposition (sections 5–7) and *spatial* decomposition (sections 8–12).

In the topological decomposition theory the underlying idea (incorrect, but fruitful) is that the C^*-algebra can be visualized as operator-valued continuous functions on the spectrum, so that we have a decomposition into irreducible C^*-algebras. The most successful outcome of the theory is the characterization of the centre of the algebra as continuous (complex-valued) functions on the spectrum.

In the measurable decomposition theory the idea (still incorrect, but nearer the truth) is to visualize the C^*-algebra as a generating set of operator-valued Borel functions on the spectrum. This gives a parametrization of the equivalence classes of separable representations of the C^*-algebra by means of classes of null sets in the spectrum equipped with a certain Borel structure.

In the spatial decomposition theory a 'local' point of view (a single functional, a single representation) is predominant, in contrast to the former theories which are essentially global in nature. It is shown that each separable representation is spatially equivalent to a 'continuous sum' of factor representations.

4.1. Spectra of C^*-algebras

4.1.1. Let A be C^*-algebra. By the *spectrum* of A we understand the set $\hat{A}$ of (spatial) equivalence classes in the set $\mathrm{Irr}(A)$ of irreducible representations of A. If A is commutative then by 3.13.2. its irreducible representations are all one-dimensional and the equivalence relation is trivial. The spectrum is therefore nothing but the set of non-zero complex homomorphisms of A so that our definition agrees with the previous one given in 1.1.5. In the general case we see that the map which to each pure state ϕ of A assign the equivalence class of (π_ϕ, H_ϕ) is a surjection from the pure state space onto $\hat{A}$. It is easy to show that this map is injective only if A is commutative (cf. 3.13.4).

4.1.2. The spectrum is by no means the only space that arises in the decomposition theory. Another useful space is the *primitive spectrum* consisting of the primitive ideals of A (see 3.13.7). We shall denote this space by $\check{A}$ (although Prim(A) or Pr(A) are common in the literature). In the commutative case a closed ideal is primitive if and only if it is maximal (cf. 3.13.8), so that $\check{A}$ and $\hat{A}$ are isomorphic. In general this is no longer true, but since equivalent representations have the same kernel there is a natural surjection of $\hat{A}$ onto $\check{A}$. In Chapter 6 we shall study the class of algebras for which $\hat{A}$ is isomorphic with $\check{A}$. In general one would tend to use $\hat{A}$ rather than $\check{A}$ since $\hat{A}$, being the larger space, may contain more information about the algebra than $\check{A}$. One advantage about $\check{A}$, however, is that it carries a natural topology.

For each set F in $\check{A}$ define a closed ideal ker(F) in A by $\ker(F) = \bigcap t,\ t \in F$. For each subset I of A define a set hull(I) in $\check{A}$ by $\mathrm{hull}(I) = \{t \in \check{A} \mid I \subset t\}$. Using the canonical maps $\mathrm{Irr}(A) \to \hat{A}$ and $\hat{A} \to \check{A}$ we define the hull of J in $\hat{A}$ and in Irr(A) as the counter-images of hull(J) in $\check{A}$.

4.1.3. THEOREM. *The class* $\{\mathrm{hull}(I) \mid I \subset A\}$ *form the closed sets for a topology on* $\check{A}$. *There is a bijective, order preserving isomorphism between the open sets in this topology and the closed ideals in* A.

Proof. Define $F^- = \mathrm{hull}(\ker(F))$. We claim that the map $F \to F^-$ satisfies the four requirements (Kuratowski's axioms) for a closure operation on the subsets of $\check{A}$.

(i) It is obvious that $\varnothing^- = \varnothing$.

(ii) If $t \in F$ then $\ker(F) \subset t$, whence $t \in F^-$, so that $F \subset F^-$.

(iii) If $F = \mathrm{hull}(I)$ for some $I \subset A$ and $t \in F^-$ then $\ker F \subset t$. However, $I \subset \ker F$ and thus $t \in F$. It follows that $F^- = F$. In particular $(F^-)^- = F^-$ for each $F \subset \check{A}$.

(iv) Take F_1 and F_2 in $\check{A}$. If $t \subset F_1^-$ then

$$t \supset \ker(F_1) \supset \ker(F_1 \cup F_2),$$

whence $t \in (F_1 \cup F_2)^-$. Conversely, if $t \in (F_1 \cup F_2)^-$ then

$$t \supset \ker(F_1 \cup F_2) = \ker(F_1) \cap \ker(F_2).$$

Since by 3.13.10, t is a prime ideal this means that $\ker(F_i) \subset t$ for $i = 1$ or $i = 2$, whence $t \in F_1^- \cup F_2^-$. Consequently $(F_1 \cup F_2)^- = F_1^- \cup F_2^-$ and the map $F \to F^-$ is the closure operation for a topology on $\check{A}$. As we saw in (iii) the closed sets in $\check{A}$ are of the form hull(I) for some $I \subset A$.

If G is an open set in $\check{A}$ define $I(G) = \ker(\check{A}\backslash G)$. If I is a closed ideal in A define $G(I) = \check{A}\backslash \text{hull}(I)$. Then the assignments $G \to I(G)$ and $I \to G(I)$ are order preserving maps between the classes of open sets in $\check{A}$ and closed ideals in A, respectively. Since G is open

$$G(I(G)) = \check{A}\backslash\text{hull}(\ker(\check{A}\backslash G)) = \check{A}\backslash(\check{A}\backslash G)^- = G.$$

Since I is a closed ideal

$$I(G(I)) = \ker(\check{A}\backslash(\check{A}\backslash\text{hull}(I))) = \ker(\text{hull}(I)) = I,$$

by 3.13.8. This proves that the maps $G \to I(G)$ and $I \to G(I)$ are the inverses of each other and the proof is complete

4.1.4. The topology on $\check{A}$ defined in 4.1.3 is called the *Jacobson topology*. Note that a point t in $\check{A}$ is closed if and only if t is a maximal ideal. We can therefore not in general expect $\check{A}$ to be a T_1-space. It is, however, always a T_0-space. For if $t_1 \neq t_2$ then either $t_1 \notin \{t_2\}^-$ or $t_2 \notin \{t_1\}^-$.

4.1.5. LEMMA. *Let B be a hereditary C^*-subalgebra of A. For each irreducible representation (π, H) of A such that $B \not\subset \ker \pi$, $(\pi | B, \pi(B)H)$ is an irreducible representation of B.*

Proof. Let $\{u_\lambda\}$ be an approximate unit for B and let p be the projection on the closure of $\pi(B)H$. Then $\{\pi(u_\lambda)\}$ is strongly convergent to p. For any pair of vectors ξ, η in pH with $\xi \neq 0$ there is by 3.13.2 an element x in A with $\pi(x)\xi = \eta$. But $u_\lambda x u_\lambda \in B$ and

$$\|\pi(u_\lambda x u_\lambda)\xi - \eta\| \to \|p\pi(x)p\xi - \eta\| = 0.$$

Consequently $\pi(B)$ acts topologically irreducibly on pH. But then it also acts algebraically irreducibly, so there must be a y in B for which $\pi(y)\xi = \eta$. In particular, $\pi(B)H$ is closed and $(\pi | B, \pi(B)H)$ is irreducible.

4.1.6. COROLLARY. *The restriction of a pure state of A to a hereditary C^*-subalgebra B is a multiple of a pure state on B (possibly zero).*

Proof. If ϕ is pure on A let $(\pi_\phi, H_\phi, \xi_\phi)$ be its associated representation. Let η_ϕ be the projection of ξ_ϕ onto $\pi_\phi(B)H_\phi$. For each x in B

$$\phi(x) = (\pi_\phi(x)\xi_\phi \mid \xi_\phi) = (\pi_\phi(x)\eta_\phi \mid \eta_\phi).$$

If $\phi | B \neq 0$ then $\eta_\phi \neq 0$ and since $(\pi_\phi | B, \pi_\phi(B)H_\phi)$ is irreducible we conclude from 3.13.2 that $\|\eta_\phi\|^{-2}\phi$ is a pure state on B.

4.1.7. LEMMA. *If B is a C^*-subalgebra of A then each pure state of B can be extended to a pure state of A.*

Proof. If ϕ is a pure state of B let

$$S_\phi = \{\psi \in S \mid \psi|B = \phi\} = \{\psi \in Q \mid \psi|B = \phi\}.$$

Then S_ϕ is a weak* closed convex subset of Q and by 3.1.6 $S_\phi \neq \emptyset$. If $\psi \in S_\phi$ and $\psi = \frac{1}{2}(\psi_1 + \psi_2)$ with ψ_1 and ψ_2 in Q then $\psi_1 \mid B = \psi_2 \mid B = \phi$ since ϕ is a pure state of B. This shows that S_ϕ is a face of Q so that any extreme point of S_ϕ will be an extreme point of Q, i.e. a pure state of A.

4.1.8. PROPOSITION. *If B is a C^*-subalgebra of A then for each irreducible representation (ρ, K) of B there is an irreducible representation (π, H) of A with a closed subspace $H_1 \subset H$ such that $(\pi \mid B, H_1)$ is spatially equivalent to (ρ, K).*

Proof. Choose a unit vector η in K and let ϕ be a pure state of A which extends the pure vector state of B determined by η. Let $(\pi_\phi, H_\phi, \xi_\phi)$ be the irreducible representation associated with ϕ and let H_1 be the closure of $\pi_\phi(B)\xi_\phi$. It follows from 3.3.7 that the representations (ρ, K) and $(\pi_\phi \mid B, H_1)$ of B are spatially equivalent.

4.1.9. PROPOSITION. *If B is a hereditary C^*-subalgebra of A the map $(\pi, H) \to (\pi \mid B, \pi(B)H)$ induces an isomorphism between $\hat{A}\backslash\text{hull}(B)$ and $\hat{B}$.*

Proof. From 4.1.5 we see that the map $(\pi, H) \to (\pi \mid B, \pi(B)H)$ takes $\text{Irr}(A)\backslash\text{hull}(B)$ into $\text{Irr}(B)$. Suppose now that (π_1, H_1) and (π_2, H_2) in $\text{Irr}(A)\backslash\text{hull}(B)$ have restrictions to B which are equivalent, and let u be a unitary from $\pi_1(B)H_1$ onto $\pi_2(B)H_2$ which effectuates this equivalence. Take a unit vector ξ_1 in $\pi_1(B)H_1$, and put $\xi_2 = u\xi_1$. Let $\{u_\lambda\}$ be an approximate unit for B. For each x in A we have $u_\lambda x u_\lambda \in B$ and thus

$$(\pi_2(x)\xi_2 \mid \xi_2) = \text{Lim}(\pi_2(u_\lambda x u_\lambda)\xi_2 \mid \xi_2)$$

$$= \text{Lim}(u^*\pi_2(u_\lambda x u_\lambda)u\xi_1 \mid \xi_1) = \text{Lim}(\pi_1(u_\lambda x u_\lambda)\xi_1 \mid \xi_1) = (\pi_1(x)\xi_1 \mid \xi_1).$$

It follows from 3.3.7 that (π_1, H_1) and (π_2, H_2) are equivalent. We have therefore an injective map from $\hat{A}\backslash\text{hull}(B)$ into $\hat{B}$. However, by 4.1.8 this map is also surjective.

4.1.10. PROPOSITION. *If B is a hereditary C^*-subalgebra of A the map $t \to t \cap B$ is a homeomorphism of $\check{A}\backslash\text{hull}(B)$ onto $\check{B}$ and we have a commutative diagram:*

$$\begin{array}{ccc} \hat{A}\backslash\text{hull}(B) & \to & \hat{B} \\ \downarrow & & \downarrow \\ \check{A}\backslash\text{hull}(B) & \to & \check{B} \end{array}$$

Proof. Take t_1, t_2 in $\check{A}\backslash\text{hull}(B)$, and write $B = L \cap L^*$, where L is a closed left ideal of A (1.5.2). If $t_1 \cap B = t_2 \cap B$ then $t_1 \cap L = t_2 \cap L$ because $x \in L$ if and only if $x^*x \in B$. Thus $t_1 L \subset t_2$ and since t_2 is prime (3.13.10) either $t_1 \subset t_2$ or $L \subset t_2$. The latter possibility is excluded, so $t_1 \subset t_2$. By a symmetric argument $t_2 \subset t_1$, so that $t_1 = t_2$.

Consider (π, H) in $\mathrm{Irr}(A)\backslash\mathrm{hull}(B)$. Then $\ker\pi \cap B = \ker(\pi \mid B)$. Since equivalent representations have the same kernel it follows that the isomorphism between $\hat{A}\backslash\mathrm{hull}(B)$ and $\hat{B}$ from 4.1.9 induces a surjective map from $\check{A}\backslash\mathrm{hull}(B)$ onto $\check{B}$ given by $t \to t \cap B$. From the first part of the proof we see that the map $t \to t \cap B$ is an isomorphism.

Suppose that F is a closed set in $\check{A}\backslash\mathrm{hull}(B)$ with image E in $\check{B}$. If $t \in \check{A}\backslash\mathrm{hull}(B)$ and $t \cap B$ belongs to the closure of E then

$$L^*\ker(F)L \subset \ker(F) \cap B = \ker(E) \subset t \cap B \subset t.$$

Since t is prime and $L \not\subset t$ we have $L^*\ker(F) \subset t$; whence also $\ker(F)L \subset t$. Again because t is prime this implies that $\ker(F) \subset t$, i.e. $t \in F$. Thus $t \cap B \in E$, so that E is closed in $\check{B}$.

Conversely, suppose that the image E of F is closed in $\check{B}$ and take t in the closure of F. Then $t \supset \ker(F)$ whence $t \cap B \supset \ker(E)$ so that $t \cap B \in E$. Since $\check{A}\backslash\mathrm{hull}(B)$ and $\check{B}$ are isomorphic this implies that $t \in F$ so that F is closed.

4.1.11. THEOREM. *Let I be a closed ideal of A.*

(i) *The maps* $(\pi, H) \to (\pi \mid I, H)$ *and* $(\pi, H) \to (\pi \bmod I, H)$ *from* $\mathrm{Irr}(A)\backslash\mathrm{hull}(I)$ *to* $\mathrm{Irr}(I)$ *and from* $\mathrm{hull}(I)$ *to* $\mathrm{Irr}(A/I)$, *respectively, induce isomorphisms of* $\hat{A}\backslash\mathrm{hull}(I)$ *onto* $\hat{I}$ *and of* $\mathrm{hull}(I)$ *onto* $(A/I)\hat{}$.

(ii) *The maps* $t \to t \cap I$ *and* $t \to t/I$ *are homeomorphisms from* $\check{A}\backslash\mathrm{hull}(I)$ *onto* $\check{I}$ *and from* $\mathrm{hull}(I)$ *onto* $(A/I)\check{}$.

(iii) *The resulting diagrams, below, are commutative:*

$$\begin{array}{ccccc} \hat{I} & \longleftarrow & \hat{A}\backslash\mathrm{hull}(I), & \mathrm{hull}(I) & \longrightarrow (A/I)\hat{} \\ \downarrow & & \downarrow & \downarrow & \downarrow \\ \check{I} & \longleftarrow & \check{A}\backslash\mathrm{hull}(I), & \mathrm{hull}(I) & \longrightarrow (A/I)\check{}. \end{array}$$

Proof. (i) follows from 4.1.9 except for the fact that $\pi(I)H$ is replaced by H in one of the formulas. This is because $\pi(I)H$ is a non-zero reducing subspace of H whenever $(\pi, H) \in \mathrm{Irr}(A)\backslash\mathrm{hull}(I)$ and therefore equal to H.

(ii) follows from 4.1.10 except for the claim that the sets $\mathrm{hull}(I)$ and $(A/I)\check{}$, which are clearly isomorphic, are also homeomorphic. But this follows immediately from the definition of the Jacobson topology.

(iii) follows from the construction of the mappings.

4.1.12. We define the *Jacobson topology* on $\hat{A}$ as the topology for which the natural map $\hat{A} \to \check{A}$ is open and continuous. Note that $\hat{A}$ will be a T_0-space if and only if $\hat{A}$ and $\check{A}$ are isomorphic. One significant benefit of this definition is that all horizontal arrows in the diagrams in 4.1.10 and 4.1.11 are homeomorphisms.

4.1.13. Notes and remarks. The Jacobson topology was introduced in [120]. For a special case in algebraic geometry is was defined by Zariski. Theorem 4.1.11 was (in essence) proved by Kaplansky [141] and 4.1.10 is found in [191].

4.2. Polish spaces

4.2.1. A topological space is said to be *Polish* if it is second countable, metrizable and complete. There exists a rich and important theory about Polish spaces and their Borel structure. We shall only need a few results, and present in this section a self-contained exposition of these.

Each second countable, compact Hausdorff space is Polish, since it can be embedded as a closed subset of $[0,1]^{\infty}$. As a consequence of 4.2.2 we see that also each second countable, locally compact Hausdorff space is Polish.

A particularly important Polish space is $\boldsymbol{N}^{\infty}$ (in the product topology). As a complete metric on $\boldsymbol{N}^{\infty}$ we recommend

$$d(a,b) = \sum 2^{-n}\delta(a(n), b(n)),$$

where $a = \{a(n)\}$ and $b = \{b(n)\}$ are elements of $\boldsymbol{N}^{\infty}$ and δ is the Kronecker symbol.

Each countable direct product of Polish spaces is Polish. Indeed, if d_n is a complete metric on the Polish space T_n then

$$d(a,b) = \sum 2^{-n}[1 + d_n(a(n), b(n))]^{-1} d_n(a(n), b(n)),$$

where a and b are elements of $\Pi\, T_n$, is a complete metric on $\Pi\, T_n$. Similarly, each countable disjoint sum of Polish spaces is Polish.

Each closed subset of a Polish space is clearly Polish. But there are many other Polish subsets:

4.2.2. LEMMA. *Each open subset and each G_δ-subset of a Polish space is Polish in its relative topology.*

Proof. Let G be an open subset of the Polish space T and let d be a complete metric for T. For each t in G we define $f(t) = [d(t, T\backslash G)]^{-1}$, and have a homeomorphism of G onto the graph $G(f)$ of f. But $G(f)$ is a closed subset of the Polish space $T \times \boldsymbol{R}$. Indeed, if $\{t_n, f(t_n)\}$ converges to (t,s) then $\{f(t_n)\}$ is bounded, which implies that $t \in G$. Since f is continuous on G we have $s = f(t)$, whence $(t,s) \in G(f)$.

Now let $\{G_n\}$ be a sequence of open subsets of T and put $E = \cap G_n$. From the first part of the proof there is for each n a Polish space T_n and a homeomorphism $f_n\colon T_n \to G_n$. Let F denote the closed subspace of the Polish space $\Pi\, T_n$ consisting of points (t_n) such that $f_n(t_n) = f_1(t_1)$ for all n. Projection onto the first coordinate gives an injective continuous map of F onto E.

However, the inverse of this map is also continuous since its composition with every coordinate projection is continuous. Consequently E is homeomorphic to F and therefore Polish.

4.2.3. LEMMA. *If a subset E of a Polish space T is Polish in its relative topology, then E is a G_δ-subset of T.*

Proof. Since each closed subset of T is a G_δ-set, we may as well assume that E is dense in T, replacing otherwise T with the closure of E.

Let d be a complete metric for E and define

$$U_n = \bigcup G, \qquad G \text{ is open in } T, \qquad \operatorname{diam}(G \cap E) \leqslant 2^{-n}.$$

Clearly $E \subset \cap U_n$. But for each t in $\cap U_n$ we can find open sets G_n in T, containing t, such that $\operatorname{diam}(G_n \cap E) \leqslant 2^{-n}$. Since E is dense in T there exists t_n in $\cap_{k=1}^{n}(G_k \cap E)$ for all n. Clearly $\{t_n\}$ is a Cauchy sequence in E, and since E is Polish, $\{t_n\}$ converges to a point t_0 in E. For each closed neighbourhood F of t there is by the same arguments an s_n in $\cap_{k=1}^{n}(G_k \cap E \cap F)$ for all n. Since $d(t_n, s_n) \leqslant 2^{-n}$, $\{s_n\}$ converges also to t_0, whence $t_0 \in F$. It follows that $t = t_0$ and thus $\cap U_n = E$.

4.2.4. LEMMA. *Each Polish space is homeomorphic to a G_δ-subset of $[0,1]^\infty$.*

Proof. Let $\{S_n\}$ be a dense sequence in the Polish space T and let d be a complete metric for T such that $d \leqslant 1$. Define $f: T \to [0,1]^\infty$ by

$$f(t)(n) = d(t, s_n).$$

Clearly f is a continuous, injective map, and since $f(t_k) \to f(t)$ if and only if $t_k \to t$ we see that f is a homeomorphism of T on $f(T)$. By 4.2.3 this implies that $f(T)$ is a G_δ-subset of $[0,1]^\infty$.

4.2.5. PROPOSITION. (Baire category theorem). *In a Polish space each countable intersection of open, dense sets is dense.*

Proof. Let $\{G_n\}$ be a decreasing sequence of open and dense sets in a Polish space T, and choose a complete metric for T. If B_0 is a closed ball in T with radius $r_0 > 0$, then by assumption we can find a closed ball $B_1 \subset B_0 \cap G_1$ with radius $r_1 < 1$. By induction we find a sequence $\{B_n\}$ of closed balls in T with radii $r_n < 2^{-n}$ such that $B_{n+1} \subset B_n \cap G_n$. Since T is complete there is a point t in T with $\{t\} = \cap B_n$. But then $t \in \cap G_n$ and $t \in B_0$. Thus $\cap G_n$ intersects every open set in T.

4.2.6. LEMMA. (Souslin's scheme). *For each Polish space T with a complete metric there is a system $\{G_{n,\lambda} \mid n \in \boldsymbol{N}, \lambda \in \boldsymbol{N}^n\}$ of non-empty, open subsets of T*

such that:

(i) $\cup_\lambda G_{1.\lambda} = T$;

(ii) $\cup_\lambda G_{n+1.\lambda} = G_{n.\mu}$ *if* $\lambda(k) = \mu(k)$ *for all* $k \leqslant n$;

(iii) *closure* $(G_{n+1.\lambda}) \subset G_{n.\mu}$ *if* $\lambda(k) = \mu(k)$ *for all* $k \leqslant n$;

(iv) $\operatorname{diam}(G_{n.\lambda}) \leqslant 2^{-n}$ *for all* n *and* λ.

Proof. Since T is separable we can find a sequence $\{G_n\}$ of open balls in T with diameters $\leqslant 1$ such that $\cup G_n = T$. Take these sets as $\{G_{1.\lambda}\}$. For each n_1 we can find a sequence $\{G_{n_1 n_2}\}$ of open balls in T with diameters $\leqslant \frac{1}{2}$ such that $\cup_{n_2} G_{n_1 n_2} = G_{n_1}$ and the closure of $G_{n_1 n_2}$ contained in G_{n_1}. Take the collection of all $G_{n_1 n_2}$ as $\{G_{2.\lambda}\}$. The proof is completed by an obvious inductive argument.

4.2.7. LEMMA. *Each family* $\{G_{n.\lambda} \mid n \in \boldsymbol{N}, \lambda \in \boldsymbol{N}^n\}$ *satisfying the conditions in 4.2.6 is a basis for the topology.*

Proof. Take t in T and let B be the open ball around t with radius $\varepsilon > 0$. Combining (i) and (ii) of 4.2.6 we see that $\cup_\lambda G_{n.\lambda} = T$ for each fixed n. Thus $t \in G_{n.\lambda}$ for some λ, and if $2^{-n} < \varepsilon$ this implies that $G_{n.\lambda} \subset B$ by (iv) of 4.2.6.

4.2.8. PROPOSITION. *For each Polish space T there is an open and continuous map h of $\boldsymbol{N}^\infty$ onto T.*

Proof. Take a system $\{G_{n.\lambda} \mid n \in \boldsymbol{N}, \lambda \in \boldsymbol{N}^n\}$ satisfying 4.2.6. If $a \in \boldsymbol{N}^\infty$ and $n \in \boldsymbol{N}$ define $\lambda(a)$ in $\boldsymbol{N}^n$ as $(a(1), a(2), \ldots, a(n))$. Now take $h(a)$ to be the unique point in $\cap_n G_{n.\lambda(a)}$.

For each λ in $\boldsymbol{N}^n$ define

$$U_{n.\lambda} = \{a \in \boldsymbol{N}^\infty \mid a(k) = \lambda(k), 1 \leqslant k \leqslant n\}.$$

It is straightforward to verify that the family $\{U_{n.\lambda} \mid n \in \boldsymbol{N}, \lambda \in \boldsymbol{N}^n\}$ satisfies 4.2.6 relative to $\boldsymbol{N}^\infty$ with the metric as in 4.2.1. The way we constructed h shows that $h(U_{n.\lambda}) \subset G_{n.\lambda}$ for all n and λ. However, if $t \in G_{n.\lambda}$ then $t \in G_{n+1.\mu}$ for some μ in $\boldsymbol{N}^{n+1}$ by (ii) of 4.2.6. Inductively we can therefore find a sequence $\{G_{m.\lambda_m}\}$, $m \geqslant n$, such that $t \in \cap_m G_{m.\lambda_m}$. But this means exactly that $t \in h(U_{n.\lambda})$. Thus $h(U_{n.\lambda}) = G_{n.\lambda}$.

Since by 4.2.7 the families $\{G_{n.\lambda}\}$ and $\{U_{n.\lambda}\}$ are bases for the topologies on T and $\boldsymbol{N}^\infty$, respectively, we conclude that h is open, continuous and surjective, as desired.

4.2.9. LEMMA. *For $i = 1, 2$, let $f_i \colon T_i \to E$ be a continuous function from the Polish space T_i into the Hausdorff space E. If $f_1(T_1) \cap f_2(T_2) = \varnothing$ there are disjoint Borel sets B_1 and B_2 in E such that $f_i(T_i) \subset B_i$.*

Proof. By 4.2.6 we can find a system $\{G^i_{n.\lambda}\}$. If for some fixed n each $f_1(G^1_{n.\lambda})$ could be separated from any $f_2(G^2_{n.\mu})$, say by disjoint Borel sets $B^1_{\lambda.\mu}$ and $B^2_{\lambda.\mu}$, then

$$f_1(T_1) = \bigcup_\lambda f_1(G^1_{n.\lambda}) \subset \bigcup_\lambda \bigcap_\mu B^1_{\lambda.\mu} \subset E \backslash \bigcap_\lambda \bigcup_\mu B^2_{\lambda.\mu},$$

$$f_2(T_2) = \bigcup_\mu f_2(G^2_{n.\mu}) \subset \bigcup_\mu \bigcap_\lambda B^2_{\lambda.\mu} = \bigcap_\lambda \bigcup_\mu B^2_{\lambda.\mu}.$$

Thus if $f_1(T_1)$ and $f_2(T_2)$ can not be separated, then for some $\lambda(n)$ and $\mu(n)$ in $\boldsymbol{N}^n$, the sets $f_1(G^1_{n.\lambda(n)})$ and $f_2(G^2_{n.\lambda(n)})$ can not be separated either. Replacing T_1 and T_2 by $G^1_{n.\lambda(n)}$ and $G^2_{n.\mu(n)}$ we can thus by induction find decreasing sequences $\{G^1_{n.\lambda(n)}\}$ and $\{G^2_{n.\mu(n)}\}$ such that $f_1(G^1_{n.\lambda(n)})$ and $f_2(G^2_{n.\mu(n)})$ can not be separated for any n. Let t_1 and t_2 be the unique points in T_1 and T_2 such that $\{t_i\} = \cap G^i_{n.\lambda(n)}$. By assumption $f_1(t_1) \neq f_2(t_2)$ and these points can therefore be separated by open sets U_1 and U_2 in E. But for large enough n, we have $G^i_{n.\lambda(n)} \subset f_i^{-1}(U_i)$; whence $f_i(G^i_{n.\lambda(n)}) \subset U_i$, a contradiction. Therefore $f_1(T_1)$ and $f_2(T_2)$ can be separated as claimed.

4.2.10. PROPOSITION. *Let $f: T \to E$ be a continuous, injective function from the Polish space T into the Hausdorff space E. Then f is a Borel isomorphism from T onto $f(T)$ in its relative Borel structure.*

Proof. We may apply 4.2.9 to any open set T_1 in T and its complement T_2. Then $f(T_1)$ and $f(T_2)$ are (relative) Borel sets in $f(T)$, which proves that f^{-1} is a Borel function, as desired.

4.2.11. We define the *lexicographic order* in $\boldsymbol{N}^\infty$ by writing $a < b$ if there is an n with $a(n) < b(n)$ and $a(k) = b(k)$ for all $k < n$. A *long interval* in $\boldsymbol{N}^\infty$ is a set

$$]-\infty, a[= \{b \in \boldsymbol{N}^\infty \mid b < a\} \qquad \text{or} \qquad]a, \infty[= \{b \in \boldsymbol{N}^\infty \mid a < b\}.$$

If $b \in]-\infty, a[$ and n is the first number for which $b(n) < a(n)$, define $\lambda = (b(1), b(2), \ldots, b(n))$. Then $\lambda \in \boldsymbol{N}^n$ and $U_{n.\lambda} \subset]-\infty, a[$, with $U_{n.\lambda}$ as in the proof of 4.2.8. This and a similar argument for $]a, \infty[$ proves that each long interval is open in $\boldsymbol{N}^\infty$. Conversely, if $U_{n.\lambda}$ is given, define a, b in $\boldsymbol{N}^\infty$ by

$$a(k) = b(k) = \lambda(k) \quad \text{for } 1 \leqslant k \leqslant n; \qquad a(k) = 1, \qquad b(k) = 17 \quad \text{for } k > n.$$

Then

$$U_{n.\lambda} \supset]a, \infty[\bigcap]-\infty, b[\neq \varnothing.$$

It follows that the long intervals form a sub-basis for the topology on $\boldsymbol{N}^\infty$.

The lexicographic order in $\boldsymbol{N}^\infty$ is a *closed well-ordering* in the sense that each closed set F in $\boldsymbol{N}^\infty$ has a first element. To find it, let $a(1)$ be the smallest first coordinate of elements from F, let $a(2)$ be the smallest second coordinate of those elements b in F with $b(1) = a(1)$, and carry on inductively. The element $a = \{a(n)\}$ is smaller than or equal to any point in F and it is clearly a limit point of F, whence $a \in F$.

4.2.12. PROPOSITION. *Let $f: T \to E$ be a function from a Polish space T onto a Borel space E such that*

(i) *$f^{-1}(s)$ is closed in T for each s in E;*

(ii) *$f(G)$ is a Borel set in E for each open set G in T.*

There is then a Borel function $g: E \to T$ such that $f \circ g$ is the identity on E.

Proof. Assume first that $T = \boldsymbol{N}^{\alpha}$. For each s in E define $g(s)$ as the first element (in the lexicographic order, see 4.2.11) of the closed set $f^{-1}(s)$. Clearly $f \circ g$ is the identity on E. We claim that

$$(*) \qquad f(]-\infty, a[) = g^{-1}(]-\infty, a[)$$

for all a in $\boldsymbol{N}^{\alpha}$. For one inclusion,

$$f(F) \supset f \circ g \circ g^{-1}(F) = g^{-1}(F)$$

for any set F in E. For the other inclusion, $g(f(b)) \leqslant b$ for any b in $\boldsymbol{N}^{\alpha}$, whence

$$g \circ f(]-\infty, a[) \subset]-\infty, a[$$

and

$$f(]-\infty, a[) \subset g^{-1} \circ g \circ f(]-\infty, a[) \subset g^{-1}(]-\infty, a[).$$

If $a \in \boldsymbol{N}^{\alpha}$ define a_n in $\boldsymbol{N}^{\alpha}$ by

$$a_n(k) = a(k) \quad \text{for } k < n; \qquad a_n(n) = a(n) + 1, \qquad a_n(k) = 1 \quad \text{for } k > n.$$

Then $b > a$ if and only if $b \geqslant a_n$ for some n. This shows that

$$]a, \infty[= \bigcup \boldsymbol{N}^{\alpha} \setminus]-\infty, a_n[.$$

Together with $(*)$ that implies that g^{-1} maps a sub-basis for the topology on $\boldsymbol{N}^{\alpha}$ into Borel subsets of E. It follows that g is a Borel function.

In the general case we have an open and continuous map h from $\boldsymbol{N}^{\alpha}$ onto T by 4.2.8. The map $f \circ h: \boldsymbol{N}^{\alpha} \to E$ satisfies the requirements (i) and (ii), so by the first part of the proof there is a Borel function $g': E \to \boldsymbol{N}^{\alpha}$ such that $f \circ h \circ g'$ is the identity on E. Take $g = h \circ g'$ and smile.

4.2.13. COROLLARY. *Let $f: T \to E$ be a surjective continuous function from a second countable, locally compact Hausdorff space T onto a Hausdorff space E. There is then a Borel function $g: E \to T$ such that $f \circ g$ is the identity on E.*

Proof. Since f is continuous and E is Hausdorff, f maps compact sets to closed (i.e. Borel) sets. Each open set in T is the countable union of compact sets, and since f preserves unions, the image of open sets are Borel sets. Clearly $f^{-1}(s)$ is closed in T for each s in E, and thus 4.2.12 can be applied.

4.2.14. Notes and remarks. This section is largely borrowed from Arveson's book [15]. We have, however, deliberately avoided the introduction of analytic sets, because every interesting set appearing in C^*-algebra theory which is analytic, is known to be a Borel set as well. The introduction of Polish spaces in operator algebra theory is one of Mackey's happy ideas, see [164]. It looks so obvious—afterwards.

4.3. Spectrum and pure states

4.3.1. For each C^*-algebra A let $P(A)$ denote the set of pure states of A equipped with the weak* topology. Thus $P(A)$ is a Hausdorff space but in general neither compact nor even locally compact.

4.3.2. PROPOSITION. *If A is a separable C^*-algebra then $P(A)$ is a G_δ-subset of the quasi-state space Q, and consequently a Polish space.*

Proof. If A is separable then Q is second countable and compact, hence Polish. Let d be a complete metric on Q. The sets

$$\{\phi \in Q \mid \phi = \tfrac{1}{2}(\phi_1 + \phi_2),\ d(\phi_1, \phi_2) \geqslant 1/n\} \cup \{0\}, \qquad n \in \boldsymbol{N},$$

are closed in Q, and $P(A)$ is the complement of their union. By 4.2.2 each G_δ-subset of a Polish space is Polish.

4.3.3. THEOREM. *For each C^*-algebra A the map $\phi \to \ker \pi_\phi$ is open and continuous from $P(A)$ onto $\check{A}$.*

Proof. If G is an open subset of $\check{A}$ let $I = \ker(\check{A}\backslash G)$ and let p be the open central projection in A'' corresponding to I (see 3.11.10). For each ϕ in $P(A)$ we have $\phi(p) = 0$ or $\phi(p) = 1$ according to whether $\pi''_\phi(p) = 0$ or $\pi''_\phi(p) = 1$. It follows that the counter-image of G in $P(A)$ is

$$\{\phi \in P(A) \mid \phi(p) = 1\} = \{\phi \in P(A) \mid \phi(p) > 0\}.$$

However, this set is open since $\hat{p}$ is lower semi-continuous on Q by 3.11.9.

To prove that the map is open, let F be a closed set in $P(A)$ such that $\phi \in F$ implies $\phi(u^* \cdot u) \in F$ for every unitary u in $\tilde{A}$. Note first that $\overline{\mathrm{Conv}(F)} \cap P(A) = F$. For if F_1 is a closed set in Q such that $F_1 \cap P(A) = F$ then each point in $\overline{\mathrm{Conv}(F_1)}$ has the form $\int_{F_1} \phi \, d\mu$ for some probability measure μ on F_1. Consequently

$$\overline{\mathrm{Conv}(F_1)} \cap P(A) = F_1 \cap P(A) = F.$$

Next, define $I = \cap \ker \pi_\phi, \phi \in F$. Then $I = \cap \phi^{-1}(0)$, $\phi \in F$ by 3.13.4, so that I is the annihilator of F. Assume for a moment that $I = \{0\}$ and let (π_F, H_F) denote the sum of irreducible representations associated with elements in F, cf. 3.7.3. By 3.7.4 (π_F, H_F) is faithful on A. If now $\psi \in P(A)\backslash F$, then $\psi \notin \overline{\mathrm{Conv}(F)}$ from the argument above, so by Hahn–Banach's theorem there is an element x in A_{sa} with $\psi(x) > 1$ and $\phi(x) \leqslant 1$ for all ϕ in F. Since F is saturated under unitary equivalence it follows from 3.13.4 that $(\pi_\phi(x)\xi \mid \xi) \leqslant 1$ for each unit vector ξ in $H_\phi, \phi \in F$, and consequently $\pi_F(x) \leqslant 1$. As (π_F, H_F) is faithful this contradicts $\psi(x) > 1$. Consequently $F = P(A)$. If $I \neq \{0\}$ let Q_I denote the invariant face of Q corresponding to the annihilator of I in Q, cf. 3.10.8. Identifying Q_I with the quasi-state space of A/I and reasoning as before we see that $F = P(A) \cap Q_I$, i.e. F is the set of non-zero extreme points in Q_I. It follows immediately that

$$F = \{\phi \in P(A) \mid \phi(I) = 0\}.$$

The image of F in $\hat{A}$ is therefore precisely the hull of I and consequently closed, cf. 4.1.12.

Now let G be an open set in $P(A)$ and define

$$\tilde{G} = \{\phi(u^* \cdot u) \mid \phi \in G, u \text{ unitary in } \tilde{A}\}.$$

Then G and $\tilde{G}$ have the same image in $\hat{A}$ and $\tilde{G}$ is open. Thus $F = P(A)\backslash\tilde{G}$ is closed and $\phi \in F$ if and only if $\phi(u^* \cdot u) \in F$. By 3.13.4 the sets $\tilde{G}$ and F have disjoint images in $\hat{A}$ and the image of F is closed as we proved above. This shows that the map from $P(A)$ to $\check{A}$ is open.

4.3.4. COROLLARY. *If A is a separable algebra then $\check{A}$ is second countable.*

4.3.5. THEOREM. *If A is a (separable) C^*-algebra then $\check{A}$ is a Baire space. (Any countable intersection of open, dense sets is dense.)*

Proof. Any Polish space is a Baire space by 4.2.5 and the image of a Baire space under a continuous, open map is a Baire space. This proves the separable case. Actually, Choquet proved that $P(A)$, being the set of extreme points of a compact, convex set, is always a Baire space; so that the result is true in general. See p. 355 of Dixmier [65].

4.3.6. PROPOSITION (cf. 3.13.10). *If A is a separable C^*-algebra then every closed prime ideal is primitive.*

Proof. Assume that 0 is a prime ideal and let $\{G_n\}$ be a basis for the topology on $\check{A}$ consisting of open, non-empty sets (use 4.3.4). By assumption any two non-zero ideals ha non-zero intersection, and since $\ker(\check{A}\backslash G_n) \cap \ker(G_n) = 0$ and $\ker(\check{A}\backslash G_n) \neq 0$ (since $G_n \neq \varnothing$) we conclude that $\ker(G_n) = 0$ so that each G_n is dense in $\check{A}$. From 4.3.5 it follows that $\cap G_n$ is non-empty, thus contains a point t which must be dense in A. But then t is an ideal contained in every primitive ideal, whence $t = 0$ by 3.13.8, and 0 is a primitive ideal.

If I is any closed prime ideal in A then 0 is a prime ideal in A/I. From the first part of the proof this implies that 0 is primitive in A/I, whence I is primitive in A.

4.3.7. Let A be a C^*-algebra with spectrum $\hat{A}$. Using 3.13.2 we choose for each t in $\hat{A}$ a pure state ϕ_t with associated representation (π_t, H_t) in t (recall that t is an equivalence class in $\mathrm{Irr}(A)$). The representation

$$\pi_a = \bigoplus_{t \in \hat{A}} \pi_t \qquad \text{on } H_a = \bigoplus_{t \in \hat{A}} H_t$$

is called the (reduced) *atomic representation* of A. Evidently the atomic representation is the result of a choice, but any other choice would give a representation $\pi'_a = \oplus \pi'_t$ where π'_t was spatially equivalent with π_t for each t in $\hat{A}$. It is, however, easy to show that two direct sums of mutually spatially equivalent representations are themselves spatially equivalent. In particular π_a and π'_a are spatially equivalent; hence *the* atomic representation (by abuse of language).

4.3.8. LEMMA.
$$\pi_a(A)'' = \prod_{t \in \hat{A}} B(H_t).$$

Proof. For each t in $\hat{A}$ we have $\pi_t(A)'' = B(H_t)$ by 3.13.2. Since the π_t's are mutually disjoint it follows from repeated use of 3.8.11 that $\pi_a(A)'' = \prod \pi_t(A)''$.

4.3.9. LEMMA. *For each pure state ϕ of A there is a unit vector ξ_ϕ in H_a, unique up to a complex factor, such that ϕ is the vector state determined by ξ_ϕ.*

Proof. There is a unique t in $\hat{A}$ such that $(\pi_\phi, H_\phi) \in t$. This means that (π_ϕ, H_ϕ) is equivalent with (π_t, H_t) and thus $\phi(x) = (\pi_t(x)\xi_\phi \mid \xi_\phi)$ for some vector ξ_ϕ in H_t by 3.13.2. Since (π_t, H_t) is irreducible, ξ_ϕ is uniquely determined up to a complex scalar of modulus one. Identifying H_t with a subspace of H_a we obtain the lemma.

4.3.10. PROPOSITION. *If x is a normal element of A and $\lambda \in \mathrm{Sp}(x) \setminus \{0\}$ there is a unit vector ξ in H_a such that $\pi_a(x)\xi = \lambda\xi$.*

Proof. Let B denote the commutative C^*-subalgebra of A generated by x. Since $B = C_0(\mathrm{Sp}(x) \setminus \{0\})$ there is by 4.1.7 a pure state ϕ of A such that $\phi \mid B$ is a complex homomorphism with $\phi(x) = \lambda$. By 4.3.9 there is a unit vector ξ in H_a such that ϕ is the vector state determined by ξ. Then

$$\|\pi_a(x)\xi - \lambda\xi\|^2 = \phi(x^*x) - \bar{\lambda}\phi(x) - \lambda\phi(x^*) + |\lambda|^2 = 0,$$

since $\phi \mid B$ is a homomorphism, so that $\pi_a(x)\xi = \lambda\xi$.

4.3.11. It is immediate from 4.3.10 that the atomic representation of a C^*-algebra A is faithful. From 3.7.7 we know that π_a has a unique extension π_a'' to a normal representation of A'', and by 4.3.8 $\pi_a''(A'') = \Pi B(H_\iota)$. This shows that π_a'' is not in general faithful on A''. Since it is often easier to work in the atomic representation than in the universal (this is the whole point of the decomposition theory), it will be convenient to single out a large class of elements in A'' on which π_a'' is still faithful, i.e. isometric.

An element x in A''_{sa} is said to be *universally measurable* if for each $\varepsilon > 0$ and each state ϕ of A there are elements a and b in $(A_{sa})^m$ such that

$$-b \leqslant x \leqslant a \qquad \text{and} \qquad \phi(a+b) < \varepsilon.$$

The set of universally measurable elements in A''_{sa} is denoted by $\mathscr{U}(A)$.

4.3.12. LEMMA. *If $x \in A''$ such that for each $\varepsilon > 0$ and each state ϕ there are elements a and b in $\mathscr{U}(A)$ with $-b \leqslant x \leqslant a$ and $\phi(a+b) < \varepsilon$ then $x \in \mathscr{U}(A)$.*

Proof. The proof is left to the reader.

4.3.13. PROPOSITION. *The space $\mathscr{U}(A)$ is a norm closed real vector space in A''_{sa} containing $((\tilde{A}_{sa})^m)^-$.*

Proof. Since $(A_{sa})^m$ is closed under addition and under multiplication with positive scalars it follows readily that $\mathscr{U}(A)$ is a real vector space.

If $x \in (A_{sa})^m$ and $\{x_i\}$ is a net in A_{sa} which increases to x then we can approximate x with itself from above and with the x_i's from below. Consequently $x \in \mathscr{U}(A)$. In particular the constants belong to $\mathscr{U}(A)$.

Let $\{x_n\}$ be a sequence in $\mathscr{U}(A)$ which is norm convergent to some x in A''_{sa}. Then

$$x_n - \varepsilon 1 \leqslant x \leqslant x_n + \varepsilon 1$$

for n large. Since $x_n \pm \varepsilon 1 \in \mathscr{U}(A)$ it follows from 4.3.12 that $x \in \mathscr{U}(A)$.

By 3.11.7 $(\tilde{A}_{sa})^m = ((A_{sa})^m)^- + \boldsymbol{R}1$ and it follows from the preceeding that $(\tilde{A}_{sa})^m \subset \mathscr{U}(A)$ and thus $((\tilde{A}_{sa})^m)^- \subset \mathscr{U}(A)$.

4.3.14. PROPOSITION. *Each element in $\mathscr{U}(A)$ can be approximated strongly from above with elements from $((A_{sa})^m)^-$ and strongly from below with elements from $((A_{sa})_m)^-$.*

Proof. Since $\mathscr{U}(A)$ is a vector space and $((A_{sa})_m)^- = -((A_{sa})^m)^-$, it suffices to prove the first statement. If $x \in \mathscr{U}(A)$ then by assumption there exists a net $\{a_i\}$ in $(A_{sa})^m$ which converges weakly to x from above. For $0 < \alpha < \|x\|^{-1}$ the net $\{(1 + \alpha a_i)^{-1}\}$ converges strongly to $(1 + \alpha x)^{-1}$ (from below) by 2.1.3. With $f_\alpha(t) = \alpha^{-1}(1 - (1 + \alpha t)^{-1})$ (as in 1.3.7) we conclude that $\{f_\alpha(a_i)\}$ converges strongly to $f_\alpha(x)$ from above. Since $1 + \alpha a_i \in (A_{sa})^m_+$ we have $f_\alpha(a_i) \in ((A_{sa})^m)^-$ by 3.11.6. As $f_\alpha(t) \to t$ uniformly on $[-\|x\|, \|x\|]$ for $\alpha \to 0$ we have

$$x \leqslant \varepsilon_\alpha 1 + f_\alpha(x) \leqslant \varepsilon_\alpha 1 + f_\alpha(a_i) \in ((A_{sa})^m)^-,$$

with $\varepsilon_\alpha \to 0$ ($\varepsilon_\alpha = \alpha\|x\|(1 - \alpha\|x\|)^{-1}$). The inequality

$$\varepsilon_\alpha 1 + f_\alpha(a_i) - x \leqslant \varepsilon_\alpha 1 + f_\alpha(a_i) - f_\alpha(x)$$

shows that a subnet of $\{\varepsilon_\alpha 1 + f_\alpha(a_i)\}$ converges strongly to x from above.

4.3.15. THEOREM. *The atomic representation is faithful on $\mathscr{U}(A)$.*

Proof. Let π_a'' denote the canonical extension of π_a to a normal representation of A'', and take x in $\mathscr{U}(A)$. If $\phi(x) < 0$ for some state ϕ of A then by definition there is an element a in $(A_{sa})^m$ such that $x \leqslant a$ and $\phi(a) < 0$. With the notation as in 3.11 this implies that the function $\hat{a}$ is not positive on S, and by 3.11.5 $\hat{a}$ is an affine, lower semi-continuous function on Q. The sets $\{\psi \in Q \mid \hat{a}(\psi) \leqslant \alpha\}$ are closed faces of Q for all real α, which implies that $\hat{a}$ attains its minimum at an extremal point of Q. Consequently $\psi(x) \leqslant \psi(a) < 0$ for some pure state ψ of A. By 4.3.9 we conclude that $\pi_a''(x) \ngeqslant 0$. It follows that $\pi_a''(x) \geqslant 0$ implies $x \geqslant 0$ for any x in $\mathscr{U}(A)$. Since

$$-\|\pi_a''(x)\|1 \leqslant \pi_a''(x) \leqslant \|\pi_a''(x)\|1,$$

and $1 \in \mathscr{U}(A)$ we see that

$$-\|\pi_a''(x)\|1 \leqslant x \leqslant \|\pi_a''(x)\|1,$$

and thus $\|x\| = \|\pi_a''(x)\|$ for each x in $\mathscr{U}(A)$.

4.3.16. If A is commutative, so that $A = C_0(\hat{A})$, then by 4.3.15 we may realize $\mathscr{U}(A)$ and $(A_{sa})^m$ as classes of functions on $\hat{A}$. It is easy to verify that $(\tilde{A}_{sa})^m$ coincides with the class of bounded, lower semi-continuous functions on $\hat{A}$.

By the Riesz representation theorem the states of A are precisely the probability measures on $\hat{A}$. It now follows from the definition of universal measurability that $\mathscr{U}(A)$ is equal to the class of bounded, universally measurable functions on $\hat{A}$.

4.3.17. Notes and remarks. Theorem 4.3.3 was in essence proved by Fell [89]. It provides a short proof of Dixmier's result (4.3.5) from [60]. No examples are known of closed prime ideals in a C^*-algebra which are not primitive, so 4.3.6 (proved in [60]) is probably true, also in the non-separable case. The result in 4.3.10 is known to every physicist, but usually in the less refined version that 'the spectrum of an operator is the set of eigenvalues'. The theory of universally measurable elements was established in [198].

4.4. The Dauns–Hofmann theorem

4.4.1. Let A be a C^*-algebra with enveloping von Neumann algebra A'' and let Z denote the centre of A''. Recall from 2.6.1 that for each x in A''_{sa} the element $c(x)$ is the smallest central element of A''_{sa} that majorizes x. If p is a minimal projection in Z then $pc(x)$ is a scalar multiple of p. If $x \geqslant 0$ then by 2.6.4

$$pc(x) = c(px) = \|c(px)\|p = \|px\|p.$$

The minimal projections in Z are in bijective correspondence with the equivalence classes of factor representations (3.8.13). In particular, the points in $\hat{A}$ correspond to minimal projections in Z. From the above we immediately deduce the following.

4.4.2. PROPOSITION. *For each x in A''_{sa} there is a bounded real function $\check{x}$ on $\hat{A}$ such that $\pi''(c(x)) = \check{x}(t)1$ whenever $(\pi, H) \in \mathrm{Irr}(A)$ and $(\pi, H) \in t$. If $x \geqslant 0$ then $\check{x}(t) = \|\pi''(x)\|$. If, moreover $x \in A$ then $\check{x}(t) = \|x/\ker \pi\|$ so that $\check{x}$ may be regarded as a function on $\check{A}$.*

4.4.3. LEMMA. *Let x be a normal element in A and E a closed subset of $\boldsymbol{C}$. Define*

$$F = \{t \in \check{A} | \mathrm{Sp}(x/t) \subset E\}$$

and

$$K = \{t \in \check{A} \,|\, \mathrm{Sp}(x/t) \cap E \neq \varnothing\}.$$

If $0 \in E$ then F is closed and if $0 \notin E$ then K is compact in $\check{A}$.

Proof. If $0 \in E$ and $\lambda \notin E$ let f be a continuous function on $\boldsymbol{C}$ which is zero on E and non-zero at λ. Since $\mathrm{Sp}(f(x)) = f(\mathrm{Sp}(x))$ we have $f(x) \in \ker(F)$. If therefore t belongs to the closure of F then $f(x) \in t$. Consequently $\lambda \notin \mathrm{Sp}(x/t)$, and thus $t \in F$. It follows that F is closed.

If $0 \notin E$ let f be a positive continuous function on $\boldsymbol{C}$ vanishing at zero which is one on E and strictly smaller than one outside E. Then with $y = f(x)$ we have

$$K = \{t \in \check{A} \,|\, \|y/t\| = 1\}.$$

If $\{F_i\}$ is a decreasing net of closed sets in $\check{A}$ such that $K \cap F_i \neq \varnothing$ for each i, let $I_i = \ker(F_i)$. Then

$$1 \geqslant \|y/I_i\| \geqslant \mathrm{Sup}\{\|y/t\| \,|\, t \in F_i\} = 1.$$

This means that the distance from y to I_i is one for all i and thus $\|y/I\| = 1$, where I is the norm closure of $\cup I_i$. In particular, $I \neq A$ so that $\mathrm{hull}(I) \neq \varnothing$. We have $\mathrm{hull}(I) \subset \cap F_i$, and since $\|y/I\| = 1$ there is by 4.3.10 (applied to A/I) a t in hull(I) such that $\|y/t\| = 1$. Consequently $t \in K$ and $K \cap (\cap F_i) \neq \varnothing$.

4.4.4. PROPOSITION. *For each x in A_+ the function $\check{x}$ is lower semi-continuous on $\check{A}$ and the sets $\{t\in\check{A}\,|\,\check{x}(t)\geqslant\alpha\}$, $x\in A_+,\alpha\geqslant 0$, are compact and form a basis for the Jacobson topology, which is therefore locally compact.*

Proof. For each $\alpha\geqslant 0$ we have

$$\{t\in\check{A}\,|\,\check{x}(t)\leqslant\alpha\}=\{t\in\check{A}\,|\,\mathrm{Sp}(x/t)\subset[0,\alpha]\}$$

so that these sets are closed by 4.4.3. This proves that $\check{x}$ is lower semi-continuous. Similarly,

$$\{t\in\check{A}\,|\,\check{x}(t)\geqslant\alpha\}=\{t\in\check{A}\,|\,\mathrm{Sp}(x/t)\cap[\alpha,\infty[\,\neq\varnothing\},$$

so that these sets are compact by 4.4.3. If G is an open set in $\check{A}$ and $t_0\in G$, let $I=\ker(\check{A}\backslash G)$ and take x in $I_+\backslash t_0$ with $\|x/t_0\|>1$. Then

$$t_0\in\{t\in\check{A}\,|\,\check{x}(t)>1\}\subset\{t\in\check{A}\,|\,\check{x}(t)\geqslant\}\subset G.$$

Since $\check{x}$ is lower semi-continuous the first of these sets is open, which proves that $\check{A}$ is a locally compact space (but not necessarily a Hausdorff space).

4.4.5. PROPOSITION. *The space $\check{A}$ is Hausdorff if and only if all functions $\check{x}$, where $x\in A_+$, are continuous on $\check{A}$.*

Proof. If $\check{A}$ is a Hausdorff space then each compact set is also closed and it follows from 4.4.4 that $\check{x}$ is continuous for all x in A_+. Conversely, if all such functions are continuous take t_1 and t_2 in $\check{A}$. If $t_1\neq t_2$ assume that $t_2\not\subset t_1$ and take $x\geqslant 0$ in $t_2\backslash t_1$. With $\alpha=\|x/t_1\|$ the sets $\{x>\frac{1}{2}\alpha\}$ and $\{x<\frac{1}{2}\alpha\}$ are disjoint neighbourhoods in $\check{A}$ of t_1 and t_2, respectively, which proves that $\check{A}$ is a Hausdorff space.

4.4.6. THEOREM. *For each x in $((\tilde{A}_{\mathrm{sa}})^m)^-$ the function $\check{x}$ is lower semi-continuous on $\hat{A}$ and the map $x\to\check{x}$ is an isometric isomorphism from the central elements in $(\tilde{A}_{\mathrm{sa}})^m$ onto the bounded, real lower semi-continuous functions on $\hat{A}$.*

Proof. If $x\in A_+$ then by 4.4.4 $\check{x}$ is lower semi-continuous on $\check{A}$ and thus on $\hat{A}$ as well. If $x\in(A_+)^m$ and $\{x_i\}$ is a net in A_+ which increases to x then $c(x_i)\nearrow c(x)$ by 2.6.5, whence $\check{x}_i\nearrow\check{x}$ by definition. Thus $\check{x}$ is the increasing limit of lower semi-continuous functions and therefore itself lower semi-continuous. Each x in $(\tilde{A}_{\mathrm{sa}})^m$ can by 3.11.7 be represented as $y+\alpha 1$ with y in $((A_{\mathrm{sa}})^m)^-$. Taking α sufficiently small (negative) we may assume that $y\geqslant 0$, whence $y+\varepsilon 1\in(A_+)^m$ by 3.11.6. Consequently $x=y+\varepsilon 1+(\alpha-\varepsilon)1$ and $c(x)=c(y+\varepsilon 1)+(\alpha-\varepsilon)1$ by 2.6.4, so that $\check{x}=(y+\varepsilon 1)^\vee+\alpha-\varepsilon$. Since the constant functions are continuous on $\hat{A}$ we conclude from the first part of the proof that $\check{x}$ is lower semi-continuous for each x in $(\tilde{A}_{\mathrm{sa}})^m$. Finally, the class of lower semi-continuous functions on $\hat{A}$ is norm closed and the map $x\to c(x)$ is norm continuous, which proves that $\check{x}$ is lower semi-continuous for all x in $((\tilde{A}_{\mathrm{sa}})^m)^-$.

Since $x = c(x)$ for each central element in A''_{sa} it is clear that the map $x \to \check{x}$ is a normal homomorphism of Z_{sa}, the centre of A''_{sa}, into the class of bounded real functions on $\hat{A}$. In fact, this map is nothing but the restriction to Z_{sa} of the atomic representation. Since the atomic representation is isometric on the class of universally measurable elements by 4.3.15, and $((\tilde{A}_{sa})^m)^- \subset \mathscr{U}(A)$ by 4.3.13 we see that $x \to \check{x}$ is an isometric isomorphism from $Z \cap ((\tilde{A}_{sa})^m)^-$ into the class of bounded real lower semi-continuous functions.

To show that the map is surjective let f be a bounded real lower semi-continuous function on $\hat{A}$ and assume first that $0 \leqslant f \leqslant 1$. Put $f_n = \Sigma_{k=1}^{2^n} 2^{-n}\chi_{nk}$, where χ_{nk} denotes the characteristic function of the set

$$\{t \in \hat{A} \mid f(t) > k2^{-n}\}.$$

Since each such set is open in $\hat{A}$ there is by definition of the Jacobson topology (see 4.1.3) a closed ideal I_{nk} in A, such that $\chi_{nk}(t) = 0$ precisely if $t \in \text{hull}(I_{nk})$. Let p_{nk} be the open projection corresponding to I_{nk} (cf 3.11.10). Then $p_{nk} \in Z \cap (A_+)^m$ and $\check{p}_{nk} = \chi_{nk}$. Define $x_n = 2^{-n}\Sigma_{k=1}^{2^n} p_{nk}$. Then $x_n \in Z \cap (A_+)^m$ and $\check{x}_n = f_n$. Since $f_n \nearrow f$ the sequence $\{x_n\}$ must increase to an element x in $Z \cap ((A_+)^m)^-$ such that $\check{x} = f$. If f is not positive we may have to modify it by adding a scalar. But by 3.11.7

$$((A_+)^m)^- + \boldsymbol{R}1 = (\tilde{A}_{sa})^m,$$

so we have established the isomorphism between $(\tilde{A}_{sa})^m \cap Z$ and the set of bounded, real lower semi-continous functions on $\hat{A}$.

4.4.7. It is easy to see that a function f which is lower semi-continuous on $\hat{A}$ must be constant on each class of elements t in $\hat{A}$ which have the same kernel in A. For if $t_1, t_2 \in \hat{A}$ and $f(t_1) > \alpha > f(t_2)$ then the set $\{f > \alpha\}$ is a neighbourhood of t_1 disjoint from $\{t_2\}$. This means that we may regard f as a lower semi-continuous function on $\check{A}$. Thus in 4.4.6 we may replace $\hat{A}$ by $\check{A}$.

4.4.8. COROLLARY (Dauns-Hofmann). *For each C^*-algebra A the complexification of the map $x \to \check{x}$ is an isomorphism of the centre of $M(A)$ onto the class of bounded continuous functions on $\check{A}$.*

Proof. By 3.12.9 the centre of $M(A)$ is $Z \cap (\tilde{A}_{sa})^m \cap (\tilde{A}_{sa})_m$. The result is now immediate from 4.4.7.

4.4.9. PROPOSITION. *For each x in $\tilde{A}_+$ the function $(\pi, H) \to \text{Tr}(\pi(x))$ from $\text{Irr}(A)$ to $[0, \infty]$ determines a lower semi-continuous function on $\check{A}$.*

Proof. Define a function $f: \check{A} \to [0, \infty]$ by

$$f = \text{Sup}(\Sigma \check{x}_k),$$

the supremum being taken over all finite sets $\{x_k\}$ in $\tilde{A}_+$ such that $\Sigma x_k \leqslant x$. Since f is the supremum of functions which by 4.4.6 are lower semi-continuous,

f is lower semi-continuous. Take (π, H) in $\mathrm{Irr}(A)$ and $\{x_k\}$ in $\tilde{A}_+$ with $\Sigma x_k \leqslant x$. Then

$$\sum \|\pi(x_k)\| \leqslant \sum \mathrm{Tr}(\pi(x_k)) \leqslant \mathrm{Tr}(\pi(x)),$$

whence $f(\ker \pi) \leqslant \mathrm{Tr}(\pi(x))$. On the other hand, given (π, H) and an orthonormal set $\{\xi_1, \xi_2, \ldots, \xi_n\}$ in H we may by 2.7.5 choose e_k in A_+ such that $\Sigma e_k^2 \leqslant 1$ and $\pi(e_k)\xi_k = \xi_k$. Then with $x_k = x^{1/2}e_k^2x^{1/2}$ we have

$$\sum \|\pi(x_k)\| = \sum \|\pi(e_k x e_k)\| \geqslant \sum (\pi(x)\xi_k \mid \xi_k),$$

whence $f(\ker \pi) = \mathrm{Tr}(\pi(x))$, and the proposition follows.

4.4.10. PROPOSITION. *The subset ${}_n\check{A}$ of $\check{A}$ corresponding to irreducible representations of A with finite dimension less than or equal to n is closed. The set $\check{A}_n = {}_n\check{A} \backslash {}_{n-1}\check{A}$ of n-dimensional representations is a Hausdorff space in its relative topology.*

Proof. Since ${}_n\check{A} = \{\ker \pi \mid \mathrm{Tr}(\pi(1)) \leqslant n\}$ we see from 4.4.9 that ${}_n\check{A}$ is closed in $\check{A}$. The set $\check{A}_n$ is by 4.1.11 homeomorphic to the primitive spectrum of the C^*-algebra $A_n = \ker({}_{n-1}\check{A})/\ker({}_n\check{A})$, which has only n-dimensional irreducible representations. If $x \in A_n$ and $0 \leqslant x \leqslant 1$ then

$$\mathrm{Tr}(\pi(x)) = n - \mathrm{Tr}(\pi(1 - x))$$

and we conclude that the trace function of x is upper semi-continuous on $\check{A}_n$, and consequently continuous. If t_0 is a limit point of the set $\{t \in \check{A}_n \mid \check{x}(t) \geqslant \alpha\}$ for some $\alpha > 0$ and $x \geqslant 0$, let f be a continuous function which is zero on $[0, \alpha - \varepsilon]$ and one on $[\alpha, \infty]$. Then the trace function of $f(x)$ is $\geqslant 1$ on the set above, and, being continuous, also $\geqslant 1$ at the point t_0. But this implies that $x(t_0) \geqslant \alpha - \varepsilon$ for all $\varepsilon > 0$; whence $x(t_0) \geqslant \alpha$ and the set above is closed. Consequently $\check{x}$ is continuous on $\check{A}_n$ for each x in $(A_n)_+$, and it follows from 4.4.5 that $\check{A}_n$ is a Hausdorff space.

4.4.11. Notes and remarks. The two results in 4.4.4 and 4.4.5 were proved by Kaplansky [141]. The proof of the Dauns–Hofmann theorem (4.4.8) given here is more complicated than necessary, but gives in addition the more general result in 4.4.6 by the author [198]. Short and direct proofs can be found in [67] and [85]. Propositions 4.4.9 and 4.4.10 combine results of Kaplansky, Fell and Dixmier, see [141, 89, 62].

4.5. Borel *-algebras

4.5.1. In the preceding sections we have explored the Jacobson topology on the spectrum of a C^*-algebra. As one of the most useful results we have obtained the Dauns–Hofmann theorem (4.4.8) which connects the topology directly with the centre of the algebra A (or $M(A)$ if $1 \notin A$).

If the Jacobson topology was our only tool, the building blocks for a decomposition theory would naturally be the simple C^*-algebras (1.2.5), because these are the ones for which the topology is trivial. As long as $\hat{A} = \check{A}$ this point of view is fruitful. However, there are simple C^*-algebras whose spectrum is uncountable. In order to handle these and, if possible, classify their representations, we must at least find a distinguished class of sets in $\hat{A}$, which separates points. This will be done by imposing a Borel structure on $\hat{A}$ (i.e. a σ-algebra of sets). In this section we introduce the concepts necessary for the construction of such a Borel structure.

4.5.2. Let H be a Hilbert space and M a subset of $\boldsymbol{B}(H)_{sa}$. The *monotone sequential closure* of M is defined as the smallest class $\mathscr{B}(M)$ in $\boldsymbol{B}(H)_{sa}$ that contains M and contains the strong limit of each monotone (increasing or decreasing) sequence of elements from $\mathscr{B}(M)$.

4.5.3. LEMMA. *Each countable subset of $\mathscr{B}(M)$ lies in the monotone sequential closure of a separable subset of M.*

Proof. The subset of $\mathscr{B}(M)$ consisting of those elements x such that $x \in \mathscr{B}(M_x)$, where M_x is a separable subset of M, is clearly monotone sequentially closed and contains M. It therefore equals M. If $\{x_n\}$ is a countable subset of $\mathscr{B}(M)$, then $\{x_n\} \subset \mathscr{B}(\cup M_{x_n})$, which proves the lemma.

4.5.4. THEOREM. *Let A be a C^*-subalgebra of $\boldsymbol{B}(H)$. Then $\mathscr{B}(A_{sa})$ is the self-adjoint part of a C^*-algebra.*

Proof. By 4.5.3 it suffices to prove the theorem under the assumption that A is separable. We may also assume that A acts non-degenerately on H. Since A is separable $(A_{sa})_\sigma = (A_{sa})^m$, and since $\tilde{A}_{sa} \subset (A_{sa})^m$ we conclude that $\mathscr{B}(A_{sa}) = \mathscr{B}(\tilde{A}_{sa})$. We may assume therefore that $1 \in A$.

Let $\mathscr{B}_1(A_{sa})$ denote the subset of elements x in $\mathscr{B}(A_{sa})$ such that $x + y \in \mathscr{B}(A_{sa})$ for each y in A_{sa}. Clearly $A_{sa} \subset \mathscr{B}_1(A_{sa})$. But $\mathscr{B}_1(A_{sa})$ is monotone sequentially closed and therefore, by the minimality condition in the definition, we must have $\mathscr{B}_1(A_{sa}) = \mathscr{B}(A_{sa})$. Thus $A_{sa} + \mathscr{B}(A_{sa}) \subset \mathscr{B}(A_{sa})$.

Now let $\mathscr{B}_2(A_{sa})$ denote the subset of elements y in $\mathscr{B}(A_{sa})$ such that $x + y \in \mathscr{B}(A_{sa})$ for each x in $\mathscr{B}(A_{sa})$. From the argument above, $A_{sa} \subset \mathscr{B}_2(A_{sa})$ and since $\mathscr{B}_2(A_{sa})$ is monotone sequentially closed we conclude that $\mathscr{B}_2(A_{sa}) = \mathscr{B}(A_{sa})$. Thus $\mathscr{B}(A_{sa}) + \mathscr{B}(A_{sa}) \subset \mathscr{B}(A_{sa})$.

The proof that $\mathscr{B}(A_{sa})$ is stable under multiplication with real scalars is similar, and thus we have shown that $\mathscr{B}(A_{sa})$ is a real vector space.

Let $\{x_n\}$ be a norm convergent sequence in $\mathscr{B}(A_{sa})$ with limit x. We may assume that $\|x_{n+1} - x_n\| \leqslant 2^{-n}$ for all n. Consider the sequence

$\{x_n - 2^{-n+1}1\}$ which belongs to $\mathscr{B}(A_{sa})$ from the above. Since

$$(x_{n+1} - 2^{-n}1) - (x_n - 2^{-n+1}1) = x_{n+1} - x_n + 2^{-n}1 \geqslant 0,$$

this sequence is increasing and therefore its limit x belongs to $\mathscr{B}(A_{sa})$. Thus $\mathscr{B}(A_{sa})$ is norm closed.

Let $\mathscr{B}_3(A_{sa})$ denote the subset of elements in $\mathscr{B}(A_{sa})$ such that $x^k \in \mathscr{B}(A_{sa})$ for all k in $\boldsymbol{N}$. We claim that $\mathscr{B}_3(A_{sa})$ is monotone sequentially closed. To see this, let $\{x_n\}$ be a monotone increasing sequence in $\mathscr{B}_3(A_{sa})$ with limit x in $\mathscr{B}(A_{sa})$. Since $\{x_n\}$ is uniformly bounded we may assume that $\|x_n\| \leqslant 1$ for all n. Then for $0 \leqslant t < 1$

$$(1 - tx_n)^{-1} = \sum_{k=0}^{\infty} t^k x_n^k \quad \text{(norm convergence).}$$

Since $x_n \in \mathscr{B}_3(A_{sa})$ it follows that $(1 - tx)^{-1} \in \mathscr{B}(A_{sa})$. The sequence $\{(1 - tx_n)^{-1}\}$ is monotone increasing (cf. 1.3.7) and therefore its limit $(1 - tx)^{-1}$ belongs to $\mathscr{B}(A_{sa})$. Since 1 and x both belong to $\mathscr{B}(A_{sa})$ we have

$$t^{-2}[(1 - tx)^{-1} - (1 + tx)] \in \mathscr{B}(A_{sa}).$$

However, the elements above converges in norm to x^2 when $t \to 0$, and since $\mathscr{B}(A_{sa})$ is norm closed, $x^2 \in \mathscr{B}(A_{sa})$. By induction we can now prove that $x^k \in \mathscr{B}(A_{sa})$ for all k, whence $x \in \mathscr{B}_3(A_{sa})$. Since therefore $\mathscr{B}_3(A_{sa})$ is monotone sequentially closed and contains A_{sa} it must equal $\mathscr{B}(A_{sa})$. From this we conclude that if x and y belong to $\mathscr{B}(A_{sa})$ then

$$xy + yx = (x + y)^2 - x^2 - y^2 \in \mathscr{B}(A_{sa}).$$

We claim that also $i(xy - yx) \in \mathscr{B}(A_{sa})$. To prove this note that

$$(*) \qquad i(xy - yx) = (x + i)^* y (x + i) - xyx - y.$$

This shows that the subset $\mathscr{B}_4(A_{sa})$ of $\mathscr{B}(A_{sa})$ consisting of the elements y in $\mathscr{B}(A_{sa})$ such that $i(xy - yx) \in \mathscr{B}(A_{sa})$ for all x in A_{sa}, is monotone sequentially closed. As $A_{sa} \subset \mathscr{B}_4(A_{sa})$ we conclude that $\mathscr{B}_4(A_{sa}) = \mathscr{B}(A_{sa})$, so that $i(xy - yx) \in \mathscr{B}(A_{sa})$ if one of the factors belong to A_{sa}. Since for all x and y in $\mathscr{B}(A_{sa})$

$$2xyx = (xy + yx)x + x(xy + yx) - (yx^2 + x^2 y) \in \mathscr{B}(A_{sa}),$$

we can use $(*)$ with x and y interchanged to show that $(y + i)^* x (y + i) \in \mathscr{B}(A_{sa})$ for all x in A_{sa} and y in $\mathscr{B}(A_{sa})$. However, the subset $\mathscr{B}_5(A_{sa})$ of $\mathscr{B}(A_{sa})$ consisting of elements x in $\mathscr{B}(A_{sa})$ such that $(y + i)^* x (y + i) \in \mathscr{B}(A_{sa})$ for all y in $\mathscr{B}(A_{sa})$ is clearly monotone sequentially closed, and since $A_{sa} \subset \mathscr{B}_5(A_{sa})$ from the above we have $\mathscr{B}_5(A_{sa}) = \mathscr{B}(A_{sa})$. Using $(*)$ we conclude that $i(xy - yx) \in \mathscr{B}(A_{sa})$ for all x and y in $\mathscr{B}(A_{sa})$, as desired.

The set $\mathscr{B}(A_{sa}) + i\mathscr{B}(A_{sa})$ is a norm closed, *-invariant complex vector space in $B(H)$. Since

$$2xy = (xy + yx) - i(i(xy - yx)),$$

we conclude from the results above that the set is actually an algebra, and thus $\mathscr{B}(A_{sa})$ is the self-adjoint part of a C^*-algebra.

4.5.5. We say that a C^*-subalgebra $\mathscr{A}$ of $B(H)$ is a *Borel *-algebra* if $\mathscr{A}_{sa}$ is monotone sequentially closed. If H is separable then by 2.4.3 any Borel *-algebra in $B(H)$ is necessarily a von Neumann algebra. For non-separable Hilbert spaces this need no longer be the case. Exactly as in 3.9.3 it can be shown that a C^*-algebra $\mathscr{A}$, which is monotone sequentially complete, has a faithful representation as a Borel *-algebra if and only if it has a separating family of sequentially normal states.

We say that a Borel *-algebra $\mathscr{A}$ is *countably generated* if there is a sequence in $\mathscr{A}$ such that no proper Borel *-subalgebra of $\mathscr{A}$ contains that sequence. It is immediate that $\mathscr{A}$ is countably generated if and only if $\mathscr{A}_{sa} = \mathscr{B}(A_{sa})$ for some separable C^*-subalgebra A of $\mathscr{A}$.

4.5.6. Let A be a C^*-algebra. We define the *enveloping Borel *-algebra* of A to be the C^*-algebra

$$\mathscr{B}(A) = \mathscr{B}(A_{sa}) + i\mathscr{B}(A_{sa}),$$

the monotone sequential closure being taken on the universal Hilbert space for A. When no confusion may arise we shall just write $\mathscr{B}$ instead of $\mathscr{B}(A)$. If A is separable $\mathscr{B}$ has a unit. This will not necessarily be the case for non-separable C^*-algebras. A much more serious obstacle to the use of non-separable C^*-algebras, is, however, that their enveloping Borel *-algebras are not countably generated.

4.5.7. PROPOSITION. *If $\mathscr{A}$ is a Borel *-algebra with unit then $f(x) \in \mathscr{A}$ for every x in $\mathscr{A}_{sa}$ and every bounded Borel function f on $\mathbf{R}$.*

Proof. Let $\mathscr{B}(\mathbf{R})$ denote the algebra of bounded Borel functions on $\mathbf{R}$. For each x in $\mathscr{A}_{sa}$ let $\mathscr{B}_x(\mathbf{R})$ denote the subset of real functions f in $\mathscr{B}(\mathbf{R})$ for which $f(x) \in \mathscr{A}$. Since $\mathscr{A}$ is a C^*-algebra $C_0(\mathbf{R}) \subset \mathscr{B}_x(\mathbf{R})$ and since the map $f \to f(x)$ is sequentially normal from $\mathscr{B}(\mathbf{R})$ to $\mathscr{A}$, by the spectral theorem, $\mathscr{B}_x(\mathbf{R})$ is a monotone sequentially closed set of functions on $\mathbf{R}$. However, $\mathscr{B}(\mathbf{R})_{sa}$ is the monotone sequential closure of $C_0(\mathbf{R})_{sa}$ (in the von Neumann algebra of all bounded functions on $\mathbf{R}$), whence $\mathscr{B}_x(\mathbf{R}) = \mathscr{B}(\mathbf{R})_{sa}$.

4.5.8. PROPOSITION. *If A is a separable C^*-algebra then $c(x) \in \mathscr{B}_{sa}$ for every x in $\mathscr{B}_{sa}$.*

Proof. The unitary group in $\tilde{A}$ is weakly dense in the unitary group of A'' by 2.3.3. Since A is separable there is a countable subgroup $\{u_n\}$ of the unitary group of $\tilde{A}$ which is norm dense in the unitary group of $\tilde{A}$, hence weakly dense in the unitary group of A''. If p is a projection in $\mathscr{B}$, then $c(p) = \vee u_n^* p u_n$ by 2.6.3. Since the least upper bound for two projections is the range projection of their sum we have

$$c(p) = \operatorname*{Lim}_n \left[\sum_{k=1}^{n} u_k^* p u_k \right] \in \mathscr{B}$$

using 4.5.7.

Take x in $\mathscr{B}_{sa}$. Using 2.6.4 we may assume that $0 \leqslant x \leqslant 1$. For fixed n let p_{nk} denote the spectral projection of x corresponding to the interval $]k2^{-n}, \infty[$ with $1 \leqslant k \leqslant 2^n$. From 4.5.7 $p_{nk} \in \mathscr{B}$ and thus $c(p_{nk}) \in \mathscr{B}$ from the first part of the proof. Put $x_n = 2^{-n} \Sigma_{k=1}^{2^n} p_{nk}$. Then by 2.6.6

$$c(x_n) = 2^{-n} \sum_{k=1}^{2^n} c(p_{nk}) \in \mathscr{B}.$$

The sequence $\{x_n\}$ is increasing and converges in norm to x. Thus by 2.6.5 $c(x_n) \nearrow c(x)$, whence $c(x) \in \mathscr{B}$.

4.5.9. THEOREM. *Let A be a C^*-algebra. For each representation (π, H) of A there is a unique sequentially normal morphism π'' of $\mathscr{B}$ that extends π, and $\pi''(\mathscr{B}_{sa})$ is the monotone sequential closure of $\pi(A_{sa})$ in $\mathbf{B}(H)$.*

Proof. The existence of a sequentially normal extension of π is clear from 3.7.7; just take the normal extension π'' of π from A to A'' and restrict it to $\mathscr{B}$. The uniqueness follows from the fact that two sequentially normal morphisms of $\mathscr{B}$ which agree on A_{sa} must agree on $\mathscr{B}_{sa}$, hence on $\mathscr{B}$. Now

$$\pi(A_{sa}) \subset \pi''(\mathscr{B}_{sa}) \subset \mathscr{B}(\pi(A_{sa})),$$

so our usual argument will yield equality in the second inclusion above, if we can show that $\pi''(\mathscr{B}_{sa})$ is monotone sequentially closed. Assume therefore that $\{x_n\}$ is an increasing sequence in $\pi''(\mathscr{B}_{sa})$ with $\|x_n\| \leqslant 1$ for all n. By successive application of 1.5.10 we can find a sequence $\{y_n\}$ in $\mathscr{B}_{sa}$ such that $\pi''(y_n) = x_n$ and $y_n \leqslant y_{n+1} \leqslant 1$ for all n. Since $\{y_n\}$ is increasing and bounded it converges to an element y in $\mathscr{B}_{sa}$, and since π'' is normal $x_n \nearrow \pi''(y)$. Thus $\pi''(\mathscr{B}_{sa})$ is monotone sequentially closed, whence $\pi''(\mathscr{B}_{sa}) = \mathscr{B}(\pi(A_{sa}))$.

4.5.10. COROLLARY. *For each separable representation (π, H) of a separable C^*-algebra A we have $\pi''(\mathscr{B}) = \pi(A)''$ and the centre of $\mathscr{B}$ maps onto the centre of $\pi(A)''$.*

Proof. By 4.5.9 $\pi''(\mathscr{B}_{sa})$ is monotone sequentially closed, whence $\pi(\mathscr{B}) = \pi(A)''$ by 2.4.3. If $x \in \mathscr{B}_{sa}$ and $\pi''(x)$ is central, then $\pi''(c(x)) = \pi''(x)$, and $c(x)$ is central in $\mathscr{B}$ by 4.5.8.

4.5.11. COROLLARY. *For each countably generated Borel *-algebra $\mathscr{A}$ there is a sequentially normal morphism of an enveloping Borel *-algebra onto $\mathscr{A}$.*

Proof. Take a separable C^*-subalgebra A which generates $\mathscr{A}$ and consider the identity map $\pi: A \to A$ as a representation of A. By 4.5.9 π'' maps $\mathscr{B}(A)$ onto $\mathscr{A}$.

4.5.12. LEMMA. *Let A be a C^*-algebra. Then $\mathscr{U}(A)$ is monotone sequentially closed.*

Proof. Let $\{x_n\}$ be a monotone increasing sequence in $\mathscr{U}(A)$ with limit x in A'', and without loss of generality assume that $0 \leqslant x_n \leqslant 1$ for all n. If ϕ is a state of A and $\varepsilon > 0$ is given then $x_n \leqslant x$ and $\phi(x - x_n) < \varepsilon$ for n large. Since $x_n \in \mathscr{U}(A)$ it suffices by 4.3.12 to approximate x from above with elements from $\mathscr{U}(A)$. To do so let $y_n = x_n - x_{n-1}$ and $x_0 = 0$. Then $y_n \in \mathscr{U}(A)_+$ and $\Sigma y_n = x$. For each n we choose a_n in $(A_{sa})^m$ such that $y_n \leqslant a_n$ and $\phi(a_n - y_n) < \varepsilon 2^{-n}$. Then with $f_\alpha(t) = (1 + \alpha t)^{-1} t$ (as in 1.3.7) we define

$$x_\alpha = \operatorname*{Lim}_m f_\alpha\left(\sum_{n=1}^m a_n\right).$$

Since f_α is operator monotone and $f_\alpha \leqslant \alpha^{-1}$ on $[0, \infty[$, the limit exists in A''. But since $\Sigma_{n=1}^m a_n \in ((A_+)^m)^-$ by 3.11.6, also $f_\alpha(\Sigma_{n=1}^m a_n) \in ((A_+)^m)^-$. It follows from 3.11.5 that $x_\alpha \in ((A_+)^m)^-$. In particular, $x_\alpha \in \mathscr{U}(A)$ by 4.3.13. We have

$$(1 + \alpha)^{-1} x \leqslant f_\alpha(x) = \operatorname*{Lim}_m f_\alpha\left(\sum_{n=1}^m y_n\right) \leqslant x_\alpha$$

Moreover,

$$\phi(x_\alpha - x) = \operatorname{Lim} \phi\left(f_\alpha\left(\sum_{n=1}^m a_n\right) - x\right) \leqslant \operatorname{Lim} \phi\left(\sum_{n=1}^m a_n - x\right)$$

$$\leqslant \sum_{n=1}^{\infty} (\phi(y_n) + \varepsilon 2^{-n}) - \phi(x) = \varepsilon,$$

using the fact that $f_\alpha(t) \leqslant t$. It follows that $\phi((1 + \alpha)x_\alpha - x) < 2\varepsilon$ for α sufficiently small, which completes the proof.

4.5.13. COROLLARY. *For each C^*-algebra A we have $\mathscr{B}_{sa} \subset \mathscr{U}(A)$ and the atomic representation is faithful on $\mathscr{B}$.*

Proof. Follows directly from 4.3.13, 4.5.12 and 4.3.15.

4.5.14. From 4.5.13 and 4.5.9 we see that $\mathscr{B}_{sa}$ is the monotone sequential closure of A_{sa} in its atomic representation. If therefore A is separable and commutative, so that $A = C_0(\hat{A})$, then $\mathscr{B} = \mathscr{B}(\hat{A})$, the algebra of bounded Borel measurable functions on $\hat{A}$ (cf. 4.3.16). This explains the notion Borel operators for the elements in $\mathscr{B}$.

The algebra of Borel operators is not the only possible generalization of the Borel functions. Considering A in its universal representation, we let $\mathscr{B}^s(A_{sa})$ denote the strong sequential closure of A_{sa} in A''_{sa}, and take $\mathscr{B}^s = \mathscr{B}^s(A_{sa}) + i\mathscr{B}^s(A_{sa})$. Furthermore, we let $\mathscr{B}^w$ denote the weak sequential closure of A in A''. It is not hard to show that both $\mathscr{B}^s$ and $\mathscr{B}^w$ are C^*-algebras. Evidently $\mathscr{B} \subset \mathscr{B}^s \subset \mathscr{B}^w$. All available information confirms the conjecture that these three algebras are equal. For commutative algebras it is well known that $\mathscr{B} = \mathscr{B}^w$. In general the problem remains open. We shall see in Chapter 6 that it has a positive solution for C^*-algebras of type I.

4.5.15. PROPOSITION. *For each C^*-algebra A the space $\mathscr{U}(A)$ is strongly sequentially closed.*

Proof. Let $\{x_n\}$ be a sequence in $\mathscr{U}(A)$ which converges strongly to x in A''_{sa}. Since $\{-x_n\}$ converges to $-x$ it suffices to show that x can be approximated weakly from above with elements from $(A_{sa})^m$, or by 4.3.12, with elements from $\mathscr{U}(A)$.

By the uniform boundedness principle the sequence $\{x_n\}$ is bounded, so we may assume that $\|x_n\| \leqslant \frac{1}{2}$ for all n. If ϕ is a state of A and $\varepsilon > 0$ is given we may further assume that $\phi((x - x_n)^2) < \varepsilon^2 4^{-n-1}$. With $y_n = x_{n+1} - x_n$ observe that $\|y_n\| \leqslant 1$ and that

$$y_n^2 \leqslant 2((x_{n+1} - x)^2 + (x_n - x)^2),$$

whence

$$\phi(|y_n|) \leqslant \phi(y_n^2)^{1/2} < \varepsilon 2^{-n}.$$

Since $y_n \in \mathscr{U}(A)$ there is by 4.3.14 a net $\{a_i\}$ in $((A_{sa})^m)^-$ converging strongly to y_n from above. From 2.3.2 it follows that we can find a_n in $((A_{sa})^m)^-$ with $y_n \leqslant a_n$ such that $\phi(|a_n|) < \varepsilon 2^{-n}$.

With $f_\alpha(t) = t(1 + \alpha t)^{-1}$ and $0 < \alpha < 1$ we set $\varepsilon_\alpha = \alpha(1 - \alpha)^{-1}$ so that $t \leqslant \varepsilon_\alpha + f_\alpha(t)$ if $t \leqslant 1$. Consequently

$$x_{n+1} - x_1 \leqslant \varepsilon_\alpha + f_\alpha(x_{n+1} - x_1) = \varepsilon_\alpha + f_\alpha\left(\sum_1^n y_k\right)$$

$$\leqslant \varepsilon_\alpha + f_\alpha\left(\sum_1^n a_k\right) \leqslant \varepsilon_\alpha + f_\alpha\left(\sum_1^n |a_k|\right).$$

Let x_α denote the strong limit of the bounded monotone increasing sequence $\{f_\alpha(\Sigma_1^n |a_k|)\}$ and note that $x_\alpha \in \mathscr{B}$ since each $a_k \in \mathscr{B}$. Passing to the limit we obtain

$$x \leqslant x_1 + \varepsilon_\alpha + x_\alpha.$$

Moreover,

$$\phi(x_1 + \varepsilon_\alpha + x_\alpha - x) = \varepsilon_\alpha + \operatorname{Lim} \phi\left(f_\alpha\left(\sum_1^n |a_k|\right) - \sum_1^n y_k\right)$$
$$\leqslant \varepsilon_\alpha + \operatorname{Lim}\left(\phi\left(\sum_1^n |a_k| + |y_k|\right)\right) \leqslant \varepsilon_\alpha + 2\varepsilon,$$

since $f_\alpha(t) \leqslant t$. As $\mathscr{B}_{sa} \subset \mathscr{U}(A)$ by 4.5.13 the proof is complete.

4.5.16. PROPOSITION. *Let $\mathscr{A}$ be a Borel *-algebra. If $x \in \mathscr{A}$ and $x = u|x|$ is the polar decomposition of x, then*

$$|x| \in \mathscr{A} \quad \textit{and} \quad u \in \mathscr{A}.$$

Proof. Since $\mathscr{A}$ is a C^*-algebra, $|x| \in \mathscr{A}$. Consider the sequence in $\mathscr{A}$ with elements

$$(x + |x|)((1/n) + |x|)^{-2}(x + |x|)^* = (u + 1)((1/n) + |x|)^{-2}|x|^2(u^* + 1).$$

The sequence is increasing with limit $(u + 1)[|x|](u^* + 1)$ in $\mathscr{A}$. Since $[|x|] = u^*u$ and $[|x^*|] = uu^*$ these two products belong to $\mathscr{A}$. Consequently

$$(u + 1)u^*u(u^* + 1) - u^*u - uu^* = u + u^* \in \mathscr{A}.$$

Applying the same argument to ix we obtain $iu - iu^* \in \mathscr{A}$ whence $u \in \mathscr{A}$.

4.5.17. PROPOSITION. *Let $\{x_n\}$ be a sequence in a Borel *-algebra $\mathscr{A}$. If $\Sigma|x_n|$ and $\Sigma|x_n^*|$ are convergent (and thus belong to $\mathscr{A}$), then Σx_n is convergent and belongs to $\mathscr{A}$.*

Proof. Let $x_n = u_n|x|$ be the polar decomposition of x_n and note that $|x_n^*| = u_n|x_n|u_n^*$. Then

$$2x_n = (1 + u_n)|x_n|(1 + u_n)^* - i(1 + iu_n)|x_n|(1 + iu_n)^* - (1 - i)(|x_n| + |x_n^*|).$$

Since

$$(1 + u_n)|x_n|(1 + u_n)^* \leqslant 2(|x_n| + u_n|x_n|u_n^*) = 2(|x_n| + |x_n^*|),$$

it follows from the assumptions that Σx_n is the linear combination of (convergent) sums of positive elements from $\mathscr{A}$. Thus Σx_n is convergent (in the strong topology) and $\Sigma x_n \in \mathscr{A}$

4.5.18. Notes and remarks. Monotone sequential closures of C^*-algebras were introduced by Kadison in [129] where most of theorem 4.5.4 was proved. Later the author undertook a more systematic study of sequential closures [193, 195, 197, 198, 201, 203]. Further results can be found in Kehlet [147] and Christensen [33]. The weak sequential closure $\mathscr{B}^w$ was studied by Davies [50, 51, 52] under the name Σ^*-algebras.

4.6. Standard algebras

4.6.1. We say that a Borel *-algebra $\mathscr{A}$ is *standard* if there is a separable C^*-algebra A with enveloping Borel *-algebra $\mathscr{B}$, and a central projection z in $\mathscr{B}$, such that $\mathscr{A}$ is isomorphic (as a C^*-algebra) to $z\mathscr{B}$. As in 2.5.2 we see that any isomorphism between sequentially complete C^*-algebras is sequentially normal, so that isomorphism is a stronger concept than it appears to be. In this section we shall see that the standard Borel *-algebras are the well behaved members of the class of Borel *-algebras.

4.6.2. LEMMA. *Let I be a closed ideal in a separable C^*-algebra A, and let z be the open central projection in A'' which supports I (cf. 3.11.10). If $\mathscr{B}$ is the enveloping Borel *-algebra for A then $z\mathscr{B}$ and $(1 - z)\mathscr{B}$ are isomorphic to the enveloping Borel *-algebras of I and A/I, respectively.*

Proof. If ϕ is a pure state of A then $\|\phi \mid I\| = 1$ if and only if $\phi(z) = 1$ and otherwise $\phi(I) = \phi(z) = 0$. It follows that if H_a is the atomic Hilbert space for A (see 4.3.7) then zH_a and $(1 - z)H_a$ are the atomic Hilbert spaces for I and A/I, respectively. We have $A/I = (1 - z)A$ and $I = zA \cap A$. Since also $z \in I^m$ we conclude that I and zA have the same monotone sequential closure on zH_a, and the result is immediate from 4.5.13.

4.6.3. THEOREM. *Let $\rho: \mathscr{A}_1 \to \mathscr{A}_2$ be a surjective sequentially normal morphism from the countably generated Borel *-algebra $\mathscr{A}_1$ onto the standard Borel *-algebra $\mathscr{A}_2$. There is a unique central projection p in $\mathscr{A}_1$ such that $\rho \mid p\mathscr{A}_1$ is an isomorphism and $\ker \rho = (1 - p)\mathscr{A}_1$. In particular, $p\mathscr{A}_1$ is standard.*

Proof. By assumption there is a separable C^*-algebra A with enveloping Borel *-algebra $\mathscr{B}$ such that $\mathscr{A}_2 = z\mathscr{B}$ for some central projection z in $\mathscr{B}$. Set $I = \{x \in A \mid zx = 0\}$ and let $\bar{z}$ be the central projection in $\mathscr{B}$ such that $1 - \bar{z}$ is the open central projection supporting I ($\bar{z}$ is the 'closure' of z). Since $\bar{z}\mathscr{B}$ is the enveloping Borel *-algebra of A/I by 4.6.2 we may assume without loss of generality that $\bar{z} = 1$, which means that the morphism $x \to zx$ is an isomorphism of A.

Choose a separable C^*-subalgebra A_1 in $\mathscr{A}_1$ such that $\rho(A_1) \supset zA$. The C^*-algebra $A_1 \cap \ker \rho$ is separable and contains therefore a countable approximate identity, which converges strongly up to a projection in $\mathscr{A}_1$. We let q denote the central cover of that projection. Note that $q \in \mathscr{A}_1$ by 4.5.8 and 4.5.11. Now $\rho(q) = 0$ since $\ker \rho$ is a monotone sequentially closed ideal in $\mathscr{A}_1$, and $\rho|(1 - q)A_1$ is an isomorphism. We can therefore define an isomorphism π_0 of A into $(1 - q)A_1$ such that $\rho(\pi_0(x)) = zx$ for all x in A. By 4.5.9 there is a unique extension of π_0 to a sequentially normal morphism π of $\mathscr{B}$ onto the Borel *-subalgebra $\mathscr{B}_1$ of $\mathscr{A}_1$ generated by $\pi_0(A)$. Moreover, $\rho(\pi(x)) = zx$ for all x in $\mathscr{B}$, since this holds on the generating set A of $\mathscr{B}$. Thus if

$z_1 = \pi(z)$ and $\mathscr{B}_0 = z_1\mathscr{B}_1$ then ρ and π are the inverses of one another when restricted to $\mathscr{B}_0$ and $z\mathscr{B}(= \mathscr{A}_2)$. It follows that $\mathscr{B}_0 \cap \ker\rho = 0$ and $\mathscr{B}_0 + \ker\rho = \mathscr{A}_1$.

Let $\{x_n\}$ be a generating sequence for $\mathscr{A}_1$. Each x_n has the form $y_n + z_n$ with y_n in $\mathscr{B}_0$ and z_n in $\ker\rho$. Let A_ρ be the separable C^*-subalgebra of $\ker\rho$ generated by $\{z_n\}$, and let $1 - p$ be the central cover in $\mathscr{A}_1$ of the strong limit of an approximate unit in A_ρ. Then p is a central projection, $\rho(1 - p) = 0$ and $pA_\rho = 0$. Thus

$$\{px_n\} = \{py_n\} \subset p\mathscr{B}_0,$$

and since $\{x_n\}$ generates $\mathscr{A}_1$ this implies that $p\mathscr{A}_1 \subset p\mathscr{B}_0$, whence $p\mathscr{A}_1 = p\mathscr{B}_0$. If $x_0 \in \mathscr{B}_0$ and $\rho(px_0) = 0$ then $z\rho(x_0) = 0$; whence $\rho(x_0) = 0$ and $x_0 = 0$. It follows that $p\mathscr{A}_1 \cap \ker\rho = 0$, and since $(1 - p)\mathscr{A}_1 \subset \ker\rho$ we conclude that $\ker\rho = (1 - p)\mathscr{A}_1$, so that $\rho \mid p\mathscr{A}_1$ is an isomorphism onto $\mathscr{A}_2(= z\mathscr{B})$.

4.6.4 PROPOSITION. *Let $\mathscr{A}_1$ and $\mathscr{A}_2$ be standard Borel *-algebras. If each can be mapped onto the other by a sequentially normal morphism, then $\mathscr{A}_1$ and $\mathscr{A}_2$ are isomorphic.*

Proof. By assumption there are surjective sequentially normal morphisms $\rho: \mathscr{A}_1 \to \mathscr{A}_2$ and $\pi: \mathscr{A}_2 \to \mathscr{A}_1$. Using 4.6.3 we find central projections p and q in $\mathscr{A}_1$ and $\mathscr{A}_2$, respectively, such that $\ker\rho = (1 - p)\mathscr{A}_1$ and $\ker\pi = (1 - q)\mathscr{A}_2$.

Set $p_0 = 1$, $p_1 = p$ and let p_2 be the central projection in $p_1\mathscr{A}_1$ such $\rho(p_2) = q$ (whence $\pi(\rho(p_2)) = p_0$). Define inductively a decreasing sequence $\{p_n\}$ of central projections in $\mathscr{A}_1$ such that $\pi(\rho(p_n)) = p_{n-2}$. Let p_∞ be the strong limit of $\{p_n\}$. Set

$$z_0 = \sum_{n=1}^{\infty} (p_{2n-1} - p_{2n}) + p_\infty,$$

$$z_1 = \sum_{n=1}^{\infty} (p_{2n} - p_{2n+1}), \qquad z_2 = \sum_{n=1}^{\infty} (p_{2n-2} - p_{2n-1}).$$

Then $z_0 + z_1 = p$ and $z_0 + z_2 = 1$. Moreover, since $\pi \circ \rho$ is an isomorphism of $(p_{2n} - p_{2n+1})\mathscr{A}_1$ onto $(p_{2n-2} - p_{2n-1})\mathscr{A}_1$ for each n, we conclude that $z_1\mathscr{A}_1$ is isomorphic to $z_2\mathscr{A}_1$; whence $p\mathscr{A}_1$ is isomorphic to $\mathscr{A}_1$. As ρ is an isomorphism of $p\mathscr{A}_1$ onto $\mathscr{A}_2$, the argument is completed.

4.6.5. LEMMA. *Let $\mathscr{A}_1$, $\mathscr{A}_2$ and $\mathscr{A}_3$ be Borel *-algebras, and let $\pi_i: \mathscr{A}_i \to \mathscr{A}_3$, $i = 1, 2$ be surjective sequentially normal morphisms. If $\mathscr{A}_1$ is standard there exists a sequentially normal morphism $\lambda: \mathscr{A}_1 \to \mathscr{A}_2$ such that $\pi_2 \circ \lambda = \pi_1$. If λ' is another such morphism, there is a projection p_2 in $\mathscr{A}_2$, with*

$\pi_2(1 - p_2) = 0$, *such that* $p_2\lambda(x) = p_2\lambda'(x)$ *for all* x *in* $\mathscr{A}_1$; *so that* λ *is 'essentially unique'.*

Proof. We may assume that $\mathscr{A}_1 = \mathscr{B}$, where $\mathscr{B}$ is the enveloping Borel *-algebra for a separable C^*-algebra A. Since π_2 is surjective, we can find a separable C^*-subalgebra B of $\mathscr{A}_2$ such that $\pi_2(B) \supset \pi_1(A)$. Let p be the strong limit in $\mathscr{A}_2$ of a countable approximate unit for the separable C^*-algebra $B \cap \ker \pi_2$. Then $\pi_2(p) = 0$ and $\pi_2|(1 - p)B(1 - p)$ is an isomorphism. For each x in A we can therefore define $\lambda_0(x)$ as the unique element in $(1 - p)B(1 - p)$ such that $\pi_2(\lambda_0(x)) = \pi_1(x)$. Clearly λ_0 is a morphism of A and so by 4.5.9 extends to a sequentially normal morphism $\lambda: \mathscr{B} \to \mathscr{A}_2$. We have $\pi_2(\lambda(x)) = \pi_1(x)$ for all x in $\mathscr{B}$, since this is true for all elements in the generating set A.

Suppose that $\lambda': \mathscr{B} \to \mathscr{A}_2$ was another sequentially normal morphism such that $\pi_2 \circ \lambda' = \pi_1$. Take a dense sequence $\{x_n\}$ in A_{sa} and for each n let $[\lambda(x_n) - \lambda'(x_n)]$ be the range projection of $\lambda(x_n) - \lambda'(x_n)$ (see 2.2.7). Let $1 - p_2 = \vee [\lambda(x_n) - \lambda'(x_n)]$. Then $\pi_2(1 - p_2) = 0$ since $\ker \pi_2$ is a monotone sequentially closed ideal in $\mathscr{A}_2$, and $p_2\lambda(x_n) = p_2\lambda'(x_n)$ for all n. However, the set $\{x \in \mathscr{B}_{sa} \mid p_2\lambda(x) = p_2\lambda'(x)\}$ is monotone sequentially closed and contains $\{x_n\}$. It therefore equals $\mathscr{B}_{sa}$, and the proof is complete.

4.6.6. THEOREM. *Let* $\mathscr{A}_1$ *and* $\mathscr{A}_2$ *be standard Borel *-algebras and let* $\pi_i: \mathscr{A}_i \to \mathscr{A}_3$, $i = 1, 2$ *be surjective sequentially normal morphisms onto a Borel *-algebra* $\mathscr{A}_3$. *Then for each* $i = 1, 2$ *there is a central projection* p_i *with* $\pi_i(1 - p_i) = 0$ *and an isomorphism* $\lambda: p_1\mathscr{A}_1 \to p_2\mathscr{A}_2$ *such that* $\pi_2 \circ \lambda = \pi_1$.

Proof. From 4.6.5 we obtain sequentially normal morphisms $\lambda_1: \mathscr{A}_1 \to \mathscr{A}_2$ and $\lambda_2: \mathscr{A}_2 \to \mathscr{A}_1$ such that $\pi_2 \circ \lambda_1 = \pi_1$ and $\pi_1 \circ \lambda_2 = \pi_2$. Set $\rho = \lambda_2 \circ \lambda_1$ and note that $\pi_1 \circ \rho = \pi_1$. Thus ρ is 'essentially' equal to the identity map, so by 4.6.5 there is a projection y_1 in $\mathscr{A}_1$, with $\pi_1(y_1) = 0$, such that $(1 - y_1)(x - \rho(x)) = 0$ for all x in $\mathscr{A}_1$. Since $\mathscr{A}_1$ is countably generated we may assume that y_1 is central, replacing it otherwise with $c(y_1)$. Similarly we can find a central projection z_1 in $\mathscr{A}_2$, with $\pi_2(z_1) = 0$, such that $(1 - z_1)(x - \lambda_1(\lambda_2(x)) = 0$ for all x in $\mathscr{A}_2$.

Set $y_2 = y_1 \vee c(\lambda_2(z_1))$ and $z_2 = z_1 \vee c(\lambda_1(y_1))$, and then inductively $y_{n+1} = y_n \vee c(\lambda_2(z_n))$, $z_{n+1} = z_n \vee c(\lambda_1(y_n))$. Since

$$\pi_1(c(\lambda_2(z_n))) = c(\pi_2(z_n)) \quad \text{and} \quad \pi_2(c(\lambda_1(y_n))) = c(\pi_1(y_n)),$$

we see that $\pi_1(y_n) = \pi_2(z_n) = 0$ for all n. Moreover, $y_{n+1} \geqslant \lambda_2(z_n)$ and $z_{n+1} \geqslant \lambda_1(y_n)$. Define $p_1 = 1 - \vee y_n$ and $p_2 = 1 - \vee z_n$. Then $\pi_1(1 - p_1) = \pi_2(1 - p_2) = 0$ and $1 - p_1 \geqslant \lambda_2(1 - p_2)$, $1 - p_2 \geqslant \lambda_1(1 - p_1)$. Furthermore,

$$p_1(x - \lambda_2(\lambda_1(x))) = 0 \quad \text{and} \quad p_2(x - \lambda_1(\lambda_2(x))) = 0,$$

for all x in $\mathscr{A}_1$ and $\mathscr{A}_2$, respectively. It follows that

$$p_1\lambda_2(p_2) = p_1\lambda_2(1-(1-p_2)) \geqslant p_1\lambda_2(1) - p_1(1-p_1)$$
$$\geqslant p_1\lambda_2(\lambda_1(1)) = p_1 1 = p_1,$$

whence $\lambda_2(p_2) \geqslant p_1$. Similarly $\lambda_1(p_1) \geqslant p_2$. Define $\lambda\, p_1\mathscr{A}_1 \to p_2\mathscr{A}_2$ by $\lambda(x) = p_2\lambda_1(x)$ and define $\mu: p_2\mathscr{A}_2 \to p_1\mathscr{A}_1$ by $\mu(x) = p_1\lambda_2(x)$. Then for each x in $p_1\mathscr{A}_1$

$$\mu(\lambda(x)) = p_1\lambda_2(p_2\lambda_1(x)) = p_1\lambda_2(p_2)\lambda_2(\lambda_1(x)) = p_1 x = x.$$

Similarly $\mu(\lambda(x)) = x$ for each x in $p_2\mathscr{A}_2$. Thus λ is an isomorphism, and clearly $\pi_2 \circ \lambda = \pi_1$.

4.6.7. COROLLARY. *Let (π_1, H) and (π_2, H) be separable representations of two separable C^*-algebras A_1 and A_2 with enveloping Borel *-algebras $\mathscr{B}_1$ and $\mathscr{B}_2$, respectively. If $\pi_1(A_1)'' = \pi_2(A_2)''$ there are central projections p_1 and p_2 in $\mathscr{B}_1$ and $\mathscr{B}_2$ with $\pi_1''(1-p_1) = \pi_2''(1-p_2) = 0$, and an isomorphism $\lambda: p_1\mathscr{B}_1 \to p_2\mathscr{B}_2$ such that $\pi_1''(x) = \pi_2''(\lambda(x))$ for all x in $p_1\mathscr{B}_1$.*

4.6.8. Any attempt to classify all C^*-algebras up to isomorphism is pointless. Already in the commutative case it would amount to a complete classification of all locally compact Hausdorff spaces up to homeomorphism, a project that no sane topologist would seriously consider. However, as we shall presently see, a classification of all Polish spaces, in particular all second countable locally compact Hausdorff spaces, up to Borel isomorphism is possible and extremely valuable.

Motivated by the commutative case we say that two separable C^*-algebras A_1 and A_2 are *Borel isomorphic* if their enveloping Borel *-algebras are isomorphic. In this case there is an affine Borel isomorphism of $S(A_1)$ onto $S(A_2)$ from which we easily conclude that $P(A_1)$ and $P(A_2)$ are Borel isomorphic and that $\hat{A}_1$ and $\hat{A}_2$ are isomorphic (as sets).

It follows from 4.6.2 that if I is a closed ideal in a separable C^*-algebra A then A is Borel isomorphic to $I \oplus (A/I)$. This produces some examples of non-isomorphic but Borel isomorphic C^*-algebras. The commutative case will give us many more.

4.6.9. Let T be a Borel space and $\mathscr{B}(T)$ the Borel *-algebra of bounded Borel functions on T (using $l^2(T)$ as a representing Hilbert space). We say that T is a *standard Borel space* if $\mathscr{B}(T)$ is a standard Borel *-algebra. Thus T is standard if $\mathscr{B}(T) = z\mathscr{B}$ where $\mathscr{B}$ is the enveloping Borel *-algebra for some separable C^*-algebra A. We may assume that A is commutative, replacing otherwise A by A/I, where $(A/I)\check{} = {}_1\check{A}$, see 4.4.10. Thus $\mathscr{B} = \mathscr{B}(\hat{A})$, and since there is a bijective correspondence between projections in $\mathscr{B}$ and Borel subsets of $\hat{A}$, we see that a Borel space T is standard if and only if T is a Borel subset of a second countable, locally compact Hausdorff space. It follows from 4.2.4 that every Borel subset of a Polish space is standard.

4.6.10. LEMMA. *Let T be a Polish space. For each Borel subset E of T there is a Polish space T_1 and a continuous bijection $f: T_1 \to E$.*

Proof. Let $\mathscr{P}$ be the class of Borel sets E in T which satisfy the condition in the lemma. By 4.2.2 $\mathscr{P}$ contains all open sets and all closed sets. Let $\{E_n\}$ be a sequence in $\mathscr{P}$ consisting of pairwise disjoint sets. It is clear that $\cup E_n \in \mathscr{P}$. Let $\{E_n\}$ be an arbitrary sequence from $\mathscr{P}$ and choose Polish spaces T_n and continuous bijections $f_n: T_n \to E_n$. The subspace F of the Polish space $\Pi\, T_n$ consisting of those points (t_n) such that $f_n(t_n) = f_1(t_1)$ for all n is closed. Projection onto the first coordinate followed by f_1 is a continuous bijection of F onto $\cap E_n$. It follows that $\cap E_n \in \mathscr{P}$. Thus $\mathscr{P}$ is closed under intersections and disjoint unions.

Let $\mathscr{P}_0$ denote the subset of $\mathscr{P}$ consisting of those sets E in $\mathscr{P}$ for which also $T\backslash E \in \mathscr{P}$. We see that $\mathscr{P}_0$ contains the open sets and that it is closed under complementation. The proof is completed when we show that $\mathscr{P}_0$ contains all Borel sets, and for this we only need to verify that $\mathscr{P}_0$ is closed under countable intersections. So take $\{E_n\}$ in $\mathscr{P}_0$. We proved, above, that $\cap E_n \in \mathscr{P}$. But

$$T\backslash \bigcap E_n = \bigcup T\backslash E_n = \bigcup_n (T\backslash E_n) \cap \left(\bigcap_{k=1}^{n-1} E_k\right),$$

which is a disjoint union of sets from $\mathscr{P}$, so that $T\backslash \cap E_n \in \mathscr{P}$ and consequently $\cap E_n \in \mathscr{P}_0$.

4.6.11. LEMMA. *Let $f: T \to E$ be a Borel map from a Polish space T into a second countable, compact space E. Then the graph of f is a Borel subset of $T \times E$, isomorphic with T.*

Proof. Define a Borel function $g: T \times E \to E \times E$ by $g(t, s) = (f(t), s)$. The graph $G(f)$ of f is then the counter image for g of the diagonal in $E \times E$. Since the latter is closed, $G(f)$ is a Borel set. By 4.6.10 $G(f)$ is the injective continuous image of a Polish space T_1 and projection onto the first coordinate gives a continuous bijection of $G(f)$ onto T. By 4.2.10, T, T_1 and $G(f)$ are Borel isomorphic.

4.6.12. PROPOSITION. *If $f: T \to E$ is an injective Borel map from a standard Borel space T into a countably generated Borel space E, then f is a Borel isomorphism from T onto $f(T)$ and $f(T)$ is a Borel set in E.*

Proof. If T_2 is compact and T_1 is a Borel space then T_1 is Borel isomorphic to a subset of T_2 (in its relative Borel structure) if and only if there is a sequentially normal morphism of $\mathscr{B}(T_2)$ onto $\mathscr{B}(T_1)$.

Using the above we see from 4.5.11 that E is Borel isomorphic to a subset of a second countable compact Hausdorff space. We may then as well assume that E itself is compact. Now the map f consists of a bijection of T onto the graph $G(f)$ of f, followed by the projection of $G(f)$ onto the second coordinate. By 4.6.11 the first part is a Borel isomorphism, and since the second is a continuous function we may as well assume that f itself is continuous. Finally by 4.6.10 we may assume that T is a Polish space. We conclude from 4.2.10 that f is a Borel isomorphism of T onto $f(T)$. By 4.6.3 the kernel of the ensuing sequentially normal morphism of $\mathscr{B}(E)$ onto $\mathscr{B}(T)$ is a projection in $\mathscr{B}(E)$, i.e. $f(T)$ is a Borel set of E.

4.6.13. THEOREM. *If two standard Borel spaces have the same cardinality* $(1, 2, \ldots, \infty$, or $2^{\infty})$ *then they are Borel isomorphic.*

Proof. The countable Borel spaces are discrete and presents no problem. We will show that any uncountable standard space T is Borel isomorphic to $\boldsymbol{N}^{\infty}$. For this we may assume, by 4.6.9 and 4.6.10 that T is a Polish space. Since by 4.2.8 and 4.2.12 we have a Borel injection of T into $\boldsymbol{N}^{\infty}$ it suffices to find a continuous injection of $\boldsymbol{N}^{\infty}$ into T, and then apply 4.6.12. After that the result is immediate from 4.6.4.

The set of points in T which have a countable neighbourhood is open and countable. Delete it.

Choose a complete metric on T and define inductively a system of closed balls $F_{n.\lambda}, n \in \boldsymbol{N}, \lambda \in \boldsymbol{N}^n$ in T such that (with $\lambda(n)$ denoting the canonical projection of $\lambda \in \boldsymbol{N}^m, n \leqslant m \leqslant \infty$, into $\boldsymbol{N}^n$):

(i) $F_{n.\lambda} \cap F_{n.\mu} = \varnothing$ if $\lambda \neq \mu$;

(ii) $F_{n+1.\lambda} \subset F_{n.\lambda(n)}$;

(iii) $0 < \operatorname{diam}(F_{n.\lambda}) \leqslant 2^{-n}$.

Then for each λ in $\boldsymbol{N}^{\infty}$ define $f(\lambda)$ as the unique point in $\cap_n F_{n.\lambda(n)}$. It is evident that f is injective and continuous. The proof is complete.

4.6.14. Notes and remarks. In measure theory a Borel space is called standard if it is Borel isomorphic to a Polish space. This is equivalent to being Borel isomorphic to a second countable (locally) compact Hausdorff space, cf. 4.6.13. Thus the direct non-commutative analogue of standard spaces are the enveloping Borel *-algebras of separable C^*-algebras. We have chosen the more general definition in 4.6.1 because it involves no extra difficulties. It is even quite possible that every standard Borel *-algebra is the enveloping Borel *-algebra for some C^*-algebra.

Proposition 4.6.12 is a celebrated result of Souslin. In the (rather perverse) proof given here it is clear that it consists of two parts: (1) $f: T \to f(T)$ is a Borel isomorphism, (2) $f(T)$ is a Borel subset. Note that the second half is a corollary of 4.6.3, i.e. belongs to the C^*-algebra theory. The first half is presumably a truly 'Polish' theorem. Corollary 4.6.7 is a generalization of von Neumann's lifting theorem: If T_1 and T_2 are standard spaces with measures μ_1, μ_2 such that $L^\infty_{\mu_1}(T_1)$ is isomorphic to $L^\infty_{\mu_2}(T_2)$, there are null sets N_1, N_2 and a Borel isomorphism $\phi: T_1 \backslash N_1 \to T_2 \backslash N_2$. The non-commutative results in this section are extracted from [201] and [203].

4.7. Borel structures on the spectrum

4.7.1. Throughout this section A denotes a separable C^*-algebra, and $\mathscr{B}$ its enveloping Borel $*$-algebra. We let $\mathscr{C}$ denote the centre of $\mathscr{B}$, so that $\mathscr{C}$ is a commutative, sequentially closed C^*-algebra. Since $\mathscr{C} \subset \mathscr{U}(A)$ we see from 4.3.15 and 4.4.2 that the map $x \to \check{x}$ is an isomorphism of $\mathscr{C}$ onto a class of functions on $\hat{A}$. We say that a subset F in $\hat{A}$ is a D-*Borel* set if $\check{x}$ is the characteristic function of F for some x in $\mathscr{C}$. In other words we set up a bijective correspondence between the class of projections in $\mathscr{C}$ and the class of D-Borel sets in $\hat{A}$. Since $\mathscr{C}$ is commutative and sequentially closed its projections form a Boolean algebra which is closed under countable unions and intersections. Thus the D-Borel sets is a σ-algebra of sets, which we call the D-*Borel structure* on $\hat{A}$ (the D is for Davies). Since $\mathscr{C}$ is generated by its projections by 4.5.7, we may now identify $\mathscr{C}$ with the algebra $\mathscr{B}(\hat{A})$ of D-Borel measurable, bounded functions on $\hat{A}$.

A word of caution concerning the above definition is in place: the D-Borel structure does not necessarily arise from a topology on $\hat{A}$. However, since $\mathscr{B}$ is so obviously the correct space free analogue of the Borel functions we feel justified in our abuse of the term Borel structure.

4.7.2. The D-Borel structure is not the only possible σ-algebra of sets on $\hat{A}$. We define the *topological Borel structure* (T-Borel) as the σ-algebra generated by the open sets in the Jacobson topology. We define the *Mackey Borel structure* (M-Borel) as the σ-algebra of sets in $\hat{A}$, whose counter-images in $P(A)$ under the canonical map from $P(A)$ onto $\hat{A}$ are Borel sets in $P(A)$ (with the weak* topology).

4.7.3. PROPOSITION. *The T-Borel structure is weaker than the D-Borel structure, which in turn is weaker than the M-Borel structure.*

Proof. Each open set in $\hat{A}$ corresponds to an element in $\mathscr{C}$ by 4.4.6. Consequently each T-Borel function is a D-Borel function.

Each D-Borel function corresponds to a bounded affine function on the quasi-state space Q of A which is obtained by sequential operations starting with the class A_{sa} of continuous affine functions on Q vanishing at zero (see

3.10.3). Consequently each element in $\mathscr{C}$ corresponds to an affine Borel function on Q, thus (by restriction) to a Borel function on $P(A)$. This proves that the canonical map $P(A) \to \hat{A}$ is measurable when $\hat{A}$ is equipped with the D-Borel structure. Since the M-Borel structure is the strongest for which the map $P(A) \to \hat{A}$ is Borel measurable, it is stronger than the D-Borel structure.

4.7.4. PROPOSITION. *Each point in $\hat{A}$ is a D-Borel set.*

Proof. Let ϕ be a pure state of A regarded as a normal functional on A''. There is then by 3.13.6 an open projection p in A'' with $\phi(p) = 0$ such that $1 - p$ is a minimal projection in A''. Let $q = c(1 - p)$. Clearly q is the minimal central projection for which $\phi(q) = 1$, whence $q = c(\pi_\phi)$ (see 3.8.1). We may identify q with the point in $\hat{A}$ corresponding to (π_ϕ, H_ϕ) by 3.8.13. But $q \in \mathscr{C}$ by 4.5.8 since $p \in (A_+)^m$. (In fact, $q \in ((\tilde{A}_+)_\delta)_{\sigma}$.) Since the map $(P(A) \to \hat{A})$ is surjective the proposition follows.

4.7.5. For each state ϕ of A let μ_ϕ be the D-Borel probability measure on $\hat{A}$ given by $\mu(F) = \phi(\check{x})$, where $\check{x}$ is the characteristic function of F and ϕ is regarded as a normal functional on A'' (thus a σ-normal functional on $\mathscr{C}$). We say that μ_ϕ is the *central measure* associated with ϕ. It is immediate from the definition that the map $\phi \to \mu_\phi$ is affine and continuous from the state space of A with the norm topology to the space of D-Borel probability measures on $\hat{A}$ with the norm topology. Under this map the pure states are mapped onto the set of point measures of $\hat{A}$.

4.7.6. PROPOSITION. *For each state ϕ of A with central measure μ_ϕ there is a natural isomorphism from $L^\infty_{\mu_\phi}(\hat{A})$ onto the centre of $\pi_\phi(A)''$, such that the diagram below is commutative.*

$$\begin{array}{ccc} \mathscr{C} & \xrightarrow{\pi''_\phi} & \pi''_\phi(\mathscr{C}) \\ \uparrow & & \uparrow \\ \mathscr{B}(\hat{A}) & \longrightarrow & L^\infty_{\mu_\phi}(\hat{A}) \end{array}$$

Proof. Since A is separable, H_ϕ is separable, thus $\pi''_\phi(\mathscr{C})$ is the centre of $\pi_\phi(A)''$ by 4.5.10. If $x \in \mathscr{C}$ then $\pi''_\phi(x) = 0$ if and only if $\phi(|x|) = 0$. Thus $\pi''_\phi(x) = 0$ if and only if $\check{x}$ is a null function for μ_ϕ. From this the isomorphism between $L^\infty_{\mu_\phi}(\hat{A})$ and $\pi''_\phi(\mathscr{C})$ is immediate.

4.7.7. Recall that a *spectral measure* is a map μ from a σ-algebra $\mathscr{S}$ of subsets of a space T into a family of mutually commuting projections on a Hilbert space H such that μ is a σ-normal homomorphism from the Boolean algebra $\mathscr{S}$ into the orthocomplemented lattice of projections in $B(H)$. A good deal of the mystery that surrounds the concept of a spectral measure evaporates when it is realized that by a simple construction μ can be extended to a σ-normal morphism from the C^*-algebra $\mathscr{B}(T)$ of bounded, $\mathscr{S}$-

measurable functions on T onto the (necessarily commutative) sequentially closed C^*-algebra generated by $\mu(\mathscr{S})$. If $f \in \mathscr{B}(T)$ we then use the suggestive notation $\int f \, d\mu$ for the image of f in $\boldsymbol{B}(H)$. Conversely, each σ-normal morphism from $\mathscr{B}(T)$ into $\boldsymbol{B}(H)$ gives by restriction to the family $\mathscr{S}$ of characteristic functions in $\mathscr{B}(T)$ a spectral measure on T (with respect to the σ-algebra $\mathscr{S}$). From these considerations we immediately deduce the following proposition.

4.7.8. PROPOSITION. *For each separable representation* (π, H) *of* A *there is a unique spectral measure* μ *on the D-Borel subsets of* $\hat{A}$ *with values in* $\boldsymbol{B}(H)$ *such that* $\pi''(x) = \int \check{x} \, d\mu$ *for each* x *in* $\mathscr{C}$.

4.7.9. Let (π, H) be a separable representation of A. We define the *null sets* of (π, H) as those D-Borel sets N in $\hat{A}$ for which $\mu(N) = 0$, where μ is the spectral measure associated with (π, H). It is clear that the null sets of (π, H) form a σ-ring on $\hat{A}$. We say that (π, H) is *concentrated* on a D-Borel set F in $\hat{A}$ if $\hat{A} \backslash F$ is a null set.

If p is a projection in $\mathscr{C}$ then, by 4.7.8, p is the characteristic function for a null set of (π, H) if and only if $\pi''(p) = 0$. Thus if $c(\pi)$ is the central cover of (π, H) (see 3.8.1) then $\{\check{p} = 1\}$ is a null set in $\hat{A}$ if and only if $c(\pi) \perp p$.

It is clear from 4.7.6 that if ϕ is a state of A with central measure μ_ϕ then the null sets of (π_ϕ, H_ϕ) are precisely the null sets for the scalar measure μ_ϕ. Since by 3.8.6 each separable representation of A is equivalent to a cyclic representation, and since equivalent representations have the same null sets (since they have the same central cover) we see that, conversely, the null sets of a separable representation are always the null sets for some D-Borel probability measure on $\hat{A}$.

4.7.10. THEOREM. *Let* A *be a separable* C^**-algebra. Two separable representations of* A *are equivalent if and only if they have the same null sets in* $\hat{A}$. *They are disjoint if they are concentrated on disjoint sets in* $\hat{A}$.

Proof. If two representations are equivalent they obviously have the same null sets. Assume now that (π_1, H_1) and (π_2, H_2) are separable representations and have the same null sets. Then $(\pi_1 \oplus \pi_2, H_1 \oplus H_2)$ is a separable representation and $c(\pi_1 \oplus \pi_2) = c(\pi_1) \vee c(\pi_2)$. By 4.5.10 there is a projection p in $\mathscr{C}$ such that

$$(\pi_1 \oplus \pi_2)''(p) = (\pi_1 \oplus \pi_2)''(c(\pi_1)(1 - c(\pi_2))).$$

This means that

$$(c(\pi_1) \vee c(\pi_2))p = (c(\pi_1) \vee c(\pi_2))c(\pi_1)(1 - c(\pi_2)) = c(\pi_1)(1 - c(\pi_2)).$$

It follows that $c(\pi_2)p = 0$. Thus by assumption $c(\pi_1)p = 0$. Consequently $(c(\pi_1) \vee c(\pi_2))p = 0$ and $c(\pi_1)(1 - c(\pi_2)) = 0$. This implies that

$c(\pi_1) \leqslant c(\pi_2)$, and by a symmetric argument $c(\pi_2) \leqslant c(\pi_1)$, so that $c(\pi_1) = c(\pi_2)$ and (π_1, H_1) is equivalent to (π_2, H_2) by 3.8.2.

If (π_1, H_1) and (π_2, H_2) are separable representations, and (π_1, H_1) is concentrated on a null set for (π_2, H_2) then there is a projection q in $\mathscr{C}$ such that $\pi_2''(q) = 0$ and $\pi_1''(1 - q) = 0$. Consequently $c(\pi_2)q = 0$ and $c(\pi_1)(1 - q) = 0$, whence $c(\pi_1) \perp c(\pi_2)$ and (π_1, H_1) and (π_2, H_2) are disjoint. Conversely, if (π_1, H_1) and (π_2, H_2) are disjoint then $(\pi_1 \oplus \pi_2, H_1 \oplus H_2)$ is a separable representation and

$$c(\pi_1 \oplus \pi_2) = c(\pi_1) + c(\pi_2)$$

(cf. 3.8.11). By 4.5.11 there is a projection q in $\mathscr{C}$ such that

$$(\pi_1 \oplus \pi_2)''(q) = (\pi_1 \oplus \pi_2)''(c(\pi_1)).$$

This means that

$$(c(\pi_1) + c(\pi_2))q = (c(\pi_1) + c(\pi_2))c(\pi_1) = c(\pi_1).$$

Consequently $c(\pi_1)q = c(\pi_1)$ and $c(\pi_2)q = 0$ so that (π_1, H_1) is concentrated on q while (π_2, H_2) is concentrated on $1 - q$.

4.7.11. Notes and remarks. Mackey originally introduced his Borel structure as a quotient structure from the set of irreducible representations on a fixed Hilbert space, see Mackey [164]. The definition used here has technical advantages and gives the same result, cf. 4.10.10 and 4.10.13. Davies introduced his Borel structure in [50], using the centre of $\mathscr{B}^w$ (cf. 4.5.14), and established 4.7.3. Proposition 4.7.8 is due to the author [198] and generalizes a result of Glimm [98]. It leads immediately to 4.7.10 which is Kadison's result [129] (more respected than beloved). It has here been coaxed in sweeter words than in the original version. The proof is taken from [198].

4.8. Disintegration on the factor spectrum

4.8.1. From the point of view of classifying representations the result obtained in 4.7.10 is most satisfying: it gives a parametrization of equivalence classes of separable representations by means of classes of null sets in the spectrum, in exact analogue with the commutative case. However, if one regards the spectral measure of a representation as a possible tool for 'composing' the representation from its irreducible parts, then the situation is much less clear. The main source of trouble is the existence of spectral measures that are extremal without corresponding to a point in the spectrum. These can of course only arise for C^*-algebras A for which the D-Borel structure on $\hat{A}$ is not countably generated. A moments reflection reveals that although $\mathscr{B}$ is countably generated (since A is separable), its centre $\mathscr{C}$ need not be. As a matter of fact for most C^*-algebras (the exception to be treated in Chapter 6) the D-Borel structure is not countably generated, and there is an abundance of extremal measures which are not point measures (corresponding to minimal central projections in A'' which do not belong to $\mathscr{C}$, cf. 3.8.13. and 4.7.4). We shall show that if the spectrum of A is enlarged so that it contains all extremal spectral measures as points, then each representation can be 'composed' from these, and in a canonical manner.

4.8.2. Let A be a separable C^*-algebra. We denote by $F(A)$ the set of *factorial states* of A, i.e. the set of states ϕ of A for which (π_ϕ, H_ϕ) is a factor representation. By 4.7.6, $\phi \in F(A)$ if and only if μ_ϕ is an extremal D-Borel measure. We denote by $\widehat{A}$ the set of equivalence classes of factor representations of A, and say that $\widehat{A}$ is the *factor spectrum* of A. (The usual notion is *quasi-spectrum*, but there is nothing 'quasi' about $\widehat{A}$). Note that $P(A) \subset F(A)$ and that $\hat{A} \subset \widehat{A}$.

The arguments in 4.4.1 show that each element x in A''_{sa} defines a bounded real function $\check{x}$ on $\widehat{A}$ and that $\check{x}$ on $\widehat{A}$ is an extension of the function $\check{x}$ on $\hat{A}$ defined in 4.4.2. In particular, the map $x \to \check{x}$ is an isomorphism of $\mathscr{C}$ onto a class of functions on $\widehat{A}$. We define the D-*Borel structure* on $\widehat{A}$ so that the D-Borel sets correspond to the projections in $\mathscr{C}$. This creates the, at first, slightly bewildering situation that $\hat{A}$ and $\widehat{A}$ are non-isomorphic spaces, but have exactly the same Borel sets. Note though that by 4.5.13 no non-empty subsets of $\widehat{A} \backslash \hat{A}$, in particular no points in $\widehat{A} \backslash \hat{A}$ are Borel sets. However, if t_1 and t_2 are distinct points in $\widehat{A}$ then by 4.7.10 there is a D-Borel set N in $\widehat{A}$ such that $t_1 \in N$ and $t_2 \notin N$.

For each state ϕ of A we shall from now on regard its central measure μ_ϕ (see 4.7.5) as a D-Borel measure on $\widehat{A}$. Now each extremal central measure on $\widehat{A}$ is a point measure.

4.8.3. PROPOSITION. *If A is a separable C^*-algebra then $F(A)$ is an $F_{\sigma\delta}$-subset of S and consequently also an $F_{\sigma\delta}$-subset of Q.*

Proof. Let $\{x_k\}$ be a dense sequence in A. If n, m, l and k are natural numbers, define G_{nmlk} as the set of states ϕ for which there is a y in A (depending on n, m, l and k) with $\|y\| \leqslant 1$ such that

$$|\phi(x_h^*(yx_i - x_iy)x_j)| < 1/l \qquad \text{and} \qquad |\phi(yx_m) - \phi(y)\phi(x_m)| > 1/n,$$

for all h, i and j between l and k. It is clear that G_{nmlk} is an open subset of S (in the weak* topology). Put $H_{nm} = \cap_{lk} G_{nmlk}$. If $\phi \in H_{nm}$ the net $\{\pi_\phi(y)\}$ (indexed by k and l) has a weak limit point z in the unit ball of $\pi_\phi(A)''$. Since

$$0 = \phi(x_h^*(zx_i - x_iz)x_j) = ((z\pi_\phi(x_i) - \pi_\phi(x_i)z)\xi_{x_j} \mid \xi_{x_h})$$

for all h, i and j, we conclude that z belongs to the centre of $\pi_\phi(A)''$. However,

$$|\phi(zx_m) - \phi(z)\phi(x_m)| \geqslant 1/n,$$

which means that z is not a scalar multiple of the identity. Consequently $\phi \notin F(A)$.

Conversely, if $\phi \in S \backslash F(A)$ there is a non-trivial central projection z in $\pi_\phi(A)''$. Since ϕ is faithful on the centre of $\pi_\phi(A)''$ (cf. 4.7.6) this implies that $\phi(z) > 0$ and $\phi(1-z) > 0$, whence $\phi(z) \neq \phi(z)^2$. Approximating z weakly with elements from $\{\pi_\phi(x_n)\}$ we therefore find an x_m and n such that

$$|\phi(z\pi_\phi(x_m)) - \phi(z)\phi(x_m)| > 1/n.$$

Since by 2.3.3 z can also be approximated weakly with elements $\pi_\phi(y)$, $\|y\| \leq 1$ we conclude that $\phi \in H_{nm}$.

We have proved that $S \backslash F(A) = \cup_{nm} H_{nm}$ so that $S \backslash F(A)$ is a $G_{\delta\sigma}$-subset of S. Thus $F(A)$ is an $F_{\sigma\delta}$-subset of S, and since S is a G_δ-subset of Q (using 3.1.4 plus the fact that A contains a countable approximate unit), $F(A)$ is an $F_{\sigma\delta}$-subset of Q as well.

4.8.4. PROPOSITION. *The canonical map: $\phi \to \pi_\phi$ induces a surjective Borel map: $\phi \to \hat{\phi}$ from $F(A)$ in its relative Borel structure onto $\hat{A}$ with the* D-Borel *structure.*

Proof. If (π, H) is a factor representation of A take a unit vector ξ in H and let ϕ be the vector state of A determined by ξ. The representation (π_ϕ, H_ϕ) of A extends canonically to a normal representation of $\pi(A)''$ (since ϕ has a normal extension to $\pi(A)''$) and since $\pi(A)''$ is a factor any normal representation is faithful by 2.5.5. Consequently (π, H) is equivalent to (π_ϕ, H_ϕ). This shows that the map $F(A) \to \hat{A}$ is surjective.

Since each counter-image of a D-Borel set is of the form $\{\phi \in F(A) \mid \phi(p) \in E\}$ for some Borel set E in $\boldsymbol{R}$ and some projection p in $\mathscr{C}$ we conclude as in 4.7.3 that the map $\phi \to \hat{\phi}$ is a Borel map from $F(A)$ to $\hat{A}$.

4.8.5. LEMMA. *For each state ϕ of A with central measure μ_ϕ there is a unique positive, sequentially normal, linear map ρ from $\mathscr{B}$ onto $L^\infty_{\mu_\phi}(\hat{A})$ such that*

$$\int \check{z}\rho(x)\,d\mu_\phi = \phi(zx), \qquad z \in \mathscr{C}, \qquad x \in \mathscr{B},$$

and ρ is an extension of the natural morphism of $\mathscr{C}$ onto $L^\infty_{\mu_\phi}(\hat{A})$.

Proof. If $z \in \mathscr{C}$ and $x \in \mathscr{B}$ then with $z = u|z|$ (2.2.9) we have

$$|\phi(zx)|^2 = |\phi(|z|^{1/2}|z|^{1/2}ux)|^2 \leqslant \phi(|z|)\phi(|z|^{1/2}x^*u^*ux|z|^{1/2})$$
$$\leqslant \phi(|z|)^2\|x\|^2$$

by the Cauchy–Schwarz inequality (3.1.3). It follows that the map $z \to \phi(zx)$ defines a unique element in $L^\infty_{\mu_\phi}(\hat{A})$ (identifying $L^\infty_{\mu_\phi}(\hat{A})$ with the set of D-Borel measures on $\hat{A}$ dominated by a multiple of μ_ϕ). We let $\rho(x)$ be that element. It is immediate that the map $x \to \rho(x)$ has the properties claimed in the lemma.

4.8.6. LEMMA. *Let ϕ, μ_ϕ and ρ be as in 4.8.5. There is a positive linear sequentially normal map λ from $\mathscr{B}$ into $\mathscr{C}$ such that $\rho(\lambda(x)) = \rho(x)$ for all x in $\mathscr{B}$. If λ' is another such map, there is a null set N for μ_ϕ in $\hat{A}$ such that the functions $\lambda(x)\check{}$ and $\lambda'(x)\check{}$ agree on $\hat{A}\backslash N$ for all x in $\mathscr{B}$; so that λ is essentially unique.*

Proof. The argument from 4.6.5 can be repeated almost verbatim. The only change is that since ρ is a positive linear map and not a morphism we must now instead of 4.5.9 appeal to the fact that any positive linear map $\lambda_0 : A \to \mathscr{B}(\hat{A})$ extends uniquely to a positive sequentially normal map $\lambda : \mathscr{B} \to \mathscr{B}(\hat{A})$. This is so because each state $x \to \lambda_0(x)(t)$, $t \in \hat{A}$ on A extends to a sequentially normal state $x \to \lambda(x)(t)$, $t \in \hat{A}$ on $\mathscr{B}$.

4.8.7. THEOREM. *Let ϕ be a state of a separable C^*-algebra A, and let μ_ϕ be its central measure on $\hat{A}$. There is then an essentially unique Borel function $t \to \phi_t$ from $\hat{A}$ into $F(A)$ with its relative Borel structure such that $\phi(x) = \int \phi_t(x)\, d\mu_\phi(t)$ for all x in $\mathscr{B}$ and such that $\phi_t(z) = \check{z}(t)$ almost everywhere for each z in $\mathscr{C}$. The set $\{t \in \hat{A} \mid \hat{\phi}_t = t\}$ has outer measure 1.*

Proof. Choose λ as in 4.8.6. For each t in $\hat{A}$ define $\phi_t(x) = \lambda(x)(t)$ for all x in $\mathscr{B}$. Then $\phi_t(z) = \check{z}(t)$ almost everywhere for each z in $\mathscr{C}$, since $\rho(\lambda(z)) = \rho(z)$. In particular $\phi_t(1) = 1$ for almost all t. Changing the map $t \to \phi_t$ on a null set for μ_ϕ we may assume that $\phi_t \in S$ for all t in $\hat{A}$.

If $x \in \mathscr{B}$ the function $t \to \phi_t(x)$ is precisely $\lambda(x)\check{}$ and consequently D-Borel measurable. Furthermore,

$$\int \phi_t(x)\, d\mu_\phi(t) = \int \lambda(x)\check{}\;\, d\mu_\phi = \phi(x).$$

Since $C(Q)$ is generated by the continuous affine functions on Q we conclude from Kadison's representation theorem (3.10.3) that the map $t \to f(\phi_t)$ is D-Borel measurable for each f in $C(Q)$. Therefore $t \to \phi_t$ is a Borel map from $\hat{A}$ into S.

If $t \to \phi'_t$ was another map satisfying the conditions above then we may define $\lambda' : \mathscr{B} \to \mathscr{C}$ by $\lambda'(x)(t) = \phi'_t(x)$ for all x in $\mathscr{B}$ (identifying $\mathscr{C}$ and $\mathscr{B}(\hat{A})$). It is immediate that λ' is a map of the form described in 4.8.6, and it follows that $\phi_t = \phi'_t$ for almost all t in $\hat{A}$.

It remains to show that the ϕ_t's are factorial states and that $\hat{\phi}_t = t$ for a thick subset of t's in $\hat{A}$. This will demand considerable efforts, and we devote most of the next section (4.9) to the completion of the proof (see 4.9.8).

4.8.8. It is appropriate to show by a simple example that the claim in 4.8.7, that the set $\{t \in \hat{A} \mid \hat{\phi}_t = t\}$ is thick in $\hat{A}$, cannot be improved.

Let ϕ be a factorial state of A such that the class $\hat{\phi}$ of (π_ϕ, H_ϕ) belongs to $\check{A}\backslash\hat{A}$. Then μ_ϕ is the point measure at $\hat{\phi}$. Define $\phi_t = \phi$ for all t in $\hat{A}$. If $z \in \mathscr{C}$ then the set

$$\{t \in \hat{A} \mid \check{z}(t) \neq \check{z}(\hat{\phi})\}$$

is a null set, and for each x in $\mathscr{B}$ we have

$$\int \phi_t(x)\, \mathrm{d}\mu_\phi(t) = \int \phi(x)\, \mathrm{d}\mu_\phi = \phi(x).$$

Thus the constant map $t \to \phi$ satisfies the conditions in 4.8.7, and is therefore, except for a negligible change, the only such map. However,

$$\{t \in \hat{A} \mid \hat{\phi}_t = t\} = \{\hat{\phi}\},$$

and $\{\hat{\phi}\}$ is not a D-Borel set in $\hat{A}$. Worse still, the inner measure (with respect to μ_ϕ) of $\{\hat{\phi}\}$ is zero, while the outer measure of $\{\hat{\phi}\}$ is one.

4.8.9. Notes and remarks. The factor spectrum was introduced by Ernest in [86] in order to decompose an arbitrary (separable) representation in factor representations (cf. 4.12.4). The origin is of course von Neumann's reduction theory [177]. Theorem 4.8.7 is new, but the hard part of the proof (section 4.9.) relies on work by Sakai and Wils.

4.9. Disintegration on the state space

4.9.1. Let Q be a second countable, compact, convex set. If $\phi \in Q$ and ν is a Radon probability measure on Q such that $\int a\, \mathrm{d}\nu = a(\phi)$ for each continuous affine function a on Q, we say that ν *represents* ϕ, or that ϕ is the *barycentre* of ν. We let $M(\phi)$ denote the set of Radon, probability measures on Q with barycentre ϕ. If ν_1 and ν_2 belongs to $M(\phi)$ we write $\nu_1 \prec \nu_2$ if $\int a^2\, \mathrm{d}\nu_1 \leqslant \int a^2\, \mathrm{d}\nu_2$ for each continuous, affine function a on Q. This clearly defines a transitive relation in $M(\phi)$.

4.9.2. LEMMA. *Let Q, ϕ and $M(\phi)$ be as in 4.9.1. If $\nu \in M(\phi)$ such that $\nu \prec \nu'$ implies $\nu' \prec \nu$ for each ν' in $M(\phi)$ then ν is concentrated on the set of extreme points of Q.*

Proof. As in the proof of 4.3.2 let d be a metric on Q compatible with the topology. Consider the closed sets

$$F_n = \{(\psi_1, \psi_2) \in Q \times Q \mid \mathrm{d}(\psi_1, \psi_2) \geqslant 1/n\},$$

$$G_n = \{\psi \in Q \mid \psi = \tfrac{1}{2}(\psi_1 + \psi_2), (\psi_1, \psi_2) \in F_n\}.$$

Choose by 4.2.13 a measure $\tilde{\nu}$ on F_n such that $\tilde{\nu} \circ m^{-1} = \nu \mid G_n$, where m denotes the 'mid-point' map of F_n onto G_n. Define a measure ν' on Q by

$$\int f\, \mathrm{d}\nu' = \int_{Q \setminus G_n} f\, \mathrm{d}\nu + \tfrac{1}{2} \int_{F_n} (f(\psi_1) + f(\psi_2))\, \mathrm{d}\tilde{\nu},$$

for each f in $C(Q)$. If a is affine then $\int a\,\mathrm{d}\nu' = \int a\,\mathrm{d}\nu$, so that $\nu' \in M(\phi)$. Furthermore, since a^2 is a convex function on Q:

$$\int a^2\,\mathrm{d}\nu' - \int_{Q\setminus G_n} a^2\,\mathrm{d}\nu = \tfrac{1}{2}\int (a^2(\psi_1) + a^2(\psi_2))\,\mathrm{d}\tilde{\nu}$$

$$\geqslant \int a^2(\tfrac{1}{2}(\psi_1 + \psi_2))\,\mathrm{d}\tilde{\nu} = \int_{G_n} a^2\,\mathrm{d}\nu$$

Thus $\nu \prec \nu'$ whence by assumption $\nu' \prec \nu$. It follows that for each continuous affine function a on Q:

$$0 = \tfrac{1}{2}\int (a^2(\psi_1) + a^2(\psi_2))\,\mathrm{d}\tilde{\nu} - \int a^2(\tfrac{1}{2}(\psi_1 + \psi_2))\,\mathrm{d}\tilde{\nu}$$

$$= \tfrac{1}{4}\int (a(\psi_1) - a(\psi_2))^2\,\mathrm{d}\tilde{\nu}.$$

Since the affine functions separates points in Q we conclude that $\tilde{\nu} = 0$, whence $\nu(G_n) = 0$. But the set of extreme points in Q is the complement of the union of all the G_n.

4.9.3. Let A be a separable C^*-algebra with quasi-state space Q. For each state ϕ of A we define the S-*measure* ν_ϕ of ϕ (S for Sakai) to be the image measure of the central measure μ_ϕ on $\hat{A}$ under the Borel map $t \to \phi_t$ from $\hat{A}$ to Q defined in 4.8.7. This Radon probability measure depends only on ϕ since the map $t \to \phi_t$ is essentially unique. With $\hat{x}$ the element in $A_0(Q)$ determined by x in A_{sa} (see 3.10) we see from 4.8.5 that ν_ϕ can also be defined as the unique Radon measure such that

$$\int (\hat{x}_1\hat{x}_2\ldots\hat{x}_n)\mathrm{d}\nu_\phi = \int \rho(x_1)\rho(x_2)\ldots\rho(x_n)\,\mathrm{d}\mu_\phi,$$

with $x_1, x_2, \ldots, x_n$ in A_{sa} (since the algebra generated by $A(Q)$ is dense in $C(Q)$). By σ-normality the same formula is valid for $x_1, x_1, \ldots, x_n$ in $\mathscr{B}_{\mathrm{sa}}$. If $x \in \mathscr{B}_{\mathrm{sa}}$ and $z \in \mathscr{C}_{\mathrm{sa}}$ then by 4.8.5

$$\int \hat{z}\hat{x}\,\mathrm{d}\nu_\phi = \int \rho(z)\rho(x)\,\mathrm{d}\mu_\phi = \phi(zx) = \int \rho(zx)\,\mathrm{d}\mu_\phi = \int (zx)\hat{}\,\mathrm{d}\nu_\phi.$$

In particular (taking $z = 1$), ν_ϕ has barycentre ϕ. Since ϕ is a state each measure in $M(\phi)$, including ν_ϕ, is concentrated on S by 3.1.4. Note that the map $\phi \to \nu_\phi$ carries no obligation to be an affine map from S into the space of probability measures on S.

4.9.4. PROPOSITION. *Let ϕ be a state of a separable C^*-algebra A such that $\pi_\phi(A)'$ is commutative. Then the S-measure of ϕ is concentrated on the set of pure states of A.*

Proof. Let ν be an element of $M(\phi)$ such that $\nu_\phi \prec \nu$. Using the Cauchy–Schwarz inequality we have for each x in $\mathscr{B}_{\mathrm{sa}}$

$$\int (\hat{x})^2\,\mathrm{d}\nu_\phi \leqslant \int (\hat{x})^2\,\mathrm{d}\nu \leqslant \int (x^2)\hat{}\,\mathrm{d}\nu = \int (x^2)\hat{}\,\mathrm{d}\nu_\phi.$$

But if $z \in \mathscr{C}_{sa}$ we know from 4.9.3 that

$$\int(\hat{z})^2 \, dv_\phi = \int(z^2)\hat{} \, dv_\phi,$$

and we conclude that

$$\int(\hat{z})^2 \, dv_\phi = \int(\hat{z})^2 \, dv = \int(z^2)\hat{} \, dv.$$

Since $(y + z)^2 = y^2 + z^2 + 2yz$ this implies that

$$(*) \qquad \int \hat{y}\hat{z} \, dv_\phi = \int \hat{y}\hat{z} \, dv = \int(yz)\hat{} \, dv$$

for all y, z in $\mathscr{C}_{sa}$.

For each x in $\mathscr{B}$ with $0 \leqslant x \leqslant 1$ define a positive functional ϕ_x by

$$\phi_x(y) = \int \hat{x}\hat{y} \, dv, \qquad y \in \mathscr{B}.$$

Since $\phi_x \leqslant \phi$ and since by assumption

$$\pi_\phi(A)' = \pi_\phi(A)' \cap \pi_\phi(A)'' = \pi_\phi(\mathscr{C})$$

(using 4.5.10), there is by 3.3.5 an element h_x in $\mathscr{C}$ with $0 \leqslant h_x \leqslant 1$ such that $\phi_x(y) = \phi(h_x y)$ for all y in $\mathscr{B}$. If now $y, z \in \mathscr{C}$ and $0 \leqslant z \leqslant 1$ then by $(*)$

$$\int \hat{z}\hat{y} \, dv = \phi(h_z y) = \int(h_z y)\hat{} \, dv = \int \hat{h}_z \hat{y} \, dv.$$

Taking $y = z - h_z$ this implies that $\hat{z} - \hat{h}_z$ is a null function for v. But the relation $v_\phi \prec v$ implies that if $a \in \mathscr{B}_{sa}$ and $\hat{a}$ is a null function for v then it is also a null function for v_ϕ. Thus for any y in $\mathscr{B}_{sa}$ we have

$$(**) \qquad \int \hat{z}\hat{y} \, dv = \phi(h_z y) = \int(h_z y)\hat{} \, dv_\phi = \int \hat{h}_z \hat{y} \, dv_\phi = \int \hat{z}\hat{y} \, dv_\phi.$$

Now take x in $\mathscr{B}$ with $0 \leqslant x \leqslant 1$. Then by $(**)$

$$\int \hat{x}\hat{y} \, dv = \phi(h_x y) = \int \hat{h}_x \hat{y} \, dv_\phi = \int \hat{h}_x \hat{y} \, dv.$$

Taking $y = x - h_z$ we see that $\hat{x} - \hat{h}_x$ is a null function for v, and thus for v_ϕ as well. This finally shows that

$$\int(\hat{x})^2 \, dv = \phi(h_x x) = \int \hat{h}_x \hat{x} \, dv_\phi = \int(\hat{x})^2 \, dv_\phi$$

whenever $0 \leqslant x \leqslant 1$. Applying a linear transformation we see that it holds for any x, i.e. $v \prec v_\phi$. By 4.9.2 this implies that v_ϕ is concentrated on the set of pure states of A.

4.9.5. LEMMA. *Let B be a separable C^*-algebra generated by two commuting C^*-subalgebras A and C, and assume that B has a unit contained in $A \cap C$. The restriction map $r: \psi \to \psi \mid A$ from $S(B)$ onto $S(A)$ satisfies $r(F(B)) \subset F(A)$ ($r(P(B)) \subset P(A)$ if C is commutative), and determines a D-Borel map $\hat{r}$ from $\hat{B}$ into $\hat{A}$. If $\phi = r(\psi)$ then $\mu_\phi = \mu_\psi \circ \hat{r}^{-1}$.*

Proof. We may regard A'' as the weak closure of A in B''. Then $\mathscr{B}(A) \subset \mathscr{B}(B)$. But as A and C commute we also have $\mathscr{C}(A) \subset \mathscr{C}(B)$. Since $\phi \in F(A)$ (respectively $F(B)$) if and only if $\phi \mid \mathscr{C}(A)$ (respectively $\phi \mid \mathscr{C}(B)$) is a complex homomorphism we see that $r(F(B)) \subset F(A)$. Moreover, two factorial states ψ_1 and ψ_2 of B correspond to the same point in $\hat{B}$ if and only if $\psi_1 \mid \mathscr{C}(B) = \psi_2 \mid \mathscr{C}(B)$, which shows that r gives rise to a map $\hat{r}$ of $\hat{B}$ into $\hat{A}$. Since $\mathscr{C}(A) \subset \mathscr{C}(B)$, $\hat{r}$ is a Borel map. It is immediate that $\mu_\phi = \mu_\psi \circ \hat{r}^{-1}$ whenever $\phi = r(\psi)$.

If C is commutative and $\psi \in P(B)$ then from the above $\psi \mid C \in F(C) = \hat{C}$, which means that $\pi_\psi(C)$ is one-dimensional in $B(H_\psi)$. Since $\pi_\psi(B)'' = B(H_\psi)$ we conclude that $\pi_\psi(A)'' = B(H_\psi)$, whence $r(\psi) \in P(A)$ by 3.13.2.

4.9.6. LEMMA. *Let A, B and C be as in 4.9.5. If $\psi \in S(B)$, with $\phi = r(\psi)$, such that the natural injection of $L^\infty_{\mu_\phi}(\hat{A})$ into $L^\infty_{\mu_\psi}(\hat{B})$ is an isomorphism (i.e. if $\pi''_\psi(\mathscr{C}(A)) = \pi''_\psi(\mathscr{C}(B))$) then $\nu_\phi = \nu_\phi \circ r^{-1}$, where ν_ϕ and ν_ψ are the S-measures of ϕ and ψ on $S(A)$ and $S(B)$, respectively.*

Proof. Consider the commutative diagram

$$\begin{array}{ccccc} \mathscr{C}(A) & \xrightarrow{\rho} & L^\infty_{\mu_\phi}(\hat{A}) & \xleftarrow{\rho} & \mathscr{B}(A) \\ \downarrow & & \downarrow & & \downarrow \\ \mathscr{C}(B) & \xrightarrow{\rho} & L^\infty_{\mu_\psi}(\hat{B}) & \xleftarrow{\rho} & \mathscr{B}(B) \end{array}$$

with ρ as in 4.8.5. If the middle inclusion map is an isomorphism then

$$\int \rho(x_1)\rho(x_2)\ldots\rho(x_n)\,d\mu_\phi = \int \rho(x_1)\rho(x_2)\ldots\rho(x_n)\,d\mu_\psi$$

for any $x_1, x_2, \ldots, x_n$ in $\mathscr{B}(A)$. By the very definition of S-measures (see 4.9.3) this implies that $\nu_\phi = \nu_\psi \circ r^{-1}$.

4.9.7. THEOREM. *If ϕ is a state of a separable C^*-algebra A the S-measure ν_ϕ of ϕ is concentrated on the set $F(A)$ of factorial states of A. Moreover, the image measure of ν_ϕ under the Borel map: $\psi \to \hat{\psi}$ is the central measure of ϕ on $\hat{A}$.*

Proof. By adjoining a unit to A we obtain exactly one more point in $F(A)$ and one in $\hat{A}$; and these are negligible sets for ν_ϕ and μ_ϕ, respectively. We may therefore assume that A has a unit.

Since for each closed ideal I of A we may identify $F(A/I)$ with a closed subset of $F(A)$ and $(A/I)^\wedge$ with a D-Borel subset of $\hat{A}$ (even a T-Borel subset), it suffices to prove the theorem under the assumption that the representation (π_ϕ, H_ϕ) is faithful.

As H_ϕ is a separable Hilbert space there is by 3.8.4 a separable C^*-algebra C which is weakly dense in $\pi_\phi(A)'$. Identifying A and $\pi_\phi(A)$ we let B denote the separable C^*-algebra generated by A and C. The identity map is a representation of B which is spatially equivalent to the cyclic representation associated with the state ψ of B given by $\psi(x) = (x\xi_\phi \mid \xi_\phi)$, and $\psi \mid A = \phi$. Since

$$B' = \pi_\phi(A)' \cap C' = \pi_\phi(A)' \cap \pi_\phi(A)''$$

we conclude from 4.9.4 that the S-measure of ψ is concentrated on the set $P(B)$. As $P(B) \subset F(B)$ it follows from 4.9.6 and 4.9.5 that the S-measure ν_ϕ of ϕ is concentrated on $F(A)$.

Now let μ be the image measure of ν_ϕ under the canonical map from $F(A)$ onto $\hat{A}$. Then for each z in $\mathscr{C}$

$$\int \check{z}\, \mathrm{d}\mu = \int \hat{z}\, \mathrm{d}\nu_\phi = \int \lambda(z)^\vee\, \mathrm{d}\mu_\phi = \int \check{z}\, \mathrm{d}\mu_\phi$$

with λ as in 4.8.6. Consequently $\mu = \mu_\phi$.

4.9.8. We can now complete the proof of 4.8.7. Define

$$N_0 = \{t \in \hat{A} \mid \phi_t \notin F(A)\}.$$

Then N_0 is the counter-image of the Borel set $Q \backslash F(A)$ in Q (cf. 4.8.3) with respect to the Borel map $t \to \phi_t$. Consequently N_0 is a D-Borel set; and since ν_ϕ is the image of μ_ϕ we have by 4.8.7

$$\mu_\phi(N_0) = \nu_\phi(Q \backslash F(A)) = 0.$$

For each x in $\mathscr{B}$ we have $\hat{x} = (\lambda(x))^\wedge$ almost everywhere with respect to ν_ϕ, whence

$$\psi(x) = \psi(\lambda(x)) = \phi_t(x), \qquad \text{where } t = \hat{\psi},$$

for all ψ in $F(A)$ except on a null set. Define

$$N = \{\psi \in F(A) \mid \psi \neq \phi_t, \qquad \text{where } t = \hat{\psi}\}.$$

Since $\mathscr{B}$ is countably generated N is a Borel set and $\nu_\phi(N) = 0$. Let T denote the image of $F(A) \backslash N$ in $\hat{A}$ under the canonical map $F(A) \to \hat{A}$. We do not claim that T is a D-Borel set. If $t \in T$ then $\hat{\phi}_t = t$. If z is a projection in $\mathscr{C}$ such that $\check{z}$ is one on T then $\hat{z}$ is one on $F(A) \backslash N$. Since by 4.9.7 μ_ϕ is the image of ν_ϕ we conclude that

$$\int \check{z}\, \mathrm{d}\mu_\phi = \int \hat{z}\, \mathrm{d}\nu_\phi \geqslant \nu_\phi(F(A) \backslash N) = 1.$$

It follows that the outer measure of T is 1 and the proof is complete.

4.9.9. From the general theory of Choquet we know that for each state ϕ of A there is at least one measure in $M(\phi)$ which is concentrated on the set of pure states of A. It is amusing to note that an apparently somewhat different proof of this result can be obtained by the method developed in this section. It suffices in the proof of 4.9.7 to take C as a separable C^*-algebra which is weakly dense in some maximal commutative von Neumann subalgebra of $\pi_\phi(A)'$. Then with B and ψ as before we have

$$B' = \pi_\phi(A)' \cap C' = C'',$$

by 2.8.1. As before we may conclude that the S-measure ν_ϕ of ψ is concentrated on $P(B)$, and since C is commutative the measure $\nu_\psi \circ r^{-1} \in M(\phi)$. But now $\nu_\psi \circ r^{-1}$ is not necessarily the S-measure of ϕ. To see this it suffices to take ϕ in $F(A)\backslash P(A)$, in which case ν_ϕ is the point measure at ϕ, so that $\nu_\phi(P(A)) = 0$.

We mention without proof that the measures ν in $M(\phi)$, concentrated on $P(A)$, which are obtained in the above manner from some maximal commutative von Neumann subalgebra of $\pi_\phi(A)'$ all satisfy $\nu_\phi \prec \nu$; and that ν_ϕ is the maximal measure in $M(\phi)$ (with respect to $\prec$) which is dominated by all such ν. This corresponds to the fact that the centre of $\pi_\phi(A)''$ is the intersection of all maximal commutative subalgebras of $\pi_\phi(A)'$.

4.9.10. Notes and remarks. Theorem 4.9.7 is due to Sakai [226]. The proof used here, which even works in the non-separable case, is largely borrowed from Wils [274, 275].

4.10. Standard measures

4.10.1. Let A be a separable C^*-algebra with factor spectrum $\hat{A}$ and let μ be a D-Borel probability measure on $\hat{A}$. It seems a natural question in connection with this theory to ask whether μ is the central measure of some state of A. To solve this problem amounts to deciding whether a given sequentially normal state on $\mathscr{C}$ can be extended to a sequentially normal state of $\mathscr{B}$. Clearly we do not expect such an extension to be unique. But the results in 4.8 and 4.9 may be interpreted to say that once an extension ϕ has been found it determines an essentially unique lifting $t \to \phi_t$ of $\hat{A}$ back into $F(A)$; therefore gives a measure ν_ϕ on $F(A)$ whose image is $\mu = \mu_\phi$ under the canonical map $\psi \to \hat{\psi}$ from $F(A)$ onto $\hat{A}$.

4.10.2. If E is a Borel space we say that a subset T of E is a *standard subset* of E, if T is a standard Borel space (see 4.6.9) in its relative Borel structure. We emphasize that T need not be a Borel subset of E. The reason why standard subsets are needed, when questions of lifting arise, is brought out by the following proposition.

4.10.3. PROPOSITION. *Let $f: F \to T$ be a surjective Borel map between standard spaces F and T. For each probability measure μ on T there is a null set N in T and a Borel map $g: T\backslash N \to F$ such that $f \circ g$ is the identity on $T\backslash N$.*

Proof. By 4.6.13 we may assume that T is compact and that F is Polish. Then by 4.6.11 we may assume that f is continuous, identifying if necessary F with the graph of f. Let K_0 be a compact Polish space in which F can be homeomorphically embedded (cf. 4.2.4), and consider the graph $G(f)$ of f as a subset of $K_0 \times T$. Since $G(f)$ is closed in $F \times T$ and F is a G_δ-subset of K_0 by

4.2.3, we see that $G(f)$ is a G_δ-subset of $K_0 \times T$. Identifying F with $G(f)$ we have established the following situation: There is a compact Polish space K $(= K_0 \times T)$ containing F as a G_δ-subset, and a continuous surjective map $f: K \to T$ (= projection on the second coordinate) whose restriction to F is the original map f.

Let $\{G_n\}$ be a decreasing sequence of open sets in K with $\cap G_n = F$. Since $f(G_n) = T$ there is by 4.2.12 for each n a Borel function $g_n: T \to G_n$ such that $f \circ g_n$ is the identity on T.

The proposition is trivial if F is countable. Assume therefore that F whence also all G_n are uncountable. By 4.6.13 we can then for each n find a topology τ_n on G_n, which generates the original Borel structure on G_n, such that (G_n, τ_n) is homeomorphic to $[0,1]$. Since $g_n(T) \subset G_1$ for all n this implies that we may regard $\{g_n\}$ as a sequence in the unit ball of $L^\infty_\mu(T)$. As the latter is weakly compact and metrizable there is a weakly convergent subsequence of $\{g_n\}$. It follows from elementary measure theory that there is a null set N_1 in T such that a further subsequence $\{g'_n\}$ of $\{g_n\}$ is pointwise convergent on $T_1 = T \backslash N_1$, to a Borel function $h_1: T_1 \to G_1$. Clearly $f \circ h_1$ is the identity on T_1. Assuming, as we may, that $g_1 \notin \{g'_n\}$ we have $g'_n(T_1) \subset G_2$ for all n. Repeating the argument from above we find a null set N_2 of T_1 and a subsequence $\{g''_n\}$ of $\{g'_n\}$ which is pointwise convergent on $T_2 = T_1 \backslash N_2$ to a Borel function $h_2: T_2 \to G_2$. However, $\{g''_n\}$ is also pointwise convergent in the τ_1-topology, and we conclude that $h_2 = h_1 | T_2$. Continuing by induction we obtain a decreasing sequence of Borel sets T_n in T such that $\mu(T \backslash T_n) = 0$ for all n, and a sequence of Borel functions $h_n: T_n \to G_n$ such that $h_{n+1} = h_n | T_{n+1}$ and $f \circ h_n$ is the identity on T_n. Set $N = T \backslash \cap T_n$ and let g be the common restriction of all h_n to $T \backslash N$. Clearly g is a Borel function on $T \backslash N$ and $f \circ g$ is the identity on $T \backslash N$, and since $g(T \backslash N) \subset G_n$ for all n we conclude that $g(T \backslash N) \subset F$.

4.10.4. THEOREM. *A D-Borel probability measure on the factor spectrum $\hat{A}$ of a separable C^*-algebra A is the central measure of some state of A if and only if there is a Borel set in $F(A)$ whose image in $\hat{A}$ under the canonical map $\psi \to \hat{\psi}$ is a standard subset of $\hat{A}$ with outer measure one.*

Proof. Let ϕ be a state of A with central measure μ_ϕ. By 4.8.7 there is a set T in $\hat{A}$ with outer measure one such that the restriction to T of the Borel map $t \to \phi_t$ is an inverse to the map $\psi \to \hat{\psi}$. Furthermore, we proved in 4.9.8 that T can be chosen as the image of a Borel set F in $F(A)$ with $\nu_\phi(F) = 1$ (set $F = F(A) \backslash N$). It follows that T in its relative Borel structure is Borel isomorphic to F. Since F is a Borel subset of the Borel subset $F(A)$ of the Polish space Q (see 4.7.3), we conclude that F is standard, and thus T is a standard subset of $\hat{A}$.

Conversely, let F be a Borel set of $F(A)$ whose image T is a standard subset of $\hat{A}$, and let μ be a D-Borel probability measure on $\hat{A}$ with $\mu^*(T) = 1$. Since the map $\psi \to \hat{\psi}$ gives a surjective Borel map between the two standard spaces F and T there is by 4.10.3 a D-Borel set N in $\hat{A}$ with $\mu(N) = 0$ and a Borel map $t \to \phi_t$ of $T\backslash N$ into F such that $\hat{\phi}_t = t$ for each t in $T\backslash N$. For each x in the enveloping Borel *-algebra $\mathscr{B}$ of A the map $t \to \phi_t(x)$ is the restriction to $T\backslash N$ of some D-Borel measurable function on $\hat{A}$. Since $\mu^*(T\backslash N) = 1$ we can thus define a unique sequentially normal state ϕ on $\mathscr{B}$ such that $\phi(x) = \int \phi_t(x)\,d\mu(t)$. If $z \in \mathscr{C}$ then

$$\phi(z) = \int \phi_t(z)\,d\mu(t) = \int \check{z}(t)\,d\mu(t),$$

whence $\mu = \mu_\phi$ and the proof is complete.

4.10.5. A measure μ on $\hat{A}$ satisfying 4.10.4 is called a *standard measure.* From the second half of the proof of 4.10.4 it follows that if μ is a standard measure, there is a Borel set F in $F(A)$, such that the map $\psi \to \hat{\psi}$ is a Borel isomorphism of F onto its image T in $\hat{A}$ (with the relative D-Borel structure) and such that $\mu^*(T) = 1$.

As in 4.7.2 we define the *Mackey–Borel structure* (M-Borel) on $\hat{A}$ as the σ-algebra of sets in $\hat{A}$ whose counter images in $F(A)$ under the canonical map $\phi \to \hat{\phi}$ are Borel sets in $F(A)$. Since each standard measure on $\hat{A}$ is the image of a Borel measure on $F(A)$ (S-measure, cf. 4.9.7) it can be extended to an M-Borel measure on $\hat{A}$. This extension facilitates some of the notions concerning standard measures.

4.10.6. PROPOSITION. *If F is a Borel subset of $F(A)$ such that the map $\phi \to \hat{\phi}$ is injective on F, then the image T of F is an M-Borel subset of $\hat{A}$ and $\phi \to \hat{\phi}$ is a Borel isomorphism of F onto T.*

Proof. Let $\{u_n\}$ be dense sequence in the unitary group of A and consider the set W in $F(A) \times F(A)$, where

$$W = \{(\phi, \psi) \mid \|\phi(u_n^* \cdot u_n) - \psi\| = 2, \forall\, n\}.$$

Since the norm is a weak* lower semi-continuous function on A_{sa}^*, we conclude that W is a Borel subset of $F(A) \times F(A)$.

If ϕ and ψ are states of A then by 3.2.3 $\|\phi - \psi\| = 2$ if and only if there is a projection p in A'' with $\phi(1-p) = 0$ and $\psi(p) = 0$. If $(\phi, \psi) \in W$ then $\phi(u_n^*(1-p_n)u_n) = 0$ and $\psi(p_n) = 0$ for a suitable sequence $\{p_n\}$. Put $p = \vee p_n$. Then $\psi(p) = 0$ and

$$\phi(u_n^*(1-p)u_n) = \phi(u_n^*(\wedge(1-p_m))u_n) = 0$$

for all n. Consequently with $z = c(1-p)$ we have $\phi(z) = 0$ and $\psi(1-z) = 0$, so that (π_ϕ, H_ϕ) and (π_ϕ, H_ψ) are disjoint by 3.8.11. Conversely, if (π_ϕ, H_ϕ) and (π_ϕ, H_ψ) are disjoint then $(\phi, \psi) \in W$. It follows from 3.8.13 that

$$(F(A) \times F(A))\backslash W = \{(\phi, \psi) \mid \hat{\phi} = \hat{\psi}\}.$$

Let V be the subset of $F(A)$ obtained by projecting the Borel set $(F(A) \times F)\backslash W$ on its first coordinate. If two points (ϕ_1, ψ_1) and (ϕ_2, ψ_2) have the same image, i.e. $\phi_1 = \phi_2$, then $\hat{\psi}_1 = \hat{\phi}_1 = \hat{\phi}_2 = \hat{\psi}_2$. By the assumption on F this implies that $\psi_1 = \psi_2$, whence $(\phi_1, \psi_1) = (\phi_2, \psi_2)$. Thus V is the injective image of a standard space under a Borel map and thus by 4.6.12 V is a Borel subset of $F(A)$, isomorphic with $(F(A) \times F)\backslash W$. But

$$V = \{\phi \in F(A) \mid \hat{\phi} \in T\},$$

whence T is an M-Borel subset of $\hat{A}$. Applying the same argument to any Borel subset F_1 of F it follows that the map $\phi \to \hat{\phi}$ is a Borel isomorphism of F onto T, as desired.

4.10.7. COROLLARY. *Each point in $\hat{A}$ is an M-Borel set.*

4.10.8. THEOREM. *A D-probability measure μ on $\hat{A}$ is standard if and only if it can be concentrated on an M-Borel set T which is a standard subset of $\hat{A}$ in the D-Borel structure. Moreover, the M-Borel and the (relative) D-Borel structures coincide on T.*

Proof. The conditions are sufficient by 4.10.4. But if μ is a standard measure we can find a Borel set F in $F(A)$ with image T in $\hat{A}$ such that $\mu^*(T) = 1$ and the map $\phi \to \hat{\phi}$ is a Borel isomorphism of F onto T. By 4.10.6 this implies that T is an M-Borel set of $\hat{A}$, M-Borel isomorphic with F. Consequently the two Borel structures on $\hat{A}$ coincide on T.

4.10.9. COROLLARY. *Each standard measure on $\hat{A}$ extends to an M-Borel probability measure on $\hat{A}$.*

4.10.10. We shall finally explain a construction which has played an important rôle in the decomposition theory. For a fixed separable Hilbert space H of dimension d, $1 \leq d \leq \infty$, and a separable C^*-algebra A, let $L_d(A)$ denote the set of bounded linear maps from A into $\boldsymbol{B}(H)$ with norm less than or equal to one. Equipped with the topology of pointwise strong convergence, i.e. a net $\{\pi_i\}$ in $L_d(A)$ converges to π if $\|\pi_i(x)\xi - \pi(x)\xi\| \to 0$ for each x in A and ξ in H, the set $L_d(A)$ becomes a Polish space. As a complete metric on $L_d(A)$ we recommend:

$$d(\rho, \pi) = \sum 2^{-n-m} \|\pi(x_n)\xi_m - \rho(x_n)\xi_m\|,$$

for some dense sequence $\{x_n\}$ in the unit ball of A and some dense sequence $\{\xi_m\}$ in the unit ball of H.

The set $\operatorname{Rep}_d(A)$ of representations of A in H can then be identified with the closed subset of $L_d(A)$ consisting of elements π such that

$$\pi(xy)\xi = \pi(x)\pi(y)\xi \qquad \text{and} \qquad (\pi(x^*)\xi \mid \eta) = (\xi \mid \pi(x)\eta)$$

for all x, y in A and all ξ, η in H.

We find it more convenient to work only with cyclic representations, and thus for a fixed unit vector ξ_0 in H we let

$$\mathrm{Cyc}_d(A) = \bigcap_{n,m} \bigcup_k \{\pi \in \mathrm{Rep}_d(A) \mid \|\pi(x_k)\xi_0 - \xi_m\| < 1/n\}.$$

Then $\mathrm{Cyc}_d(A)$ is precisely the set of representations of A on H for which ξ_0 is a cyclic vector and being a G_δ-subset of $\mathrm{Rep}_d(A)$ it is a Polish space in the relative topology by 4.2.2.

4.10.11. LEMMA. *The map* $\pi \to (\pi(\cdot)\xi_0 \mid \xi_0)$ *is continuous from* $\mathrm{Cyc}_d(A)$ *onto the Borel set* S_d *of states* ϕ *of* A *such that* $\dim H_\phi = d$.

Proof. (cf. 4.4.10). Let ${}_dS$ denote the closed subset of states ϕ of A such that

$$\mathrm{Sup}\left\{\sum \phi(x_n^* x_n)\right\} \leqslant d,$$

the supremum being taken over all finite subsets $\{x_n\}$ of A for which $\phi(x_n^* x_m) = \delta_n^m$. Then ${}_dS$ consists precisely of those states ϕ for which $\dim H_\phi \leqslant d$. Since

$$S_d = {}_dS \setminus \bigcup_{n<d} {}_nS \cdot$$

we have shown that S_d is a Borel set in S.

It is clear from the definition of the topology on $\mathrm{Cyc}_d(A)$ that the map $\pi \to (\pi(\cdot)\xi_0 \mid \xi_0)$ is continuous from $\mathrm{Cyc}_d(A)$ into S_d with the weak* topology. To see that the map is surjective, take ϕ in S_d and consider the representation $(\pi_\phi, H_\phi, \xi_\phi)$. Let u be an isometry of H_ϕ onto H such that $u\xi_\phi = \xi_0$, and define $\pi = u\pi_\phi(\cdot)u^*$. Then $\pi \in \mathrm{Cyc}_d(A)$ and $(\pi(\cdot)\xi_0 \mid \xi_0) = \phi$.

4.10.12. COROLLARY. *The subsets* $\mathrm{Irr}_d(A)$ *and* $\mathrm{Fac}_d(A)$ *of* $\mathrm{Cyc}_d(A)$, *consisting of the irreducible and factorial elements, respectively, are both Borel subsets of* $\mathrm{Cyc}_d(A)$, *and consequently standard spaces.*

4.10.13. PROPOSITION. *The map which to each representation associates its equivalence class, induces surjective maps from* $\mathrm{Irr}_d(A)$ *to* $\hat{A}_d$ *and from* $\mathrm{Fac}_d(A)$ *onto* $\check{A}_d$. *The resulting quotient Borel structures on* $\hat{A}_d$ *and* $\check{A}_d$ *are precisely the M-Borel structures.*

Proof. If $F_d(A) = F(A) \cap S_d$ we have a commuting diagram:

$$\begin{array}{ccc} \mathrm{Fac}_d(A) & \longrightarrow & F_d(A) \\ & \searrow \quad \swarrow & \\ & \check{A}_d & \end{array}$$

Since by 4.10.11 the horizontal arrow is a continuous map, we see that the counter image in $\mathrm{Fac}_d(A)$ of an M-Borel set in $\hat{A}_d$, is a Borel set in $\mathrm{Fac}_d(A)$. Conversely, if E is a subset of $\hat{A}_d$ whose counter image G in $\mathrm{Fac}_d(A)$ is a Borel set then the same holds for the complements E' and G'. Since each representation in G is disjoint (see 3.8.13) from any representation in G' we conclude that the images T and T' of G and G' in $F_d(A)$ are the complements of each other. However, both G and G' are continuous images of Polish spaces by 4.6.10 and it follows from 4.2.9 that T (and T') is a Borel set of $F_d(A)$. But T is the counter image of E, which shows that E is an M-Borel set.

The proof for the triple $\mathrm{Irr}_d(A)$, $P_d(A)$ and $\hat{A}_d$ is the same.

4.10.14. Notes and remarks. Theorem 4.10.4 was proved by Effros [76] in the form given in 4.10.8. Proposition 4.10.6 was established by Ernest [86], but with $\mathrm{Fac}(A)$ (cf. 4.10.12) instead of $F(A)$. By 4.10.13 this is equivalent with 4.10.6. The results in 4.10.10–4.10.13 are our excuse for working exclusively with the state space, instead of using the customary approach to decomposition theory via representations on a concrete Hilbert space.

4.11. Direct integrals of von Neumann algebras

4.11.1. Let T be a Borel space and $\{H_t \mid t \in T\}$ a family of separable Hilbert spaces. A *vector field* on T is a function $\eta: T \to \cup H_t$ such that $\eta(t) \in H_t$ for each t in T. We say that $\{H_t\}$ is a *Borel field of Hilbert spaces* if there exists a sequence $\{\eta_n\}$ of vector fields on T such that

(i) For each t in T the set $\{\eta_n(t) \mid n \in \boldsymbol{N}\}$ is total in H_t.

(ii) Each function $t \to (\eta_n(t) \mid \eta_m(t))$ is a Borel function on T.

A vector field η on T is *Borel measurable* (with respect to $\{\eta_n\}$) if $t \to (\eta(t) \mid \eta_n(t))$ is a Borel function for all n. A sequence $\{\eta'_n\}$ of Borel vector fields on T satisfying (i) and (ii) is *equivalent* to $\{\eta_n\}$ if it produces the same space of Borel vector fields.

Using the obvious 'pointwise' version of the Gram–Schmidt orthonormalization method, followed by an addition of vector fields with disjoint supports, we may construct from a given sequence $\{\eta_n\}$ an equivalent sequence $\{\eta'_n\}$ of Borel vector fields on T such that

(iii) For each t in T the set $\{\eta'_n(t) \mid n < \dim(H_t) + 1\}$ is an orthonormal basis for H_t;

(iv) $\eta'_n(t) = 0$ if $n > \dim(H_t)$.

If ξ_1 and ξ_2 are Borel vector fields on T then

$$(\xi_1(t)\,|\,\xi_2(t)) = \sum (\xi_1(t)\,|\,\eta'_n(t))(\eta'_n(t)\,|\,\xi_2(t)),$$

which shows that the function $t \to (\xi_1(t)\,|\,\xi_2(t))$ is a Borel function on T.

4.11.2. Let μ be a bounded Borel measure on T, and $\{H_t\,|\,t \in T\}$ a Borel field of Hilbert spaces. A Borel field on T is *square integrable* if the (Borel) function $t \to \|\xi(t)\|^2$ is integrable with respect to μ. Similarly ξ is a *null field* if $t \to \|\xi(t)\|$ is a null function for μ. We define an inner product on the vector space of square integrable Borel fields modulo null fields by setting

$$(\xi_1\,|\,\xi_2) = \int_T (\xi_1(t)\,|\,\xi_2(t))\,\mathrm{d}\mu(t).$$

Since each square integrable ξ has the form

$$(*) \qquad \xi(t) = \sum (\xi(t)\,|\,\eta'_n(t))\eta'_n(t) = \sum f_n(t)\eta'_n(t),$$

where $f_n \in L^2_\mu(T)$ and $\{\eta'_n\}$ is a sequence satisfying (iii) and (iv) of 4.11.1 it is easy to prove that the inner product space described above is in fact a Hilbert space. We call it the *direct integral* of the Borel field $\{H_t\}$ with respect to μ, and denote it by

$$\int_T^{\oplus} H_t\,\mathrm{d}\mu(t).$$

If $L^2_\mu(T)$ is separable, a situation which occurs if μ is concentrated on a standard subset of T (see 4.10.2), then the formula $(*)$ shows that $\int^{\oplus} H_t\,\mathrm{d}\mu(t)$ is separable as well. If T is countable then $\int^{\oplus} H_t\,\mathrm{d}\mu(t)$ is the direct sum of the H_t's.

4.11.3. Let T be a Borel space and $\{H_t\,|\,t \in T\}$ a Borel field of Hilbert spaces on T. An *operator field* on T is a bounded function $x: T \to \cup \boldsymbol{B}(H_t)$ such that $x(t) \in \boldsymbol{B}(H_t)$ for each t in T. An operator field x on T is *Borel measurable* if $t \to x(t)\xi(t)$ is a Borel vector field on T for each Borel vector field ξ on T. This is equivalent to the condition that $t \to (x(t)\eta_n(t)\,|\,\eta_m(t))$ is a Borel function on T for all η_n and η_m in a sequence defining the Borel structure on $\{H_t\}$. Suppose that for each t in T is given a von Neumann algebra $\mathcal{M}_t$ in $\boldsymbol{B}(H_t)$. We denote by $\mathscr{B}(T, \{\mathcal{M}_t\})$ the Borel $*$-algebra of Borel operator fields x on T such that $x(t) \in \mathcal{M}_t$ for each t in T. The largest of these algebras is $\mathscr{B}(T, \{\boldsymbol{B}(H_t)\})$ consisting of all Borel operator fields on T. The smallest is $\mathscr{B}(T, \{\boldsymbol{C}1_t\})$, which may be identified with $\mathscr{B}(T)$. We say that a family $\{\mathcal{M}_t\,|\,t \in T\}$ of von Neumann algebras is a *Borel field of von Neumann algebras* on T (with respect to a given Borel field of Hilbert spaces) if there is a $*$-invariant sequence $\{x_n\}$ in $\mathscr{B}(T, \{\mathcal{M}_t\})$ such that $\mathcal{M}_t$ is generated by $\{x_n(t)\}$ for each t in T. Evidently $\{\boldsymbol{C}1_t\}$ is a Borel field. To prove that also $\{\boldsymbol{B}(H_t)\}$ is a Borel field, note first that if $\{\eta'_n\}$ is a sequence of Borel vector fields on T satisfying (iii) and (iv) of 4.11.1 then the set T_d of points t such that $\dim(H_t) = d$, $1 \leqslant d \leqslant \infty$, is evidently a Borel subset of T. Next, on each T_d define a countable number of Borel operator fields $\{x_{ij} \mid 1 \leqslant i < d+1,\ 1 \leqslant j < d+1\}$:

$$x_{ij}(t)\eta'_j(t) = \eta'_i(t), \qquad x_{ij}(t)\eta'_k(t) = 0 \quad \text{if } k \neq j,$$

(these are *constant operator fields* on T_d). Arranging these Borel operator fields in a sequence $\{x_n\}$ it is clear that $\{x_n(t)\}$ generates $\boldsymbol{B}(H_t)$ for each t in T.

4.11.4. If $\{H_t \mid t \in T\}$ is a Borel field of Hilbert spaces on T and μ is a bounded Borel measure on T we define a sequentially normal representation π_μ of $\mathscr{B}(T, \{\boldsymbol{B}(H_t)\})$ on $\int^\oplus H_t \, d\mu(t)$ by

$$(\pi_\mu(x)\xi)(t) = x(t)\xi(t), \qquad \xi \in \int^\oplus H_t \, d\mu(t), \qquad x \in \mathscr{B}(T, \{\boldsymbol{B}(H_t)\}).$$

The kernel of π_μ consists precisely of those x in $\mathscr{B}(T, \{\boldsymbol{B}(H_t)\})$ for which the (Borel) function $t \to \|x(t)\|$ is a null function for μ.

For any family $\{\mathscr{M}_t \mid t \in T\}$ of von Neumann algebras we say that $\pi_\mu(\mathscr{B}(T, \{\mathscr{M}_t\}))$ is the *direct integral* of the von Neumann algebras $\{\mathscr{M}_t\}$ with respect to μ and we denote it by

$$\int_T^\oplus \mathscr{M}_t \, d\mu(t).$$

4.11.5. LEMMA. *With the notations as in 4.11.4,*

$$\int^\oplus \boldsymbol{C}1_t \, d\mu(t) \qquad \textit{and} \qquad \int^\oplus \boldsymbol{B}(H_t) \, d\mu(t)$$

are the commutants of one another.

Proof. It is clear that $\int^\oplus \boldsymbol{C}1_t \, d\mu(t)$ is (monotone) sequentially closed and isomorphic to $L^\infty_\mu(T)$. Choosing a sequence $\{\eta'_n\}$ of Borel vector fields on T satisfying (iii) and (iv) of 4.11.1 we see that μ on $\mathscr{B}(T, \{\boldsymbol{C}1_t\})$ is the vector functional determined by η'_1 in $\int^\oplus H_t \, d\mu(t)$. Since μ is faithful on $L^\infty_\mu(T)$, each increasing net in $\int^\oplus \boldsymbol{C}1_t \, d\mu(t)$ has the same limit as some increasing sequence from the net (cf. 3.4.2). Since $\int^\oplus \boldsymbol{C}1_t \, d\mu(t)$ is sequentially closed, it is therefore monotone closed and thus a von Neumann algebra.

Take x' in $(\int^\oplus \boldsymbol{C}1_t \, d\mu(t))'$ and choose a sequence $\{\eta_n\}$ of bounded Borel vector fields on T that form a vector space over the rational numbers such that $\{\eta_n(t)\}$ is dense in H_t (take e.g. $\{\eta_n\}$ as the set of rational combinations of the sequence $\{\eta'_n\}$). For each n there is a square integrable Borel field ξ_n whose image in $\int^\oplus H_t \, d\mu(t)$ is $x'\eta_n$. If $f \in \mathscr{B}(T)$ then by assumption

$$x'(f\eta_n) = (\pi_\mu f)x'\eta_n = \pi_\mu(f)\xi_n = f\xi_n,$$

and $\|f\xi_n\| \leqslant \|x'\|\|f\eta_n\|$. Since f is arbitrary this implies that $\|\xi_n(t)\| \leqslant \|x'\|\|\eta_n(t)\|$ almost everywhere. Changing all η_n on a null set for μ we may assume that $\|\xi_n(t)\| \leqslant \|x'\|\|\eta_n(t)\|$ and that $\alpha\xi_n(t) + \beta\xi_m(t) = \xi_k(t)$ for all t in T whenever α and β are rationals such that $\alpha\eta_n + \beta\eta_m = \eta_k$. There is then precisely one operator $x(t)$ in $\boldsymbol{B}(H_t)$ such that $x(t)\eta_n(t) = \xi_n(t)$ and $\|x(t)\| \leqslant \|x'\|$. The operator field $x: t \to x(t)$ is Borel measurable since $x\eta_n = \xi_n$ and $\{\eta_n\}$ determines the Borel structure on $\{H_t\}$. Finally, as $\mathscr{B}(T, \boldsymbol{B}(H_t)\})$ commutes with $\mathscr{B}(T, \{\boldsymbol{C}1_t\})$, we have for each f in $\mathscr{B}(T)$,

$$\pi_\mu(x)(f\eta_n) = \pi_\mu(f)\pi_\mu(x)\eta_n = \pi_\mu(f)\xi_n$$

$$= \pi_\mu(f)x'\eta_n = x'\pi_\mu(f)\eta_n = x'(f\eta_n).$$

Since vector fields of the form $f\eta_n$ constitute a total set in $\int^\oplus H_t \, d\mu(t)$ (cf. (*) in 4.11.2) we conclude that $\pi_\mu(x) = x'$, whence $(\int^\oplus \boldsymbol{C}1_t \, d\mu(t))' = \int^\oplus \boldsymbol{B}(H_t) \, d\mu(t)$.

The converse statement follows from von Neumann's bicommutant theorem, since we proved that $\int^\oplus \boldsymbol{C}1_t \, d\mu(t)$ was weakly closed.

4.11.6. We say that the operators in $\int^\oplus \boldsymbol{C}1_t \, d\mu(t)$ are *diagonalizable* on $\int^\oplus H_t \, d\mu(t)$ (with respect to the Borel field $\{H_t\}$, which plays the rôle of 'basis'). We say that the operators in $\int^\oplus \boldsymbol{B}(H_t) \, d\mu(t)$ are *decomposable* on $\int^\oplus H_t \, d\mu(t)$.

4.11.7. PROPOSITION. *Let $\{H_t \mid t \in T\}$ be a Borel field of Hilbert spaces on a standard space T. If $\{\mathscr{M}_t \mid t \in T\}$ is a Borel field of von Neumann algebras then $\{\mathscr{M}'_t \mid t \in T\}$ is also a Borel field of von Neumann algebras on T.*

Proof. We may consider separately the Borel subsets T_d of T for which $\dim(H_t) = d$, $1 \leqslant d \leqslant \infty$, and thus assume that $T_d = T$. Fixing a Hilbert space H of dimension d we may then realize the Borel vector fields on T as the set of weakly Borel measurable vector valued functions from T into H. Furthermore we may identify $\mathscr{B}(T, \{\boldsymbol{B}(H_t)\})$ with the set of bounded, weakly Borel measurable operator valued functions from T into $\boldsymbol{B}(H)$.

By assumption there is a *-invariant sequence $\{x_n\}$ in $\mathscr{B}(T, \{\mathscr{M}_t\})$ such that $\{x_n(t)\}$ generates $\mathscr{M}_t$ for each t in T. We may assume that $\|x_n\| \leqslant 1$ for all n. Let $\boldsymbol{B}(H)_s^1$ denote the unit ball of $\boldsymbol{B}(H)$ in the strong topology. Then $\boldsymbol{B}(H)_s^1$ is a Polish space and $\{x_n\}$ consists of Borel functions from T to $\boldsymbol{B}(H)_s^1$. Take a dense sequence $\{\xi_i\}$ in H and let A be the separable commutative C^*-subalgebra of $\boldsymbol{B}(T)$ generated by the functions $t \to \|x_n(t)\xi_i\|$, together with a sequence in $\mathscr{B}(T)$ that separates points in T and contains the function 1. Identifying A with $C(\hat{A})$ we have a natural Borel injection of T into $\hat{A}$. Since both T and $\hat{A}$ are standard Borel spaces we may by 4.6.12 identify T with a dense Borel subset of the second countable, compact space $\hat{A}$. Evidently each x_n extends uniquely to a strongly continuous function from $\hat{A}$ into $\boldsymbol{B}(H)_s^1$, and for each t in $\hat{A}$ we let $\mathscr{M}_t$ denote the von Neumann algebra generated by $\{x_n(t)\}$.

Let $B(H)^1_w$ denote the unit ball of $B(H)$ equipped with the weak topology, so that $B(H)^1_w$ is a second countable, compact space. Denote by E the compact subset of $\hat{A} \times B(H)^1_w$ consisting of points (t, x') such that $x'x_n(t) = x_n(t)x'$ for all n in $\boldsymbol{N}$. If $\{F_n\}$ is a basis for the topology on $B(H)^1_w$ consisting of compact sets, let $E_n = E \cap (\hat{A} \times F_n)$ and let T_n be the projection of E_n on $\hat{A}$. Since both E_n and T_n are compact and second countable and the projection map is continuous, there is by 4.2.13 a Borel function $g_n: T_n \to E_n$ such that $g_n(t) = (t, f_n(t))$, where f_n is a Borel function from T_n into F_n. Let x'_n be the Borel operator field on $\hat{A}$ such that $x'_n(t) = f_n(t)$ if $t \in T_n$ and $x'_n(t) = 0$ if $t \notin T_n$. By construction $x_n(t) \in \mathcal{M}'_t$ for each t in $\hat{A}$. Moreover, for each fixed t in $\hat{A}$ we have $x'_n(t) \in F_n$ which shows that $\{x'_n(t)\}$ is weakly dense in the unit ball of $\mathcal{M}'_t$. It follows that $\{\mathcal{M}'_t \mid t \in \hat{A}\}$ is a Borel field of von Neumann algebras on $\hat{A}$, hence $\{\mathcal{M}'_t \mid t \in T\}$ is a Borel field on the Borel subset T of $\hat{A}$.

4.11.8. THEOREM. *Let $\{H_t \mid t \in T\}$ be a Borel field of Hilbert spaces on a standard Borel space T, and let $\{\mathcal{M}_t \mid t \in T\}$ be a Borel field of von Neumann algebras on T. If μ is a bounded Borel measure on T then $\int^\oplus \mathcal{M}_t \, d\mu(t)$ and $\int^\oplus \mathcal{M}'_t \, d\mu(t)$ are the commutants of one another.*

Proof. It is clear that $\int^\oplus \mathcal{M}_t \, d\mu(t)$ and $\int^\oplus \mathcal{M}'_t \, d\mu(t)$ commute. Take y' in the commutant of $\int^\oplus \mathcal{M}_t \, d\mu(t)$. Since $\int^\oplus \mathcal{M}_t \, d\mu(t)$ contains the algebra of diagonalizable operators we conclude from 4.11.5 that $y' = \pi_\mu(x')$, where x' is a Borel operator field on T and π_μ is as in 4.11.4.

Let $\{x_n\}$ be a generating sequence for $\{\mathcal{M}_t\}$. Since $\pi_\mu(x_n x' - x' x_n) = 0$ for all n we have $x_n(t)x'(t) = x'(t)x_n(t)$ for all n and almost all t in T. Setting $x'(t) = 0$ on a null set we may thus assume that $x'(t)$ commutes with $\{x_n(t)\}$ for all t in T. But this implies that $x'(t) \in \mathcal{M}'_t$, whence $y' \in \int^\oplus \mathcal{M}'_t \, d\mu$. Thus

$$\left(\int^\oplus \mathcal{M}_t \, d\mu(t)\right)' = \int^\oplus \mathcal{M}'_t \, d\mu(t).$$

The other equation follows from a symmetric argument, since $\{\mathcal{M}'_t \mid t \in T\}$ is a Borel field of von Neumann algebras by 4.11.7.

4.11.9. THEOREM. *For each $i = 1, 2$ let $\{H^i_t \mid t \in T_i\}$ be a Borel field of Hilbert spaces on a standard space T_i, and let μ_i be a bounded Borel measure on T_i. If v is an isometry of $\int^\oplus H^1_t \, d\mu_1(t)$ onto $\int^\oplus H^2_t \, d\mu_2(t)$, such that*

$$\int^\oplus \boldsymbol{C}1_t \, d\mu_2(t) = v\left(\int^\oplus \boldsymbol{C}1_t \, d\mu_1(t)\right)v^*$$

then there are null sets N_1 and N_2 and a Borel isomorphism τ of $T_2 \backslash N_2$ onto $T_1 \backslash N_1$ such that $\mu_2 \circ \tau^{-1}$ is equivalent to μ_1. Moreover, with $z = d(\mu_2 \circ \tau^{-1})/d\mu_1$ there is a family $\{v_t \mid t \in T_1 \backslash N_1\}$ of isometries such that $v_t H^1_t = H^2_{\tau^{-1}(t)}$ for all t in $T_1 \backslash N_1$ and such that if $v\xi = \eta$ for some square integrable Borel vector field ξ in $\{H^1_t \mid t \in T_1 \backslash N_1\}$, then $z^{-1/2}(t)v_t\xi(t) = \eta(\tau^{-1}(t))$.

Proof. Since T_i is standard, $\mathscr{B}(T_i)$ is a standard Borel *-algebra. The assumptions imply that the sequentially normal representations π_1 and π_2 of $\mathscr{B}(T_1)$ and $\mathscr{B}(T_2)$ obtained form μ_1 and μ_2, respectively, satisfy $\pi_2(\mathscr{B}(T_2)) = v\pi_1(\mathscr{B}(T_1))v^*$. It follows from 4.6.6 that we can find null sets N_1 and N_2, corresponding to projections $1 - p_1$ and $1 - p_2$, and an isomorphism λ of $p_1\mathscr{B}(T_1)$ onto $p_2\mathscr{B}(T_2)$ such that $v\pi_1(x)v^* = \pi_2(\lambda(p_1 x))$ for each x in $\mathscr{B}(T_1)$.

This defines a unique Borel isomorphism τ of $T_2\backslash N_2$ onto $T_1\backslash N_1$ such that $x(\tau(s)) = \lambda(p_1 x)(s)$ for all x in $\mathscr{B}(T_1)$ and each s in $T_2\backslash N_2$. Clearly the measures $\mu_2 \circ \tau^{-1}$ and μ_1 are equivalent and we put $z = \mathrm{d}(\mu_2 \circ \tau^{-1})/\mathrm{d}\mu_1$.

Let $\{\eta'_n\}$ be a sequence of vector fields on T_1 satisfying (iii) and (iv) of 4.11.1, and choose a sequence of square integrable Borel vector fields $\{\eta_n\}$ on T_2 such that $v\eta'_n = \eta_n$ in $\int^{\oplus} H_s^2 \, \mathrm{d}\mu_2(s)$. Set $\eta''_n(s) = \eta_n(s)z(\tau(s))^{1/2}$. Then for each n, m and each x in $\mathscr{B}(T_1)$ we have

$$\begin{aligned}
\int x(t)(\eta'_n(t) \mid \eta'_m(t)) \, \mathrm{d}\mu_1(t) &= (\pi_1(x)\eta'_n \mid \eta'_m) = (v\pi_1(x)v^*\eta_n \mid \eta_m) \\
&= (\pi_2(\lambda(p_1 x))\eta_n \mid \eta_m) = \int x(\tau(s))(\eta_n(s) \mid \eta_m(s)) \, \mathrm{d}\mu_2(s) \\
&= \int x(t)(\eta_n(\tau^{-1}(t)) \mid \eta_m(\tau^{-1}(t))z(t) \, \mathrm{d}\mu_1(t) \\
&= \int x(t)(\eta''_n(\tau^{-1}(t) \mid \eta''_m(\tau^{-1}(t)) \, \mathrm{d}\mu_1(t).
\end{aligned}$$

From this we conclude that except on a null set (which we may assume is equal to N_2) we have $\|\eta''_n(s)\| = 0$ or 1 with $\|\eta''_n(s)\| \geqslant \|\eta''_m(s)\|$ and $(\eta''_n(s) \mid \eta''_m(s)) = 0$ if $n > m$. Thus $\{\eta''_n\}$ satisfies (iii) and (iv) of 4.11.1 except that the span of $\{\eta''_n(s)\}$ may not always be H_s^2. However, if $p_n(s)$ denotes the projection on the subspace spanned by $\eta''_n(s)$ then $p_n(s)\xi(s) = (\xi(s) \mid \eta''_n(s))\eta''_n(s)$ for each $\xi(s)$ in H_s^2. Consequently p_n is a Borel operator field on $T_2\backslash N_2$ and so is $p = \Sigma\, p_n$. Since $\{\eta'_n\}$ is a total set in $\int^{\oplus} H_t^1 \, \mathrm{d}\mu_1(t)$ we have $\pi_2(p) = 1$, whence $p(s) = 1$ almost everywhere. Enlarging N_2 we may assume that $p(s) = 1$ for all s in $T_2\backslash N_2$. Now define v_t from H_t^1 onto $H_{\tau^{-1}(t)}^2$ by $v_t\eta'_n(t) = \eta''_n(\tau^{-1}(t))$. Clearly v_t is an isometry and

$$\begin{aligned}
(v\eta'_n)(\tau^{-1}(t)) = \eta_n(\tau^{-1}(t)) &= z^{-1/2}(t)\eta''_n(\tau^{-1}(t)) \\
&= z^{-1/2}(t)v_t\eta'_n(t),
\end{aligned}$$

for all n and all t in $T_1\backslash N_1$. By 4.11.2 each square integrable Borel vector field ξ on $T_1\backslash N_1$ has the form $\xi(t) = \Sigma\, f_n(t)\eta'_n(t)$, where $\{f_n\} \subset L^2_{\mu_1}(T_1)$. If the f_n's are bounded Borel functions we have

$$\begin{aligned}
(v\xi)(\tau^{-1}(t)) &= \sum (v\pi_1(f_n)\eta'_n)(\tau^{-1}(t)) \\
&= \sum (\pi_2(f_n \circ \tau)v\eta'_n)(\tau^{-1}(t)) \\
&= \sum f_n(t)z^{-1/2}(t)v_t\eta'_n(t) = z^{-1/2}(t)v_t\xi(t),
\end{aligned}$$

Consequently the same formula holds also for unbounded functions in $L^2_{\mu_1}(T_1)$, and the proof is complete.

4.11.10. COROLLARY. *Let* $\{H_t \mid t \in T\}$ *be a Borel field of Hilbert spaces on a standard space* T, *and let* μ *be a bounded Borel measure on* T. *If* $\{\xi_n\}$ *is a sequence of square integrable vector fields which is dense in* $\int^{\oplus} H_t \, d\mu(t)$, then $\{\xi_n(t)\}$ *is dense in* H_t *for almost all* t.

4.11.11. Notes and remarks. The contents of this section are the classic elements in von Neumann's reduction theory (see von Neumann [177] or Dixmier [59]). The authors preference of Borel functions to (classes of) measurable functions has dictated a setting in which Borel fields are the basic objects. The real difference from [177] is, however, slight: although 4.11.7 is probably new.

4.12. Direct integrals of representations

4.12.1. Let $\{H_t \mid t \in T\}$ be a Borel field of Hilbert spaces on a Borel space T. We say that a family $\{\pi_t \mid t \in T\}$ of non-degenerate representations of a separable C^*-algebra A is a *Borel field of representations* if $t \to \pi_t(x)$ is a Borel operator field on T for each x in A. If $\mathscr{B}$ is the enveloping Borel *-algebra associated with A then $t \to \pi_t''(x)$ is also a Borel operator field for each x in $\mathscr{B}$. Moreover, the family $\{\pi_t''(\mathscr{B}) \mid t \in T\}$ is a Borel field of von Neumann algebras on T, since A is separable.

4.12.2. If μ is a bounded Borel measure on T and $\{H_t \mid t \in T\}$, $\{\pi_t \mid t \in T\}$ are as in 4.12.1, then we define a representation π of A (and its sequentially normal extension π'' to $\mathscr{B}$) on $\int^{\oplus} H_t \, d\mu(t)$ by

$$(\pi''(x)\xi)(t) = \pi_t''(x)\xi(t), \qquad \xi \in \int^{\oplus} H_t \, d\mu(t), \qquad x \in \mathscr{B}.$$

The kernel of π'' consists precisely of those x in $\mathscr{B}$ for which the (Borel) function $t \to \|\pi_t''(x)\|$ is a null function for μ. We say that π is the *direct integral* of the Borel field $\{\pi_t\}$ with respect to μ and write

$$\pi = \int_T^{\oplus} \pi_t \, d\mu(t).$$

4.12.3. THEOREM. *Let* ϕ *be a state of a separable* C^**-algebra* A *and let* μ_ϕ *be the central measure on* $\hat{A}$ *corresponding to* ϕ. *There exists a family* $\{(\pi_t, H_t) \mid t \in \hat{A}\}$ *of factor representations of* A *such that* $\pi_t \in t$ *on a D-standard, M-Borel subset* T *of* $\hat{A}$ *with outer measure* 1. *Moreover,* $\{H_t \mid t \in \hat{A}\}$ *is a D-Borel field of Hilbert spaces,* $\{\pi_t \mid t \in \hat{A}\}$ *is a D-Borel field of representations of* A, *and* (π_ϕ, H_ϕ) *is spatially equivalent to* $(\int^{\oplus} \pi_t \, d\mu_\phi(t), \int^{\oplus} H_t \, d\mu_\phi(t))$. *If* $\{(\rho_t, K_t) \mid t \in \hat{A}\}$ *satisfies the same conditions there is a null set* N *and a family* $\{v_t \mid t \in T\backslash N\}$ *of isometries such that* $v_t H_t = K_t$ *and* $v_t^* \rho_t v_t = \pi_t$ *for all* t *in* $T\backslash N$.

Proof. Let $\phi = \int \phi_t \, d\mu_\phi(t)$ be the canonical disintegration of ϕ in factorial states described in 4.8.7. For each t in $\hat{A}$ let (π_t, H_t, ξ_t) be the cyclic factor representation associated with ϕ_t. Then $\mu_\phi^*\{t \in \hat{A} \mid \pi_t \in t\} = 1$. With $\{x_n\}$ a dense sequence in A we define a sequence $\{\xi_n\}$ of vector fields on $\{H_t\}$ by $\xi_n(t) = \pi_t(x_n)\xi_t$. This gives $\{H_t \mid t \in \hat{A}\}$ the structure of a Borel field of Hilbert spaces, and at the same time assures that $\{\pi_t \mid t \in \hat{A}\}$ becomes a Borel field of representations. Set $H = \int^\oplus H_t \, d\mu_\phi(t)$ and $\pi = \int^\oplus \pi_t \, d\mu_\phi(t)$. Since vector fields of the form $\Sigma f_n \xi_n$, where $f_n \in \mathscr{B}(\hat{A})$, are dense in H (by (*) in 4.11.2) and $\mathscr{B}(\hat{A}) = \mathscr{C}$, it follows that the Borel vector field $\xi_0 : t \to \xi_t$ is cyclic for $\pi(A)$ in H. Moreover, for each x in A

$$
\begin{aligned}
(\pi(x)\xi_0 \mid \xi_0) &= \int (\pi_t(x)\xi_t \mid \xi_t) \, d\mu_\phi(t) \\
&= \int \phi_t(x) \, d\mu_\phi(t) = \phi(x).
\end{aligned}
$$

We conclude from 3.3.7 that (π_ϕ, H_ϕ) is spatially equivalent to (π, H).

Suppose now that the family $\{(\rho_t, K_t) \mid t \in \hat{A}\}$ satisfies the same conditions and set $K = \int^\oplus K_t \, d\mu_\phi(t)$, $\rho = \int^\oplus \rho_t \, d\mu_\phi(t)$. Let v be an isometry of H onto K such that $v\pi(x)v^* = \rho(x)$ for each x in A. Then in particular $v\pi''(x)v^* = \rho''(x)$ for each x in $\mathscr{C}$. Moreover, by 4.10.4 and 4.10.8 we may restrict attention to a D-standard, M-Borel subset T of $\hat{A}$ with outer measure 1 such that $\pi_t \in t$ and $\rho_t \in t$ for all t in T. The assumptions in 4.11.9 are satisfied and we conclude that there is a null set N and a family $\{v_t \mid t \in T\backslash N\}$ of isometries such that $v_t H_t = K_t$ for each t in $T\backslash N$ such that $(v\xi)(t) = v_t\xi(t)$ for each square integrable Borel vector field ξ in $\{H_t \mid t \in T\backslash N\}$. Thus for each x in A

$$
\begin{aligned}
v_t^*\rho_t(x)v_t\xi(t) &= v_t^*\rho_t(x)(v\xi)(t) = v_t^*(\rho(x)v\xi)(t) \\
&= v_t^*(v\pi(x)\xi)(t) = v_t^*v_t(\pi(x)\xi)(t) = \pi_t(x)\xi(t);
\end{aligned}
$$

whence $v_t^*\rho_t v_t = \pi_t$ for all t in $T\backslash N$.

4.12.4. THEOREM. *Let (π, H) be a non-degenerate separable representation of a separable C^*-algebra A. There exists*:

(i) *a standard measure μ on $\hat{A}$*;

(ii) *a Borel field $\{H_t \mid t \in \hat{A}\}$ of Hilbert spaces*;

(iii) *an isometry u from $\int^\oplus H_t \, d\mu(t)$ onto H*;

(iv) *a Borel field $\{\pi_t \mid t \in \hat{A}\}$ of factor representations of A such that $\pi_t \in t$ on a standard subset T of $\hat{A}$ which is M-Borel and has outer measure 1, and such that. $u^*\pi u = \int^\oplus \pi_t \, d\mu(t)$.-*

If ν, $\{K_t \mid t \in \hat{A}\}$, v and $\{\rho_t \mid t \in \hat{A}\}$ satisfy the same conditions, then ν is equivalent to μ and there is a D-Borel set N in $\hat{A}$ with $\mu(N) = 0$ and a family $\{v_t \mid t \in T\backslash N\}$ of isometries such that $v_t H_t = K_t$ and $v_t^*\rho_t v_t = \pi_t$ for all t in $T\backslash N$.

Proof. We may assume that (π, H) is faithful and that $1 \in A$. We may also assume that there is a unit vector ξ_0 in H which is cyclic for $(\pi(A) \cup \pi(A)')''$. Indeed, the projection on $[\pi(A)\pi(A)'\xi_0]$ belongs to the centre of $\pi(A)''$ for each unit vector ξ_0 in H. Since H is separable, there can be at most countably many of these. This gives a partition of $\hat{A}$ in a sequence of D-Borel subsets, and it suffices to prove the theorem on each of these.

Choose a separable C^*-algebra C containing 1, which is weakly dense in $\pi(A)'$, and let B be the C^*-algebra generated by $\pi(A)$ and C. There is then a unique representation π' of B on H such that $\pi' | A = \pi$ and $\pi' | C =$ identity. Moreover, ξ_0 is cyclic for $\pi'(B)$ and determines a vector state ψ on B with restriction ϕ to A. Consider the commutative diagram:

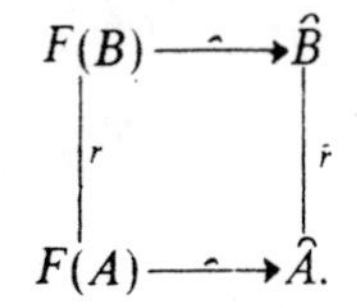

Since $\pi'(A)''$ and $\pi'(B)''$ have the same centre we conclude from 4.9.5 and 4.9.6 that $\mu_\phi = \mu_\psi \circ \hat{r}^{-1}$ and $\nu_\phi = \nu_\psi \circ r^{-1}$ (ν_ϕ and ν_ψ denoting the S-measures of ϕ and ψ, respectively). We want to restrict all maps to sets with (outer) measure 1, on which they are isomorphisms. To do so, note that $L^\infty_{\nu_\psi}(F(B)) = L^\infty_{\nu_\phi}(F(A))$ and since $F(B)$ is standard, it follows readily that r is injective outside a null set for ν_ψ. This means that we can define a Borel map $f: F(A) \to F(B)$ such that $r(f(\omega)) = \omega$ almost everywhere (with respect to ν_ϕ). For the canonical maps $\omega \to \hat{\omega}$ on $F(A)$ and $F(B)$ we have by 4.10.4 Borel maps $t \to \phi_t$ on $\hat{A}$ and $t \to \psi_t$ on $\hat{B}$, which are Borel inverses to $\omega \to \hat{\omega}$ on standard subsets T_ϕ and T_ψ, respectively. Combining these results we see that by deleting null sets from T_ϕ and T_ψ we may assume that $\hat{r}(T_\psi) = T_\phi$ and $f(\phi_{\hat{r}(t)})$ for each t in T_ψ.

Let $\mu = \mu_0$ and $T = T_\phi$. For each t in A let (π_t', H_t) be the cyclic representation of B associated with $f(\phi_t)$ and put $\pi_t = \pi_t' | A$. Then $\{H_t \,|\, t \in \hat{A}\}$ is a Borel field of Hilbert spaces (the Borel structure being generated by vector fields $\{\xi_x \,|\, x \in B\}$) and $\{\pi_t \,|\, t \in \hat{A}\}$ is a Borel field of factor representations of A by 4.9.5 such that $\pi_t \in t$ for each t in T. Finally, since $f(\phi_{\hat{r}(t)}) = \psi_t$ for each t in T_ψ we conclude that $(\int^\oplus \pi_t' \, d\mu(t), \int^\oplus H_t \, d\mu(t))$ is the canonical disintegration of (π_ψ, H_ψ) given in 4.12.2. From this the existence part of the theorem follows since (π_ψ, H_ψ) is spatially equivalent to (π, H).

Suppose that ν, $\{K_t\}$, v and $\{\rho_t\}$ satisfy the same conditions. Then, as in the proof of 4.12.3, μ and ν are equivalent and there is a null set N in $\hat{A}$ and a family of isometries $\{v_t \,|\, t \in T \backslash N\}$ such that $v_t H_t = K_t$ for each t in $T \backslash N$. Moreover, if $z = d\nu/d\mu$ then for each Borel vector field ξ in $\{H_t \,|\, t \in T \backslash N\}$ that is square integrable with respect to μ, we have $(v^* u\xi)(t) = z^{-1/2}(t) v_t \xi(t)$. As in 4.12.3 it follows that $v_t^* \rho_t v_t = \pi_t$ for each t in $T \backslash N$.

4.12.5. COROLLARY. *Let $\mathcal{M}$ be a von Neumann algebra on a separable Hilbert space H. There exists:*

(i) *A probability measure μ on a standard Borel space T;*

(ii) *A Borel field $\{H_t \mid t \in T\}$ of Hilbert spaces;*

(iii) *An isometry u from $\int^{\oplus} H_t \, d\mu(t)$ onto H;*

(iv) *A Borel field $\{\mathcal{M}_t \mid t \in T\}$ of factors such that $u^*\mathcal{M}u = \int^{\oplus} \mathcal{M}_t \, d\mu(t)$ and $u^*(\mathcal{M} \cap \mathcal{M}')u$ is the algebra of diagonalizable operators on $\int^{\oplus} H_t \, d\mu(t)$.*

If E, v, $\{K_t \mid t \in E\}$, v and $\{\mathcal{N}_t \mid t \in E\}$ satisfy the same conditions there are null sets N_1 and N_2 and a Borel isomorphism τ of $E\backslash N_2$ onto $T\backslash N_1$ such that $v \circ \tau^{-1}$ is equivalent to μ. Moreover, there is a family of isometries $\{v_t \mid t \in T\backslash N_1\}$ such that $v_t H_t = K_{\tau^{-1}(t)}$ and $v_t^* \mathcal{N}_{\tau^{-1}(t)} v_t = \mathcal{M}_t$ for each t in $T\backslash N_1$.

Proof. The existence of a disintegration for $\mathcal{M}$ follows from 4.12.4 by choosing a separable, weakly dense C^*-subalgebra A of $\mathcal{M}$. The uniqueness of the disintegration follows from 4.11.9 just as in 4.12.3.

4.12.6. Notes and remarks. Theorems 4.12.3 and 4.12.4 are the work of many hands. Without any claim of completeness we mention Effros, Ernest, Godement, Guichardet, Mackey, Mautner, Naimark, Segal and Tomita. The versions given here are borrowed from Dixmier, see 8.4.2 of [65]. The proof, however, rests on Sakai's disintegration of states from 4.8.7. Corollary 4.12.5 is von Neumann's original result from [177].

Chapter 5

Weights and Traces

In this chapter we shall study unbounded functionals on C^*-algebras. They arise in many contexts, but the commutative theory probably offers the best introduction to the subject.

If T is a locally compact Hausdorff space, σ-compact for convenience, we consider the familiar C^*-algebra $C_0(T)$ and its enveloping Borel *-algebra $\mathscr{B}(T)$. A positive *Borel measure* on T is (or can be identified with) a positive homogeneous affine function $\phi:\mathscr{B}(T)_+ \to [0, \infty]$, which is sequentially normal in the sense that $\operatorname{Lim} \phi(x_n) = \phi(x)$, whenever $\{x_n\}$ is a bounded increasing sequence in $\mathscr{B}(T)_+$ with limit x (Lebesgue's monotone convergence theorem). Recall that ϕ is said to be *σ-finite* if there is an increasing sequence $\{x_n\}$ in $\mathscr{B}(T)_+$ with limit 1 such that $\phi(x_n) < \infty$ for all n. Further, ϕ is a *Radon measure* if $\phi(x) < \infty$ for all positive x in the set $K(T)$ of continuous functions on T with compact supports. Of course, all bounded Borel measures (i.e. $\phi(1) < \infty$) are Radon measures, but for unbounded measures the distinction has to be made.

The best excuse for unbounded measures is Lebesgue measure: If T is a locally compact, non-compact group (viz. $T = \boldsymbol{R}$) then there are no translation invariant bounded measures on T. However, the Haar measure on T is a translation invariant, unbounded Radon measure. We shall have similar uses of unbounded functionals on C^*-algebras, functionals which are invariant under some group of automorphisms of the algebra, but where no such bounded functionals exist.

5.1. Weights

5.1.1. Let A be a C^*-algebra. A *weight* on A is a function $\phi: A_+ \to [0, \infty]$ such that

(i) $\phi(\alpha x) = \alpha\phi(x)$ if $x \in A_+$ and $\alpha \in \boldsymbol{R}_+$;

(ii) $\phi(x + y) = \phi(x) + \phi(y)$ if x and y belong to A_+.

Consider the set

$$A_+^\phi = \{x \in A_+ \mid \phi(x) < \infty\}.$$

We say that ϕ is *densely defined* if A_+^ϕ is dense in A_+. If $\mathscr{M}$ is a von Neumann algebra we say that ϕ is *semi-finite* if $\mathscr{M}_+^\phi$ is weakly dense in $\mathscr{M}_+$. If $\mathscr{A}$ is a Borel *-algebra we say that ϕ is *σ-finite* if $\mathscr{A}_+^\phi$ contains an increasing sequence with limit 1. In the case where $\mathscr{M}$ is a σ-finite von Neumann algebra (see 3.8.3) and thus, *a fortiori*, a Borel *-algebra the notions semi-finite and σ-finite coalesce: In one direction this is easy to see, since $e_n \nearrow 1$ and $\{e_n\} \subset \mathscr{M}_+^\phi$ implies that the set $\cup e_n \mathscr{M}_+ e_n$ is contained in $\mathscr{M}_+^\phi$ and weakly dense in $\mathscr{M}_+$. The other implication follows from the fact that the open unit ball of $\mathscr{M}_+$ intersects $\mathscr{M}_+^\phi$ in an increasing net (cf. the proof of 1.4.2). If therefore $\mathscr{M}$ is σ-finite then $\mathscr{M}_+^\phi$ contains a countable approximate unit $\{e_n\}$ for the C^*-algebra generated by $\mathscr{M}_+^\phi$. Since $\mathscr{M}_+^\phi$ is weakly dense this implies that $e_n \nearrow 1$.

If ϕ is a weight on a C^*-algebra A, then ϕ is *lower semi-continuous* if for each α in $\boldsymbol{R}_+$ the set $\{x \in A_+ \mid \phi(x) \leqslant \alpha\}$ is closed. If we are dealing with weights on von Neumann algebras or Borel *-algebras then it is more relevant to ask for normality or sequential normality of the weight. There are several ways of formulating this normality condition and though they are all equivalent this is not at all apparent. We shall use here the strongest form of the normality condition: If ϕ is a weight on a Borel *-algebra $\mathscr{A}$ we say that ϕ is a *σ-normal* weight if there exists a sequence $\{\phi_n\}$ of sequentially normal positive functionals on $\mathscr{A}$ such that $\phi(x) = \Sigma\, \phi_n(x)$ for every x in $\mathscr{A}_+$. It is clear that each sequentially normal positive functional is σ-normal and that any countable sum of σ-normal weights is a σ-normal weight.

We shall be mostly concerned with weights on Borel *-algebras which are both σ-finite and σ-normal. These we shall just call *σ-weights*. Note that the sum of two σ-weights is not necessarily a σ-weight, since it need not be σ-finite.

5.1.2. LEMMA. *For each weight ϕ on a C^*-algebra A the linear span A^ϕ of A_+^ϕ is a hereditary *-subalgebra of A with $(A^\phi)_+ = A_+^\phi$, and there is a unique extension of ϕ to a positive linear functional on A^ϕ. Moreover, the set $A_2^\phi = \{x \in A \mid x^*x \in A_+^\phi\}$ is a left ideal of A such that $y^*x \in A^\phi$ for any x, y in A_2^ϕ. Finally, for any x, y in A_2^ϕ we have $|\phi(y^*x)|^2 \leqslant \phi(y^*y)\phi(x^*x)$.*

Proof. Since A_+^ϕ is a hereditary cone in A_+ we see as in the proof of 1.5.2 that A_2^ϕ is a left ideal of A. From the *polarization identity*:

$$y^*x = \frac{1}{4}\sum_{k=0}^{3} i^k(x + i^k y)^*(x + i^k y)$$

it follows immediately that the *-algebra $(A_2^\phi)^*(A_2^\phi)$ is equal to A^ϕ, and that $(A^\phi)_+ = A_+^\phi$. It is clear that ϕ extends uniquely to a positive linear functional on A^ϕ, and the proof of the Cauchy–Schwarz inequality for weights is identical with the one given in 3.1.3.

5.1.3. PROPOSITION. *For each lower semi-continuous weight ϕ on a C^*-algebra A there is a non-degenerate representation (π_ϕ, H_ϕ) of A and a linear map $x \to \xi_x$ from A_2^ϕ to a dense subspace of H_ϕ such that $(\pi_\phi(x)\xi_y \mid \xi_z) = \phi(z^*xy)$ for all x in A and y, z in A_2^ϕ.*

Proof. Let H_ϕ be the completion of the pre-Hilbert space $A_2^\phi - L_\phi$, where L_ϕ is the left kernel of ϕ, where $x \to \xi_x$ is the quotient map of A_2^ϕ onto $A_2^\phi - L_\phi$ and where $(\xi_x \mid \xi_y) = \phi(y^*x)$ is the inner product (by 5.1.2). Taking $\pi_\phi(x)\xi_y = \xi_{xy}$ we obtain exactly as in 3.3.3 a *-representation of A on H_ϕ. If $\{u_\lambda\}$ is an approximate unit for A and $x \in A_2^\phi$ then

$$\|\xi_x - \pi_\phi(u_\lambda)\xi_x\|^2 \leqslant \phi(x^*(1 - u_\lambda)x) \to 0$$

since ϕ is lower semi-continuous; whence (π_ϕ, H_ϕ) is a non-degenerate representation.

5.1.4. LEMMA. *Let ϕ be a σ-weight on a Borel *-algebra $\mathscr{A}$. For each σ-normal positive functional ψ dominated by ϕ there is a vector ξ in H_ϕ such that $\psi(x) = (\pi_\phi(x)\xi \mid \xi)$ for all x in $\mathscr{A}$, and an a in $\pi_\phi(\mathscr{A})'$ with $0 \leqslant a \leqslant 1$ such that $a^{1/2}\xi_x = \pi_\phi(x)\xi$ for all x in $\mathscr{A}_2^\phi$.*

Proof. Exactly as in 3.3.5 we can show that there is an element a in $\pi_\phi(\mathscr{A})'$ with $0 \leqslant a \leqslant 1$ such that $\psi(x^*x) = (a\xi_x \mid \xi_x)$ for each x in $\mathscr{A}_2^\phi$. Let $\{e_n\}$ be an increasing sequence in $\mathscr{A}_+^\phi$ with limit 1. Then with $\xi_n = \xi_{e_n^{1/2}}$ we have

$$\|a^{1/2}\xi_n - a^{1/2}\xi_m\|^2 = \psi((e_n^{1/2} - e_m^{1/2})^2) \to 0,$$

and we define ξ as the limit of the Cauchy sequence $\{a^{1/2}\xi_n\}$. For each x in $\mathscr{A}$,

$$(\pi_\phi(x)\xi \mid \xi) = \operatorname{Lim}(\pi_\phi(x)a^{1/2}\xi_n \mid a^{1/2}\xi_n) = \operatorname{Lim}\psi(e_n^{1/2}xe_n^{1/2}) = \psi(x),$$

since ψ is sequentially normal. Moreover, for each x in $\mathscr{A}_2^\phi$,

$$\begin{aligned}\|a^{1/2}\xi_x - \pi_\phi(x)\xi\|^2 &= \operatorname{Lim}\|a^{1/2}(\xi_x - \pi_\phi(x)\xi_n)\|^2 \\ &= \operatorname{Lim}\psi((x - xe_n^{1/2})^*(x - xe_n^{1/2})) \\ &= \operatorname{Lim}\psi((1 - e_n^{1/2})x^*x(1 - e_n^{1/2})) = 0,\end{aligned}$$

and the proof is complete.

5.1.5. THEOREM. *For each σ-weight ϕ on a countably generated Borel *-algebra $\mathscr{A}$ the representation (π_ϕ, H_ϕ) is separable and π_ϕ is a sequentially normal morphism of $\mathscr{A}$ onto the von Neumann algebra $\pi_\phi(\mathscr{A})$. Moreover, there is a* *sequence* $\{\xi_n\}$ *of vectors in* H_ϕ *such that* $\phi(x) = \Sigma(\pi_\phi(x)\xi_n \mid \xi_n)$ *for each* x *in* $\mathscr{A}_+$.

Proof. By assumption $\phi = \Sigma\,\phi_n$, and since the ϕ_n's are norm continuous, ϕ is lower semi-continuous. Thus the result from 5.1.3 can be applied. If $\{x_n\}$ is an increasing sequence in $\mathscr{A}_{sa}$ with limit x and if $y \in \mathscr{A}_2^\phi$ then

$$((\pi_\phi(x) - \pi_\phi(x_n))\xi_y \mid \xi_y) = \phi(y^*(x - x_n)y) \searrow 0,$$

since ϕ is σ-normal. This shows that (π_ϕ, H_ϕ) is a sequentially normal representation.

For each n we have $\phi_n \leqslant \phi$ and thus there is a sequence $\{\xi_n\}$ in H_ϕ and a sequence $\{a_n\}$ in $\pi_\phi(\mathscr{A})'$ satisfying 5.1.4. Consequently

$$\phi(x) = \sum \phi_n(x) = \sum(\pi_\phi(x)\xi_n \mid \xi_n)$$

for each x in $\mathscr{A}_+$. Now

$$\sum(a_n\xi_x \mid \xi_x) = \sum(\pi_\phi(x^*x)\xi_n \mid \xi_n) = \phi(x) = \|\xi_x\|^2$$

for each x in $\mathscr{A}_2^\phi$, which shows that $\Sigma\,a_n = 1$. Since $\mathscr{A}$ is countably generated, $a_n^{1/2}H_\phi$, being the closure of $\pi_\phi(\mathscr{A})\xi_n$, is separable, and consequently H_ϕ itself is separable. It follows from 2.4.3 that $\pi_\phi(\mathscr{A})$ is a von Neumann algebra.

5.1.6. LEMMA. *Let* ϕ *be a* σ*-weight on a von Neumann algebra* $\mathscr{M}$. *The densely defined map* $\xi: \mathscr{M} \to H_\phi$ *given by* $x \to \xi_x, x \in \mathscr{M}_2^\phi$ *(cf. 5.1.3) is a closed operator from* $\mathscr{M}$ *equipped with the* σ*-weak topology to* H_ϕ *with the weak topology.*

Proof. Let $\{\phi_n\}$ be an increasing sequence of positive normal functionals on $\mathscr{M}$ such that $\phi(x) = \operatorname{Lim}\phi_n(x)$ for each x in $\mathscr{M}_+$. By 5.1.4 there is for each n a vector ξ_n in H_ϕ and an element a_n in $\pi_\phi(\mathscr{M})'_+$ such that $\phi_n(x) = (\pi(x)\xi_n \mid \xi_n)$ for all x; and $\pi_\phi(x)\xi_n = a_n^{1/2}\xi_x$ if $x \in \mathscr{M}_2^\phi$. Since $\phi_n \nearrow \phi$ it follows that $a_n \nearrow 1$.

Suppose that $\{x_i\}$ is a net in $\mathscr{M}_2^\phi$ such that $x_i \to x$ σ-weakly and $\xi_{x_i} \to \xi$ weakly for some x in $\mathscr{M}$ and ξ in H_ϕ. Then

$$\operatorname*{Lim}_i \pi_\phi(x_i)\xi_n = \pi_\phi(x)\xi_n, \qquad \operatorname*{Lim}_i a_n^{1/2}\xi_{x_i} = a_n^{1/2}\xi,$$

weakly. Consequently $\pi_\phi(x)\xi_n = a_n^{1/2}\xi$, whence

$$\phi_n(x^*x) = \|\pi_\phi(x)\xi_n\|^2 = (a_n\xi \mid \xi).$$

In the limit we obtain $\phi(x^*x) = \|\xi\|^2 < \infty$ so that $x \in \mathscr{M}_2^\phi$. But then

$$a_n^{1/2}\xi_x = \pi_\phi(x)\xi_n = a_n^{1/2}\xi$$

for every n, whence $\xi_x = \xi$, as desired.

5.1.7. PROPOSITION. *Let* ϕ *and* ψ *be* σ*-weights on a von Neumann algebra* $\mathscr{M}$ *and let* $\xi^\phi: \mathscr{M} \to H_\phi$ *and* $\xi^\psi: \mathscr{M} \to H_\psi$ *be the associated maps (cf. 5.1.3). If there* *is a set M in $\mathcal{M}_2^\phi \cap \mathcal{M}_2^\psi$ such that $\{(x, \xi_x^\phi) \mid x \in M\}$ and $\{(x, \xi_x^\psi) \mid x \in M\}$ are dense in the graphs of ξ^ϕ and ξ^ψ, respectively; and an isometry $u: H_\phi \to H_\psi$ such that $u\xi_x^\phi = \xi_x^\psi$ for every x in M, then $\phi = \psi$.*

Proof. If $x \in \mathcal{M}_2^\phi$ choose a net $\{x_i\}$ in M such that $\{(x_i, \xi_{x_i}^\phi)\}$ converges to (x, ξ_x^ϕ) in the graph of ξ^ϕ. Thus $x_i \to x$, σ-weakly and $\xi_{x_i}^\phi \to \xi_x^\phi$, weakly. Since $u\xi_{x_i}^\phi = \xi_{x_i}^\psi$ we see that $\xi_{x_i}^\psi \to u\xi_x^\phi$, weakly. It follows from 5.1.6 that $x \in \mathcal{M}_2^\psi$ and that $\xi_x^\psi = u\xi_x^\phi$. Consequently

$$\phi(x^*x) = \|\xi_x^\phi\|^2 = \|\xi_x^\psi\|^2 = \psi(x^*x).$$

Thus $\phi = \psi$ on $\mathcal{M}_+^\phi$, whence $\phi \leq \psi$. By a symmetric argument $\psi \leq \phi$, i.e. $\phi = \psi$.

5.1.8. Notes and remarks. Weights were introduced by Dixmier (see 1.4.2 of [59]) and many of their properties were predicted by Tomita (cf. [265] p. 75). However, the first serious studies were made by Combes [34] and the author [190]. The main results 5.1.3 and 5.1.5 are due to Combes. The author studied a special class of weights (C^*-integrals) on C^*-algebras and carried that theory to its utter limit in [200].

When the Tomita–Takesaki theory (see 8.13) swept through operator algebra theory in the early seventies and Combes [37] showed that a left Hilbert algebra structure (the basic ingredient in the old-fashioned T–T theory) was equivalent with the existence of a faithful, normal, semi-finite weight, the theory of weights on von Neumann algebras became fashionable. Some basic research was carried out by Pedersen and Takesaki [210], and Haagerup [111] was able to settle Dixmier's old problem from 1.4.2 of [59] in the affirmative: every weight on a von Neumann algebra which is normal (in the sense that it preserves limits of monotone increasing nets) can be expressed as a sum of positive normal functionals. Thanks to this result our definition of σ-normal weights in 5.1.1 is justifiable.

5.2. Traces

5.2.1. A *trace* on a C^*-algebra A is a weight ϕ such that $\phi(u^*xu) = \phi(x)$ for all x in A_+ and all unitary u in $\tilde{A}$. Thus the traces are exactly the weights which are invariant under the group of inner automorphisms of A (strictly speaking, the inner automorphisms of $\tilde{A}$). The motivating example is of course the canonical trace Tr on $B(H)$, cf. 3.5.1.

If ϕ is a trace and $x \in A_+^\phi$, then for each unitary u in $\tilde{A}$

$$u^*x = (x^{1/2}u)^*x^{1/2} = \frac{1}{4}\sum_{k=0}^{3} i^k(1 + i^k u)^*x(1 + i^k u)$$

(polarization identity), and $(1 + i^k u)^* x (1 + i^k u) \leqslant 2(x + u^* x u)$ for each k. By assumption $u^* x u \in A_+^\phi$, and thus $u^* x \in A^\phi$. Since each element in $\tilde{A}$ is the linear combination of unitaries (see 1.1.9) it follows that A^ϕ is an ideal of A. This implies that also A_2^ϕ is an ideal.

5.5.2. PROPOSITION. *If ϕ is a trace on a C*-algebra A then $\phi(yx) = \phi(xy)$ for each x in A^ϕ and y in $\tilde{A}$. If moreover ϕ is lower semi-continuous then $\phi(x^*x) = \phi(xx^*)$ for all x in A and $\phi(xy) = \phi(yx)$ for all x and y in A_2^ϕ.*

Proof. If $x \in A$ and u is unitary in $\tilde{A}$, then $ux \in A^\phi$ by 5.2.1 and

$$\phi(ux) = \phi(u^*(ux)u) = \phi(xu).$$

Since each element in $\tilde{A}$ is a linear combination of unitaries it follows that $\phi(yx) = \phi(xy)$ for each y in $\tilde{A}$.

Assume now that ϕ is lower semi-continuous and take x in A_2^ϕ. Then $xx^* \in A_2^\phi$ and if $\{u_\lambda\}$ is an approximate unit for A^ϕ then since A^ϕ is dense in A_2^ϕ (they have the same closure cf. 1.5.3) we have

$$\phi(xx^*) \leqslant \operatorname{Lim\,inf} \phi(u_\lambda x x^* u_\lambda) = \operatorname{Lim\,inf} \phi(x^* u_\lambda^2 x) \leqslant \phi(x^*x).$$

Consequently $x^* \in A_2^\phi$ and $\phi(xx^*) = \phi(x^*x)$. If both x and y belong to $(A_2^\phi)_{sa}$ then since

$$(x + iy)^*(x + iy) - (x + iy)(x + iy)^* = 2i(xy - yx),$$

we conclude that $\phi(xy) = \phi(yx)$. As A_2^ϕ is a *-ideal this equality then holds for all x and y in A_2^ϕ.

5.2.3. PROPOSITION. *If ϕ is a σ-finite trace on a countably generated Borel *-algebra $\mathscr{A}$, then ϕ is σ-normal if and only if $\phi(x) = \operatorname{Lim} \phi(x_n)$ for each increasing sequence $\{x_n\}$ in $\mathscr{A}_+$ with limit x.*

Proof. By assumption there is an increasing sequence $\{e_n\}$ in $\mathscr{A}_+^\phi$ with limit 1. For each x in $\mathscr{A}$ we have $\phi(e_n x) = \phi(e_n^{1/2} x e_n^{1/2})$ by 5.2.2, which shows that each functional $\phi(e_n \cdot)$ is positive, bounded and sequentially normal. The same is therefore true for the functionals

$$\phi_n = \phi((e_{n+1} - e_n)\cdot), \qquad n \in \boldsymbol{N};$$

and for each x in $\mathscr{A}_+$

$$\sum \phi_n(x) = \sum \phi(x^{1/2}(e_{n+1} - e_n)x^{1/2}) = \phi\left(\sum x^{1/2}(e_{n+1} - e_n)x^{1/2}\right) = \phi(x),$$

which shows that ϕ is σ-normal.

5.2.4. LEMMA. *If ϕ is a σ-weight on a countably generated Borel *-algebra $\mathscr{A}$, and if $\phi(u_n^* \cdot u_n) \leqslant \phi$ for a sequence $\{u_n\}$ of unitaries which is norm dense in the unitary group of a C*-algebra A generating $\mathscr{A}$, then ϕ is a trace on $\mathscr{A}$.*

Proof. By 5.1.5 we may assume that $\mathscr{A}$ is a von Neumann algebra, that A is strongly dense in $\mathscr{A}$, and that ϕ is weakly lower semi-continuous on $\mathscr{A}_+$. *A fortiori*, ϕ is norm lower semi-continuous so that $\phi(u^*xu) \leqslant \phi(x)$ for all unitaries u in A and all x in $\mathscr{A}_+^\phi$. Since the unitary group of A is strongly dense in the unitary group of $\mathscr{A}$ by 2.3.3 it follows that $\phi(u^*xu) \leqslant \phi(x)$ for every unitary u in $\mathscr{A}$ and every x in $\mathscr{A}_+^\phi$. But this implies that ϕ is a trace on $\mathscr{A}$.

5.2.5. LEMMA. (cf. 1.4.10). *If $\{x_n\}$ and $\{y_m\}$ are finite sets in a C*-algebra A such that $\Sigma x_n^* x_n = \Sigma y_m y_m^*$, then there is a set of elements $\{z_{nm}\}$ in A such that*

$$\sum z_{nm} z_{nm}^* = x_n x_n^*, \qquad \sum z_{nm}^* z_{nm} = y_m^* y_m.$$

Proof. Let $a = \Sigma x_n^* x_n (= \Sigma y_m y_m^*)$ and define z_{nm} as the limit of the sequence with elements

$$z_{nmk} = x_n[(1/k) + a]^{-1/2} y_m,$$

which is norm convergent by 1.4.5. Then

$$\sum_m z_{nm} z_{nm}^* = \operatorname{Lim}_k x_n[(1/k) + a]^{-1/2} \left(\sum y_m y_m^*\right) [(1/k) + a]^{-1/2} x_n^* = x_n x_n^*$$

arguing as in the proof of 1.4.10. Similarly $\Sigma_n z_{nm}^* z_{nm} = y_m^* y_m$ and the proof is complete.

5.2.6. We define an equivalence relation in A_+ by setting $x \approx y$ if there is a finite set $\{z_n\}$ in A such that $x = \Sigma z_n^* z_n$ and $y = \Sigma z_n z_n^*$. As an immediate consequence of 5.2.5 we see that if x, y_1, y_2 belong to A_+ and $x \approx y_1 + y_2$ then $x = x_1 + x_2$ with $x_i \approx y_i$ for $i = 1, 2$. Since the equivalence relation is clearly additive this shows that the equivalence classes form a partially ordered cone which satisfy the Riesz decomposition property (cf. 1.4.10).

If ϕ is a lower semi-continuous trace on A and x, y are elements of A_+ with $x \approx y$ then $\phi(x) = \phi(y)$ by 5.2.2. Conversely, if $A = C(H)$ and Tr is the canonical trace then for positive operators x, y of finite rank $x \approx y$ if and only if $\operatorname{Tr}(x) = \operatorname{Tr}(y)$.

If $\mathscr{A}$ is a Borel *-algebra one can formulate a countable analogue of 5.2.5: if $\Sigma x_n^* x_n = \Sigma y_m y_m^*$ there is a double sequence $\{z_{nm}\}$ in $\mathscr{A}$ such that $\Sigma z_{nm} z_{nm}^* = x_n x_n^*$ and $\Sigma z_{nm}^* z_{nm} = y_m^* y_m$. The proof is virtually the same.

We use the notation $y \precsim x$ to mean $y \approx x_1, x_1 \leqslant x$.

5.2.7. THEOREM. *Let B be a hereditary C*-subalgebra of a C*-algebra A, and let ϕ be a lower semi-continuous weight on B. For each x in A_+ define*

$$\tilde{\phi}(x) = \operatorname{Sup}\{\phi(y) \mid y \in B_+, y \precsim x\}.$$

Then $\tilde{\phi}$ is a lower semi-continuous trace on A and $\tilde{\phi} \mid B_+$ is the smallest trace dominating ϕ.

Proof. It is clear that $\tilde{\phi}$ is positive homogeneous. Take x_1 and x_2 in A_+ and given $\varepsilon > 0$ choose y_1 and y_2 in B_+ such that $\tilde{\phi}(x_i) \leqslant \varepsilon + \phi(y_i)$, and $y_i \precsim x_i$, $i = 1, 2$. Then $y_1 + y_2 \precsim x_1 + x_2$ whence

$$\tilde{\phi}(x_1) + \tilde{\phi}(x_2) \leqslant 2\varepsilon + \phi(y_1 + y_2) \leqslant 2\varepsilon + \tilde{\phi}(x_1 + x_2).$$

Conversely, choose y in B_+ such that $\tilde{\phi}(x_1 + x_2) \leqslant \varepsilon + \phi(y)$ and $y \precsim x_1 + x_2$. Then by 5.2.6 $y = y_1 + y_2$ with $y_i \precsim x_i$, $i = 1, 2$, and since B is hereditary, $y_i \in B_+$. Thus

$$\tilde{\phi}(x_1 + x_2) \leqslant \varepsilon + \phi(y_1) + \phi(y_2) \leqslant \varepsilon + \tilde{\phi}(x_1) + \tilde{\phi}(x_2).$$

It follows that $\tilde{\phi}$ is a weight on A_+. However, if u is unitary in $\tilde{A}$ and $x \in A_+$, $y \in B_+$, then $x \approx y$ if and only if $u^*xu \approx y$. Consequently $\tilde{\phi}(u^*xu) = \tilde{\phi}(x)$ so that $\tilde{\phi}$ is a trace on A_+.

Clearly $\tilde{\phi} \mid B_+$ dominates ϕ. On the other hand, if ψ is a trace on B dominating ϕ then for all x and y in B_+ with $y \precsim x$ we have $\phi(y) \leqslant \psi(y) \leqslant \psi(x)$. It follows that $\tilde{\phi}(x) \leqslant \psi(x)$, so that $\tilde{\phi} \mid B_+$ is indeed the smallest trace dominating ϕ.

Assume now that ϕ is lower semi-continuous and let $\{x_n\}$ be a sequence in A_+ which converges to x, such that $\tilde{\phi}(x_n) \leqslant 1$ for all n. If $y \in B_+$ and $y \precsim x$, say $y = \Sigma z_k z_k^*$, $x \geqslant \Sigma z_k^* z_k$ with $1 \leqslant k \leqslant m$ then for each $\varepsilon > 0$ we have $x \leqslant x_n + \varepsilon$ for n sufficiently large. It follows from 5.2.5 that $y = y_n + y_\varepsilon$ where $y_n = \Sigma u_k^* u_k$, $\Sigma u_k u_k^* \leqslant x_n$, $y_\varepsilon = \Sigma v_k^* v_k$, $\Sigma v_k v_k^* \leqslant \varepsilon$ and $1 \leqslant k \leqslant m$. Consequently

$$\|y_\varepsilon\| < \Sigma \|v_k\|^2 \leqslant m\varepsilon,$$

which shows that $\{y_n\}$ converges to y. Moreover,

$$\phi(y_n) \leqslant \tilde{\phi}(y_n) \leqslant \tilde{\phi}(x_n) \leqslant 1,$$

and since ϕ is lower semi-continuous we conclude that $\phi(y) \leqslant 1$. It follows that $\tilde{\phi}(x) \leqslant 1$, which proves that $\tilde{\phi}$ is lower semi-continuous.

5.2.8. PROPOSITION. *Let p be a projection in a countably generated Borel *-algebra $\mathscr{A}$. If ϕ is a σ-weight on $p\mathscr{A}p$ such that $\tilde{\phi}$ (as defined in 5.2.7) is σ-finite, then $\tilde{\phi}$ is a σ-trace on $\mathscr{A}$, and if ϕ is faithful on $p\mathscr{A}p$ then $\tilde{\phi}$ is faithful on $c(p)\mathscr{A}$. In particular, if ϕ is a σ-trace on $p\mathscr{A}p$ then $\tilde{\phi}$ is a σ-trace on $\mathscr{A}$ which extends ϕ.*

Proof. If $\tilde{\phi}$ is σ-finite then by 5.2.3 it is also σ-normal if $\Sigma\tilde{\phi}(x_n) = \tilde{\phi}(x)$ for each sequence $\{x_n\}$ in $\mathscr{A}_+$ with $\Sigma x_n = x$. But if $y \in p\mathscr{A}p$ and $y \precsim x$ then by the countable analogue of 5.2.5 (see 5.2.6) $y = \Sigma y_n$ with $y_n \precsim x_n$ for each n. Consequently,

$$\phi(y) = \Sigma \phi(y_n) \leqslant \Sigma \tilde{\phi}(x_n),$$

which shows that $\tilde{\phi}(x) \leqslant \Sigma \tilde{\phi}(x_n)$. The other inequality is evident, so $\tilde{\phi}$ is σ-trace.

No element in $(1 - c(p))\mathscr{A}$ is equivalent to any non-zero element in $p\mathscr{A}_+p$, which implies that $\tilde{\phi}$ vanishes on $(1 - c(p))\mathscr{A}$. We may therefore as well assume that $c(p) = 1$. If now ϕ is faithful on $p\mathscr{A}p$ and $L_{\tilde{\phi}}$ denotes the left kernel of $\tilde{\phi}$ then $L_{\tilde{\phi}}$ is a sequentially closed ideal of $\mathscr{A}$ since $\tilde{\phi}$ is a σ-trace. If $L_{\tilde{\phi}} \neq 0$ then $L_{\tilde{\phi}} \cap p\mathscr{A}p \neq 0$ since $c(p) = 1$. But $\phi \leqslant \tilde{\phi}$ on $p\mathscr{A}p$ and ϕ is faithful; whence $L_{\tilde{\phi}} \cap p\mathscr{A}p = 0$ and $\tilde{\phi}$ is faithful, since $L_{\tilde{\phi}} = 0$.

If ϕ is a σ-trace on $p\mathscr{A}p$ then from the definition in 5.2.7 it follows that $\tilde{\phi}(x) = \phi(x)$ for each x in $p\mathscr{A}_+p$. To show that $\tilde{\phi}$ is a σ-trace it suffices by the first part of the proof to show that $\tilde{\phi}$ is σ-finite. By assumption there is an increasing sequence $\{e_n\}$ in $p\mathscr{A}_+p$ with limit p such that $\phi(e_n) < \infty$ for all n. With $\{u_k\}$ a countable generating subgroup of unitaries in $\mathscr{A}$ we set $e = \Sigma\, \alpha_k u_k^* e_k u_k$, where the α_k's are chosen so that $\alpha_k > 0$ and $\Sigma\, \alpha_k < \infty$, $\Sigma\, \alpha_k \phi(e_k) < \infty$. Then $[e] = 1$ $(= c(p))$ and since $\tilde{\phi}$ is lower semi-continuous (in norm) by 5.2.7.

$$\tilde{\phi}(e) = \sum \alpha_k \tilde{\phi}(u_k^* e_k u_k) = \sum \alpha_k \phi(e_k) < \infty.$$

With p_n as the spectral projection of e corresponding to $[1/n, \infty[$ we have $p_n \nearrow 1$ and $\tilde{\phi}(p_n) \leqslant n\tilde{\phi}(e) < \infty$, whence $\tilde{\phi}$ is σ-finite.

5.2.9. Notes and remarks. General traces were introduced by Murray and von Neumann [169]. Since then they have formed an indispensable ingredient in von Neumann algebra theory. Much later Dixmier and Guichardet [64, 106] made systematic use of traces on C^*-algebras, leading to Dixmier's and Fell's theory [64, 91] of C^*-algebras with continuous trace (cf. 6.1.9). The extension result in 5.2.7 is due to the author, see [193].

Deviating slightly from 5.2.6 one may define an equivalence relation $\sim$ in A_+ by setting $x \sim y$ if $x = \Sigma\, z_n^* z_n$, $y = \Sigma\, z_n z_n^*$, the sums being infinite but norm convergent. The set of differences $x - y$, where $x \sim y$ in A_+ becomes a closed subspace A_0 of A_{sa} and the quotient space $A^q = A_{sa}/A_0$ is a partially ordered vector space satisfying the Riesz decomposition property. The dual space of A^q is isomorphic with the set of finite tracial functionals on A, so that information about A^q leads to information about traces on A. Moreover a notion of finite elements (relative to $\sim$) leads to a type theory somewhat resembling the von Neumann algebra type theory (cf. 5.5.12). A self-contained account of these results is found in Cuntz and Pedersen [48].

5.3. The Radon–Nikodym theorem for traces

5.3.1. All non-commutative versions of the Radon–Nikodym theorem are of the form: if ψ and ϕ are positive functionals on a C^*-algebra and if ψ is 'small' relative to ϕ, then there is an operator h such that ψ can be expressed in terms of ϕ and h. One such theorem is 3.3.5 where h is an element in the commutant.

In this section we shall obtain Radon–Nikodym theorems where h is in the algebra itself, or rather in its weak closure. The most satisfying results obtains when ϕ is a trace. However, for later use (modular theory in Chapter 8) we prove also a result for general functionals.

5.3.2. PROPOSITION. *If ϕ and ψ are normal functionals on a von Neumann algebra $\mathcal{M}$ such that $0 \leqslant \psi \leqslant \phi$, then for each complex λ with $\operatorname{Re}\lambda \geqslant \frac{1}{2}$ there is an element h in $\mathcal{M}_+^1$ such that $\psi = \lambda\phi(h\cdot) + \bar{\lambda}\phi(\cdot h)$. If ϕ is faithful then h is unique.*

Proof. The set

$$M = \{\lambda\phi(h\cdot) + \bar{\lambda}\phi(\cdot h) \mid h \in \mathcal{M}_+^1\}$$

is convex and compact in $(\mathcal{M}_*)_{\text{sa}}$, since $\mathcal{M}_+^1$ is convex and σ-weakly compact. If $\psi \notin M$ then by the Hahn–Banach theorem there is an element a in $(\mathcal{M}_*)_{\text{sa}}^*$ $(= \mathcal{M}_{\text{sa}}$ by 3.6.5) and a real t such that

$$\psi(a) > t, \qquad M(a) \leqslant t.$$

Write $a = a_+ - a_-$ and take $h = [a_+]$. Then

$$\psi(a_+) \geqslant \psi(a_+ - a_-) > t \geqslant 2\operatorname{Re}\lambda\phi(a_+) \geqslant \phi(a_+)$$

in contradiction with $\psi \leqslant \phi$.

If ϕ is faithful and if $\psi = \lambda\phi(k\cdot) + \bar{\lambda}\phi(\cdot k)$ for some k in $\mathcal{M}_{\text{sa}}$ then since

$$(\lambda + \bar{\lambda})(h - k)^2 = \lambda h(h - k) + \bar{\lambda}(h - k)h - \lambda k(h - k) - \bar{\lambda}(h - k)k$$

we conclude that

$$2\operatorname{Re}\lambda\phi((h - k)^2) = \psi(h - k) - \psi(h - k) = 0,$$

whence $h = k$.

5.3.3. THEOREM. *Let ϕ be a σ-trace on a countably generated Borel *-algebra $\mathcal{A}$ and let (π_ϕ, H_ϕ) be the sequentially normal representation associated with ϕ. There is a sequentially normal anti-representation ρ_ϕ of $\mathcal{A}$ on H_ϕ and a conjugate linear isometry j of H_ϕ onto itself such that j^2 is the identity and $j\pi_\phi(x)j = \rho_\phi(x^*)$ for all x in $\mathcal{A}$. Moreover, $\pi_\phi(\mathcal{A})' = \rho_\phi(\mathcal{A})$ and $\rho_\phi(\mathcal{A})' = \pi_\phi(\mathcal{A})$.*

Proof. For each x in $\mathcal{A}$ and y in $\mathcal{A}_2^\phi$ define $\rho_\phi(x)\xi_y = \xi_{yx}$. By 5.2.2

$$\begin{aligned}\|\xi_{yx}\|^2 &= \phi(x^*y^*yx) = \phi(yxx^*y^*) \leqslant \|x\|^2\phi(yy^*)\\ &= \|x\|^2\phi(y^*y) = \|x\|^2\,\|\xi_y\|^2.\end{aligned}$$

Therefore $\rho_\phi(x)$ extends to a bounded operator on H_ϕ. It is straightforward to verify that the map $x \to \rho_\phi(x)$ satisfies all the requirements for being a sequentially normal representation of $\mathcal{A}$, except that we now have $\rho_\phi(x_1x_2) = \rho_\phi(x_2)\rho_\phi(x_1)$ (anti-homomorphism). Define j on H_ϕ as the unique extension of the conjugate linear isometry (involution) defined by $j\xi_y = \xi_{y*}$ for each y in $\mathscr{A}_2^\phi$. Elementary computations show that $j\pi_\phi(x)j = \rho_\phi(x^*)$ for all x in $\mathscr{A}$. Moreover, $\rho_\phi(x_1)\pi_\phi(x_2) = \pi_\phi(x_2)\rho_\phi(x_1)$ for x_1 and x_2 in $\mathscr{A}$, whence $\pi_\phi(\mathscr{A}) \subset \rho_\phi(\mathscr{A})'$ and $\rho_\phi(\mathscr{A}) \subset \pi_\phi(\mathscr{A})'$.

We claim that $\pi_\phi(e)x\pi_\phi(e) \in \pi_\phi(\mathscr{A})$ for each e in $\mathscr{A}_+^\phi$ and x in $\rho_\phi(\mathscr{A})'_{\text{sa}}$. To show this choose a sequence $\{x_n\}$ in $\mathscr{A}_2^\phi$ such that $\pi_\phi(e)x\xi_e = \operatorname{Lim} \xi_{x_n}$. For each y in $\mathscr{A}_2^\phi$ we have

$$
\begin{aligned}
(j\pi_\phi(e)x\xi_e \mid \xi_y) &= (j\xi_y \mid \pi_\phi(e)x\xi_e) \\
&= (\pi_\phi(e)\xi_{y^*} \mid x\xi_e) = (\rho_\phi(y^*)\xi_e \mid x\xi_e) \\
&= (x\xi_e \mid \rho_\phi(y)\xi_e) = (\pi_\phi(e)x\xi_e \mid \xi_y).
\end{aligned}
$$

This shows that $j\pi_\phi(e)x\xi_e = \pi_\phi(e)x\xi_e$, whence

$$
\pi_\phi(e)x\xi_e = \operatorname{Lim} \tfrac{1}{2}(\xi_{x_n} + j\xi_{x_n}) = \operatorname{Lim} \tfrac{1}{2}\xi_{x_n + x_n^*}.
$$

We may therefore assume that $x_n = x_n^*$ for all n. For all y, z in $\mathscr{A}_2^\phi$ and a in $\pi_\phi(\mathscr{A})'$ we now compute

$$
\begin{aligned}
(a\pi_\phi(e)x\pi_\phi(e)\xi_y \mid \xi_z) &= (a\rho_\phi(y)\pi_\phi(e)x\xi_e \mid \xi_z) \\
= \operatorname{Lim}(a\rho_\phi(y)\xi_{x_n} \mid \xi_z) &= \operatorname{Lim}(a\pi_\phi(x_n)\xi_y \mid \xi_z) \\
= \operatorname{Lim}(a\xi_y \mid \pi_\phi(x_n)\xi_z) &= \operatorname{Lim}(a\xi_y \mid \rho_\phi(z)\xi_{x_n}) \\
= (a\xi_y \mid \rho_\phi(z)\pi_\phi(e)x\xi_e) &= (\pi_\phi(e)x\pi_\phi(e)a\,\xi_y \mid \xi_z).
\end{aligned}
$$

This shows that

$$
a\pi_\phi(e)x\pi_\phi(e) = \pi_\phi(e)x\pi_\phi(e)a,
$$

whence

$$
\pi_\phi(e)x\pi_\phi(e) \in \pi_\phi(\mathscr{A})'' = \pi_\phi(\mathscr{A}),
$$

as desired.

Since ϕ is a σ-trace there is an increasing sequence $\{e_n\}$ in $\mathscr{A}_+^\phi$ such that $e_n \nearrow 1$. Consequently

$$
x = \operatorname{Lim} \pi_\phi(e_n)x\pi_\phi(e_n) \in \pi_\phi(\mathscr{A})
$$

for every x in $\rho_\phi(\mathscr{A})'_{\text{sa}}$ from the above, whence $\rho_\phi(\mathscr{A})' \subset \pi_\phi(\mathscr{A})$. Thus $\rho_\phi(\mathscr{A})' = \pi_\phi(\mathscr{A})$ and $\pi_\phi(\mathscr{A})' = \rho_\phi(\mathscr{A})$, completing the proof.

5.3.4. PROPOSITION. *Let ϕ be a σ-trace on a countably generated Borel *-algebra $\mathscr{A}$. For each σ-weight ψ on $\mathscr{A}$ which is dominated by ϕ there is an element h in $\mathscr{A}$, with $0 \leqslant h \leqslant 1$, such that $\psi(x) = \phi(h^{1/2}xh^{1/2})$ for all x in $\mathscr{A}_+$. The weight ψ is bounded if and only if $\phi(h) < \infty$ and it is a trace if and only if h is central.*

Proof. By 5.1.4 there is an element a in $\pi_\phi(\mathscr{A})'$ with $0 \leqslant a \leqslant 1$ such that $\psi(x^*x) = (a\xi_x \mid \xi_x)$ for each x in $\mathscr{A}_2^\phi$. Choose by 5.3.3 an h in $\mathscr{A}$ with $0 \leqslant h \leqslant 1$ such that $\rho_\phi(h) = a$. Then

$$\psi(x^*x) = (\rho_\phi(h)\xi_x \mid \xi_x) = \phi(x^*xh) = \phi(h^{1/2}x^*xh^{1/2}).$$

Let $\{e_n\}$ be a sequence in $\mathscr{A}_+^\phi$ with $e_n \nearrow 1$. Then for each x in $\mathscr{A}_+$

$$\psi(x) = \operatorname{Lim} \psi(x^{1/2}e_nx^{1/2}) = \operatorname{Lim} \phi(h^{1/2}x^{1/2}e_nx^{1/2}h^{1/2}) = \phi(h^{1/2}xh^{1/2}).$$

Clearly $\phi(h) < \infty$ if and only if ψ is bounded, and ψ is a trace if h is central. But if ψ is a trace then for each x in $\mathscr{A}_2^\phi$ and each unitary u in $\mathscr{A}$

$$(\rho_\phi(uhu^*)\xi_x \mid \xi_x) = \phi(x^*xuhu^*) = \phi(u^*x^*xuh)$$
$$= \psi(u^*x^*xu) = \psi(x^*x) = (\rho_\phi(h)\xi_x \mid \xi_x).$$

Thus $\rho_\phi(h)$ is central in $\rho_\phi(\mathscr{A})$ and we may replace h by $c(h)$ to obtain the desired result.

5.3.5. PROPOSITION. *Let ϕ be a σ-weight on a countably generated Borel *-algebra $\mathscr{A}$ and assume that there is an increasing sequence of projections $\{p_n\}$ in $\mathscr{A}_+^\phi$ with $p_n \nearrow 1$ such that $\phi \mid p_n\mathscr{A}p_n$ is a trace for every n. Then ϕ is a σ-trace on $\mathscr{A}$.*

Proof. By 5.1.5 we may assume that $\mathscr{A}$ is a von Neumann algebra and that ϕ is weakly lower semi-continuous on $\mathscr{A}_+$. Define positive normal functionals ϕ_n on $\mathscr{A}$ by $\phi_n(x) = \phi(p_nxp_n)$. Since $n \leqslant m$ implies that $p_n \leqslant p_m$, and ϕ is a trace on $p_m\mathscr{A}p_m$, it follows that $\phi_n \leqslant \phi_m$. We define a σ-weight $\bar\phi$ on $\mathscr{A}$ by

$$\bar\phi(x) = \operatorname{Lim} \phi_n(x), \qquad x \in \mathscr{A}_+.$$

For every x in $\mathscr{A}$ we have

$$\bar\phi(x^*p_nx) = \operatorname{Lim} \phi(p_mx^*p_nxp_m) = \operatorname{Lim} \phi(p_nxp_mx^*p_n) = \phi(p_nxx^*p_n)$$

whence in the limit $\bar\phi(x^*x) = \bar\phi(xx^*)$; so that $\bar\phi$ is a σ-trace. Moreover, if $x \in \mathscr{A}_+^{\bar\phi}$

$$\bar\phi(x) = \operatorname{Lim} \phi(p_nxp_n) \geqslant \phi(x),$$

since ϕ is lower semi-continuous, so that $\phi \leqslant \bar\phi$. By 5.3.4 we have $\phi = \bar\phi(h^{1/2} \cdot h^{1/2})$ for some h in $\mathscr{A}$ with $0 \leqslant h \leqslant 1$. However, $\phi(x) = \bar\phi(x)$ for every x in $\cup p_n\mathscr{A}p_n$, whence $h = 1$ and $\phi = \bar\phi$.

5.3.6. LEMMA. *If ϕ is a σ-trace and ψ is a σ-weight on a Borel *-algebra $\mathscr{A}$, then $\phi + \psi$ is a σ-weight.*

Proof. By assumption there exists sequences $\{x_n\}$ and $\{y_n\}$ in $\mathscr{A}_+^\phi$ and $\mathscr{A}_+^\psi$, respectively, such that $\Sigma x_n = \Sigma y_m = 1$. Define $z_{nm} = y_m^{1/2}x_ny_m^{1/2}$. Then

$\psi(z_{nm}) < \infty$ since $z_{nm} \leqslant y_m$ and $\phi(z_{nm}) = \phi(x_n^{1/2} y_m x_n^{1/2}) < \infty$. As $\Sigma z_{nm} = 1$ this implies that the weight $\phi + \psi$ which is evidently σ-normal, is σ-finite as well.

5.3.7. COROLLARY. *If $\mathcal{M}$ is a factor on a separable Hilbert space then all normal, semi-finite traces on $\mathcal{M}$ are proportional.*

Proof. If ϕ and ψ are normal, semi-finite traces on $\mathcal{M}$ then $\phi + \psi$ is semi-finite by 5.3.6 and the result is immediate from 5.3.4.

5.3.8. PROPOSITION (cf. 5.1.7). *If ϕ and ψ are σ-traces on a countably generated Borel *-algebra $\mathscr{A}$ such that $\phi(x) = \psi(x) < \infty$ for all x in the positive part of a *-algebra A which generates $\mathscr{A}$, then $\phi = \psi$.*

Proof. For each x in A_+ the two σ-normal positive functionals $\phi(x\cdot)$ and $\psi(x\cdot)$ coincide on the norm closure $\bar{A}$ of A. Since $\mathscr{A}_{sa} = \mathscr{B}(\bar{A}_{sa})$ by 4.5.4 it follows that $\phi(x\cdot) = \psi(x\cdot)$. If therefore $y \in \mathscr{A}_+^{\phi}$ then $\phi(\cdot y)$ is bounded and coincides with $\psi(y^{1/2}\cdot y^{1/2})$ on $\bar{A}$ and therefore on $\mathscr{A}$. It follows that $\psi(y) < \infty$ and that $\phi(\cdot y) = \psi(\cdot y)$, in particular $\phi(y) = \psi(y)$. Consequently $\phi = \psi$.

5.3.9. LEMMA. *Let ϕ be a faithful, normal, semi-finite trace on a von Neumann algebra $\mathcal{M}$. For each normal state ψ on $\mathcal{M}$ there is a sequence $\{e_n\}$ of pairwise orthogonal projections in $\mathcal{M}$ with sum 1 such that $0 \leqslant \psi(e_n\cdot) \leqslant n\phi(e_n\cdot e_n)$ for each n.*

Proof. We take e_0 as the maximal projection in $\mathcal{M}$ with $\psi(e_0) = 0$. Then ψ is faithful on $(1 - e_0)\mathcal{M}(1 - e_0)$. Let $\{e_n\}$ be a maximal family of non-zero projections in $\mathcal{M}$ such that $\Sigma e_n \leqslant 1 - e_0$ and such that $0 \leqslant \psi(e_n\cdot) \leqslant m_n\phi(e_n\cdot e_n)$ for each n, where $m_n \in \boldsymbol{N}$. Since $\Sigma\psi(e_n) \leqslant 1$ the family $\{e_n\}$ is countable, and by a reordering we may assume that $\psi(e_n\cdot) \leqslant n\phi(e_n\cdot e_n)$. Set $p = 1 - e_0 - \Sigma e_n$. If $p \neq 0$ then by restricting to $p\mathcal{M}p$ we may as well assume that $p = 1$.

Take $\alpha > 0$ such that $\alpha\phi(1) > 1$. The function $x \to \psi(x) - \alpha\phi(x)$ is bounded above and weakly upper semi-continuous on the weakly compact set $\mathcal{M}_+^1$. The function therefore attains its maximum $\beta \geqslant 0$. Since it is affine and $\mathcal{M}_+^1$ is convex with the projections as extreme points, we may assume that $\psi(q) - \alpha\phi(q) = \beta$ for some projection q in $\mathcal{M}$. For any unitary u in $\mathcal{M}$ we have

$$\psi(u^*qu) - \alpha\phi(u^*qu) \leqslant \beta = \psi(q) - \alpha\phi(q),$$

whence $\psi(u^*qu) \leqslant \psi(q)$. If therefore $x \in \mathcal{M}_{sa}$ then

$$\psi(q) \geqslant \psi(\exp(-i\varepsilon x)q\exp(i\varepsilon x)) = \psi(q + i\varepsilon(qx - xq) + \varepsilon y(\varepsilon)),$$

where $\|y(\varepsilon)\| \to 0$ as $\varepsilon \to 0$. Consequently $i\psi(qx - xq) \leqslant 0$ and since this holds also for $-x$ we conclude that $\psi(qx) = \psi(xq)$ for all x in $\mathcal{M}$. Thus with $e = 1 - q$ and $x \in \mathcal{M}_+$ we have $\psi(ex) = \psi(exe) \geqslant 0$, whence $\psi(e\cdot) \geqslant 0$.

Moreover, $\psi(exe) - \alpha\phi(exe) \leq 0$ for all x in $\mathcal{M}_+$, since otherwise the value at $q + exe$ would exceed β. Thus $\psi(e \cdot e) \leq \alpha\phi(e \cdot e)$. Finally, if $e = 0$ then $q = 1$ and $\psi(1) - \alpha\phi(1) = \beta$, i.e. $\alpha\phi(1) \leq 1$ a contradiction. Thus $e \neq 0$ which contradicts the maximality of the family $\{e_n\}$ in the first part of the proof. Consequently $p = 0$ and $e_0 + \Sigma e_n = 1$, as desired.

5.3.10. If $\mathcal{M}$ is a von Neumann algebra on a Hilbert space H and h is a not necessarily bounded, self-adjoint operator on H we say that h is *affiliated* with $\mathcal{M}$ if $u^*hu = h$ for every unitary u in $\mathcal{M}'$. From the bicommutant theorem it follows that h is affiliated with $\mathcal{M}$ if and only if $f(h) \in \mathcal{M}$ for every bounded Borel function on $\mathrm{Sp}(h)$. In particular, if $h \geq 0$ then h is affiliated with $\mathcal{M}$ if and only if $(1 + \varepsilon h)^{-1}h \in \mathcal{M}$ for some $\varepsilon > 0$ (and hence any $\varepsilon > 0$).

Let ϕ be a normal semi-finite trace on $\mathcal{M}$. For each operator $h \geq 0$, affiliated with $\mathcal{M}$ define $\phi(h\cdot)$ on $\mathcal{M}_+$ by

$$\phi(hx) = \mathrm{Lim}\,\phi(x^{1/2}(1 + \varepsilon h)^{-1}hx^{1/2}).$$

5.3.11. THEOREM. *Let ϕ be a σ-trace on a countably generated Borel *-algebra $\mathcal{A}$. For each σ-weight ψ on $\mathcal{A}$ such that $\phi(x) = 0$ implies $\psi(x) = 0$ for all x in $\mathcal{A}_+$ there is a unique positive operator h on H_ϕ, affiliated with $\pi_\phi(\mathcal{A})$, such that $\psi(x) = \phi(h\pi_\phi(x))$ for all x in $\mathcal{A}_+$. Conversely, for any positive operator h on H_ϕ affiliated with $\pi_\phi(\mathcal{A})$ the function $x \to \phi(h\pi_\phi(x))$ is a σ-weight on $\mathcal{A}$ which is 'absolutely continuous' with respect to ϕ.*

Proof. If h is a positive operator affiliated with $\mathcal{M} = \pi_\phi(\mathcal{A})$, let p_n be the spectral projection of h corresponding to the interval $]n-1, n]$. Then $p_n h \in \mathcal{M}_+$ for each n and for x in $\mathcal{M}_+$ we have

$$\phi(hx) = \mathrm{Lim}\,\phi(x^{1/2}(1 + \varepsilon h)^{-1}hx^{1/2}) = \mathrm{Lim}\sum \phi(x^{1/2}(1 + \varepsilon h)^{-1}hp_n x^{1/2})$$
$$= \sum \phi(x^{1/2}hp_n x^{1/2}) = \sum \phi(p_n h^{1/2}xh^{1/2}p_n).$$

Each function $\phi(p_n h^{1/2} \cdot h^{1/2}p_n)$ is a σ-weight on $\mathcal{M}$ (cf. 5.3.4) and since the p_n's are pairwise orthogonal the sum, i.e. $\phi(h\cdot)$, is a σ-weight on $\mathcal{M}$; hence by composition with $\pi_\phi(\cdot)$ gives a σ-weight on $\mathcal{A}$.

Assume now that ψ is a σ-weight on $\mathcal{A}$ such that $\phi(x) = 0$ implies $\psi(x) = 0$ for each x in $\mathcal{A}_+$. Since $\ker \pi_\phi = L_\phi$ this implies that we may regard ψ (and ϕ) as a σ-weight on the von Neumann algebra $\mathcal{M} = \pi_\phi(\mathcal{A})$.

Consider first the case where ψ is bounded. By 5.3.9 we have a sequence $\{e_n\}$ of projections in $\mathcal{M}$ with sum 1 such that $0 \leq \psi(e_n \cdot) \leq n\phi(e_n \cdot e_n)$. Consequently $\psi(e_n \cdot) = \phi(h_n^{1/2} \cdot h_n^{1/2})$ for some operator h_n in $e_n\mathcal{M}_+e_n$ with $\|h_n\| \leq n$ by 5.3.4. Since $h_n h_m = 0$ for $n \neq m$ we may define $h = \Sigma h_n$ and h will be a positive operator on H_ϕ affiliated with $\mathcal{M}$. Finally, if $x \in \mathcal{M}_+$ then

$$\psi(x) = \sum \psi(e_n x) = \sum \phi(h_n^{1/2}xh_n^{1/2})$$
$$= \sum \phi(x^{1/2}h_n x^{1/2}) = \mathrm{Lim}\,\phi(x^{1/2}(1 + \varepsilon h)^{-1}hx^{1/2}) = \phi(hx).$$

In the general case $\psi = \Sigma\,\psi_n$ where each ψ_n is bounded. Thus from what we proved above $\psi_n = \phi(k_n\,\cdot)$ for some positive operator k_n affiliated with $\mathscr{M}$. Set

$$a_m = 1 + \sum_{n=1}^{m} [1 + (1/m)k_n]^{-1}k_n.$$

Then $\{a_m\}$ is an increasing sequence in $\mathscr{M}_+$, and for each x in $\mathscr{A}_2^\phi$:

$$(a_m\xi_x \,|\, \xi_x) = \phi(x^*a_m x) \to \phi(x^*x) + \sum \psi_n(xx^*) = \phi(xx^*) + \psi(xx^*).$$

Since $\phi + \psi$ is σ-finite by 5.3.6 there is an increasing sequence $\{q_n\}$ of projections in $\mathscr{A}$ with $q_n \nearrow 1$ such that $\phi(q_n) + \psi(q_n) < \infty$ for each n. From the computations above it follows that

$$\operatorname*{Lim}_m (a_m\xi_{q_nx} \,|\, \xi_{q_nx}) = \phi(q_nxx^*q_n) + \psi(q_nxx^*q_n) < \infty.$$

Set $b = \operatorname{Lim} a_m^{-1}$. Clearly $b \in \mathscr{M}_+$, and if $\xi_0 \in H_\phi$ such that $b\xi_0 = 0$ then

$$|(\xi_0 \,|\, \xi_{q_nx})|^2 = |(a_m^{-1/2}\xi_0 \,|\, a_m^{1/2}\xi_{q_nx})|^2 \leqslant (a_m^{-1}\xi_0 \,|\, \xi_0)(a_m\xi_{q_nx} \,|\, \xi_{q_nx}) \to 0.$$

But $\cup\, \pi_\phi(q_n)H_\phi$ is dense in H_ϕ, whence $\xi_0 = 0$. It follows that $h = b^{-1} - 1$ is a positive self-adjoint operator affiliated with $\mathscr{M}$, and for x in $\mathscr{M}_+$

$$\begin{aligned}
\phi(hx^2) &= \operatorname*{Lim}_\varepsilon \phi(x(1 + \varepsilon h)^{-1}hx) \\
&= \operatorname*{Lim}_\varepsilon \phi(x(1 + \varepsilon(b^{-1} - 1))^{-1}(b^{-1} - 1)x) \\
&= \operatorname*{Lim}_\varepsilon \phi(x(\varepsilon + (1 - \varepsilon)b)^{-1}(1 - b)x) \\
&= \operatorname*{Lim}_{\varepsilon,m} \phi(x(\varepsilon + (1 - \varepsilon)a_m^{-1})^{-1}(1 - a_m^{-1})x) \\
&= \operatorname*{Lim}_{\varepsilon,m} \phi(x(1 + \varepsilon(a_m - 1))^{-1}(a_m - 1)x) \\
&\leqslant \operatorname*{Lim}_m \phi(x(a_m - 1)x) = \sum \phi(k_nx^2) = \sum \psi_n(x^2) = \psi(x^2).
\end{aligned}$$

Thus $\phi(h\,\cdot) \leqslant \psi$. However, since $a_m^{-1} \geqslant b$ we have $a_m - 1 \leqslant h$, whence $\Sigma_{n=1}^m \psi_n \leqslant \phi(h\,\cdot)$ for all m, and thus $\phi(h\,\cdot) = \psi$.

Finally, to show that the correspondence $h \to \phi(h\,\cdot)$ is an isomorphism, assume that h_1 and h_2 are positive operators affiliated with $\mathscr{M}$ such that $\phi(h_1\,\cdot) = \psi = \phi(h_2\,\cdot)$. This means that

$$\operatorname{Lim} \phi(x^*(1 + \varepsilon h_1)^{-1}h_1x) = \operatorname{Lim} \phi(x^*(1 + \varepsilon h_2)^{-1}h_2x)$$

for every x in $\mathcal{M}$. If now x, y belong to $\mathcal{A}_\phi^2$ and p_1 and p_2 are spectral projections of h_1 and h_2, respectively, corresponding to bounded intervals then

$$\begin{aligned}(p_1 h_1 p_2 \xi_x \mid \xi_y) &= \phi(\pi_\phi(y^*) p_1 h_1 p_2 \pi_\phi(x)) \\ &= \psi(p_2 \pi_\phi(xy^*) p_1) = \phi(h_2 p_2 \pi_\phi(xy^*) p_1) \\ &= (p_1 h_2 p_2 \xi_x \mid \xi_y).\end{aligned}$$

It follows that $p_1 h_1 p_2 = p_1 h_2 p_2$. Adding $\varepsilon p_1 p_2$ to this equation and multiplying on the left with $(\varepsilon + h_1)^{-1}$ and on the right with $(\varepsilon + h_2)^{-1}$ we obtain

$$p_1 p_2 (\varepsilon + h_2)^{-1} = (\varepsilon + h_1)^{-1} p_1 p_2.$$

Since p_1 and p_2 are arbitrarily close to 1 this implies that

$$(\varepsilon + h_2)^{-1} = (\varepsilon + h_1)^{-1}$$

for any $\varepsilon > 0$, whence $h_1 = h_2$ by spectral theory.

5.3.12. Notes and remarks. Proposition 5.3.2 was found by Sakai in 1962, see p. 77 of [231]. Later he showed in [225] that it is also possible to write $\psi = \phi(t \cdot t)$ for some t in $\mathcal{M}_+$. The linear version (5.3.2) of the Radon–Nikodym theorem is valid also if ϕ and ψ are weights (cf. [272]), but the quadratic version is not (cf. proposition 7.7 of [210]). We return to general Radon–Nikodym theorems in Section 8.14.

Theorem 5.3.3 is due to Murray and von Neumann [170], but the proof is new. Theorem 5.3.11 was proved by Dye [73] for finite traces and extended to σ-finite traces by Segal [240]. The result is valid for any von Neumann algebra (not necessarily countably generated), but we have chosen the setting of Borel *-algebras in order to show the close relation with the ordinary (commutative) Radon–Nikodym theorem. As shown by Christensen [33] all of Segal's non-commutative integration theory can be formulated within a Borel *-algebraic framework.

5.4. Semi-finite von Neumann algebras

5.4.1. Let $\mathcal{M}$ be a von Neumann algebra on a separable Hilbert space (although this restriction is not essential). We say that $\mathcal{M}$ is *finite* (respectively *semi-finite*), if there exists a faithful, normal, finite (respectively semi-finite) trace on $\mathcal{M}$. We say that $\mathcal{M}$ is *properly infinite* (respectively *purely infinite*), if there are no non-zero normal, finite (respectively semi-finite) traces on $\mathcal{M}$.

5.4.2. PROPOSITION. *Each von Neumann algebra $\mathcal{M}$ has a unique decomposition $\mathcal{M} = \mathcal{M}_1 \oplus \mathcal{M}_2 \oplus \mathcal{M}_3$ such that $\mathcal{M}_1$ is finite, $\mathcal{M}_2$ is semi-finite but properly infinite, and $\mathcal{M}_3$ is purely infinite.*

Proof. Let ϕ be a normal trace on $\mathcal{M}$. Then the set $\mathcal{N}_\phi = \{x \in \mathcal{M} \mid \phi(x^*x) = 0\}$ is a weakly closed ideal of $\mathcal{M}$ since ϕ is weakly lower semi-continuous. Consequently $\mathcal{N}_\phi = (1 - p)\mathcal{M}$ for some central projection p in $\mathcal{M}$ (2.5.4) and ϕ is faithful on $p\mathcal{M}$. By 5.2.1 the weak closure of $\mathcal{M}^\phi$ is also an ideal of $\mathcal{M}$, and thus there is a central projection q such that ϕ is semi-finite on $q\mathcal{M}$ and purely infinite on $(1 - q)\mathcal{M}$. Consequently ϕ is faithful and semi-finite on $pq\mathcal{M}$.

Let $\{\phi_n, p_n\}$ be a maximal family of normal finite traces ϕ_n and pairwise orthogonal central projections p_n, such that ϕ_n is faithful on $p_n\mathcal{M}$ for each n. Since $\mathcal{M}$ is σ-finite the family $\{\phi_n, p_n\}$ is countable, and we define $\phi(x) = \Sigma\, 2^{-n}\phi_n(1)^{-1}\phi_n(p_n x)$ for each x in $\mathcal{M}$. If $p = \Sigma\, p_n$ then ϕ is faithful normal and finite on $p\mathcal{M}$, and by the maximality of the family $\{\phi_n, p_n\}$ we see that $(1 - p)\mathcal{M}$ is properly infinite.

Let $\{\psi_n, q_n\}$ be a maximal family of normal semi-finite traces ψ_n and pairwise orthogonal projections $q_n \leqslant 1 - p$ such that ψ_n is faithful on $q_n\mathcal{M}$. As before, set $\psi(x) = \Sigma\, \psi_n(q_n x)$ and $q = \Sigma\, q_n$. Then $q \perp p$, ψ is faithful, normal and semi-finite on $q\mathcal{M}$ and $(1 - q - p)\mathcal{M}$ is purely infinite by the maximality of the family $\{\psi_n, q_n\}$.

5.4.3. THEOREM. *If $\mathcal{M}$ is a semi-finite von Neumann algebra on a separable Hilbert space H then $\mathcal{M}'$ is semi-finite.*

Proof. If $\mathcal{M}'$ is not semi-finite we may assume by 5.4.2 that $\mathcal{M}'$ is purely infinite (and $\mathcal{M}$ is semi-finite). Let ϕ be a faithful, normal, semi-finite trace on $\mathcal{M}$. There is then a non-zero, positive, normal functional ψ on $\mathcal{M}$ dominated by ϕ; and since ψ is a sum of a sequence of vector functionals by 3.6.6 there is a non-zero vector ξ in H such that $(x\xi \mid \xi) \leqslant \phi(x)$ for each x in $\mathcal{M}$. By 5.3.3 there is then a (unique) element h^2 in $\mathcal{M}^\phi_+$ such that

$$(x\xi \mid \xi) = \phi(h^2x) = \phi(hxh) = (\pi_\phi(x)\xi_h \mid \xi_h).$$

Denote by p the projection on $[\mathcal{M}\xi]$ and by q the projection on $[\pi_\phi(\mathcal{M})\xi_h]$. Then $p \in \mathcal{M}'$ and $q \in \pi_\phi(\mathcal{M})'$, and by 3.3.7 there is an isometry u of pH onto qH_ϕ which implements an isomorphism between $p\mathcal{M}$ and $q\pi_\phi(\mathcal{M})$. It follows that

$$u(p\mathcal{M}'p)u^* = q\pi_\phi(\mathcal{M})'q.$$

Since $\pi_\phi(\mathcal{M})'$ is semi-finite by 5.3.3 (as trace on $\pi_\phi(\mathcal{M})'$ use $\phi'(x) = \phi(jxj)$) we conclude that $c(p)\mathcal{M}'$ is semi-finite by 5.2.8; for if ψ is a faithful normal semi-finite trace on $p\mathcal{M}'p$, then $\tilde{\psi}$ is faithful on $c(p)\mathcal{M}'$. This contradicts the assumption that $\mathcal{M}'$ was purely infinite, and it follows that $\mathcal{M}'$ is semi-finite as desired.

5.4.4. Let p and q be projections in a C^*-algebra B. If there exists a partial isometry v in B such that $v^*v = p$ and $vv^* = q$, we say that p is *equivalent* to q and write $p \sim q$. It is easy to verify that this is indeed an equivalence relation.

If $B = B(H)$ then two projections are equivalent if and only if pH and qH have the same dimensions. Thus the equivalence classes for projections in a general von Neumann algebra $\mathscr{M}$ may be visualized as 'generalized dimension'.

It is clear that if $p \sim q$ then in particular $p \approx q$ (with $\approx$ as defined in 5.2.6). In fact the two relations $\sim$ and $\approx$ coincide on the set of projections in a von Neumann algebra, but we shall not need this result.

5.4.5. PROPOSITION. *If p and q are projections in a von Neumann algebra $\mathscr{M}$ and $p \sim q$, then $c(p) = c(q)$ and $zp \sim zq$ for each central projection z. If $\{p_i\}$ and $\{q_i\}$ are orthogonal families of projections such that $p_i \sim q_i$ for each i, then $\Sigma\, p_i \sim \Sigma\, q_i$.*

Proof. Only the last assertion needs proof. Assume then that $v_i^* v_i = p_i$ and $v_i v_i^* = q_i$ for each i. Since the v_i's have pairwise orthogonal ranges and supports the sum $\Sigma\, v_i$ is strongly convergent to an element v in $\mathscr{M}$. We have $v^*v = \Sigma\, v_i^* v_i$ and $vv^* = \Sigma\, v_i v_i^*$, whence $\Sigma p_i \sim \Sigma\, q_i$.

5.4.6. We say that a projection p in a von Neumann algebra $\mathscr{M}$ is *finite* if $\phi(p) < \infty$ for some faithful normal semi-finite trace ϕ on $c(p)\mathscr{M}$. By 5.2.8, p is finite if and only if $p\mathscr{M}p$ is a finite von Neumann algebra. Clearly any projection which is majorized by a finite projection is finite, and it is not hard to show, using 5.2.8 that a von Neumann algebra $\mathscr{M}$ is semi-finite if and only if it contains a finite projection p with $c(p) = 1$; or, equivalently, if and only if each non-zero projection in $\mathscr{M}$ majorizes a non-zero finite projection.

If p is a finite projection in a von Neumann algebra $\mathscr{M}$ and $q \leqslant p$ with $q \sim p$ (or $q \approx p$) then with ϕ a faithful, normal, finite trace on $p\mathscr{M}p$ we have $\phi(q) = \phi(p)$, whence $\phi(p - q) = 0$ and $p = q$. Thus the finite projections are really 'finite' with respect to the equivalence relation $\sim$ (cf. the definition of finite cardinal numbers). It is interesting that this 'finiteness' property actually characterizes finite projections (5.4.10).

5.4.7. PROPOSITION. (Comparison). *Let p and q be projections in a von Neumann algebra $\mathscr{M}$. There are then projections p_1 and q_1 in $\mathscr{M}$ such that $p_1 \leqslant p$, $q_1 \leqslant q$, $p_1 \sim q_1$ and $c(p - p_1) \perp c(q - q_1)$.*

Proof. Let $\mathscr{M}^1$ denote the unit ball of $\mathscr{M}$ in the weak topology. Then $q\mathscr{M}^1 p$ is convex and compact, and contains therefore an extreme point v. By 1.4.8 this implies that v^*v and vv^* are projections with

$$(1 - vv^*)q\mathscr{M}p(1 - v^*v) = 0.$$

Put $p_1 = v^*v$ and $q_1 = vv^*$. Then $p_1 \leqslant p$, $q_1 \leqslant q$ and $p_1 \sim q_1$. Moreover, $(q - q_1)\mathscr{M}(p - p_1) = 0$ which by 2.6.3 implies that $c(q - q_1)c(p - p_1) = 0$.

5.4.8. COROLLARY. *If p and q are projections in $\mathscr{M}$ there is a central projection z in $\mathscr{M}$ such that*

$$zp \sim zq_1 \leqslant zq \quad \textit{and} \quad (1 - z)q \sim (1 - z)p_1 \leqslant (1 - z)p.$$

Proof. Put $z = c(q - q_1)$ so that $zp = zp_1$ and $(1 - z)q = (1 - z)q_1$; and apply 5.4.7.

5.4.9. COROLLARY. *If $\mathcal{M}$ is a factor then either $p \sim q_1 \leqslant q$ or $q \sim p_1 \leqslant p$.*

5.4.10. THEOREM. *If p is a projection in a von Neumann algebra $\mathcal{M}$ such that $q \sim p$ implies $q = p$ for all projections $q \leqslant p$, then p is finite.*

Proof. Working in $p\mathcal{M}p$ we may assume that $p = 1$. Since $\mathcal{M}$ operates on a separable Hilbert space we can find a faithful, normal state ϕ of $\mathcal{M}$. By Zorn's lemma there is a maximal family (under inclusion) of pairs of equivalent, non-zero projections p_i, q_i with $\phi(p_i) > 2\phi(q_i)$, such that $\Sigma\, p_i \leqslant 1$ and $\Sigma\, q_i \leqslant 1$. Set $p' = 1 - \Sigma\, p_i$ and $q' = 1 - \Sigma\, q_i$. By 5.4.5, $\Sigma\, p_i \sim \Sigma\, q_i$ and consequently neither $p' = 0$ nor $q' = 0$. For if $p' = 0$, then $q' = 0$; otherwise $\Sigma\, p_i \sim 1$ but $\Sigma\, p_i \neq 1$. And if $p' = q' = 0$, then $\phi(1) > 2\phi(1)$, a contradiction. We claim that $p' \sim q'$. Otherwise, by 5.4.8 we would have $zp' \sim zq'' \lneqq zq'$ for some central projection z. But then by 5.4.5

$$1 = 1 - z + z(1 - p') + zp' \sim 1 - z + z(1 - q') + zq'' \lneqq 1,$$

a contradiction.

By Zorn's lemma there is a maximal family of pairs of equivalent, non-zero projections e_j, f_j with $\alpha\phi(e_j) < \phi(f_j)$, such that $\Sigma\, e_j \leqslant p'$ and $\Sigma f_j \leqslant q'$. With $\alpha = \phi(p')^{-1}\phi(q')$ and $e' = p' - \Sigma\, e_j$, $f' = q' - \Sigma f_j$ we have $\Sigma\, e_j \sim \Sigma f_j$ and consequently neither $e' = 0$ nor $f' = 0$ (proof as above). Moreover, $e' \sim f'$ (again proof as above), and so $e' = v^*v$, $f' = vv^*$.

For any pair e, f of equivalent non-zero projections in $e'\mathcal{M}e'$ we have $e \sim vev^*$ and thus $\alpha\phi(e) \geqslant \phi(vev^*)$ by the maximality of $\{e_j, f_j\}$. But also $f \sim vev^*$, whence $\phi(f) \leqslant 2\phi(vev^*)$ by the maximality of $\{p_i, q_i\}$. Put together this gives $\phi(f) \leqslant 2\alpha\phi(e)$. If now $x \in e'\mathcal{M}e'$ and $x = u|x|$ is its polar decomposition; then with $\lambda_n = nm^{-1}\|x\|^2$, $0 \leqslant n \leqslant m$, and p_n the spectral projection of x^*x corresponding to the interval $]\lambda_{n-1}, \lambda_n]$ we have $\|\Sigma\, \lambda_n p_n - x^*x\| \leqslant m^{-1}\|x\|^2$. Since $xx^* = u(x^*x)u^*$ this implies that $\|\Sigma\, \lambda_n u p_n u^* - xx^*\| \leqslant m^{-1}\|x\|^2$. But $p_n \sim up_nu^*$, whence $\phi(up_nu^*) \leqslant 2\alpha\phi(p_n)$, and from the linearity of ϕ it follows that $\phi(xx^*) \leqslant 2m^{-1}\|x\|^2 + 2\alpha\phi(x^*x)$ for each m, whence $\phi(xx^*) \leqslant 2\alpha\phi(x^*x)$ for every x in $e'\mathcal{M}e'$. From 5.2.8 we have a trace $\tilde{\phi}$ on $e'\mathcal{M}e'$ and from the definition of $\tilde{\phi}$ we see that $\phi \leqslant \tilde{\phi} \leqslant 2\alpha\phi$. Thus $\tilde{\phi}$ is a faithful, normal, finite trace on $e'\mathcal{M}e'$, and again by 5.2.8 extends to a faithful, normal, semi-finite trace on $c(e')\mathcal{M}$. If $c(e') \neq 1$ we repeat the whole proof with $(1 - c(e'))\mathcal{M}$ instead of $\mathcal{M}$. Applying the standard maximality argument we obtain in this way a faithful, normal, semi-finite trace on $\mathcal{M}$. To complete the proof we need just the following lemma.

5.4.11. LEMMA. *If $\mathcal{M}$ is a semi-finite von Neumann algebra such that no proper subprojection in $\mathcal{M}$ is equivalent to 1, then $\mathcal{M}$ is finite.*

Proof. Let ϕ be a faithful, normal, semi-finite trace on $\mathcal{M}$ and let $\{z_n\}$ be a maximal family of orthogonal central projections with $\phi(z_n) < \infty$ for all n. If $\Sigma z_n = 1 - z_0$ with $z_0 \neq 0$ take a non-zero projection $p_0 \leqslant z_0$ with $\phi(p_0) < \infty$. Choose a maximal orthogonal family $\{p_n\}$ of projections in $z_0\mathcal{M}$ such that $p_n \sim p_0$ for each n, and set $q = 1 - \Sigma p_n$. If $\{p_n\}$ was an infinite set (necessarily countable) then

$$q + \sum p_{2n} \lneqq q + \sum p_n = 1.$$

But by 5.4.5 the projections above are equivalent contrary to our assumption. Thus $\{p_n\}$ is finite. By 5.4.8 there is a central projection z such that $zp_0 \sim zq_1 \leqslant zz_0q$ and $(1-z)z_0q \sim (1-z)p' \leqslant (1-z)p_0$. The second equation gives

$$\phi((1-z)z_0) = \phi\left((1-z)z_0\left(q + \sum p_n\right)\right) \leqslant \phi(p_0) + \sum \phi(p_n) < \infty,$$

in contradiction with the maximality of $\{z_n\}$ unless $(1-z)z_0 = 0$. But then the first equation gives $p_0 \sim q_1 \leqslant z_0q$ in contradiction with the maximality of $\{p_n\}$. Consequently the assumption $z_0 \neq 0$ can not hold, whence $\Sigma z_n = 1$. Define

$$\psi = \sum 2^{-n}\phi(z_n)^{-1}\phi(z_n\,\cdot)$$

and note that ψ is a faithful, normal and finite trace on $\mathcal{M}$.

5.4.12. PROPOSITION. *A projection p in a von Neumann algebra $\mathcal{M}$ is finite if and only if for every net $\{x_i\}$ in the unit ball of $\mathcal{M}$ such that $x_ip \to 0$ strongly, we have $px_i^* \to 0$ strongly.*

Proof. We may assume without loss of generality that $c(p) = 1$. If p is finite there is a faithful normal semi-finite trace ϕ on $\mathcal{M}$ such that $\phi(p) < \infty$. It follows from 5.3.10 that the set of positive functionals ψ in $\mathcal{M}_*$ dominated by a multiple of ϕ is norm dense in $(\mathcal{M}_*)_+$. If $x_ip \to 0$ strongly then $px_i^*x_ip \to 0$ weakly, whence

$$\phi(x_ipx_i^*) = \phi(px_i^*x_ip) \to 0.$$

Thus $\psi(x_ipx_i^*) \to 0$ for a norm dense set of functionals ψ in $(\mathcal{M}_*)_+$, whence $x_ipx_i^* \to 0$ weakly since the net $\{x_ipx_i^*\}$ is bounded; i.e. $px_i^* \to 0$ strongly.

If p is not finite there is by 5.4.10 a partial isometry u in $\mathcal{M}$ such that $uu^* = p$ and $u^*u \lneqq p$. Let q be the limit projection of the decreasing sequence $\{u^{n*}u^n\}$ and define $u_n = u^n(1-q) = u^n - q$. Then

$$pu_n^*u_np = (1-q)u^{n*}u^n(1-q) \searrow 0;$$

$$u_npu_n^* = u^n(1-q)u^{n*} = p - q \neq 0.$$

Thus $u_np \to 0$ strongly but $pu_n^* \nrightarrow 0$ strongly.

5.4.13. COROLLARY. *A von Neumann algebra $\mathcal{M}$ is finite if and only if the involution is strongly continuous on bounded subsets of $\mathcal{M}$.*

5.4.14. Notes and remarks. Murray and von Neumann originally defined finite and semi-finite algebras in terms of finite projections (finite in the sense of equivalence, cf. 5.4.10), and established the existence of a trace in a semi-finite factor in [169]. This result was then generalized to arbitrary von Neumann algebras by Dixmier [53]. It must be recalled that Murray and von Neumann worked almost exclusively with factors, and [53] was one of the first papers that showed this restriction to be unnecessary.

As noted later by Dixmier it is actually more natural to take the trace as the basic concept and postpone the thorny characterization in terms of finite projections (5.4.10) as long as possible. This is the point of view taken in [59] and also here.

The proof of 5.4.10 is an attempt to improve an already quite simple proof by Kadison (see [126]) of the dreaded Murray–von Neumann trace result; it appeared in [202]. Corollary 5.4.13 is due to Sakai, see [222]. It can be used to prove Tomiyama's theorem [269], that if $\pi: \mathcal{M} \to \mathcal{N}$ is a projection of a von Neumann algebra $\mathcal{M}$ onto one of its von Neumann subalgebras $\mathcal{N}$ (i.e. π is a normal positive, linear, idempotent map of norm one) then the type of $\mathcal{N}$ is no larger than that of $\mathcal{M}$.

5.5. Von Neumann algebras of type I

5.5.1. A positive element x in a C^*-algebra A is *abelian* if the hereditary C^*-subalgebra of A containing x is commutative. In particular a projection p is *abelian* if pAp is commutative. In a von Neumann algebra $\mathcal{M}$ with centre $\mathcal{Z}$ the centre of $p\mathcal{M}p$ is $\mathcal{Z}p$. If therefore p is abelian then $p\mathcal{M}p = \mathcal{Z}p$.

If p is a minimal projection then it is clearly abelian. Conversely, if p is an abelian projection and $\mathcal{M}$ is a factor then p is minimal because $p\mathcal{M}p = \mathbf{C}p$. Thus abelian projections are the analogues of minimal projections in factors.

We say that a von Neumann algebra $\mathcal{M}$ is of *type I* if each non-zero projection in $\mathcal{M}$ majorizes a non-zero abelian projection.

5.5.2. LEMMA. *Let p and q be abelian projections in a von Neumann algebra $\mathcal{M}$. If $c(p) = c(q)$ then $p \sim q$.*

Proof. Assume first that $p \sim q_1 \leqslant q$. Then $q_1 = eq$ for some central projection e in $\mathcal{M}$, since q is abelian. Consequently

$$c(p) = c(q_1) = ec(q) = ec(p)$$

by 2.6.4, whence $e \geqslant c(q)$ and $q_1 = q$.

In the general case we apply 5.4.8 to obtain a central projection z such that $z \sim zq_1 \leqslant zq$ and $(1-z)q \sim (1-z)p_1 \leqslant (1-z)p$. It follows from the above that $zp \sim zq$ and $(1-z)q \sim (1-z)p$, whence $p \sim q$.

5.5.3. PROPOSITION. *A von Neumann algebra $\mathcal{M}$ is of type I and if and only if there is an abelian projection p in $\mathcal{M}$ with $c(p) = 1$.*

Proof. If p is abelian and $c(p) = 1$, let q be a non-zero projection in $\mathcal{M}$. By 5.4.7 p and q have equivalent subprojections p_1 and q_1 such that $c(p - p_1) \perp c(q - q_1)$. If $p_1 = 0$ then $c(q - q_1) = 0$ whence $q = q_1$; and $q_1 \sim 0$ implies $q_1 = 0$, a contradiction. Thus p_1 and q_1 are non-zero and since p_1 is abelian, q_1 is also abelian. Thus $\mathcal{M}$ is of type I.

Assume now that $\mathcal{M}$ is of type I and choose a maximal family $\{p_i\}$ of abelian non-zero projections such that $\{c(p_i)\}$ is an orthogonal family. Put $p = \Sigma p_i$ and note that $c(p) = \Sigma c(p_i)$. If $c(p) \neq 1$ there is by assumption a non-zero abelian projection $p_0 \leqslant 1 - c(p)$. This contradicts the maximality of $\{p_i\}$ and consequently $c(p) = 1$. Since

$$p\mathcal{M}p = \prod p_i \mathcal{M} p_i,$$

we conclude that p is an abelian projection.

5.5.4. COROLLARY. *Let $\mathcal{M}$ be a von Neumann algebra of type I on a separable Hilbert space, and p an abelian projection with $c(p) = 1$. There is then a faithful, normal, semi-finite trace $\tilde{\phi}$ on $\mathcal{M}$ such that $\tilde{\phi}(p) = 1$.*

Proof. Since $p\mathcal{M}p$ is isomorphic to $L_\mu^\infty(T)$ by 3.4.4, there is a faithful normal finite trace ϕ on $p\mathcal{M}p$ (take any finite measure on T which is equivalent to μ) and by normalization we may assume that $\phi(p) = 1$. By 5.2.8 ϕ extends to a normal, semi-finite trace $\tilde{\phi}$ on $\mathcal{M}$ and since $c(p) = 1$, $\tilde{\phi}$ is faithful on $\mathcal{M}$.

5.5.5. LEMMA. *If $\{p_i\}$ and $\{q_j\}$ are families of pairwise orthogonal, equivalent, abelian projections with sum 1 in a von Neumann algebra $\mathcal{M}$ on a separable Hilbert space, then they have the same cardinality.*

Proof. Assume that $\text{card}\{p_i\} = n < \infty$ and take p_1 in $\{p_i\}$ and q_1 in $\{q_j\}$. Since the p_i's are equivalent we have $c(p_1) \geqslant \Sigma p_i = 1$, whence $c(p_1) = 1$. Likewise $c(q_1) = 1$. But then $p_1 \sim q_1$ by 5.5.2, so if ϕ is a faithful, normal, semi-finite trace on $\mathcal{M}$ with $\phi(p_1) = 1$ (cf. 5.5.4), then $\phi(q_1) = 1$. Now

$$n = \sum \phi(p_i) = \phi(1) = \sum \phi(q_j) = \text{card}\{q_j\}\phi(q_1) = \text{card}\{q_j\},$$

and the lemma follows.

5.5.6. Let $\mathcal{M}$ be a von Neumann algebra of type I. We say that $\mathcal{M}$ is *homogeneous of degree n* if the unit of $\mathcal{M}$ can be written as a sum of n orthogonal, equivalent, abelian projections. By 5.5.5 the degree is well defined.

5.5.7. PROPOSITION. *Each von Neumann algebra $\mathcal{M}$ of type I on a separable Hilbert space has a unique decomposition $\mathcal{M} = \Pi\mathcal{M}_n$, $1 \leqslant n \leqslant \infty$, where the $\mathcal{M}_n$'s are homogeneous of degree n.*

Proof. For each cardinal $n \leqslant \infty$ let $\{z_j\}$ be a maximal family of orthogonal central projections each of which is a sum of n orthogonal, equivalent, abelian projections $\{p_{ji}\}$. Since $c(p_{ji}) = z_j$ for all i and j the projection $p_i = \Sigma_j p_{ji}$ is abelian for each i, $1 \leqslant i < n + 1$. With $e_n = \Sigma z_j$ we have $c(p_i) = e_n$ for each i, so that $\{p_i\}$ is a family of orthogonal, equivalent, abelian projections by 5.5.2. Moreover,

$$\sum_i p_i = \sum_{ij} p_{ji} = \sum_j z_j = e_n,$$

which shows that the algebra $\mathcal{M}_n = \mathcal{M}e_n$ is homogeneous of degree n.

If $n \neq m$ then $e_n e_m = 0$ by 5.5.5. Let $e = \Sigma e_n$. We must prove that $e = 1$. Note that by the maximality of the family $\{z_j\}$ the algebra $\mathcal{M}(1 - e)$ contains no homogeneous central summand. However, $\mathcal{M}(1 - e)$ is evidently of type I so we need only show that each von Neumann algebra $\mathcal{M}$ of type I contains a non-zero, homogeneous, central summand.

For this, let $\{q_i\}$ be a maximal family of orthogonal abelian projections in $\mathcal{M}$ with $c(q_i) = 1$. By 5.5.2, these are all equivalent, and by 5.5.3 the family is nonvoid. Put $z = 1 - c(1 - \Sigma q_i)$. If $z = 0$ there would be an abelian projection $q \leqslant 1 - \Sigma q_i$ with $c(q) = c(1 - \Sigma q_i) = 1$, in contradiction with the maximality of $\{q_i\}$. Consequently $z \neq 0$ and since

$$c(z - \sum zq_i) = zc(1 - \sum q_i) = 0,$$

we have $z = \Sigma zq_i$, whence $\mathcal{M}z$ is homogeneous.

5.5.8. COROLLARY. *Let $\mathcal{M}$ be a factor of type I. Then $\mathcal{M}$ is isomorphic to $\boldsymbol{B}(H)$ where $\dim(H)$ is the degree of homogeneity for $\mathcal{M}$.*

Proof. Take a non-zero abelian projection p in $\mathcal{M}$. Since $\mathcal{M}$ is a factor p is minimal and $c(p) = 1$. Let ϕ be a normal state of $\mathcal{M}$ with $\phi(p) = 1$. Then ϕ is evidently a pure state of $\mathcal{M}$ and consequently (π_ϕ, H_ϕ) is irreducible. It follows that $\pi_\phi(\mathcal{M}) = \boldsymbol{B}(H_\phi)$. Clearly the degree of homogeneity for $\mathcal{M}$ and $\pi_\phi(\mathcal{M})$ are the same, namely $\dim(H_\phi)$.

5.5.9. LEMMA. *Let $\mathcal{M}$ be a von Neumann algebra of type I on a Hilbert space H. Then $\mathcal{M}'$ is isomorphic to a von Neumann algebra with abelian commutant.*

Proof. Take by 5.5.3 an abelian projection p in $\mathcal{M}$ with $c(p) = 1$. Then $\mathcal{M}'$ is isomorphic with $\mathcal{M}'p$ by 2.6.7, and the commutant of $\mathcal{M}'p$ on pH is $p\mathcal{M}p$.

5.5.10. LEMMA. *Let $\mathcal{M}$ be a commutative von Neumann algebra on a Hilbert space H. Then $\mathcal{M}'$ is of type I.*

Proof. Let q be a non-zero projection in $\mathcal{M}'$ and choose a unit vector ξ in qH. Let p be the cyclic projection on $[\mathcal{M}\xi]$. Then $p \in \mathcal{M}'$ and $p \leqslant q$. Since $\mathcal{M}p$ is commutative with a cyclic vector, it is maximal commutative by 2.8.3 (on pH). Hence

$$\mathcal{M}p = (\mathcal{M}p)' = p\mathcal{M}'p,$$

so that p is an abelian projection, and consequently $\mathcal{M}'$ is of type I (cf. 5.5.1).

5.5.11. THEOREM. *Let $\mathcal{M}$ be a von Neumann algebra. The following conditions are equivalent:*

(i) *$\mathcal{M}$ is of type I;*

(ii) *$\mathcal{M}'$ is of type I;*

(iii) *$\mathcal{M}$ is isomorphic to a von Neumann algebra with abelian commutant;*

(iv) *$\mathcal{M}'$ is isomorphic to a von Neumann algebra with abelian commutant.*

Proof. (i) ⇒ (iv) 5.5.9. (iv) ⇒ (ii) 5.5.10. (ii) ⇒ (iii) 5.5.9. (iii) ⇒ (i) 5.5.10.

5.5.12. The type I theory cuts into the classification scheme given in 5.4.1 and makes another distinction desirable. We say that a von Neumann algebra is of *type II* if it is semi-finite, but contains no non-zero abelian projections. We say that the algebra is of *type III* if it is purely infinite. It follows easily from 5.4.2 and 5.5.1 that each von Neumann algebra $\mathcal{M}$ has a unique decomposition

$$\mathcal{M} = \mathcal{M}_{\mathrm{I}} \oplus \mathcal{M}_{\mathrm{II}} \oplus \mathcal{M}_{\mathrm{III}}$$

into central summands of distinct types.

Of course we have still to distinguish between finite and properly infinite von Neumann algebras both in type I and type II. Note, however, that by 5.5.4 a von Neumann algebra which is homogeneous of degree n is finite if n is finite and properly infinite if n is infinite. We say that the algebra is of *type I_n*.

For a von Neumann algebra $\mathcal{M}$ of type II we say that it is of *type II_1* if it is finite and of *type II_∞* if it is properly infinite. Combining our results we have the following.

5.5.13. THEOREM. *Each von Neumann algebra $\mathcal{M}$ on a separable Hilbert space has a unique decomposition*

$$\mathcal{M} = \mathcal{M}_{\mathrm{I}_1} \oplus \mathcal{M}_{\mathrm{I}_2} \oplus \cdots \oplus \mathcal{M}_{\mathrm{I}_\infty} \oplus \mathcal{M}_{\mathrm{II}_1} \oplus \mathcal{M}_{\mathrm{II}_\infty} \oplus \mathcal{M}_{\mathrm{III}}$$

into central summands of distinct types.

5.5.14. Notes and remarks. Abelian projections were introduced by Dixmier [53] under the name 'variétés irréductibles'. Abelian projections in AW^*-algebras appear in [140]. We shall later use abelian elements (not necessarily projections) in C^*-algebras in order to define C^*-algebras of type I, see section 6.1.

5.6. The minimal dense ideal

5.6.1. THEOREM. *For each C^*-algebra A there is a dense hereditary ideal $K(A)$, which is minimal among all dense ideals.*

Proof. Let $K(]0,\infty[)$ denote the set of continuous functions on $]0,\infty[$ with compact support and define

$$K(A)_0 = \{f(x) \mid x \in A_+, f \in K(]0,\infty[)_+\}.$$

Let

$$K(A)_+ = \left\{x \in A_+ \mid x \leqslant \sum_{k=1}^{n} x_k, x_k \in K(A)_0\right\},$$

so that $K(A)_+$ is the smallest hereditary cone in A_+ containing $K(A)_0$. If $K(A)$ denotes the linear span of $K(A)_+$ we conclude as in 5.1.3 that $K(A)$ is a hereditary $*$-algebra with $(K(A))_+ = K(A)_+$. Since $u^*f(x)u = f(u^*xu)$ for any unitary u in $\tilde{A}$ we have $u^*K(A)_0u = K(A)_0$, and as in 5.2.1 this implies that $K(A)$ is an ideal. Let $\{f_n\}$ be a sequence in $K(]0,\infty[)_+$ such that $|t - f_n(t)| < n^{-1}$ for all t in $[0,n]$. Then $\|f_n(x) - x\| \to 0$ for each x in A_+, which shows that $K(A)_+$ is dense in A_+, whence $K(A)$ is dense in A.

Let I be a dense ideal of A and define

$$J = \{x \in I_+ \mid y^*y \leqslant x \Rightarrow yy^* \in I, \forall y \in A\}.$$

If x_1, x_2 belong to J and $y^*y \leqslant x_1 + x_2$ then by 1.4.9 $yy^* = y_1y_1^* + y_2y_2^*$ with $y_i^*y_i \leqslant x_i$ for $i = 1,2$. Consequently $yy^* \in I$, whence $x_1 + x_2 \in J$. It follows that J is a hereditary cone in A_+. Since $u^*Ju = J$ for each unitary u in $\tilde{A}$ we conclude that $y^*xy \in J$ for each x in J and y in A. We claim that J is dense in I_+ (thus dense in A_+ also). To see this take x in I_+ and $\varepsilon > 0$. Choose f in $K(]0,\infty[)_+$ such that $|f(t) - t| \leqslant \varepsilon$ for all t in $[0,\|x\|]$ and let g be a function in $K(]0,\infty[)_+$ which is 1 on the support of f. Then $\|f(x) - x\| \leqslant \varepsilon$ and $f(x)g(x) = f(x)$. Moreover, since the function $h: t \to t^{-1}g(t)$ is continuous we have $g(x) = h(x)x \in I$. If $y^*y \leqslant f(x)$ then $yg(x) = y$, whence $yy^* = yg(x)^2y^* \in I$. Consequently $f(x) \in J$, and J is dense in I_+.

Take x_0 in $K(A)_0$. Then $x_0 = f(x)$ for some x in A_+ and f in $K(]0,\infty[)_+$. Choose another function g in $K(]0,\infty[)_+$ which is 1 on the support of f and put $x_1 = g(x)$. Then $x_0x_1 = x_0$. With I and J as above we have $\|x_1 - y\| \leqslant \frac{1}{2}$ for some y in J, since J is dense in A_+. Then

$$x_0 = x_0^{1/2}x_1x_0^{1/2} = x_0^{1/2}(x_1 - y)x_0^{1/2} + x_0^{1/2}yx_0^{1/2} \leqslant \tfrac{1}{2}x_0 + x_0^{1/2}yx_0^{1/2};$$

whence $x_0 \leqslant 2x_0^{1/2}yx_0^{1/2} \in J$. Since J is a hereditary cone in A_+ this proves that $K(A)_+ \subset J$. It follows that $K(A) \subset I$ so that $K(A)$ is the smallest dense ideal of A.

5.6.2. PROPOSITION. *If $\{x_k\}$ is a finite set in $K(A)$ then the hereditary C^*-subalgebra generated by $\{x_k\}$ is contained in $K(A)$. If therefore $x \in K(A)_+$ then $x^\alpha \in K(A)_+$ for each $\alpha > 0$ and $y^*y \in K(A)$ whenever $yy^* \in K(A)$.*

Proof. Since each x_k is a linear combination of elements in $K(A)_+$, each of which are dominated by a finite sum of elements from $K(A)_0$, we may assume that $\{x_k\} \subset K(A)_0$. As in the last part of the proof of 5.6.1 there is for each k an element y_k in $K(A)_0$ with $x_k y_k = x_k$. Put $x = \Sigma x_k$ and let B denote the hereditary C^*-subalgebra of A generated by x. Then

$$B_+ = \{z \in A_+ \mid \operatorname{Lim} z(x + n^{-1})^{-1}x = z\}.$$

If $z \in B_+$ put $z_k = \operatorname{Lim} z(x + n^{-1})^{-1}x_k$ (this limit exists in A since $x_k \leqslant x$). Then $z_k = z_k y_k \in K(A)$, and since $z = \Sigma z_k$ we conclude that $B_+ \subset K(A)_+$, as desired.

It is immediate from the above that $x^\alpha \in K(A)_+$ for each x in $K(A)_+$ and $\alpha > 0$. If therefore $x^*x \in K(A)$ then it follows that $|x| \in K(A)$; and since $x = u|x|^{1/2}$ by 1.4.6, for some u in A, we conclude that $xx^* = u|x|u^* \in K(A)$.

5.6.3. It is obvious from the definition that $K(A)$ contains every element x in A_+ such that $\operatorname{Sp}(x)\backslash\{0\}$ is closed. Thus all projections in A belong to $K(A)$. In particular, $K(A) = A$ whenever $1 \in A$. It may however happen that $K(A) = A$ even though $1 \notin A$. If A is commutative then $K(A)$ is isomorphic to the set $K(\hat{A})$ of continuous functions on the spectrum with compact support. If $A = C(H)$, the compact operators on H, then $K(A)$ is the set of operators with finite rank. This result is generalized in 5.6.6 below. In most other cases the computation of $K(A)$ is quite difficult, if at all possible.

5.6.4. LEMMA. *If p and q are projections in a von Neumann algebra $\mathscr{M}$, then $p \vee q - q \sim p - p \wedge q$.*

Proof. We may assume that $\mathscr{M}$ is generated by p and q. In that case $1 = p \vee q$, and $p\mathscr{M}p$ and $(1-q)\mathscr{M}(1-q)$ are generated by $\{p, pqp\}$ and $\{1-q, (1-q)p(1-q)\}$, respectively. In particular, $p\mathscr{M}p$ and $(1-q)\mathscr{M}(1-q)$ are commutative, so that p and $1-q$ are abelian projections.

The largest central projection in $\mathscr{M}$ which is orthogonal to p is $(1-p) \wedge q + (1-p) \wedge (1-q)$. Since $p \vee q = 1$ we have

$$(1-p) \wedge (1-q) = 0.$$

Consequently

$$c(p) = 1 - (1-p) \wedge q = p \vee (1-q).$$

Similarly,

$$c(1-q) = 1 - q \wedge p - q \wedge (1-p) = (1-q) \vee p - q \wedge p.$$

It follows from 5.5.2 that

$$p - p \wedge q \sim 1 - q = p \vee q - q.$$

5.6.5. COROLLARY. *If p and q are projections in a countably generated Borel $*$-algebra $\mathscr{A}$, and ϕ is a σ-trace on $\mathscr{A}$, then*

$$\phi(p \vee q) + \phi(p \wedge q) = \phi(p) + \phi(q).$$

5.6.6. PROPOSITION. *Let A be a norm closed ideal in a countably generated Borel $*$-algebra $\mathscr{A}$, and denote by $\mathscr{P}(A)$ the set of projections in A. If $p \in \mathscr{P}(A)$ and either $q \leqslant p$ or $q \sim p$ then $q \in \mathscr{P}(A)$. Moreover, if p and q belong to $\mathscr{P}(A)$ then $p \vee q \in \mathscr{P}(A)$. Conversely, if $\mathscr{P}$ is a class of projections in $\mathscr{A}$ satisfying the conditions above, then $\mathscr{P} = \mathscr{P}(A)$, where A is the smallest norm closed ideal of $\mathscr{A}$ containing $\mathscr{P}$; and $K(A)$ is precisely the smallest ideal of $\mathscr{A}$ containing $\mathscr{P}$.*

Proof. If $p \in \mathscr{P}(A)$ and $q = u^*u$, $uu^* = p$, then $q = u^*pu \in \mathscr{P}(A)$. If $q \leqslant p \in \mathscr{P}(A)$ then $q \in \mathscr{P}(A)$ since A_+ is hereditary in $\mathscr{A}_+$. That $p \vee q \in \mathscr{P}(A)$ for all p and q in $\mathscr{P}(A)$ follows from 5.6.4.

The ideal I of $\mathscr{A}$ generated by a set $\mathscr{P}$ satisfying the above conditions is equal to the set of elements x in $\mathscr{A}$ such that $px = x$ for some p in $\mathscr{P}$. It is clearly hereditary, and its norm closure is an ideal A of $\mathscr{A}$. The projections of A all belong to I, and the projections in I all belong to $\mathscr{P}$. Thus $\mathscr{P}(A) = \mathscr{P}$. Since I is hereditary and dense, $K(A) \subset I$. On the other hand $\mathscr{P}(A) \subset K(A)$, whence $I = K(A)$, as desired.

5.6.7. PROPOSITION. *For each separable C^*-algebra A with enveloping Borel $*$-algebra $\mathscr{B}$ there is a bijective correspondence between the following sets:*

(i) *The unitarily invariant, positive linear functionals on $K(A)$;*

(ii) *The densely defined, lower semi-continuous traces on A;*

(iii) *The σ-traces ϕ on $\mathscr{B}$ such that $K(A) \subset \mathscr{B}^\phi$;*

(iv) *The σ-traces ϕ on $\mathscr{B}$ such that $\mathscr{B}^\phi \cap A$ is dense in A.*

Proof. If ϕ belongs to class (i) let $\{u_\lambda\}$ be the canonical approximate unit for $K(A)$, consisting of the elements in the positive part of the open unit ball of $K(A)$ (cf. 1.4.3). Put $\phi_\lambda(x) = \phi(u_\lambda x)$ for each x in A. If $x \geqslant 0$ then by 5.2.2

$$\phi(u_\lambda x) = \phi(u_\lambda^{1/2} x u_\lambda^{1/2}) = \phi(x^{1/2} u_\lambda x^{1/2}),$$

which shows that $\{\phi_\lambda\}$ is an increasing net of positive linear functionals on A.

For each x in A_+ define $\bar{\phi}(x) = \operatorname{Lim} \phi_\lambda(x)$. Then $\bar{\phi}$ is a lower semi-continuous weight on A, and if $x \in K(A)_+$ then $\bar{\phi}(x) \leqslant \phi(x)$. However, for each x in $K(A)_0$ there is a y in $K(A)_0$ with $xy = x$ and we may assume that $\|y\| = 1$, whence $(1 - \varepsilon)y \in \{u_\lambda\}$. Thus

$$\bar{\phi}(x) \geqslant \phi((1 - \varepsilon)yx) = (1 - \varepsilon)\phi(x),$$

and since ε is arbitrary $\bar{\phi}(x) = \phi(x)$. Since $\bar{\phi} \leqslant \phi$ the set $\{x \in K(A)_+ \mid \phi(x) = \bar{\phi}(x)\}$ is a hereditary cone, and as it contains $K(A)_0$ we conclude that $\bar{\phi}$ is an extension of ϕ. If v is unitary in $\tilde{A}$ then $vu_\lambda v^* = u_\mu$ for some μ, whence

$$\phi_\lambda(v^*xv) = \phi(u_\lambda v^*xv) = \phi(vu_\lambda v^*x) = \phi_\mu(x),$$

which shows that $\bar{\phi}(v^*xv) = \bar{\phi}(x)$ for all x in A_+, so that $\bar{\phi}$ is a densely defined, lower semi-continuous trace.

If conversely ϕ belongs to class (ii), then A^ϕ is a dense, hereditary ideal, whence $K(A) \subset A^\phi$ by 5.6.1. Put $\phi_1 = \phi | K(A)$. Since ϕ is lower semi-continuous and $\{u_\lambda\}$ is an approximate unit for A we have, for each x in A_+

$$\phi(x) = \text{Lim}\, \phi(x^{1/2}u_\lambda x^{1/2}) = \text{Lim}\, \phi(u_\lambda x) = \text{Lim}\, \phi_{1\lambda}(x) = \bar{\phi}_1(x).$$

This proves the bijective correspondence between the classes (i) and (ii). The correspondence between the classes (iii) and (iv) is immediate from 5.6.1 since $\mathscr{B}^\phi \cap A$ is always a hereditary ideal of A.

Now choose a countable group $\{u_k\}$ which is dense in the unitary group of $\tilde{A}$. Then choose a countable approximate unit $\{e_n\}$ in $K(A)$ such that $e_1 = 0$ and for each n and k, $u_k e_n u_k^* \leqslant e_m$ for some m. If ϕ belongs to class (i) define

$$\phi_n(x) = \phi((e_{n+1} - e_n)x)$$

for each x in A. Then $\{\phi_n\}$ is a sequence of positive functionals on A which all extends to normal functionals on A''. It follows that $\tilde{\phi} = \Sigma\, \phi_n$ is a σ-weight on A'' which extends ϕ, and consequently $\tilde{\phi} | \mathscr{B}$ is a σ-weight on $\mathscr{B}$. If $x \in A_+$ then

$$\sum_{i=1}^{n-1} \phi_i(v_k^* x v_k) = \phi(e_n v_k^* x v_k) = \phi(v_k e_n v_k^* x) \leqslant \phi(e_m x) = \sum_{i=1}^{m-1} \phi_i(x),$$

and consequently $\tilde{\phi}(v_k^* \cdot v_k) \leqslant \tilde{\phi}$. By 5.2.4 this implies that $\tilde{\phi}$ is a trace on $\mathscr{B}$. Finally, if φ is an element in class (iii) then with $\{e_n\}$ as before we conclude from the σ-normality of ϕ that since $\{e_n\} \subset \mathscr{B}^\phi$,

$$\phi(x) = \text{Lim}\, \phi(e_{n+1}x) = \sum \phi((e_{n+1} - e_n)x)$$

for each x in $\mathscr{B}_+$. This shows that if $\phi_1 = \phi | K(A)$ then $\tilde{\phi}_1 = \phi$, and the proof is complete.

5.6.8. *Notes and remarks.* Theorem 5.6.1 is due to Laursen and Sinclair [160]. A slightly weaker result ($K(A)$ is minimal among all dense hereditary ideals) was proved by the author [190]. The minimal dense ideal was used in [190] (see also [200]) to define an analogue of Radon measures (C^*-integrals). Lazar and Taylor used the multipliers of $K(A)$ to define a non-commutative analogue of the algebra $C(T)$ of (unbounded) continuous functions on a locally compact Hausdorff space T, see [161]. The difficulties involved in determining $K(A)$ for some specific C^*-algebra A can be realized by looking at the (counter) examples in [208] and [94]. In this book we shall only use the minimal dense ideal as a common domain of definition for all densely defined traces, cf. 5.6.7.

5.7. Borel sets in the factor spectrum

5.7.1. Let A be a separable C^*-algebra with enveloping Borel $*$-algebra $\mathscr{B}$. With α as any one of the symbols I_n $(1 \leqslant n \leqslant \infty)$, $\mathrm{II}_1, \mathrm{II}_\infty$ and III, we say that a separable representation (π, H) of A is of *type* α if the von Neumann algebra $\pi''(\mathscr{B})$ is of type α. Since a factor has precisely one of these types we obtain in this way a partition of the factor spectrum $\hat{A}$ in disjoint subsets $\hat{A}_\alpha$. Note that it follows from 5.5.8 that the sets $\hat{A}_n$ and $\hat{A}_{I_n}$ are canonically isomorphic for each n $(1 \leqslant n \leqslant \infty)$.

5.7.2. LEMMA. *The set of states ϕ of A such that $\pi_\phi(A)'$ contains a non-zero abelian projection is a $G_{\delta\sigma}$-subset of S.*

Proof. Let $\{x_n\}$ be a sequence which is dense in A and $\{y_n\}$ a sequence which is dense in A^1_+. Choose $\varepsilon > 0$ such that $64(1-\varepsilon)^{-2}\varepsilon^{1/2} < 1$. For each z in A^1_+ consider the set of states ϕ such that

$$(*) \qquad \phi(x_i^*(y_j z y_k - y_k z y_j)x_i)^2 < 16\phi(x_i^* x_i)\phi(x_i^*(1-z)x_i) + 1/n$$

for all i, j, k between 1 and n, and moreover

$$(**) \qquad (1+\varepsilon)\phi(x_m^* z x_m) > \phi(x_m^* x_m) > 1.$$

Clearly each such set is open in S and we denote by G_{mn} their union over all z in A^1_+.

If $\phi \in \cap_n G_{mn}$ we can find a sequence $\{z_n\}$ in A^1_+ such that ϕ and z_n satisfy $(*)$ and $(**)$. Put $\mathscr{M} = \pi_\phi(A)''$. Since $\mathscr{M}^1_+$ is weakly compact we may assume that $\{\pi_\phi(z_n)\}$ is weakly convergent to an element a in $\mathscr{M}^1_+$. From $(*)$ we have

$$((\pi_\phi(y_j)a\pi_\phi(y_k) - \pi_\phi(y_k)a\pi_\phi(y_j))\xi_{x_i} \mid \xi_{x_i})^2 \leqslant 16\|\xi_{x_i}\|^2((1-a)\xi_{x_i} \mid \xi_{x_i})$$

for all i, j and k. Since $\{\pi_\phi(y_j)\}$ is strongly dense in $\mathscr{M}^1_+$ and $\{\xi_{x_i}\}$ is dense in H_ϕ we conclude that

$$((bac - cab)\xi \mid \xi)^2 \leqslant 16\|\xi\|^2((1-a)\xi \mid \xi).$$

for all b, c in $\mathscr{M}^1_+$ and all ξ in H_ϕ. From $(**)$ we have

$$(1+\varepsilon)(a\xi_{x_m} \mid \xi_{x_m}) \geqslant \|\xi_{x_m}\|^2 \geqslant 1,$$

which implies that the spectral projection p of a corresponding to the interval $[1-\varepsilon, 1]$ is non-zero. Replacing ξ with $a^{1/2}p\xi$ and b, c with $(1-\varepsilon)a^{-1/2}pbpa^{-1/2}$, $(1-\varepsilon)a^{-1/2}pcpa^{-1/2}$ we obtain the inequality

$$(1-\varepsilon)^4((pbpcp - pcpbp)\xi \mid \xi)^2 \leqslant 16\varepsilon\|p\xi\|^4$$

for all b, c in $\mathcal{M}_+^1$ and all ξ in H_ϕ. It follows that for all b, c in $p\mathcal{M}_+^1 p$ we have $\|bc - cb\| \leq (1-\varepsilon)^{-2}4\varepsilon^{1/2}$. However, each element in $p\mathcal{M}^1 p$ is a combination of 4 positive elements, whence $\|bc - cb\| \leq 64(1-\varepsilon)^{-2}\varepsilon^{1/2}$ for all b, c in $p\mathcal{M}^1 p$. If $p\mathcal{M}p$ was non-commutative then for some projection $q \leq p$ there would be an element v of norm 1 in $q\mathcal{M}(p-q)$. But then $\|vq - qv\| = \|0 - v\| = 1$, in contradiction with $64(1-\varepsilon)^{-2}\varepsilon^{1/2} < 1$. Consequently $\pi_\phi(A)''$ contains a non-zero abelian projection (viz. p) for every ϕ in $\cup_m \cap_n G_{mn}$.

Conversely, take a state ϕ such that the algebra $\mathcal{M} = \pi_\phi(A)''$ contains a non-zero abelian projection p. There is then by Kaplansky's density theorem a sequence $\{z_n\}$ in A_+^1 such that $\{\pi_\phi(z_n)\}$ is strongly convergent to p. Since $p \neq 0$ there is a vector ξ_{x_m} such that $\|\xi_{x_m}\|^2 > 1$ and $(1+\varepsilon)(p\xi_{x_m} \mid \xi_{x_m}) > \|\xi_{x_m}\|^2$. We may assume therefore that all z_n satisfy (**) for some fixed m. Since p is abelian we have for all b, c in $\mathcal{M}_+^1$ and all ξ in H_ϕ,

$$
\begin{aligned}
((bpc - cpb)\xi \mid \xi) &= ((bpc - cpb)(1-p)\xi \mid \xi) + ((bpc - cpb)p\xi \mid \xi) \\
&= ((bpc - cpb)(1-p)\xi \mid \xi) + ((bpc - cpb)p\xi \mid (1-p)\xi) \\
&\leq 2\|(1-p)\xi\|\,\|\xi\| + 2\|\xi\|\,\|(1-p)\xi\| = 4\|\xi\|((1-p)\xi \mid \xi)^{1/2}.
\end{aligned}
$$

From this inequality it follows that (passing to a subsequence) we may assume that each z_n satisfies (*), whence $\phi \in \cap_n G_{mn}$. We have shown that the set described in the lemma is $\cup_m \cap_n G_{mn}$, and consequently a $G_{\delta\sigma}$-subset of S.

5.7.3. LEMMA. *The set of states ϕ of A such that $\pi_\phi(A)''$ contains a non-zero finite projection is a $G_{\delta\sigma}$-subset of S.*

Proof. Let $\{x_n\}$ be a sequence which is dense in A and take $\varepsilon > 0$ such that $\varepsilon(1-\varepsilon)^{-2} < 1$. Let G_{kmn} denote the open subset of states ϕ for which there is a pair, y, z in A_+^1 with $(1-\varepsilon)z \leq y \leq z$ such that

$$
\begin{aligned}
(*)\quad & |\phi(y(x_i x_j - x_j x_i))| \\
& < |\phi(yx_i x_i^*)\phi(x_j^*(1-z)x_j)|^{1/2} + |\phi(yx_j x_j^*)\phi(x_i^*(1-z)x_i|^{1/2} + 1/n,
\end{aligned}
$$

$$(**)\qquad |\phi(yx_i - x_i y)| < 1/n \quad \text{and} \quad |\phi(zx_i - x_i z)| < 1/n,$$

$$(***)\qquad (1+\varepsilon)|\phi(zx_k^* x_k)| > \phi(x_k^* x_k) > 1/m,$$

for all i, j between 1 and n.

If $\phi \in \cap_n G_{kmn}$ we can find sequences $\{y_n\}$ and $\{z_n\}$ in A_+^1 such that ϕ, y_n and z_n satisfy (*), (**) and (***) for each n. With $\mathcal{M} = \pi_\phi(A)''$ we may further assume that $\{\pi_\phi(y_n)\}$ and $\{\pi_\phi(z_n)\}$ are weakly convergent to elements h and a in $\mathcal{M}_+^1$ with $(1-\varepsilon)a \leq h \leq a$. Regarding ϕ as a normal state of $\mathcal{M}$ we let $\mathcal{M}_\phi$ denote the von Neumann subalgebra of elements c in $\mathcal{M}$ for which $\phi(bc) = \phi(cb)$ for all b in $\mathcal{M}$. From (**) we see that both h and a belong to $\mathcal{M}_\phi$.

Let p be the spectral projection of a corresponding to the interval $[1-\varepsilon, 1]$. If $\phi(p) = 0$ then from (***) and the Cauchy–Schwarz inequality

$$1/m < \phi(x_k^* x_k) \leqslant (1+\varepsilon)\phi(a x_k^* x_k) = (1+\varepsilon)\phi((1-p)a x_k^* x_k)$$
$$\leqslant (1+\varepsilon)(1-\varepsilon)\phi(x_k^* x_k) = (1-\varepsilon^2)\phi(x_k^* x_k)$$

a contradiction. Thus $\phi(p) > 0$. Let $\psi = \phi(h\cdot)$. Then ψ is a normal functional and since $h \in \mathcal{M}_\phi$, $\psi \geqslant 0$. (If $b \geqslant 0$ then $\phi(hb) = \phi(h^{1/2} b h^{1/2}) \geqslant 0$.) Moreover, $\psi \mid p\mathcal{M}p$ is non-zero since $\psi(p) = \phi(php) \geqslant (1-\varepsilon)\phi(p)$. From (*) we have for all b, c in $\mathcal{M}$

$$|\psi(bc - cb)| \leqslant (\psi(bb^*)\phi(c^*(1-a)c))^{1/2} + (\psi(cc^*)\phi(b^*(1-a)b))^{1/2}.$$

If b, c belong to $p\mathcal{M}p$ then $c^*(1-a)c \leqslant \varepsilon c^* c$ and since furthermore $p \leqslant (1-\varepsilon)^{-2} php$ we have for all b, c in $p\mathcal{M}p$

$$|\psi(bc - cb)| \leqslant \varepsilon^{1/2}(1-\varepsilon)^{-1}((\psi(bb^*)\psi(c^*c))^{1/2} + (\psi(cc^*)\psi(b^*b))^{1/2}).$$

With $c = b^*$ this gives

$$\psi(bb^*) - \psi(b^*b) \leqslant \varepsilon^{1/2}(1-\varepsilon)^{-1}(\psi(bb^*) + \psi(b^*b)),$$

and consequently $\psi(bb^*) \leqslant \alpha\psi(b^*b)$ for all b in $p\mathcal{M}p$, where $\alpha = (1 - \varepsilon^{1/2}(1-\varepsilon)^{-1})^{-1}(1 + \varepsilon^{1/2}(1-\varepsilon)^{-1})$. If $c \in p\mathcal{M}_+ p$ and $\Sigma b_i^* b_i \leqslant c$ then we see from this that $\psi(\Sigma b_i b_i^*) \leqslant \alpha \Sigma \psi(b_i^* b_i) \leqslant \alpha\psi(c)$. It follows that the trace $\tilde{\psi}$ as defined in 5.2.7 satisfies $\tilde{\psi} \leqslant \alpha\psi$ on $p\mathcal{M}p$. Since $\tilde{\psi}$ is normal by 5.2.8 we conclude that $\pi_\phi(A)''$ contains a non-zero finite projection for each ϕ in $\cup_{k,m} \cap_n G_{kmn}$.

Conversely, take a state ϕ such that the algebra $\mathcal{M} = \pi_\phi(A)''$ contains a non-zero finite projection. This is equivalent to the existence of a normal trace τ on $\mathcal{M}$ which is faithful and semi-finite on $q\mathcal{M}$, where q is a non-zero central projection in $\mathcal{M}$ (cf. the proof of 5.4.2). It follows from 5.3.11 that $\phi(q\cdot) = \tau(k\cdot)$ for some positive operator k affiliated with $q\mathcal{M}$. Note that $\phi(q) > 0$ since ϕ is faithful on the centre of $\mathcal{M}$ (see e.g. 4.7.6). Thus if p is the spectral projection of k corresponding to an interval $[(1-\varepsilon)\beta, \beta]$, with ε as before, then for some $\beta > 0$ we must have $\phi(p) > 1/m$ for a suitable m. Moreover, since $\{\pi_\phi(x_k^* x_k)\}$ is a strongly dense sequence in $\mathcal{M}_+$ we may assume that

$$(1+\varepsilon)\phi(p x_k^* x_k) > \phi(x_k^* x_k) > 1/m$$

for some k. Now let $h = (1-\varepsilon)\beta k^{-1} p$, whence $(1-\varepsilon)p \leqslant h \leqslant p$. Since $\mathcal{M}_\phi \cap q\mathcal{M} = \{k\}' \cap q\mathcal{M}$ we see that both h and p belong to $\mathcal{M}_\phi$. Choose sequences $\{y_n'\}$ and $\{z_n\}$ in A_+^1 such that $\{\pi_\phi(y_n')\}$ and $\{\pi_\phi(z_n)\}$ are strongly convergent to $\varepsilon^{-1}(h - (1-\varepsilon)p)$ and p, respectively. Then with $y_n = (1-\varepsilon)z_n + \varepsilon z_n y_n' z_n$ we have $(1-\varepsilon)z_n \leqslant y_n \leqslant z_n$ and $\{\pi_\phi(y_n)\}$ is strongly convergent to h. Passing to a subsequence if necessary we may assume that y_n and z_n satisfy (**), and (***) for each n. Now

$$\phi(h\cdot) = \tau(kh\cdot) = (1-\varepsilon)\beta\tau(p\cdot),$$

so that $\phi(h\cdot)$ is a trace on $p\mathcal{M}p$. Consequently, for all b, c in $\mathcal{M}$

$$\begin{aligned}|\phi(h(bc-cb))| &= |\phi(hp(bc-cb)p)| = |\phi(hp(b(1-p)c - c(1-p)b)p)| \\ &\leqslant |\phi(hb(1-p)c)| + |\phi(hc(1-p)b)| \\ &\leqslant (\phi(hbb^*h)\phi(c^*(1-p)c))^{1/2} + (\phi(hcc^*h)\phi(b^*(1-p)b))^{1/2} \\ &\leqslant (\phi(hbb^*)\phi(c^*(1-p)c))^{1/2} + (\phi(hcc^*)\phi(b^*(1-p)b))^{1/2},\end{aligned}$$

using the facts that p and h belong to $\mathcal{M}_\phi$ and $h^2 \leqslant h$. It follows that we may assume that y_n and z_n satisfy (*) as well, whence $\phi \in \cap_n G_{kmn}$. We have shown that the set described in the lemma is $\cup_{k,m} \cap_n G_{kmn}$, and consequently a $G_{\delta\sigma}$-subset of S.

5.7.4. THEOREM. *Let A be a separable C*-algebra with factor spectrum $\hat{A}$. Each of the sets $\hat{A}_\alpha, \alpha = I_n (1 \leqslant n \leqslant \infty)$, II_1, II_∞ and III, is an M-Borel set in $\hat{A}$. Moreover, $\hat{A}$ and $\hat{A}_I$ are Borel isomorphic in the respective M-Borel structures.*

Proof. Let S_1 and S_2 denote the subsets of S described in 5.7.2 and 5.7.3, respectively. In a factor each non-zero projection has central cover 1. It follows that the counter-image of $\hat{A}_I$ (with respect to the canonical surjection $\phi \to \hat{\phi}$ of $F(A)$ onto $\hat{A}$) is precisely $S_1 \cap F(A)$, and the counter image of $\hat{A}_I \cup \hat{A}_{II}$ is precisely $S_2 \cap F(A)$. Thus by the definition of the M-Borel structure $\hat{A}_I$, $\hat{A}_{II}$ and $\hat{A}_{III} (= \hat{A} \backslash (\hat{A}_I \cup \hat{A}_{III}))$ are Borel sets.

Clearly $\hat{A}$ and $\hat{A}_I$ are isomorphic as sets by 5.5.8 and since the set $P(A)$ of pure states is a Borel subset of $S_1 \cap F(A)$, the isomorphism $\iota: \hat{A} \to \hat{A}_I$ is an M-Borel map. Let W be as in the proof of 4.10.6. If T is an M-Borel set in $\hat{A}$ with counter image F in $P(A)$, then the counter image F_1 of $\iota(T)$ in $F(A)$ is obtained by projecting the set $(F \times (S_1 \cap F(A))) \backslash W$ on its second coordinate. However, the counter-image of $\hat{A}_I \backslash \iota(T)$ in $F(A)$ is the projection of the set $((P(A)\ F) \times (S_1 \cap F(A))) \backslash W$. It follows from 4.2.9 that F_1 (and $S_1 \cap F(A) \backslash F_1$) is a Borel subset of $F(A)$. Thus $\iota(T)$ is an M-Borel set in $\hat{A}$ and ι is a Borel isomorphism.

By 4.4.10 each set $\hat{A}_n$ is a Borel set of $\hat{A}$ for $1 \leqslant n \leqslant \infty$ (even a T-Borel set). Thus $\hat{A}_{I_n}$ is an M-Borel subset of $\hat{A}$ for $1 \leqslant n \leqslant \infty$.

Let $FT(A)$ denote the closed subset of $F(A)$ consisting of states ϕ such that $\phi(xy) = \phi(yx)$ for all x, y in A. Thus $FT(A)$ consists of the factorial states which are also traces. By 5.3.7 the restriction of the map $\phi \to \hat{\phi}$ is a bijection of $FT(A)$ on $\hat{A}_{II_1} \cup_{n<\infty} \hat{A}_{I_n}$. It follows from 4.10.6 that $\hat{A}_{II_1}$ is an M-Borel set in A (even a standard subset). This completes the proof.

5.7.5. In the course of the proof of 5.7.4 we showed that $\hat{A}_{\mathrm{I}_n}$ for $n < \infty$ and $\hat{A}_{\mathrm{II}_1}$ are standard spaces in the M-Borel structure. It remains to see whether similar techniques can give the same information about $\hat{A}_{\mathrm{I}\,\infty}$ and $\hat{A}_{\mathrm{II}\,\infty}$.

Let $T(\mathscr{B})$ denote the cone of all σ-traces on $\mathscr{B}$, equipped with the *weak* Borel structure*, which is the weakest Borel structure for which each function $\phi \to \phi(x)$, $x \in \mathscr{B}_+$ is Borel measurable. The weak* Borel structure separates points, but is probably not always countably generated.

Let $T(A)$ denote the subset of $T(\mathscr{B})$ consisting of σ-traces ϕ such that $K(A) \subset \mathscr{B}^{\phi}$; i.e. $T(A)$ is the set described in 5.6.7.

5.7.6. LEMMA. *The set $T(A)$ is a Borel subset of $T(\mathscr{B})$ and the weak* Borel structure is the weakest Borel structure for which every function $\phi \to \phi(x)$, $x \in K(A)_+$, is Borel measurable. Moreover, $T(A)$ is a standard Borel space.*

Proof. Let $\{e_n\}$ be a countable approximate unit for A contained in $K(A)_+$. Using 5.6.7 we see that

$$T(A) = \{\phi \in T(\mathscr{B}) \,|\, \forall n: \phi(e_n) < \infty\},$$

and consequently a Borel subset of $T(\mathscr{B})$. If $T(A)$ is equipped with the weakest Borel structure for which each function $\phi \to \phi(x)$, $x \in K(A)_+$, is measurable, then for each x in A_{sa} the function $\phi \to \phi(e_n^{1/2} x e_n^{1/2})$ is a bounded Borel function on $T(A)$. Since $\mathscr{B}_{\mathrm{sa}}$ is the monotone sequential closure of A_{sa}, the same is true with x in $\mathscr{B}_{\mathrm{sa}}$. Finally, if $x \in \mathscr{B}_+$

$$\phi(x) = \mathrm{Lim}\, \phi(x^{1/2} e_n x^{1/2}) = \mathrm{Lim}\, \phi(e_n^{1/2} x e_n^{1/2}),$$

whence $\phi \to \phi(x)$ is a Borel function. Consequently this Borel structure coincides with the weak* Borel structure defined in 5.7.5.

For each n let A_n denote the hereditary C^*-subalgebra of A generated by e_n (i.e. $A_n = (e_n A e_n)^-$) and let I_n denote the smallest closed ideal containing A_n. By 5.6.2, $A_n \subset K(A)$ and by 5.6.1, $K(A) \subset \cup I_n$. Let E_n denote the cone of finite traces on A_n and define a map

$$f: T(A) \to \prod E_n$$

by $(f(\phi))_n = \phi \,|\, A_n$. Note that f is injective; for if $\phi \,|\, A_n = \psi \,|\, A_n$ for all n, then $\phi(e_n x) = \psi(e_n x)$ for all n and all x in A_+, whence $\phi(x) = \psi(x)$ by lower semi-continuity of ϕ and ψ. The set of traces of norm less than or equal to one on a separable C^*-algebra is weak* compact. It follows that each E_n is a standard Borel space in the weak* Borel structure, whence ΠE_n is standard with the product Borel structure. Let

$$E = \{(\phi_n) \in \prod E_n \,|\, \forall n: \phi_{n+1} \,|\, A_n = \phi_n\}.$$

Then E is a Borel subset of ΠE_n, hence standard, and f is a Borel map from $T(A)$ into E. However, if $(\phi_n) \in E$ define, for each n and each x in $(I_n)_+$,

$$\tilde{\phi}_n(x) = \mathrm{Sup}\{\phi_n(y) \mid y \in A_n, y \precsim x\}.$$

By 5.2.7, $\tilde{\phi}_n$ is an extension of ϕ_n to a lower semi-continuous trace on I_n. Furthermore, $\tilde{\phi}_{n+1} \mid I_n = \tilde{\phi}_n$ since $\tilde{\phi}_{n+1}$ and $\tilde{\phi}_n$ coincide on A_n. We can therefore define a unique trace ϕ on $\cup I_n$ such that $\phi \mid A_n = \phi_n$ for every n. Since $K(A) \subset \cup I_n$ and since ϕ is finite on $\cup A_n$, hence on $K(A)$, we see that $\phi \in T(A)$. This proves that $f(T(A)) = E$. At the same time we see that f^{-1} is a Borel map. Indeed, if $\phi = f^{-1}(\phi_k)$, then for each x in A_+ and each n we have

$$\phi(x^{1/2}e_n x^{1/2}) = \phi(e_n^{1/2} x e_n^{1/2}) = \phi_n(e_n^{1/2} x e_n^{1/2}),$$

so that the map $(\phi_k) \to \phi(x^{1/2}e_n x^{1/2})$ is Borel measurable. It follows that the map $(\phi_k) \to \phi(x)$ is measurable for every x in A_+, whence f^{-1} is a Borel map. This completes the proof.

5.7.7. THEOREM. *Let A be a separable C^*-algebra. The set $\hat{A}_K$ of equivalence classes of semi-finite factor representations associated with traces in $T(A)$ is an M-Borel set in $\hat{A}$ and is standard in the relative Borel structure.*

Proof. With $\{e_n\}$ as a countable approximate unit for A contained in $K(A)$ define

$$F_n = \{\phi \in T(A) \mid \phi(e_n) = 1, \phi(1 - c[e_n]) = 0, \phi(e_n \cdot) \in F(A)\}.$$

Then F_n is a Borel subset of $T(A)$. Indeed, the first condition ensures that we have a map $\phi \to \phi(e_n \cdot)$ into the state space of A, a map which is clearly a Borel map, and F_n is then the counter image under this map of the Borel set $F(A)$.

We claim that F_n is the set of traces ϕ such that $\phi(e_n) = 1$ and (π_ϕ, H_ϕ) is a factor representation. It is rather obvious that this set is contained in F_n (if (π_ϕ, H_ϕ) is factorial then $\pi_\phi(c[e_n]) = 1$). Conversely, if $\phi \in F_n$ then it follows from the second condition that the representations associated with ϕ and $\phi(e_n \cdot)$ have the same kernel in $\mathscr{B}$, and thus are equivalent by 4.7.10. Since $\phi(e_n \cdot) \in F(A)$, (π_ϕ, H_ϕ) is factorial.

Consider the Borel map $\phi \to \phi(e_n \cdot)$ from F_n into $F(A)$. If ϕ, ψ belong to F_n and $\phi(e_n \cdot) = \psi(e_n \cdot)$ then (π_ϕ, H_ϕ) and (π_ϕ, H_ϕ) are equivalent from the above, whence ϕ and ψ are proportional by 5.3.7. Since $\phi(e_n) = \psi(e_n)$ this implies that $\phi = \psi$. Thus $\phi \to \phi(e_n \cdot)$ is an injective Borel map from the standard Borel space F_n. It follows from 4.6.12 that the image E_n in $F(A)$ is also Borel. We have a commutative diagram:

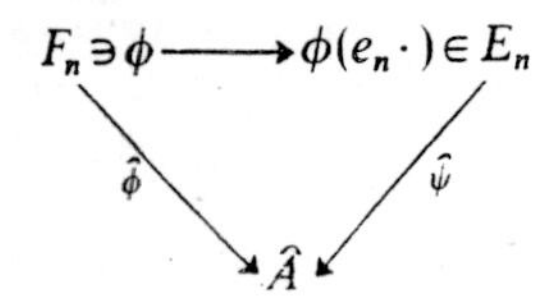

Since $\hat{\phi} \neq \hat{\psi}$ implies $\phi \neq \psi$ for all ϕ, ψ in F_n, 4.10.6 applies to show that the subset of $\hat{A}_K$ coming from semi-finite factor representations with a trace ϕ for which $\phi(e_n) > 0$, is a standard M-Borel subset of $\hat{A}$. But $\hat{A}_K$ is the union of these sets and the proof is complete.

5.7.8. PROPOSITION. *Let A be a separable C^*-algebra with enveloping Borel $*$-algebra $\mathscr{B}$. For each countable set $\{a_n\}$ which generates $\mathscr{B}$ let $\hat{A}(a_n)$ denote the set of equivalence classes of semi-finite factor representations associated with σ-traces ϕ such that $\phi(|a_n|) < \infty$ for all n. Then $\hat{A}(a_n)$ is a standard M-Borel subset of $\hat{A}$.*

Proof. Let A_1 denote the C^*-subalgebra of $\mathscr{B}$ generated by $\{a_n\}$. Then by 4.5.9 the canonical injection $A_1 \to \mathscr{B}$ extends to a sequentially normal morphism $\rho: \mathscr{B}_1 \to \mathscr{B}$, and since A_1 generates $\mathscr{B}$ we have $\rho(\mathscr{B}_1) = \mathscr{B}$. But then by 4.6.3 there is a central projection p in $\mathscr{B}_1$ such that $\ker \rho = (1-p)\mathscr{B}_1$. We may therefore identify $\mathscr{B}$ and $p\mathscr{B}_1$. In particular we may embed the set of σ-traces ϕ on $\mathscr{B}$ such that $\phi(|a_n|) < \infty$ for all n, as a Borel subset of the set $\{\phi \in T(A_1) \mid \phi(1-p) = 0\}$. It follows that $\hat{A}(a_n)$ is isomorphic to a Borel subset of the set $\{t \in (\hat{A}_1)_K \mid \check{p}(t) = 1\}$. By 5.7.7 this is a standard M-Borel set in $\hat{A}_1$ and consequently $\hat{A}(a_n)$ is a standard M-Borel subset of $\hat{A}$.

5.7.9. THEOREM. *Let A be a separable C^*-algebra, and denote by $\hat{A}_N$ the set of equivalence classes of semi-finite factor representations associated with σ-traces in $T(\mathscr{B})$ such that $0 < \phi(x) < \infty$ for some x in A_+. Then $\hat{A}_N$ is an M-Borel set in $\hat{A}$ and is standard in the relative Borel structure.*

Proof. By 4.3.4 and 4.1.3 there is a sequence $\{I_n\}$ of closed, non-zero ideals of A such that for every closed ideal of A we have

$$I = \{\textstyle\sum I_n \mid I_n \subset I\}^-.$$

In particular, each non-zero I contains some I_n. Let ι_n denote the canonical injection of $\hat{I}_n$ into $\hat{A}$, and note that ι_n is an M-Borel isomorphism and that $\iota_n(\hat{I}_n)$ is an M-Borel subset of $\hat{A}$ (even a D-Borel subset).

Now take a semi-finite factor representation (π, H) of A associated with a factorial trace ϕ in $T(\mathscr{B})$ such that $0 < \phi(x) < \infty$ for some x in A_+. Then A^ϕ is a non-zero $*$-ideal in A, whence $K(I_n) \subset A^\phi$ for some n, by 5.6.1. Furthermore we may assume that ϕ is non-zero on $K(I_n)$. Thus $\phi \mid I_n \in T(I_n)$; whence $(\pi \mid I_n, H)$ belongs to $(\hat{I}_n)_K$. It follows that

$$\hat{A}_N = \bigcup \iota_n((\hat{I}_n)_K),$$

and therefore a standard Borel space.

5.7.10. *Notes and remarks.* The computational lemmas 5.7.2 and 5.7.3 are taken from unpublished work (1974) by the author. However, theorem 5.7.4 is older.

The isomorphism between $\hat{A}_{\mathrm{I}}$ and $\hat{A}$ was established by Dixmier [63] and the fact that $\hat{A}_{\mathrm{III}}$ (hence also $\hat{A}_{\mathrm{II}}$) is an M-Borel set was proved by Nielsen [178]. (See also the work Schwartz [234].) That $\hat{A}_{\mathrm{II}_1}$ is a standard space was shown by Guichardet [106] and the generalization to $\hat{A}_N$ (5.7.9) is due to Halpern, see [112]. The proof of 5.7.9 given here utilizes the minimal dense ideal (cf. 5.7.7) and gives at the same time 5.7.8, which may have applications in the future.

Factor representations associated with $\hat{A}_N$ were called 'normal' by Guichardet [106], following a terminology introduced by Godement [102]. Cuntz [46] gives a series of perfectly normal and readily available C^*-algebras for which no normal representations exist.

5.8. Disintegration of weights and traces

5.8.1. Let A be a separable C^*-algebra with enveloping Borel $*$-algebra $\mathscr{B}$. We denote by $W(\mathscr{B})$ the set of σ-weights on $\mathscr{B}$. As mentioned before, the set of all σ-normal weights on $\mathscr{B}$ form a cone, while the set of σ-finite weights do not. Thus $W(\mathscr{B})$ is not a cone, as the sum is only a partially defined operation. We say that an element ϕ in $W(\mathscr{B})$ is *factorial* if (π_ϕ, H_ϕ) is a factor representation of A.

5.8.2. LEMMA. *The following conditions on an element ϕ in $W(\mathscr{B})$ are equivalent*:

(i) *ϕ is factorial;*

(ii) *$\phi(x^*\cdot x)$ is a factorial state for each x in $\mathscr{B}$ with $\phi(x^*x) = 1$;*

(iii) *$\phi(z\cdot)$ is proportional to ϕ for each z in $\mathscr{C}_+$.*

Proof. (i) $\Rightarrow$ (iii). If $z \in \mathscr{C}_+$ then $\pi_\phi(z) = \lambda 1$, whence

$$\phi(zx) = \sum (\pi_\phi(zx)\xi_n \mid \xi_n) = \lambda\phi(x)$$

by 5.1.5 for each x in $\mathscr{B}_+$.

(iii) $\Rightarrow$ (ii). If $\phi(x^*x) = 1$ and z_1, z_2 are elements of $\mathscr{C}_+$ then $\phi(z_1\cdot) = \lambda\phi$, where $\lambda = \phi(z_1 x^*x)$. Consequently,

$$\phi(x^*z_1z_2x) = \lambda\phi(x^*z_2x) = \phi(x^*z_1x)\phi(x^*z_2x).$$

Thus $\phi(x^*\cdot x)$ is multiplicative on $\mathscr{C}$, hence factorial (see 4.8.2).

(ii) $\Rightarrow$ (i). If z is a projection in $\mathscr{C}$ such that $\pi_\phi(z)$ is neither 0 or 1, then there is an element x in $\mathscr{B}_2^\phi$, with $\|\xi_x\| = 1$, such that

$$0 < (\pi_\phi(z)\xi_x \mid \xi_x) < 1.$$

However,

$$\begin{aligned}(\pi_\phi(z)\xi_x \mid \xi_x) &= \phi(x^*zx) = \phi(x^*z^2x)\\ &= \phi(x^*zx)^2 = (\pi_\phi(z)\xi_x \mid \xi_x)^2,\end{aligned}$$

a contradiction. Consequently $\pi_\phi(z)$ is a multiple of 1 and $\pi_\phi(\mathscr{B})$ is a factor.

5.8.3. THEOREM. *Let A be a separable C^*-algebra with enveloping Borel $*$-algebra $\mathscr{B}$. For each σ-weight ϕ on $\mathscr{B}$ there is a sequence $\{x_n\}$ in $\mathscr{B}_2^\phi$, a standard measure μ on $\hat{A}$ and a function $t \to \phi_t$ from $\hat{A}$ into the set of factorial σ-weights on $\mathscr{B}$ such that:*

(i) $\{x_n\} \subset \mathscr{B}_2^{\phi_t}$ *and* $\{\xi_{x_n}\}$ *is dense in* H_{ϕ_t} *for almost all* t;

(ii) $t \to \phi_t(x)$ *is an extended valued Borel function for each* x *in* $\mathscr{B}_+$;

(iii) $\int \phi_t(x)\,d\mu(t) = \phi(x)$ *for each* x *in* $\mathscr{B}_+$;

(iv) $\phi_t(zx) = \check{z}(t)\phi_t(x)$ *almost everywhere for each* z *in* $\mathscr{C}_+$ *and* x *in* $\mathscr{B}_+$;

(v) *the set* $\{t \in \hat{A} \mid \hat{\phi}_t = t\}$ *has outer measure* 1.

Proof. Let μ, $\{H_t\}$ and $\{\pi_t\}$ be the components in the canonical disintegration of the separable representation (π_ϕ, H_ϕ) obtained from 4.12.4. Choose a sequence $\{x_n\}$ *in* $\mathscr{B}_2^\phi$ such that $\{\xi_{x_n}\}$ is dense in H_ϕ. For each n there is a square integrable Borel vector field η_n on $\hat{A}$ whose image in H_ϕ is ξ_{x_n}. It follows from 4.11.10 that $\{\eta_n(t)\}$ is dense in H_t for almost all t. By 5.1.5 there is a sequence $\{\xi_n\}$ of square integrable Borel vector fields such that

$$\phi(x) = \sum (\pi_\phi(x)\xi_n \mid \xi_n) = \sum \int (\pi_t(x)\xi_n(t) \mid \xi_n(t))\,d\mu(t)$$

for each x in $\mathscr{B}_+$. Defining

$$\phi_t(x) = \sum (\pi_t(x)\xi_n(t) \mid \xi_n(t)),$$

we obtain a function $t \to \phi_t$ satisfying (ii) and (iii), and also (iv) since the π_t's are factor representations. If $x \in \mathscr{B}_+^\phi$ then $\phi_t(x) < \infty$ almost everywhere, so we may assume that each ϕ_t is a σ-weight with $\{x_n\} \subset \mathscr{B}_2^{\phi_t}$. If we can show that (π_t, H_t) is spatially equivalent with $(\pi_{\phi_t}, H_{\phi_t})$ almost everywhere, then conditions (i) and (v) will be satisfied.

Now, for each x in $\mathscr{B}$ and z in $\mathscr{C}_+$

$$\begin{aligned}\int (\pi_{\phi_t}(x)\xi_{x_n} \mid \xi_{x_m})\check{z}(t)\,d\mu(t) &= \int \phi_t(x_m^* x x_n)\check{z}(t)\,d\mu(t) = \phi(z x_m^* x x_n)\\ &= (\pi_\phi(zx)\xi_{x_n} \mid \xi_{x_m}) = \int (\pi_t(x)\eta_n(t) \mid \eta_m(t))\check{z}(t)\,d\mu(t).\end{aligned}$$

It follows that

$$(\pi_{\phi_t}(x)\xi_{x_n} \mid \xi_{x_m}) = (\pi_t(x)\eta_n(t) \mid \eta_m(t))$$

almost everywhere, and since $\mathscr{B}$ is countably generated this implies the existence, almost everywhere, of isometries $u_t \colon H_t \to H_{\phi_t}$ such that $u_t\pi_t(x)\eta_n(t) = \pi_{\phi_t}(x)\xi_{x_n}$ for all n and all x in $\mathscr{B}$. Thus π_t is at least spatially equivalent to a subrepresentation of π_{ϕ_t}. To show that $u_t H_t = H_{\phi_t}$ consider the functional

$$\phi_{t,n}(x) = \sum_{k=1}^{n} (\pi_t(x)\eta_k(t) \mid \eta_k(t)).$$

Since $\phi_{t_n} \leqslant \phi_t$ there exists a_{tn} in $\pi_{\phi_t}(\mathscr{B})'$ and ξ'_{nt} in H_{ϕ_t} satisfying the conditions in 5.1.4. Put $\phi_n = \int \phi_{tn}\,\mathrm{d}\mu(t)$. Then ϕ_n is a positive functional on A with central measure μ and factorial disintegration $\{\phi_{tn} | t \in \hat{A}\}$ (see 4.8.7). Since $\phi_n \leqslant \phi$ there is a square integrable Borel vector field ξ'_n on $\hat{A}$ such that $\phi_n(x) = (\pi_\phi(x)\xi'_n | \xi'_n)$. Arguing exactly as before we show that $u_t\xi'_n(t) = \xi'_{nt}$ almost everywhere. Thus for each fixed t outside a null set we have for each x in $\mathscr{B}_2^{\phi_t}$

$$u_t\pi_t(x)\xi'_n(t) = \pi_{\phi_t}(x)\xi'_{nt} = a_{tn}^{1/2}\xi_x \to \xi_x \quad \text{as} \quad n \to \infty,$$

whence $u_tH_t = H_{\phi_t}$.

5.8.5. COROLLARY. *If ϕ is a σ-weight on $\mathscr{B}$ such that $\phi | A$ is densely defined and if $\phi = \int \phi_t\,\mathrm{d}\mu(t)$ is the canonical disintegration of ϕ on $\hat{A}$ then $\phi_t | A$ is densely defined almost everywhere.*

5.8.4. COROLLARY. *If ϕ is a σ-trace on $\mathscr{B}$ and if $\phi = \int \phi_t\,\mathrm{d}\mu(t)$ is the canonical disintegration of ϕ on $\hat{A}$ then almost all ϕ_t's are σ-traces.*

Proof. Choose a sequence of projections $\{p_n\}$ in $\mathscr{B}^\phi$ such that $p_n \nearrow 1$. Then $\phi(p_n \cdot p_n) = \int \phi_t(p_n \cdot p_n)\,\mathrm{d}\mu(t)$ is the canonical disintegration of a positive functional (cf. 4.8.7), and from the uniqueness of this it follows that $\phi_t(p_n \cdot p_n)$ is a trace for almost all t. By 5.3.5 this implies that ϕ_t is a trace for almost all t.

5.8.6. LEMMA. *Let μ be a probability measure on a standard Borel space T, and let $\{H_t | t \in T\}$ and $\{\mathscr{M}_t | t \in T\}$ be Borel fields of Hilbert spaces and von Neumann algebras, respectively. If all the $\mathscr{M}_t$'s are semi-finite factors, then the von Neumann algebra $\int^\oplus \mathscr{M}_t\,\mathrm{d}\mu(t)$ is semi-finite.*

Proof. We may consider separately the Borel subsets T_d of T for which $\dim(H_t) = d$, $1 \leqslant d \leqslant \infty$, and thus assume that $T_d = T$. Fixing a Hilbert space H of dimension d we may then realize the Borel vector fields on T as the set of Borel measurable vector valued functions from T into H. Furthermore, we may identify $\mathscr{B}(T, \{\mathscr{M}_t\})$ with the set of bounded, strongly Borel measurable operator valued functions x from T into $\boldsymbol{B}(H)$ such that $x(t)x'_i(t) = x'_i(t)x(t)$, where $\{x'_i\}$ is a generating sequence for the Borel field $\{\mathscr{M}'_t | t \in T\}$ (see 4.11.7).

Let P denote the set of projections in $\boldsymbol{B}(H)$, equipped with the strong topology. Then P as a closed subset of the Polish space $B(H)^1$ is Polish. Let E denote the positive part of the unit ball of $\boldsymbol{B}(H)_*$ in the norm topology. Then E is Polish, since $\boldsymbol{B}(H)_*$ is a separable Banach space (isomorphic with $\boldsymbol{T}(H)$ for a separable H, cf. 3.5). If $\{x_i\}$ is a generating sequence for the Borel field $\{\mathscr{M}_t | t \in T\}$ consider the Borel subset F of the standard space $T \times P \times E$ consisting of elements (t, p, ϕ) such that

(i) $x'_i(t)p = px'_i(t)$ for all i;

(ii) $\phi(p) = 1$;

(iii) $\phi(px_i(t)px_j(t)p) = \phi(px_j(t)px_i(t)p)$ for all i and j.

Then $(t, p, \phi) \in F$ if and only if $p \in \mathcal{M}_t$ and $\phi \mid p\mathcal{M}_t p$ is a normal state and a trace (since (iii) implies that $\phi(xy) = \phi(yx)$ for all x, y in $p\mathcal{M}_t p$). For each fixed t in T there is by assumption a faithful, normal, semi-finite trace ψ_t. We can therefore find a projection p_t in $\mathcal{M}_t$ such that $\psi_t(p_t) = 1$ (changing if necessary ψ_t by a scalar). There is then by 3.6.6 an element ϕ in E such that $\phi \mid \mathcal{M}_t = \phi_t(p_t \cdot p_t)$. It follows that the projection of F on its first coordinate is a surjection on T. By 4.10.3 there is then a null set N and Borel maps $t \to p(t)$ and $t \to \phi_t$ fom $T\backslash N$ into P and E, respectively, such that $(t, p(t), \phi_t) \in F$ for all t in $T\backslash N$.

Let p be the image of the operator field $\{p(t) \mid t \in T\backslash N\}$ in the von Neumann algebra $\mathcal{M} = \int^{\oplus} \mathcal{M}_t \, d\mu(t)$, and let ϕ be the normal state on $\mathcal{M}$ given by $\phi(x) = \int \phi_t(x(t)) \, d\mu(t)$ for each x in $\mathcal{B}(T, \{\mathcal{M}_t\})$. Then ϕ is a trace on $p\mathcal{M}p$ by construction; and ϕ is faithful on $p\mathcal{M}p$ since each ϕ_t is faithful on $p(t)\mathcal{M}_t p(t)$. Since $p(t) \neq 0$ for almost all t we have $c(p) = 1$ and by 5.2.8 we can extend ϕ to a faithful, normal semi-finite trace on $\mathcal{M}$. Thus $\mathcal{M}$ is semi-finite.

5.8.7. LEMMA. *Let $\mu, T, \{H_t\}$ and $\{\mathcal{M}_t\}$ be as in* 5.8.6. *If all the $\mathcal{M}_t$'s are finite factors, then $\int^{\oplus} \mathcal{M}_t \, d\mu(t)$ is finite.*

Proof. In the proof of 5.8.6 we may replace P by the single element 1. The Borel space F will then be a subset of $T \times E$. Since each $\mathcal{M}_t$ has a finite normalized trace the projection of F on its first coordinate is still a surjection on T. Thus by 4.10.3 there is a Borel field $t \to \phi_t$ of normalized traces on $T\backslash N$, whence $\phi = \int \phi_t d\mu(t)$ is a faithful, normal finite trace on $\int^{\oplus} \mathcal{M}_t \, d\mu(t)$.

5.8.8. LEMMA. *Let $\mu, T, \{H_t\}$ and $\{\mathcal{M}_t\}$ be as in 5.8.6. If all the $\mathcal{M}_t$'s are factors of type I_n, then $\int^{\oplus} \mathcal{M}_t \, d\mu(t)$ is of type I_n.*

Proof. In the proof of 5.8.6 we replace P by the n-fold product ΠP_m of copies of P. Then F is defined as the Borel subset of $T \times \Pi P_m$ consisting of elements $(t, \{p_m\})$ such that:

(i′) $x_i'(t)p_m = p_m x_i'(t)$ for all i and $m < n + 1$;

(ii′) $\Sigma p_m = 1$;

(iii′) $p_m x_i(t) p_m x_j(t) p_m = p_m x_j(t) p_m x_i(t) p_m$ for all i, j and $m < n + 1$.

Since each $\mathcal{M}_t$ contains exactly n orthogonal, abelian (= minimal) non-zero projections, we still have a surjective Borel map of F onto T. An application of 4.10.3 therefore produces n projection-valued Borel operator fields $t \to p_m(t)$ which are pointwise non-zero, abelian and pairwise orthogonal. If p_m denotes the image in the von Neumann algebra $\mathcal{M} = \int^{\oplus} \mathcal{M}_t \, d\mu(t)$ then each p_m is abelian, and $\Sigma p_m = 1$. Since $c(p_m) = 1$, all the p_m's are equivalent by 5.5.2, and it follows that $\mathcal{M}$ is of type I_n.

5.8.9. THEOREM. *Let $\mathcal{M}$ be a von Neumann algebra on a separable Hilbert space and let $\mathcal{M} = \int^{\oplus} \mathcal{M}_t \, d\mu(t)$ be the canonical disintegration obtained from 4.12.5. If α denotes one of the symbols $I_n (1 \leqslant n \leqslant \infty)$, II_1, II_∞ and III, then $\mathcal{M}$ is of type α if and only if almost all the $\mathcal{M}_t$'s are of type α.*

Proof. If $\mathcal{M}$ is finite or semi-finite it follows from 5.8.5 that almost all the $\mathcal{M}_t$'s are finite or semi-finite, respectively. Also, if $\mathcal{M}$ contains an abelian projection with central cover 1 then its central disintegration will produce non-zero abelian projections in almost all the $\mathcal{M}_t$'s. Since furthermore equivalence is preserved under disintegration it follows readily that if $\mathcal{M}$ is of type I_n then almost all of the $\mathcal{M}_t$'s are of type I_n. We have shown that if $\mathcal{M}$ is of type α, $\alpha \neq \mathrm{III}$ then almost all of the $\mathcal{M}_t$'s are of type α. The converse of this is contained in 5.8.6, 5.8.7 and 5.8.8. Since the types are mutually exclusive and exhaust all possibilities we conclude that $\mathcal{M}$ is of type III if and only if almost all the $\mathcal{M}_t$'s are of type III. This completes the proof.

5.8.10. COROLLARY. *Let $\mathcal{M}$ be a von Neumann algebra of type I_n on a separable Hilbert space. There is a probability measure μ on a standard space T and a Hilbert space H of dimension n such that $\mathcal{M}$ is isomorphic to the Borel *-algebra of bounded strongly (or weakly) Borel measurable functions from T to $\boldsymbol{B}(H)$ modulo the ideal of μ-null functions.*

Proof. By 5.5.11 we may assume that $\mathcal{M}'$ is commutative. We then disintegrate $\mathcal{M} = \int^{\oplus} \mathcal{M}_t \, d\mu(t)$, where μ is a probability measure on a standard Borel space T (4.12.5), and by 5.8.7 we may assume that all the $\mathcal{M}_t$'s are factors of type I_n. Moreover, since $\mathcal{M}' = \int^{\oplus} \mathcal{M}_t' \, d\mu(t)$ we have $\mathcal{M}_t' = \boldsymbol{C}1_t$, whence $\mathcal{M}_t = \boldsymbol{B}(H_t)$ almost everywhere. Since $\dim(H_t) = n$ the result is immediate.

5.8.11. Notes and remarks. Theorem 5.8.3 is extracted from Sutherland's thesis [253], which also contains a uniqueness statement similar to the one in 4.8.7. One obvious difficulty in proving the uniqueness of the disintegration is to determine when two weights are equal, cf. 5.1.7. To do this, Sutherland uses the theory of left Hilbert algebras (the Tomita–Takesaki theory) as developed in [259]. An alternative approach to direct integrals of left Hilbert algebras is presented in [157]. The result on disintegration of traces (5.8.5) was obtained by Guichardet [106]. Part of theorem 5.8.9 was known to von Neumann, but the complete version is due to Schwartz, see [234].

Chapter 6

Type I C^*-algebras

A C^*-algebra is either extremely well behaved (type I) or totally misbehaved (antiliminary). That is the content of this chapter, exemplified in the last theorem (6.8.7). Thus there is a natural temptation to concentrate on type I C^*-algebras and forget about the rest. As long as the theory is applied to group representations this point of view is quite fruitful, because a large number of interesting groups (among them all compact groups) give rise to C^*-algebras of type I. For the applications in theoretical physics, however, the situation is not so easy. As a matter of fact all the relevant algebras are antiliminary.

For these reasons we have set a course from Calypso to Circe: we first delve into the many beautiful properties of the type I C^*-algebras. But then we present in some detail a class of antiliminary algebras (the Glimm algebras) that are deceptively simple in appearance, so that the reader may familiarize himself with the fact that such algebras exist. We then show that the Glimm algebras are the worst that can occur: If a C^*-algebra is not of type I then it has a subalgebra, some quotient of which is a Glimm algebra.

6.1. Abelian elements

6.1.1. Recall from 5.5.1 that a positive element x in a C^*-algebra A is *abelian* if the hereditary C^*-subalgebra generated by x, i.e. the norm closure of xAx, is commutative. Just as for von Neumann algebras the existence of sufficiently many abelian elements in a C^*-algebra is desirable since it makes possible a very detailed structure theory.

We say that a C^*-algebra A is *of type I* if each non-zero quotient of A contains a non-zero abelian element. If A is even generated (as a C^*-algebra) by its abelian elements we say that it is *of type* I_0. Finally we say that A is *antiliminary* if it contains no non-zero abelian elements. For various reasons, mainly historical, C^*-algebras of type I are also called *GCR-*, *postliminary* or *smooth* algebras in the literature.

6.1.2. A word of caution concerning the above definition: A von Neumann algebra of type I is not in general of type I as a C^*-algebra. It suffices to consider the von Neumann algebra $B(H)$, with H infinite-dimensional, and verify that the quotient $B(H)/C(H)$—the so-called *Calkin algebra*—contains no abelian elements.

6.1.3. LEMMA. *A positive element x in a C^*-algebra A is abelian if and only if $\dim \pi(x) \leqslant 1$ for every irreducible representation (π, H) of A.*

Proof. If x is abelian and (π, H) is an irreducible representation then $\pi(x)$ is abelian in $\pi(A)$, whence $\pi(x)B(H)\pi(x)$ is commutative and thus $\dim \pi(x) \leqslant 1$.

Conversely, if $x \in A_+$ and $\dim \pi(x) \leqslant 1$ for each irreducible representation, then xAx is commutative in the atomic representation (see 4.3.7). Since that representation is faithful by 4.3.11, x is abelian.

6.1.4. LEMMA. *If A is a C^*-algebra of operators acting irreducibly on a Hilbert space H such that $A \cap C(H) \neq 0$ then $C(H) \subset A$ and each faithful irreducible representation of A is unitarily equivalent with the identity representation.*

Proof. Since $A \cap C(H) \neq 0$ we can by spectral theory find a finite-dimensional projection in $A \cap C(H)$. By 2.7.5 there is then a one-dimensional projection p in A. If ξ is a unit vector in pH then for any vector η in H there is by 3.13.2 an element x in A such that $x\xi = \eta$. Consequently x^*px is the projection on $\mathbf{C}\eta$. It follows that A contains all one-dimensional projections, whence $C(H) \subset A$.

To prove the second half of the lemma it suffices by 3.13.2 and 3.3.7 to show that each pure state ϕ of A for which (π_ϕ, H_ϕ) is faithful, is determined by a vector in H. Now $\phi \,|\, C(H)$ is non-zero (since (π_ϕ, H_ϕ) is faithful) and therefore a pure state of $C(H)$ by 4.1.6. Using 3.5.4 this implies that $\phi(x) = (x\xi \,|\, \xi)$ for some unit vector ξ in H and all x in $C(H)$. However, the extension of a state from an ideal to the algebra is unique by 3.1.6, whence $\phi(x) = (x\xi \,|\, \xi)$ for all x in A and the proof is complete.

6.1.5. THEOREM. *Let A be a C^*-algebra of type I. Then*

(i) *$C(H) \subset \pi(A)$ for each irreducible representation (π, H) of A;*

(ii) *$\hat{A} = \check{A}$.*

Proof. If (π, H) is an irreducible representation of A let x be an abelian element in $\pi(A)$ with norm one. By 6.1.3 x is a one-dimensional projection on H, and thus $C(H) \subset \pi(A)$ by 6.1.4.

(i) $\Rightarrow$ (ii). If (π_1, H_1) and (π_2, H_2) are irreducible representations of A with the same kernel and if $C(H_i) \subset \pi_i(A)$ for $i = 1, 2$, then the representations are unitarily equivalent by 6.1.4. It follows that $\hat{A} = \check{A}$.

6.1.6. COROLLARY. *If A is of type I_0 then $\pi(A) = C(H)$ for each irreducible representation (π, H) of A.*

Proof. By 6.1.5 we have $C(H) \subset \pi(A)$. On the other hand, $\pi(A)$ is generated by its abelian elements, i.e. by one-dimensional operators, cf. 6.1.3, and therefore $\pi(A) \subset C(H)$.

6.1.7. PROPOSITION. *For each C^*-algebra A the C^*-subalgebra A_0 generated by the abelian elements is the largest ideal of A of type I_0.*

Proof. Let J denote the smallest hereditary cone in A_+ containing the abelian elements. Thus $x \in J$ if and only if there is a finite set $\{x_i\}$ of abelian elements such that $x \leqslant \Sigma x_i$. It follows from 1.4.9 that there is a set $\{y_i\}$ in A such that $x = \Sigma y_i^* y_i$ and $y_i y_i^* \leqslant x_i$ for each i. From 6.1.3 we see that any positive element dominated by an abelian element is abelian and that yy^* being abelian implies that y^*y is abelian. It follows that x is a sum of abelian elements, whence $J \subset A_0$. A similar argument shows that $u^*Ju = J$ for each unitary u in $\tilde{A}$. Consequently the norm closure $\bar{J}$ of J is the positive part of an ideal of A. Since $\bar{J} \subset A_0$ and $\bar{J}$ contains all abelian elements of A we conclude that $\bar{J} = (A_0)_+$ so that A_0 is an ideal of A. Since an abelian element of an ideal of A is abelian in A we see that A_0 is the largest ideal of A of type I_0.

6.1.8. COROLLARY. *A C^*-algebra of type I_0 has an approximate unit each of whose elements is a sum of abelian elements.*

Proof. From the proof of 6.1.7 we see that the set of elements of the form Σx_i, x_i abelian, is a dense, invariant, hereditary cone in A_+, and therefore contains an approximate unit for A.

6.1.9. For each x in A_+ the (canonical) trace of $\pi(x)$ depends only on the equivalence class of an irreducible representation (π, H) of A, so that we may define a function $\hat{x}: \hat{A} \to [0, \infty]$ by $\hat{x}(t) = \mathrm{Tr}(\pi(x))$ whenever $(\pi, H) \in t$. From 4.4.9 we see that $\hat{x}$ is a lower semi-continuous on $\hat{A}$ in the Jacobson topology. Moreover, the map $x \to \hat{x}$ is affine and faithful from A_+ into the cone of extended valued lower semi-continuous functions on $\hat{A}$.

The preference of the transformation $x \to \hat{x}$ to the transformation $x \to \check{x}$ introduced in 4.4.2 depends on the possibility of finding elements x in A such that $\hat{x}$ is finite. Note that if x is abelian then $\hat{x} = \check{x}$; in particular $\hat{x}$ is bounded. If conversely $x \in A_+$ such that $\hat{x} = \check{x}$ then $\dim \pi(x) \leqslant 1$ for each irreducible representation, whence x is abelian by 6.1.3.

We say that an element x in A_+ has *continuous trace* if $\hat{x} \in C^b(\hat{A})$. We say that the C^*-algebra A has *continuous trace* if the set of elements with continuous trace is dense in A_+.

6.1.10. LEMMA. *If A is a C^*-algebra with continuous trace there is for each t in $\hat{A}$ an element a in A_+ and a neighbourhood G of t such that $\hat{a}(s) = \check{a}(s) = 1$ for every s in G.*

Proof. If (π, H) is an irreducible representation associated with t there is by assumption an element x in A_+ such that $\hat{x} \in C^b(\hat{A})$ and $\|\pi(x)\| = 1$. If y is any element in A_+ dominated by x then $\hat{y}$ and $(x-y)^\wedge$ are lower semi-continuous functions whose sum is continuous; and consequently $\hat{y}$ (and $(x-y)^\wedge$) belongs to $C^b(\hat{A})$. Applying the function $t \to t \wedge 1$ to x we may therefore assume that $\|x\| = 1$. Now $\pi(x)$ is an operator of trace class so that

$$H_1 = \{\xi \in H \mid \pi(x)\xi = \xi\}$$

is a finite dimensional subspace of H. Applying a suitable continuous function to x we may assume that $\pi(x)$ is the projection on H_1. From 2.7.3 there is an element y in A_+ such that $\pi(y)\xi_1 = \xi_1$ for some unit vector ξ_1 in H_1 and $\pi(y) = 0$ on the orthogonal complement of ξ_1 in H_1. Replacing x with $x^{1/2}yx^{1/2}$ we may assume that $\hat{x}(t) = \check{x}(t) = 1$.

For $0 < \varepsilon < \frac{1}{3}$ define

$$G = \{s \in \hat{A} \mid \hat{x}(s) < 1 + \varepsilon \quad \text{and} \quad \check{x}(s) > 1 - \varepsilon\}.$$

Then G is a neighbourhood of t and for each irreducible representation (ρ, K) associated with an s in G the largest eigenvalue of $\rho(x)$ is $\geqslant 1 - \varepsilon$ and the second largest is $\leqslant (1 + \varepsilon) - (1 - \varepsilon) = 2\varepsilon$. Thus if f is a continuous function such that $f(\alpha) = 0$ for $\alpha \leqslant 2\varepsilon$ and $f(\alpha) = 1$ for $\alpha \geqslant 1 - \varepsilon$ then with $a = f(x)$ we have $\hat{a}(s) = \check{a}(s)$ for every s in G.

6.1.11. THEOREM. *Let A be a C^*-algebra with continuous trace. Then*

(i) *A is of type I_0;*

(ii) *$\hat{A}$ is a locally compact Hausdorff space;*

(iii) *For each t in $\hat{A}$ there is an abelian element x such that $\hat{x} \in K(\hat{A})$ and $\hat{x}(t) = 1$.*

The last condition is also sufficient for A to have continuous trace.

Proof. For a fixed but arbitrary t in $\hat{A}$ let G and a be objects satisfying 6.1.10. Let y be a positive element in the closed ideal corresponding to G (cf. 4.1.3) such that $(aya)^\vee(t) = 1$. The element $z = aya$ is abelian since $\hat{z} = \check{z}$ on G and $\hat{z} = 0 = \check{z}$ on $\hat{A}\backslash G$. Since $\check{z}(t) \neq 0$ it follows that the set of abelian elements is not contained in any primitive ideal of A. Thus by 3.13.8 it is not contained in any proper ideal of A, so that A is of type I_0 by 6.1.7.

To show that $\hat{A}$ is a Hausdorff space take $t_1 \neq t_2$ in $\hat{A}$ and choose an element x in A_+ such that $\check{x}(t_1) = 1$ and $\check{x}(t_2) = 0$. This is possible since $\hat{A} = \check{A}$ by 6.1.5. By assumption there is an element y in A_+ with continuous trace such that $\|x - y\| \leqslant \frac{1}{2}$. Put $z = y^{1/2}xy^{1/2}$ and observe that z has continuous trace since $z \leqslant \|x\|y$. Clearly $\hat{z}(t_2) = 0$, since

$$\check{z}(t_2) \leqslant \|y\|\, \check{x}(t_2) = 0;$$

but

$$1 = (x^2)^\vee(t_1) \leqslant (x^{1/2}(x - y)x^{1/2})^\vee(t_1) + (x^{1/2}yx^{1/2})^\vee(t_1)$$
$$\leqslant \tfrac{1}{2}\check{x}(t_1) + \check{z}(t_1) \leqslant \tfrac{1}{2} + \hat{z}(t_1)$$

so that $\hat{z}(t_1) \geqslant \frac{1}{2}$. We can therefore separate t_1 and t_2 using the continuous function $\hat{z}$.

To prove (iii) it suffices to take the element a from the first part of the proof together with a positive function f in $K(G)$ such that $f(t) = 1$. By 4.4.8 f corresponds to a multiplier of A so the element $x = f \cdot a$ belongs to A and satisfies the conditions in (iii).

Now let A be a C^*-algebra satisfying (iii) and consider the set M of elements in A_+ with continuous trace. As we observed in the proof of 6.1.10, M is hereditary and since $(u^*xu)^\wedge = \hat{x}$ for every x in A_+ and every unitary u in $\tilde{A}$ we see that the closure of M is the positive part of an ideal I. Condition (iii) implies that I is not contained in any primitive ideal of A, whence $I = A$ by 3.13.8 and A has continuous trace.

6.1.12. PROPOSITION. *If A is a C^*-algebra with continuous trace there is for each x in $K(A)_+$ an $n < \infty$ such that $\dim \pi(x) \leqslant n$ for every irreducible representation (π, H) of A. Moreover, the map $x \to \hat{x}$ is a faithful, positive linear surjection of $K(A)$ onto $K(\hat{A})$.*

Proof. As we saw in the proof of 6.1.11 the set of elements in A_+ with continuous trace is the positive part of a dense ideal. Since $K(A)$ is the minimal dense ideal (5.6.1) it follows that $\hat{x} \in C^b(\hat{A})$ for each x in $K(A)_+$.

Assume now that $x_0 \in K(A)_0$ (notations as in the proof of 5.6.1). Then $x_0 = x_0 x_1$ for some x_1 in $K(A)_+$ and for each irreducible representation (π, H) we have

$$\dim \pi(x_0) \leqslant \operatorname{Tr} \pi(x_1) \leqslant \operatorname{Sup} \hat{x}_1 < \infty.$$

Thus the dimension of $\pi(x_0)$ is bounded above, and since each x in $K(A)_+$ is dominated by a finite sum of elements from $K(A)_+$ the same is true for each x in $K(A)_+$. Furthermore, $\check{x}_0(t) = 0$ if $\check{x}_1(t) < 1$ and since the set $\{t \in \hat{A} \mid \check{x}_1(t) \geqslant \frac{1}{2}\}$ is compact by 4.4.4 we conclude that $\check{x}_0$, and therefore also $\hat{x}_0$, has compact support. Again this is enough to ensure that $\hat{x}$ has compact support for each x in $K(A)$.

We have shown that the map $x \to \hat{x}$ is a faithful positive linear map of $K(A)$ into $K(\hat{A})$. However, for each compact set F in $\hat{A}$ there is by 6.1.10 an element x

in $K(A)_+$ such that $\hat{x} > 0$ on F. Applying a suitable multiplier from $C^b(\hat{A})$ to x we can obtain for any given f in $C_0(F)$ an element in $K(A)$ such that $\hat{y} = f$.

6.1.13. PROPOSITION. *Let A be a C^*-algebra with continuous trace. If x, y are elements of $K(A)_+$ then $\hat{x} = \hat{y}$ if and only if there is a finite set $\{z_n\}$ in A such that $x = \Sigma z_n^* z_n$ and $y = \Sigma z_n z_n^*$* (i.e. $x \approx y$, cf. *5.2.6*).

Proof. The condition $x \approx y$ is clearly sufficient to imply $\hat{x} = \hat{y}$. To prove the necessity, let F denote the support of $\hat{x}$ and fix t in F. By 6.1.11 there is an abelian element a with continuous trace such that $\hat{a} > \frac{1}{2}$ in a neighbourhood G of t. If I denotes the closed ideal of A corresponding to G we claim that $f \cdot z_1 \in K(I)$ for each z_1 in $K(A)$ and each f in $K(G)$. It suffices to show this for z_1 in $K(A)_0$, i.e. we may assume that $z_1 = z_1 z_2$ for some z_2 in $K(A)$. Then choose g in $K(G)$ such that $f = fg$ and note that

$$f \cdot z_1 = (f \cdot z_1)(g \cdot z_2);$$

whence $f \cdot z_1 \in K(I)$, since $g \cdot z_2 \in I$.

Consider the set

$$M = \{z \in I_+ \mid \exists b \in I_+, \alpha \geqslant 0 : z \approx b, b \leqslant \alpha a\}.$$

It is easy to verify that M is the positive part of an ideal in I, and since $\hat{a} > 0$ on G, this ideal is dense in I, whence $K(I)_+ \subset M$. Combining this with the argument above we obtain for each f in $K(G)_+$ elements b, c in I_+ such that

$$f \cdot x \approx b, \qquad f \cdot y \approx c$$

and b, c are both dominated by some multiple of a. But then b and c are abelian elements, and since $\hat{b} = \hat{c}$ because $\hat{x} = \hat{y}$ we conclude that actually $b = c$. Since $\approx$ is an equivalence relation it follows that $f \cdot x \approx f \cdot y$.

To complete the proof take a finite covering (G_i) of F and let (f_i) be a partition of unity subordinate to the covering, such that $f_i \in K(G_i)$ for each i. Then

$$x = \sum f_i \cdot x \approx \sum f_i \cdot y = y.$$

6.1.14. Notes and remarks. The spiritual fathers of the type I C^*-algebra theory are Kaplansky and Mackey. The material presented in this section belongs, however, to a later period, and is largely borrowed from the papers by Fell [91] and Dixmier [64]. The author is responsible for the emphasis on abelian elements and the introduction of type I_0 algebras.

Fell proved that C^*-algebras satisfying conditions (i), (ii) and (iii) of 6.1.11 have continuous trace. The converse was established by Dixmier, who also made several simplifications in the other half of the proof. Propositions 6.1.12 and 6.1.13 are found in Pedersen and Petersen [208].

6.2. Composition series

6.2.1. A C^*-algebra A is called *liminary* (or *CCR*) if $\pi(A) = C(H)$ for each irreducible representation (π, H) of A. From 6.1.6 we see that a C^*-algebra of type I_0 is liminary. The converse is false.

6.2.2. LEMMA. *Each C^*-algebra A contains a largest liminary ideal (possibly zero).*

Proof. Let I be the set of elements x in A such that $\pi(x) \in C(H)$ for all irreducible representations (π, H) of A. Since $\pi(A) \cap C(H)$ is an ideal of $\pi(A)$ we see that I is a closed ideal of A. Each irreducible representation (ρ, H) of I extends by 4.1.11 to an irreducible representation (π, H) of A. By definition $\rho(I) = \pi(I) \subset C(H)$ which proves that I is liminary. It is clear from the construction that I is the largest liminary ideal contained in A.

6.2.3. THEOREM. *A liminary C^*-algebra is of type I.*

Proof. Since each quotient of a liminary C^*-algebra is liminary it suffices to show that a liminary C^*-algebra $A \neq 0$ contains a non-zero abelian element. Let x and y be non-zero elements of A_+^1 with $x = xy$. For each irreducible representation (π, H) of A we have $\pi(y) \in C(H)$ and $\pi(y)$ is a unit for $\pi(x)$. It follows that $\dim \pi(x) < \infty$ for all irreducible representations of A. The function: $(\pi, H) \to \dim \pi(x)$ depends only on the equivalence class of (π, H) and defines therefore a function $\dim x: \hat{A} \to \mathbf{N}$ which is lower semicontinuous in the Jacobson topology since

$$\{t \in \hat{A} \mid \dim x(t) \leqslant n\} = \bigcap_m \{t \in \hat{A} \mid x^{1/m})\hat{}\,(t) \leqslant n\}.$$

An abelian element in a hereditary C^*-subalgebra of A is also abelian in A. We may therefore assume without loss of generality that xAx is dense in A. We claim that there is an m in $\mathbf{N}$ and a non-empty open set G in $\hat{A}$ such that $\dim x \leqslant m$ on G. Otherwise each of the open sets

$$G_n = \{t \in \hat{A} \mid \dim x(t) > n\}$$

is dense in $\hat{A}$. But this is impossible since $\cap G_n = \varnothing$ and $\hat{A}$ is a Baire space by 4.3.5. Since G corresponds to a closed ideal of A we may assume again without loss of generality that $G = \hat{A}$. Since xAx is dense in A it then follows that $\dim(H) \leqslant m$ for all irreducible representations (π, H) of A. Assuming as we may that m is the lowest bound for the dimensions of irreducible representations of A we have $\hat{A} = {}_m\hat{A}$ (notations as in 4.4.10) and $\hat{A}_m = \hat{A} \backslash {}_{m-1}\hat{A}$ is a non-empty open set in $\hat{A}$ corresponding to a closed ideal A_m of A having only m-dimensional irreducible representations.

For each element y in $(A_m)_+^1$ the functions $\hat{y}$ and $(1-y)^\wedge$ are lower semicontinuous. However,

$$\hat{y}(t) + (1-y)^\wedge(t) = \hat{1}(t) = m$$

for all t in $\hat{A}_m$ whence $\hat{y}$ (and $(1-y)^\wedge$) is continuous. It follows that A_m has continuous trace and therefore contains abelian elements by 6.1.11.

6.2.4. By a *composition series* for a C^*-algebra A we mean a strictly increasing family of closed ideals $\{I_\alpha\}$ indexed by a segment $\{0 \leqslant \alpha \leqslant \beta\}$ of the ordinals such that $I_0 = 0$, $I_\beta = A$; and for each limit ordinal γ we have

$$I_\gamma = \left(\bigcup_{\alpha<\gamma} I_\alpha\right)^- \qquad \text{(norm closure).}$$

If each I_α is an essential ideal in $I_{\alpha+1}$ we say that the composition series is *essential*. It is clear that a composition series for a separable C^*-algebra can be at most countable.

6.2.5. As an easy example of a composition series let A be a C^*-algebra having an upper bound $d < \infty$ for the dimensions of its irreducible representations. For each n, $0 < n < d$, let I_n be the kernel of all irreducible representations (π, H) of A with $\dim(H) \leqslant d - n$. Then $\{I_n\}$ is a finite composition series for A with the property that I_{n+1}/I_n is a *homogeneous algebra* having only irreducible representations of dimension $d - n$, see 4.4.10.

6.2.6. THEOREM. *The following conditions on a C^*-algebra A are equivalent*:

(i) *A is of type I;*

(ii) *A has a composition series $\{I_\alpha \mid 0 \leqslant \alpha \leqslant \beta\}$ such that $I_{\alpha+1}/I_\alpha$ is of type I_0 for each $\alpha < \beta$;*

(iii) *A has a composition series $\{I_\alpha \mid 0 \leqslant \alpha \leqslant \beta\}$ such that $I_{\alpha+1}/I_\alpha$ is liminary for each $\alpha < \beta$;*

(iv) *A has a composition series $\{I_\alpha \mid 0 \leqslant \alpha \leqslant \beta\}$ such that $I_{\alpha+1}/I_\alpha$ is of type I for each $\alpha < \beta$.*

Proof. (i) $\Rightarrow$ (ii). If A is of type I put $I_0 = 0$. Then, proceeding by induction, if α is not a limit ordinal define I_α as the closed ideal of A generated by the elements whose images in $A/I_{\alpha-1}$ are abelian; if α is a limit ordinal define I_α as the closure of the union of the I_γ's with $\gamma < \alpha$. By assumption there is an ordinal β such that $I_\beta = A$ and thus the family $\{I_\alpha \mid 0 \leqslant \alpha \leqslant \beta\}$ is a composition series for A with the desired properties.

(ii) $\Rightarrow$ (iii) $\Rightarrow$ (iv). Obvious from 6.1.6 and 6.2.3.

(iv) $\Rightarrow$ (i). If $\{I_\alpha \mid 0 \leqslant \alpha \leqslant \beta\}$ is a composition series for A satisfying (iv) and I

is a proper closed ideal of A let γ be the first ordinal such that $I \not\supset I_\gamma$. Since $I_\alpha \subset I$ for all $\alpha < \gamma$ we see that γ is not a limit ordinal. Therefore I_γ/I is a quotient of the C^*-algebra $I_\gamma/I_{\gamma-1}$ of type I, and thus itself of type I. It follows that the quotient A/I contains a non-zero abelian element (in I_γ/I), and since this holds for any I we conclude that A is of type I.

6.2.7. PROPOSITION. *Each C^*-algebra A has a largest ideal of type I and A/I is antiliminary.*

Proof. Do the construction in the proof of (i) $\Rightarrow$ (ii) of 6.2.6. There is then a first ordinal β for which the procedure terminates, i.e. $I_{\beta+1} = I_\beta$. This means that A/I_β contains no abelian elements. Put $I = I_\beta$. Then I is of type I by 6.2.6 (with composition series $\{I_\alpha \mid 0 \leqslant \alpha \leqslant \beta\}$) and A/I is antiliminary. Clearly I is the largest ideal of type I in A.

6.2.8. COROLLARY. *If a C^*-algebra A is antiliminary then for each non-zero x in A there is an irreducible representation (π, H) of A such that $\pi(x) \notin C(H)$.*

Proof. By 6.2.6 A is antiliminary if and only if its liminary ideal is zero.

6.2.9. PROPOSITION. *Each C^*-subalgebra and each quotient of a C^*-algebra which is liminary or of type I is liminary or of type I, respectively.*

Proof. If B is a C^*-subalgebra of the liminary algebra A it follows directly from 4.1.8 that $\rho(B) \subset C(K)$ for each irreducible representation (ρ, K) of B; whence B is liminary by 6.1.4.

Assume now that A is of type I and let $\{I_\alpha \mid 0 \leqslant \alpha \leqslant \beta\}$ be a composition series for A satisfying 6.2.6 (iii). Put $J_\alpha = I_\alpha \cap B$. Then except for a simple relabelling arising from the α's for which $J_\alpha = J_{\alpha+1}$ the set $\{J_\alpha \mid 0 \leqslant \alpha \leqslant \beta\}$ is a composition series for B. Since $J_{\alpha+1}/J_\alpha$ is isomorphic with a C^*-subalgebra of the liminary algebra $I_{\alpha+1}/I_\alpha$ we conclude from the first part of the proof that $J_{\alpha+1}/J_\alpha$ is liminary. It follows from 6.2.6 that B is of type I.

The quotient results are obvious.

6.2.10. PROPOSITION. *Each hereditary C^*-subalgebra and each quotient of a C^*-algebra which is of type I_0 or has continuous trace is of type I_0 or has continuous trace, respectively.*

Proof. If B is a hereditary C^*-subalgebra of the C^*-algebra A of type I_0 then for each non-zero x in B_+ there is by 6.1.8 for each $\varepsilon > 0$ an element $u = \Sigma x_i$, x_i abelian, such that $\|x^{1/2}(1-u)x^{1/2}\| < \varepsilon$. But each element $x^{1/2}x_ix^{1/2}$ belongs to B and is abelian by 6.1.3; and

$$\|x - \sum x^{1/2}x_ix^{1/2}\| = \|x^{1/2}(1-u)x^{1/2}\| < \varepsilon.$$

It follows that B is of type I_0.

Assume now that A has continuous trace and take x in B_+. There is then for each $\varepsilon > 0$ an element u in A_+ with continuous trace such that $\|x^{1/2}(1-u)x^{1/2}\| < \varepsilon$. The element $x^{1/2}ux^{1/2}$ belongs to B and has continuous trace in A. But $\hat{B}$ is homeomorphic to an open subset of $\hat{A}$ by 4.1.10 and we conclude from 4.1.9 that $x^{1/2}ux^{1/2}$ has continuous trace in B. Since $\|x - x^{1/2}ux^{1/2}\| < \varepsilon$ we see that B has continuous trace.

The quotient results are obvious.

6.2.11. THEOREM. *Let A be a C^*-algebra of type I. Then A contains an essential ideal which has continuous trace. Moreover, A has an essential composition series $\{I_\alpha \mid 0 \leqslant \alpha \leqslant \beta\}$ such that $I_{\alpha+1}/I_\alpha$ has continuous trace for each $\alpha < \beta$.*

Proof. Let x be a non-zero abelian element in A and let B denote the hereditary C^*-subalgebra of A generated by x. If I denotes the smallest closed ideal of A containing B then

$$\hat{I} = \{t \in \hat{A} \mid \check{x}(t) > 0\}.$$

From 4.1.10 we see that $\hat{I}$ is homeomorphic to $\hat{B}$. However, B is commutative, whence $\hat{B}$ is a Hausdorff space. Consequently $\hat{I}$ is a Hausdorff space so that $\check{x}$ is continuous by 4.4.5. As $\check{x} = \hat{x}$ we conclude from 6.1.11 (iii) that I has continuous trace.

Let $\{I_i\}$ be a maximal family of pairwise orthogonal non-zero ideals of A with continuous trace and put $I_1 = \oplus I_i$ (direct sum). Then clearly I_1 has continuous trace and it suffices to show that it is an essential ideal. If not, there is a non-zero ideal I_0 orthogonal to I_1. But then I_0, being an ideal of A, is of type I by 6.2.9, and thus by the first part of the proof contains a non-zero ideal with continuous trace, in contradiction with the maximality of the family $\{I_i\}$. It follows that I_1 is an essential ideal of A.

The composition series is now easy to construct. Consider A/I_1 which is of type I and apply the previous result to obtain a larger ideal I_2 such that I_2/I_1 has continuous trace; and proceed by transfinite induction.

6.2.12. Note that the composition series for a C^*-algebra of type I constructed in 6.2.6 (ii) and (iii) are unique and essential if at each stage one takes $I_{\alpha+1}$ such that $I_{\alpha+1}/I_\alpha$ is the largest ideal of A/I_α which is of type I_0 or liminary, respectively (cf. 6.1.7 and 6.2.2). By contrast the composition series in 6.2.11 is the result of a choice (selecting a maximal family) and there is in general no canonical procedure for its construction.

6.2.13. Notes and remarks. The condition in 6.2.1 was invented by Kaplansky [141]. The original name CCR means 'completely continuous representations' (completely continuous operators being another name for $C(H)$). The next layer in the hierarchy, 'GCR' indicates a generalization of the CCR condition. The modern names 'liminary' and 'postliminary' do not mean anything, which may be more aesthetic. In any case the Anglo-Saxon Habit of

Condensing Every Formula into its Leading Characters (abbreviated ASHCEFLC) should not be tolerated in mathematics. Theorem 6.2.3 was proved by Kaplansky [142]. Theorem 6.2.6 together with 6.2.11 give an amalgamated version of results by Kaplansky [141] and Fell [91]. Proposition 6.2.9 is due to Kaplansky (cf. [141]), and 6.2.10 is found in [208].

Homogeneous C^*-algebras (cf. 6.2.5) were introduced by Fell in [91], where he showed that if $n < \infty$, each n-homogeneous C^*-algebra A is equal to the set of continuous cross sections vanishing at infinity on a fibre bundle with base space $\hat{A}$ and fibre space M_n. The extension of this characterization in terms of fibre bundles, but now with variable dimensions on the fibres ('collapsing cross sections') was pursued by Dixmier and Douady [68]. Eventually this road into the realm of topology must be widened, if we are to learn about the detailed structure of type I C^*-algebras.

6.3. Borel *-algebras of type I

6.3.1. Throughout this section A will denote a separable C^*-algebra and $\mathscr{B}$ its enveloping Borel *-algebra. Note that if $\{I_\alpha \mid 0 \leqslant \alpha \leqslant \beta\}$ is a composition series (necessarily countable) for A, then this corresponds to a well ordered, increasing family $\{p_\alpha \mid 0 \leqslant \alpha \leqslant \beta\}$ of open central projections in $\mathscr{B}$. If α is not a limit ordinal put $e_\alpha = p_\alpha - p_{\alpha-1}$. Then $\mathscr{B}$ is the direct sum of the Borel *-algebras $e_\alpha\mathscr{B}$, and by 4.6.2 each of these is the enveloping Borel *-algebra for the quotient $I_\alpha/I_{\alpha-1}$. This observation, together with 6.2.11 often permits us to extend results for $\mathscr{B}$ when A has continuous trace to the case where A is only of type I.

6.3.2. PROPOSITION. *If A is a separable C^*-algebra of type I, the T-Borel structure on $\hat{A}$ is standard, and the T-Borel, D-Borel and M-Borel structures all coincide.*

Proof. Using the previous remarks we may assume that A has continuous trace. Then $\hat{A}$ is a second countable locally compact Hausdorff space by 6.1.11, and consequently the T-Borel structure is standard. Let $\hat{A}_T$ and $\hat{A}_M$ denote $\hat{A}$ equipped with the T-Borel and the M-Borel structure, respectively; and let $P(A)$ denote the space of pure states of A. The natural surjection $\phi \to \hat{\phi}$ of $P(A)$ onto $\hat{A}_T$ is open and continuous by 4.3.3 and admits therefore a right Borel inverse $g: \hat{A}_T \to P(A)$ by 4.2.12. However, regarding $\phi \to \hat{\phi}$ as a map from $P(A)$ onto $\hat{A}_M$, it is Borel (by definition) and composing this with g we obtain a bijective Borel map from $\hat{A}_T$ onto $\hat{A}_M$. Since $\hat{A}_T$ has a weaker Borel structure than $\hat{A}_M$ by 4.7.3, they coincide. The D-Borel structure is trapped between the two others, and therefore coalesce with these.

6.3.3. LEMMA. *If A has continuous trace there is a sequence $\{p_n\}$ of pairwise orthogonal abelian projections in $\mathscr{B}$ such that* $1 = c(p_1) \geqslant c(p_2) \geqslant \cdots$, *and* $\Sigma\, p_n = 1$.

Proof. Using 6.1.11 and the fact that $\hat{A}$ is second countable hence σ-compact we can find a sequence $\{G_n\}$ of open sets in $\hat{A}$ and a sequence $\{a_n\}$ of abelian elements in A such that $\hat{a}_n = 1$ on G_n for each n. Let f_n be the characteristic function corresponding to the set $G_n \backslash \cup_{m<n} G_m$. Identifying the T-Borel functions with the centre of $\mathscr{B}$ (cf. 4.7.2 and 6.3.2), we define

$$p_1 = \sum f_n \cdot a_n.$$

Then $p_1 \in \mathscr{B}$ and since for each t in $\hat{A}$

$$\check{p}_1(t) = \sum f_n(t) = 1 = \sum f_n(t)\hat{a}_n(t) = \hat{p}_1(t),$$

we see that p_1 is an abelian projection with $c(p_1) = 1$.

Let $\{u_k\}$ be a dense sequence in the unitary group of $\tilde{A}$ and for convenience assume that $u_1 = 1$. Suppose that we have already constructed pairwise orthogonal abelian projections $p_1, p_2, \ldots, p_n$ in $\mathscr{B}$ such that $\hat{p}_m(t) = 0$ only if $\hat{1}(t) < m$. Put $q_0 = \Sigma\, p_i$ and define inductively a sequence $\{q_m\}$ in $\mathscr{B}$ by

$$q_{m+1} = (u_{m+1}^* p_1 u_{m+1}) \vee (q_0 + c(q_1 + \cdots + q_m)) - (q_0 + c(q_1 + \cdots + q_m)).$$

By construction $\{q_m\}$ is a sequence of abelian projections with pairwise orthogonal central covers. Moreover, they are all orthogonal to q_0. Therefore the projection $p_{n+1} = \Sigma\, q_m$ is abelian and orthogonal to q_0. If $\hat{p}_{n+1}(t) = 0$ then

$$((u_{m+1}^* p_1 u_{m+1}) \vee q_0)\hat{}\,(t) = \hat{q}_0(t)$$

for all m, and since

$$1 = c(p_1) = \bigvee u_m^* p_1 u_m$$

this implies that $\hat{q}_0(t) \geqslant \hat{1}(t)$. However, $\hat{q}_0(t) \leqslant n$ and thus we have established that $\hat{p}_{n+1}(t) = 0$ only if $\hat{1}(t) < n + 1$. In particular $c(p_{n+1}) \leqslant c(p_n)$.

We claim that the sequence $\{p_n\}$ constructed above satisfies

$$u_n^* p_1 u_n \leqslant p_1 + \cdots + p_n$$

for each n. Suppose that this has been established for all $m \leqslant n$ and let $q_0, q_1, \ldots$ be as before. Then $q_m = 0$ for all $m \leqslant n$, whence

$$q_{n+1} = (u_{n+1}^* p_1 u_{n+1}) \vee q_0 - q_0$$

which implies that

$$u_{n+1}^* p_1 u_{n+1} \leqslant q_0 + q_{n+1} \leqslant p_1 + \cdots + p_n + p_{n+1},$$

as desired. It follows immediately from this that

$$1 = c(p_1) = \bigvee u_n^* p_1 u_n \leqslant \sum p_n \leqslant 1.$$

6.3.4. THEOREM. *Let A be a separable C^*-algebra of type I. For each d, with $1 \leqslant d \leqslant \infty$ there is a standard Borel space $\hat{A}_d$ and a Hilbert space H_d of dimension d such that (notations as in 4.11.3)*

$$\mathscr{B} = \prod \mathscr{B}(\hat{A}_d, \boldsymbol{B}(H_d)).$$

Proof. From the remarks in 6.3.1 we see that it suffices to prove the theorem under the assumption that A has continuous trace. There is then a sequence $\{p_n\}$ satisfying 6.3.3. For $d < \infty$ define

$$\hat{A}_d = \{t \in \hat{A} \mid \hat{p}_d(t) = 1, \hat{p}_{d+1}(t) = 0\}$$

and note that $\hat{A}_d$ (as in 4.4.10) is precisely the classes of d-dimensional irreducible representations of A. With $\hat{A}_\infty = \hat{A} \backslash \cup \hat{A}_d$ we have a partition $\{\hat{A}_d \mid 1 \leqslant d \leqslant \infty\}$ of $\hat{A}$ consisting of Borel subsets, and thus each $\hat{A}_d$ is a standard Borel space. Let e_d be the central projection in $\mathscr{B}$ corresponding to $\hat{A}_d$, and note that $e_d = c(p_d) - c(p_{d+1})$ for $d < \infty$.

Take a dense sequence $\{x_k\}$ in A and for each k and each $n \leqslant d$ let

$$p_n x_k p_1 e_d = u_{nk} |p_n x_k p_1 e_d|$$

be the polar decomposition. By 4.5.16 $u_{nk} \in \mathscr{B}$ and we note that $(u_{nk}^* u_{nk})\check{}(t)$ is either zero or one, and that for each t in $\hat{A}_d$ there is a least k for which $(u_{nk}^* u_{nk})\check{}(t) = 1$. It follows that if $y_n = \Sigma\, 3^{-k} u_{nk}$ and $y_n = u_n |y_n|$ is its polar decomposition then $u_n^* u_n = p_1 e_d$ and $u_n u_n^* = p_n e_d$, because this is valid at each point t in $\hat{A}_d$. From this we obtain a $d \times d$-dimensional set of matrix units $\{v_{nm}\}$ by defining $v_{nm} = u_n u_m^*$. Note the equations:

(i) $v_{nm}^* = v_{mn}$,

(ii) $v_{nm} v_{kp} = \delta(m, k) v_{np}$,

(iii) $v_{nm}^* v_{nm} = p_m e_d$.

Let H_d be a d-dimensional Hilbert space and choose an orthonormal basis $\{\xi_n \mid 1 \leqslant n < d + 1\}$. There is then a faithful representation of the system $\{v_{nm}\}$ on H_d given by

$$v_{nm} \xi_k = \delta(k, m) \xi_n.$$

We wish to extend this to a faithful representation of $e_d \mathscr{B}$ into $\mathscr{B}(\hat{A}_d, \boldsymbol{B}(H_d))$—the bounded weakly Borel measurable functions from $\hat{A}_d$ into $\boldsymbol{B}(H_d)$. To do so note that for each x in $e_d \mathscr{B}$ we have $p_n x p_m = f \cdot v_{nm}$, where f is the bounded Borel function on $\hat{A}_d$ (identified with a central element of $\mathscr{B}$) such that $f(t) = (p_n x v_{nm})\hat{}(t)$. That f is a Borel function follows from 4.5.8 and the facts that $(p_n x p_n)^\vee = (p_n x p_n)^\wedge$ if $x \in \mathscr{B}_+$ and that $(p_n x p_n)^\wedge$ depends linearly on x. For each t in A_d we now define $x(t)$ on H_d by

$$(x(t)\xi_m \mid \xi_n) = f(t),$$

and have a faithful representation of $e_d\mathscr{B}$ into $\mathscr{B}(\hat{A}_d, \boldsymbol{B}(H_d))$.

To show that the representation is surjective take x in $\mathscr{B}(\hat{A}_d, \boldsymbol{B}(H_d))$. Then as above $p_n x p_m = f \cdot v_{nm}$ with f in $\mathscr{B}(\hat{A}_d)$. Thus $p_n x p_m \in e_d\mathscr{B}$. But $x = \Sigma\, p_n x p_m$ and we conclude from 4.5.17 that $x \in e_d\mathscr{B}$.

6.3.5. COROLLARY. *If A is a C^*-algebra of type I then $\mathscr{B}$ is sequentially weakly closed.*

6.3.6. We say that a Borel *-algebra $\mathscr{A}$ is *of type* I if there is a sequence $\{T_d \mid 1 \leqslant d \leqslant \infty\}$ of standard Borel spaces such that

$$\mathscr{A} = \prod \mathscr{B}(T_d, \boldsymbol{B}(H_d)).$$

In particular we say that $\mathscr{A}$ is *homogeneous of degree d* if $\mathscr{A} = \mathscr{B}(T_d, \boldsymbol{B}(H_d))$. With this definition 6.3.4 simply says that if A is a C^*-algebra of type I then $\mathscr{B}$ is a Borel *-algebra of type I. Conversely, each Borel *-algebra $\mathscr{A} = \Pi\mathscr{B}(T_d, \boldsymbol{B}(H_d))$ of type I is the enveloping Borel *-algebra for a separable C^*-algebra A. It suffices to choose a compact Hausdorff space topology for each T_d which generate the Borel structure; and then let A be the direct sum of the C^*-algebras $C(T_d, \boldsymbol{C}(H_d))$. Of course there are many other C^*-algebras which has the same enveloping Borel *-algebra, just as in the commutative case. However, as there are only a countable number of standard Borel spaces the following observation is useful.

6.3.7. COROLLARY. *If A and B are separable C^*-algebras of type I they are Borel isomorphic if and only if $\hat{A}_d$ and $\hat{B}_d$ are Borel isomorphic for each d.*

6.3.8. PROPOSITION. *If A is a separable C^*-algebra of type I then A'' is a von Neumann algebra of type I.*

Proof. It follows from 6.3.4 that $\mathscr{B}$ contains an abelian projection p with central cover 1. But then p is also an abelian projection in A'' with central cover 1 (cf. 4.5.8), whence A'' is of type I by 5.5.3.

6.3.9. COROLLARY. *If A is a C^*-algebra of type I then $\check{A} = \hat{A}$.*

Proof. Each factor representation (π, H) of A extends to a normal morphism of A'', whence $\pi(A)''$ is of type I and (π, H) is equivalent to an irreducible representation by 5.5.9

6.3.10. Notes and remarks. Proposition 6.3.2 was (in essence) proved by Fell [90]. Theorem 6.3.4 was proved by Davies [50] with $\mathscr{B}^w$ instead of $\mathscr{B}$ (cf. 4.5.14). The author's extension to $\mathscr{B}$ (from [195]) is marginal, but it does provide us with a proof of the equation $\mathscr{B} = \mathscr{B}^w$; at least in the type I case. Proposition 6.3.8 (in the form 6.3.9) was established by Kaplansky [141].

6.4. Glimm algebras

6.4.1. It is appropriate at this point to give examples of antiliminary algebras. Although we have already hinted in 6.1.2 that the Calkin algebra is antiliminary, we must obviously produce separable examples to keep within the spirit of this treatise.

The examples mentioned in section 1.2 are all of type I, and from 6.2.9 we see that all known operations which generate new C^*-algebras from old, will produce type I algebras if the input is of type I. There is, however, one operation which does not respect type, namely the C^*-completion of an inductive limit of a sequence of C^*-algebras. Using this operation with a sequence of finite-dimensional algebras we shall produce our examples: the Glimm algebras (also known as the uniformly hyperfinite C^*-algebras or UHF algebras).

The Glimm algebras are in their own right a very interesting class. One of them (in some sense the simplest) appears as a model in quantum statistical mechanics for particles obeying Fermi statistics, and is therefore known as the *Fermion algebra.* We shall use the Glimm algebras to give examples (hitherto conspicuously missing) of von Neumann algebras of type II and III. Finally, in section 6.8 we show that the Glimm algebras are the universal obstruction for a C^*-algebra to be of type I, and prove the converse of 6.3.9.

6.4.2. As in section 1.2 let $\boldsymbol{M}_m$ denote the C^*-algebra of $m \times m$ matrices identified with $\boldsymbol{B}(H_m)$. If $v_{ij}^{(m)}$ is the matrix with 1 in the (i,j) position and zeros elsewhere then the set $\{v_{ij}^{(m)} \mid 1 \leqslant i \leqslant m, 1 \leqslant j \leqslant m\}$ is a complete system of matrix units for $\boldsymbol{M}_m$ and we have the relations

(i) $v_{ij}^{(m)} v_{kl}^{(m)} = v_{il}^{(m)} \delta(j,k)$,

(ii) $v_{ij}^{(m)*} = v_{ji}^{(m)}$,

(iii) $\Sigma_i\, v_{ii}^{(m)} = 1$.

From (i) and (ii) we see that it suffices to know the elements $\{v_{i1}^{(m)} \mid 1 \leqslant i \leqslant m\}$ in order to generate $\boldsymbol{M}_m$ as a $*$-algebra. This will be used in 6.6.

Suppose now that $\iota: \boldsymbol{M}_m \to \boldsymbol{M}_n$ is a morphism of $\boldsymbol{M}_m$ into $\boldsymbol{M}_n$ such that $\iota(1) = 1$. Since $\boldsymbol{M}_m$ is simple, ι is an injection. Let d be the dimension (trace) of $\iota(v_{11}^{(m)})$. Since all $v_{ii}^{(m)}$ are equivalent in $\boldsymbol{M}_m$ with sum 1 we conclude that $md = n$. It is then possible to find a complete set of matrix units $\{v_{ij}^{(n)}\}$ for $\boldsymbol{M}_n$ such that

$$(*) \qquad \iota(v_{ij}^{(m)}) = \sum_{k=1}^{d} v_{i+(k-1)m,\, j+(k-1)m}^{(n)}.$$

Whenever we have an embedding of M_m in M_n we shall assume that the matrix units satisfy (*). Conversely, if m and n are given and $md = n$ for some d in N then the relations (*) determine an embedding $\iota: M_m \to M_n$.

Given a sequence $\{m(n) \mid n \in N\}$ of natural numbers greater than one let $m(n)! = \Pi_{k=1}^n m(k)$. Then consider the inductive system

$$M_{m(1)!} \xrightarrow{\iota} M_{m(2)!} \xrightarrow{\iota} \cdots \xrightarrow{\iota} M_{m(n)!} \xrightarrow{\iota} \cdots .$$

The inductive limit $M_\infty = \cup M_{m(n)!}$ satisfies all axioms for a C^*-algebra except that of completeness. Denote by A_∞ the completion of M_∞. Then A_∞ is a C^*-algebra which we shall call the *Glimm algebra of rank* $\{m(n)\}$. The *Fermion algebra* is the Glimm algebra for which $m(n) = 2$ for all n.

6.4.3. PROPOSITION. *Every Glimm algebra is a separable simple C^*-algebra with unit, and has a unique tracial state.*

Proof. Since each $M_{m(n)!}$ is separable and M_∞ is dense in A_∞, A_∞ is separable. Furthermore, the embedding of $M_{m(n)!}$ into $M_{m(n+1)!}$ preserves the unit, so M_∞ has a unit, whence A_∞ has a unit.

If π is a non-zero morphism of A_∞ then $\pi \mid M_{m(n)!}$ is an isometry for each n, since $1 \in M_{m(n)!}$ and $M_{m(n)!}$ is simple. Consequently $\pi \mid M_\infty$ is an isometry, whence by continuity π is an isometry of A_∞, which is therefore simple.

For each n let τ_n denote the normalized trace on $M_{m(n)!}$ (i.e. $\tau_n = (m(n)!)^{-1}\mathrm{Tr}$). From (*) in 6.4.2 we see that $\tau_{n+1} \circ \iota = \tau_n$ for each n so that there is a unique functional τ on M_∞ such that $\tau \mid M_{m(n)!} = \tau_n$ for all n. Since τ is a tracial state on M_∞ it extends by continuity to a tracial state on A_∞. Conversely, if ϕ is a tracial state on A_∞ then from the uniqueness of the trace on $B(H_{m(n)!})$ we conclude that $\phi \mid M_{m(n)!} = \tau_n$ for all n, whence $\phi = \tau$.

6.4.4. COROLLARY. *There exists a factor of type II_1.*

Proof. The von Neumann algebra $\mathscr{M} = \pi_\tau(A_\infty)''$ has a non-zero finite normal trace, viz. the normal extension of τ. The kernel of τ is a central projection, so we may assume that τ is faithful on $z\mathscr{M}$ for some central projection z. Since τ is faithful on M_∞, $z \neq 0$ (actually $z = 1$, cf. 6.5.8). Since τ is the unique tracial state on A_∞, the centre of $z\mathscr{M}$ is trivial, so $z\mathscr{M}$ is a factor. This factor is finite but not finite-dimensional, since $z\pi_\tau(A_\infty)$ is isomorphic to A_∞, and therefore $z\mathscr{M}$ is of type II_1.

6.4.5. LEMMA. *If p and q are projections in a C^*-algebra A such that $\|p - q\| < 1$ then p and q are equivalent in A.*

Proof. Since $\|p - pqp\| < 1$ the element pqp is invertible in the subalgebra pAp. There is therefore an element x in $(pAp)_+$ such that $x^2pqp = p$. Define $u = qx$. Then $u \in A$ and since $x = xp$, $u^*u = xpqpx = p$. Furthermore, $uu^* = qx^2q$. However, since $x^2pqp = p$ we have $qx^2q\,qpq = qpq$, so that

qx^2q is a unit for qpq. Since qpq is invertible in qAq it follows that $qx^2q = q$, whence $uu^* = q$.

6.4.6. THEOREM. *Let A be a Glimm algebra of rank $\{m(n)\}$ and for each prime number p let $\varepsilon(p)$ be the supremum of exponents ε for which p divides some $m(n)!$ (so that $0 \leqslant \varepsilon(p) \leqslant \infty$). If A'_∞ is a Glimm algebra isomorphic to A_∞ then $\varepsilon'(p) = \varepsilon(p)$ for each p.*

Proof. For each n consider a minimal projection e in $\boldsymbol{M}_{m(n)!}$. Then $\tau(e) = (m(n)!)^{-1}$. If $\pi: A_\infty \to A'_\infty$ is an isomorphism there is an n' and an element x in $\boldsymbol{M}_{m'(n')!}$ with $0 \leqslant x \leqslant 1$ such that $\|\pi(e) - x\| < 2^{-4}$. Since e is a projection we have $\|x - x^2\| < 3 \cdot 2^{-4}$ so that the spectrum of x is contained in $[0, \frac{1}{4}] \cup [\frac{3}{4}, 1]$. With q as the spectral projection of x corresponding to the interval $[\frac{3}{4}, 1]$ we therefore have $\|x - q\| \leqslant \frac{1}{4}$, whence $\|\pi(e) - q\| < 1$. By 6.4.3 we have $\tau' \circ \pi = \tau$, so from 6.4.5 it follows that $\tau'(q) = (m(n)!)^{-1}$. However, the only possible values of τ' on projections in $\boldsymbol{M}_{m'(n')!}$ are multiples of $(m'(n')!)^{-1}$. It follows that $m'(n')! = dm(n)!$ for some d in $\boldsymbol{N}$. If therefore p^ε divides $m(n)!$ it also divides $m'(n')!$, whence $\varepsilon(p) \leqslant \varepsilon'(p)$. By a symmetric argument $\varepsilon'(p) \leqslant \varepsilon(p)$ and the theorem follows.

6.4.7. COROLLARY. *There are uncountably many non-isomorphic Glimm algebras.*

6.4.8. Notes and remarks. Glimm algebras first appeared in the thesis of Glimm [95]. The corresponding notion for von Neumann algebras (where a factor $\mathscr{M}$ is called hyperfinite if there is an increasing sequence $\{\boldsymbol{M}_{m(n)}\}$ of matrix subalgebras whose union is dense in $\mathscr{M}$) goes back to Murray and von Neumann, see [171]. Contrary to the C^*-case, all hyperfinite factors are isomorphic, so that 6.4.4 gives only one example of a II_1 von Neumann algebra. To each Glimm algebra we can assign a sequence of powers (possibly infinite) of primes as in 6.4.6. Such an assignment is sometimes called a 'supernatural number'. Theorem 6.4.6 then says that two Glimm algebras that are isomorphic have the same supernatural number. Actually Glimm proved that the converse is true as well, so that the supernatural number is a complete invariant for the algebra. A characterization of Glimm algebras among all separable C^*-algebras with units also appeared in [95]. The demand is that for each $\varepsilon > 0$ and each finite set $x_1, \ldots, x_k$ in A there is some matrix subalgebra $\boldsymbol{M}_m \subset A$ and elements $y_1, \ldots, y_k$ in $\boldsymbol{M}_m$, such that $\|x_k - y_k\| < \varepsilon$ for all k. The corresponding approximation, with (generalized) Hilbert–Schmidt norm instead of $\|\cdot\|$ characterizes the hyperfine factor among all finite II_1 factors.

Dixmier [66] considered inductive limits of matrix algebras without the demand that the embeddings preserve units. The resulting C^*-algebras were called matroid and would typically include alg bras of the form $A \otimes C(H)$, A a Glimm algebra. Later Bratteli [25] undertook a study of C^*-algebras which are inductive limits of arbitrary finite-dimensional C^*-algebras with arbitrary embeddings. The resulting class of approximately finite-dimensional C^*-algebras (or AF-algebras) displays a rich structure and is particularly valuable as testing ground for various hypotheses about general C^*-algebras, since the AF-algebras lend themselves to detailed analysis. Bratteli gave a classification of AF-algebras in terms of (equivalence classes of) embedding schemes: The Bratteli diagrams. Later Elliott [81] found a more algebraic formulation of the classification invariants in terms of abelian local semi-groups.

6.5. Product states of Glimm algebras

6.5.1. Let m, n and d be natural numbers with $n = md$ and let $\iota: M_m \to M_n$ be the embedding determined by $(*)$ in 6.4.2. The commutant of $\iota(M_m)$ in M_n is isomorphic to M_d. In fact, if

$$u_{kl} = \sum_{i=1}^{m} v^{(n)}_{i+(k-1)m,\, i+(l-1)m} \qquad \text{for } 1 \leqslant k \leqslant d,\ 1 \leqslant l \leqslant d,$$

then $\{u_{kl}\}$ is a complete system of matrix units for $\iota(M_m)'$. Since for all i, j, k, l we have

$$v^{(n)}_{i+(k-1)m,\, j+(l-1)m} = u_{kl}\iota(v^{(m)}_{ij}),$$

we see that M_n is generated by $\iota(M_m)'$ and $\iota(M_m)$ in such a way that $M_n = \iota(M_m)' \otimes \iota(M_m)$.

In matrix form we can represent each element x in M_n as a $d \times d$-matrix (x_{ij}) with x_{ij} in M_m. Then $x \in \iota(M_m)$ if all x_{ii} are equal and $x_{ij} = 0$ for $i \neq j$; while $x \in \iota(M_m)'$ if all x_{ij} are scalar multiples of the unit in M_m.

If A_∞ is a Glimm algebra of rank $\{m(n)\}$ then by successive applications of the formula above we obtain

$$M_{m(n)!} = M_{m(1)} \otimes M_{m(2)} \otimes \cdots \otimes M_{m(n)}.$$

Since the tensor product is associative (and commutative) we have for each $k < n$

$$\iota(M_{m(k)!}) = M_{m(k)} \otimes 1_{m(k+1)} \otimes \cdots \otimes 1_{m(n)},$$

$$\iota(M_{m(k)!})' = 1_{m(k)!} \otimes M_{m(k+1)} \otimes \cdots \otimes M_{m(n)}.$$

With the proper definition A_∞ becomes the infinite C^*-tensor product of the algebras $\{M_{m(n)}\}$.

6.5.2. LEMMA. *Let A_∞ be a Glimm algebra of rank $\{m(n)\}$ and let Λ be a sequence* *of convex combinations* $\{\Lambda^n\}$ *each of length* $m(n)$. *There is a unique state* ϕ_Λ *on* A_∞ *such that with* $\boldsymbol{M}_{0!} = \boldsymbol{C}$, $\phi_0 = \iota$ *and* $\phi_n = \phi_\Lambda | \boldsymbol{M}_{m(n)!}$ *we have*

$$(**) \qquad \phi_n(x) = \sum_{i=1}^{m(n)} \Lambda_i^n \phi_{n-1}(x_{ii}),$$

for each x *in* $\boldsymbol{M}_{m(n)!}$ *represented in the form* $x = (x_{ij})$, $x_{ij} \in \boldsymbol{M}_{m(n-1)!}$, $1 \leqslant i \leqslant m(n)$, $1 \leqslant j \leqslant m(n)$ *and all* n *in* $\boldsymbol{N}$.

Proof. The conditions $\phi_0 =$ identity and $(**)$ define by induction a sequence $\{\phi_n\}$ of functionals on the algebras $\boldsymbol{M}_{m(n)!}$. Since $x \geqslant 0$ implies that $x_{ii} \geqslant 0$ for all $i \leqslant m(n)$ we see that $\phi_{n-1} \geqslant 0$ implies $\phi_n \geqslant 0$. By induction we conclude that $\phi_n \geqslant 0$ for all n. Since furthermore $\phi_n(1) = 1$ for all n, we have a sequence of states. Now $\phi_n | \boldsymbol{M}_{m(n-1)!} = \phi_{n-1}$ so there is a unique positive functional ϕ_∞ of norm one on $\boldsymbol{M}_\infty$ such that $\phi_\infty | \boldsymbol{M}_{m(n)!} = \phi_n$ for all n; and ϕ_∞ extends by continuity to a state ϕ_Λ of A. Clearly ϕ_Λ is the unique state on A_∞ satisfying $(**)$ for all n.

6.5.3. Since $\boldsymbol{M}_{m(n)!} = \boldsymbol{M}_{m(1)} \otimes \cdots \otimes \boldsymbol{M}_{m(n)}$, each element in $\boldsymbol{M}_{m(n)!}$ can be written as a linear combination of elements of the form $x = x^{(1)} \otimes \cdots \otimes x^{(n)}$, with $x^{(k)}$ in $\boldsymbol{M}_{m(k)}$ for $1 \leqslant k \leqslant n$. We claim that

$$(***) \qquad \phi_\Lambda(x) = \prod_{k=1}^{n} \left(\sum_{i=1}^{m(k)} \Lambda_i^k x_{ii}^{(k)} \right).$$

Suppose that we have proved $(***)$ for all tensors of length n and take x in $\boldsymbol{M}_{m(n+1)!}$ of the form $x^{(1)} \otimes \cdots \otimes x^{(n+1)}$. Then with $y = x^{(1)} \otimes \cdots \otimes x^{(n)}$ we have by $(**)$ in 6.5.2 and our assumption

$$\begin{aligned} \phi_\Lambda(x) &= \sum_{i=1}^{m(n+1)} \Lambda_i^{n+1} \phi_\Lambda(y x_{ii}^{(n+1)}) \\ &= \phi_\Lambda(y) \left(\sum_{i=1}^{m(n+1)} \Lambda_i^{n+1} x_{ii}^{(n+1)} \right) = \prod_{k=1}^{n+1} \left(\sum_{i=1}^{m(k)} \Lambda_i^k x_{ii}^{(k)} \right). \end{aligned}$$

Thus $(***)$ can be verified by induction.

Note that $(***)$ defines an n-linear form on $\boldsymbol{M}_{m(1)} \times \cdots \times \boldsymbol{M}_{m(n)}$ and thus extends uniquely to a linear functional on $\boldsymbol{M}_{m(1)} \otimes \cdots \otimes \boldsymbol{M}_{m(n)}$; so that $(***)$ is actually an equivalent definition of ϕ_Λ. The formula $(***)$ also explains the name *product state* of A_∞, applied to ϕ_Λ. We see that the tracial state on A_∞ defined in 6.4.3 is the product state with $\Lambda_i^n = m(n)^{-1}$ for all $i \leqslant m(n)$ and all n in $\boldsymbol{N}$.

6.5.4. LEMMA. *If* $x, y \in A_\infty$, $x \in \boldsymbol{M}_{m(n)!}$ *and* $y \in (\boldsymbol{M}_{m(n)!})'$ *for some* n, *then* $\phi_\Lambda(xy) = \phi_\Lambda(x)\phi_\Lambda(y)$ *for each product state* ϕ_Λ *on* A_∞.

Proof. Suppose first that $y \in M_{m(k)}$ with $k > n$. We can then write $x = \tilde{x} \otimes 1_{m(n+1)} \otimes \cdots \otimes 1_{m(k)}$ and $y = 1_{m(n)!} \otimes \tilde{y}$ (cf. 6.5.1). Applying (***) from 6.5.3 we get

$$\begin{aligned}\phi_\Lambda(xy) &= \phi_\Lambda(\tilde{x} \otimes \tilde{y}) \\ &= \phi_\Lambda(\tilde{x} \otimes 1_{m(n+1)} \otimes \cdots \otimes 1_{m(k)})\phi_\Lambda(1_{m(n)!} \otimes \tilde{y}) \\ &= \phi_\Lambda(x)\phi_\Lambda(y).\end{aligned}$$

For an arbitrary y in A_∞ commuting with $M_{m(n)!}$ choose a sequence $\{y_k\}$ converging to y such that $y_k \in M_{m(k)!}$ for each k. Let $U_{m(n)!}$ denote the unitary group of $M_{m(n)!}$. This is a compact group with Haar measure du. Put

$$z_k = \int_{U_{m(n)!}} u y_k u^* \, du.$$

The $z_k \in \mathcal{M}_{m(k)!}$ for $k \geq n$; and from the invariance of Haar measure we see that z_k commutes with $U_{m(n)!}$ hence with $M_{m(n)!}$. Since y commutes with $M_{m(n)!}$ we have

$$\begin{aligned}\|y - z_k\| &= \left\| \int_{U_{m(n)!}} (uyu^* - uy_ku^*)\, du \right\| \\ &\leq \int_{U_{m(n)!}} \|u(y - y_k)u^*\| \, du = \|y - y_k\| \to 0.\end{aligned}$$

From the first part of the proof we know that $\phi_\Lambda(xz_k) = \phi_\Lambda(x)\phi(z_k)$ and by continuity we conclude that $\phi_\Lambda(xy) = \phi_\Lambda(x)\phi(y)$.

6.5.5. Lemma. *If for all n and i, Λ_i^n is either 0 or 1, then ϕ_Λ is a pure state of A_∞.*

Proof. Take a positive functional ψ on A_∞ such that $\psi \leq \phi_\Lambda$. Suppose that we have established that $\psi \mid M_{m(n)!} = \psi(1)\phi_n$ for some $n \geq 0$. Assuming, as we may without loss of generality, that $\Lambda_1^{n+1} = 1$ and $\Lambda_i^{n+1} = 0$ for $1 < i \leq m(n+1)$, we write each x in $M_{m(n+1)!}$ in the form $x = (x_{ij})$, $x_{ij} \in M_{m(n)!}$. Then $\phi_{n+1}(x) = \phi_n(x_{11})$. Since $\psi \leq \phi_\Lambda$ this implies that

$$\psi(x) = \psi(x_{11}) = \psi(1)\phi_n(x_{11}) = \psi(1)\phi_{n+1}(x),$$

so that $\psi \mid M_{m(n+1)!} = \psi(1)\phi_{n+1}$. By induction we obtain $\psi = \psi(1)\phi_\Lambda$, whence ϕ_Λ is pure.

6.5.6. Proposition. *Let ϕ_Λ and ϕ_Δ be pure product states of A_∞. The irreducible representations associated with ϕ_Λ and ϕ_Δ are equivalent if and only if $\Delta^n = \Lambda^n$ for all but a finite number of indices.*

Proof. If $\Lambda^n = \Delta^n$ whenever $n > k$ consider the two pure states $\phi_\Lambda \mid M_{m(k)!}$ and $\phi_\Delta \mid M_{m(k)!}$ on $M_{m(k)!}$. All irreducible representations of $M_{m(k)!}$ are equivalent so by 3.13.4 there is a unitary u in $M_{m(k)!}$ such that $\phi_\Lambda(u^*\cdot u) = \phi_\Delta$ on $M_{m(k)!}$. Since $\Lambda^n = \Delta^n$ for all $n > k$ it follows from (**) in 6.5.2 that $\phi_\Lambda(u_*\cdot u) = \phi_\Lambda$ on A_∞, and thus their associated representations are equivalent.

Conversely, if the representations associated with ϕ_Λ and ϕ_Δ are equivalent there is by 3.13.4 a unitary u in A_∞ such that $\phi_\Lambda(u^*\cdot u) = \phi_\Delta$. For sufficiently large n there is an element x in $M_{m(n)!}$ with $\|u - x\| < \frac{1}{2}$. If now $\Lambda^k \neq \Delta^k$ for some $k > n$, say $\Lambda_1^k = 0$ and $\Delta_1^k = 1$, consider the element e in $M_{m(k)!}$, represented in the form $e = e_{ij}$, $e_{ij} \in M_{m(k-1)!}$, such that $e_{11} = 1$, $e_{ij} = 0$ otherwise. Then e commutes with $M_{m(k-1)!}$, hence with $M_{m(n)!}$, so by 6.5.4

$$
\begin{aligned}
1 &= 1 + \phi_\Lambda(x^*x)\phi_\Lambda(e) = 1 + \phi_\Lambda(x^*xe) = 1 + \phi_\Lambda(x^*ex)\\
&> \phi_\Lambda((u^* - x^*)eu) + \phi_\Lambda(x^*e(u - x)) + \phi_\Lambda(x^*ex)\\
&= \phi_\Lambda(u^*eu) = \phi_\Delta(e) = 1,
\end{aligned}
$$

a contradiction. Consequently $\Lambda^k = \Delta^k$ for all $k > n$.

6.5.7. THEOREM. *Every Glimm algebra A_∞ is antiliminary. Each irreducible representation (π, H) of A_∞ is faithful with $\pi(A_\infty) \cap C(H) = 0$, and A_∞ has uncountably many of these, each disjoint from the other.*

Proof. The existence of uncountably many disjoint irreducible representations follows directly from 6.5.6. Since A_∞ is simple (6.4.3) each irreducible representation (π, H) is faithful. If $\pi(A_\infty) \cap C(H) \neq 0$ then by 6.1.4 $C(H) \subset \pi(A_\infty)$. However, as A_∞ is simple this would mean that A_∞ was isomorphic with $C(H)$ which contradicts 6.1.5 (ii) and the first part of the proof (it also contradicts 6.4.3, since $C(H)$ has no tracial state when H is infinite-dimensional).

6.5.8. PROPOSITION. *Each product state of a Glimm algebra is factorial.*

Proof. Let ϕ_Λ be a product state of a Glimm algebra A_∞ of rank $\{m(n)\}$, and let $(\pi_\Lambda, H_\Lambda, \xi_\Lambda)$ denote the cyclic representation associated with ϕ_Λ. If z belongs to the centre of $\pi_\Lambda(A_\infty)''$ there is by Kaplansky's density theorem a sequence $\{y_k\}$ in A_∞ with $\|y_k\| \leqslant \|z\|$ for all k, such that $\pi_\Lambda(y_k) \to z$ weakly. Fix n, and with $U_{m(n)!}$ as in the proof of 6.5.4 construct z_k from y_k as before. Then z_k commutes with $M_{m(n)!}$. Also, for each u in $U_{m(n)!}$ we have

$$\pi_\Lambda(uy_ku^*) \to \pi_\Lambda(u)z\pi_\Lambda(u^*) = z \qquad \text{(weakly)},$$

whence $\pi(z_k) \to z$ weakly by Lebesgue's dominated convergence theorem.

Thus for all x, y in $M_{m(n)!}$ we have by 6.5.4

$$\begin{aligned}(z\xi_x \mid \xi_y) &= \operatorname{Lim}(\pi(z_k)\xi_x \mid \xi_y)\\ &= \operatorname{Lim} \phi_\Lambda(y^* z_k x) = \operatorname{Lim} \phi_\Lambda(z_k)\phi_\Lambda(y^* x)\\ &= (z\xi_\Lambda \mid \xi_\Lambda)(\xi_x \mid \xi_y).\end{aligned}$$

This holds for all n and we conclude that $z = (z\xi_\Lambda \mid \xi_\Lambda)1$ so that $\pi_\Lambda(A_\infty)''$ is a factor.

6.5.9. PROPOSITION. *Each product state ϕ_Λ of a Glimm algebra A_∞, such that $\Lambda_i^n > 0$ for all i and n, is faithful as a normal functional on $\pi_\Lambda(A_\infty)''$.*

Proof. If $x \in M_{m(n)!}$ represented in the form $x = (x_{ij})$, $x_{ij} \in M_{m(n-1)!}$, then $x \geqslant 0$ and $x_{ii} = 0$ for $1 \leqslant i \leqslant m(n)$ imply $x = 0$. Using (**) in 6.5.2 it follows by induction that ϕ_Λ is faithful on $M_{m(n)!}$ for every n. Since the positive unit sphere of $M_{m(n)!}$ is compact there is therefore a constant α_n such that $\|x\| \leqslant \alpha_n \phi_\Lambda(x)$ for each positive x in $M_{m(n)!}$. Thus for every unitary u in $M_{m(n)!}$ we have

$$\phi_\Lambda(u^* \cdot u) \leqslant \alpha_n \phi_\Lambda$$

on $M_{m(n)!}$. Suppose that we have proved this inequality also on $M_{m(k)!}$ with $k \geqslant n$. Then take x in $(M_{m(k+1)!})_+$ represented in the form $x = (x_{ij}), x_{ij} \in M_{m(k)!}$. Using (**) in 6.5.2 we obtain

$$\phi_\Lambda(u^* x u) = \sum \Lambda_i^{k+1} \phi_\Lambda(u^* x_{ii} u) \leqslant \sum \Lambda_i^{k+1} \alpha_n \phi_\Lambda(x_{ii}) = \alpha_n \phi_\Lambda(x).$$

By induction we conclude that $\phi_\Lambda(u^* \cdot u) \leqslant \alpha_n \phi_\Lambda$ on A_∞. But then by weak continuity we have

$$\phi_\Lambda(\pi_\Lambda(u^*) x \pi_\Lambda(u)) \leqslant \alpha_n \phi_\Lambda(x)$$

for each unitary u in $M_{m(n)!}$ and each positive x in $\pi_\Lambda(A_\infty)''$. It follows that if p is a projection in $\pi_\Lambda(A_\infty)''$ with $\phi_\Lambda(p) = 0$ then $\phi_\Lambda(u^* p u) = 0$ for every unitary u in $\pi_\Lambda(A_\infty)$. Therefore $\phi_\Lambda(c(p)) = 0$ whence $c(p) = 0$. Consequently $p = 0$ and ϕ_Λ is faithful on $\pi_\Lambda(A_\infty)''$.

6.5.10. In the rest of this section we specialize to the Fermion algebra F defined in 6.4.2. Actually all our results would go through for any Glimm algebra whose rank $\{m(n)\}$ satisfied $m(n) = a^n$ for all n and some fixed number a. We concentrate on $a = 2$ to ease the notation. For each λ in $[0, \frac{1}{2}]$ let Λ be the sequence of convex combinations of length 2 such that $\Lambda_1^n = \lambda$ and $\Lambda_2^n = 1 - \lambda$ for all n. We shall denote by ϕ_λ the product state of F arising from Λ and by $(\pi_\lambda, H_\lambda, \xi_\lambda)$ the cyclic representation of F associated with ϕ_λ. We already know that $\pi_0(F)'' = B(H_0)$ and that $\pi_{1/2}(F)''$ is a factor of type II_1 (cf. 6.5.5, 6.4.4 and 6.5.8). We shall determine the type of the von Neumann algebras $\mathcal{M}_\lambda = \pi_\lambda(F)''$ for $0 < \lambda < \frac{1}{2}$.

6.5.11. Let Π denote the group of permutations of $\boldsymbol{N}$ which leave all but a finite number of elements fixed. Thus Π is the inductive limit of all groups Π_k of permutations of the first k numbers. Given t in Π_k and $n \geqslant k$ let u_t be the unique unitary operator on $H_{2^n} = H_2 \otimes \cdots \otimes H_2$ such that

$$u_t(\xi_1 \otimes \cdots \otimes \xi_n) = \xi_{t(1)} \otimes \cdots \otimes \xi_{t(n)}$$

for any n-tuple $(\xi_1, \ldots, \xi_n)$ of vectors in H_2. We have $u_t \in M_{2^n}$ and it is clear from the construction that the map $t \to u_t$ is an isomorphism of Π into the unitary group of $\boldsymbol{M}_\infty$.

If $x \in \boldsymbol{M}_{2^n}$ of the form

$$x = x_1 \otimes \cdots \otimes x_n, \qquad x_i \in \boldsymbol{M}_2,$$

and $t \in \Pi_k$, $k \leqslant n$, then

$$(\#) \qquad u_t x u_t^* = x_{t(1)} \otimes \cdots \otimes x_{t(n)},$$

since for each vector ξ in H_{2^n} of the form $\xi_1 \otimes \cdots \otimes \xi_n$ we have

$$\begin{aligned} u_t x u_t^* \xi &= u_t(x_1 \otimes \cdots \otimes x_n)(\xi_{t^{-1}(1)} \otimes \cdots \otimes \xi_{t^{-1}(n)}) \\ &= u_t(x_1 \xi_{t^{-1}(1)} \otimes \cdots \otimes x_n \xi_{t^{-1}(n)}) = x_{t(1)}\xi_1 \otimes \cdots \otimes x_{t(n)}\xi_n. \end{aligned}$$

6.5.12. Fix once and for all a sequence $\{u_n\}$ in $\{u_t \mid t \in \Pi\}$ such that the permutation t_n corresponding to u_n satisfies $t_n(i) > n$ for all $i \leqslant n$. Then from $(\#)$ in 6.5.11 we see that for each x in $\boldsymbol{M}_{2^k}$ the elements $\{u_n x u_n^*\}$ commute with $\boldsymbol{M}_{2^k}$ when $n \geqslant k$.

From (***) in 6.5.3 and $(\#)$ in 6.5.11 we see that the product states ϕ_λ defined in 6.5.10 satisfy $\phi_\lambda(u_t x u_t^*) = \phi_\lambda(x)$ for each x in $\boldsymbol{M}_{2^n}$ and each t in Π_k, $k \leqslant n$. It follows immediately that

$$\phi_\lambda(u_t x u_t^*) = \phi_\lambda(x)$$

for all x in $\boldsymbol{F}$ and all t in Π.

6.5.13. LEMMA. *If* $x \in \boldsymbol{F}$ *then* $\{\pi_\lambda(u_n x u_n^*)\}$ *is weakly convergent to* $\phi_\lambda(x)1$.

Proof. If x, y, z belong to $\boldsymbol{M}_{2^k}$ and $n \geqslant k$ then from 6.5.12 together with 6.5.4 we have

$$\phi_\lambda(z^* u_n x u_n^* y) = \phi_\lambda(u_n x u_n^* z^* y) = \phi_\lambda(u_n x u_n^*)\phi_\lambda(z^* y) = \phi_\lambda(x)\phi_\lambda(z^* y).$$

Since this is true for all n and k with $n \geqslant k$ it follows by continuity that

$$\phi_\lambda(z^* u_n x u_n^* y) \to \phi_\lambda(x)\phi_\lambda(z^* y)$$

for all x, y, z in $\boldsymbol{F}$. However, this means that

$$(\pi_\lambda(u_n x u_n^*)\xi_y \mid \xi_z) \to \phi_\lambda(x)(\xi_y \mid \xi_z).$$

Since the sequence $\{\pi_\lambda(u_n x u_n^*)\}$ is bounded and weakly convergent on a dense set of vectors we conclude that $\pi_\lambda(u_n x u_n^*) \to \phi_\lambda(x)1$, weakly.

6.5.14. LEMMA. *If ψ is a positive functional on* F *which has a normal extension to* $\pi_\lambda(F)''$ *for some* λ *in* $[0,\frac{1}{2}]$ *and satisfies* $\psi(u_t x u_t^*) = \psi(x)$ *for all* x *in* F *and* t *in* Π, *then* ψ *is a scalar multiple of* ϕ_λ.

Proof. Since ψ is weakly continuous on bounded sets in $\pi_\lambda(F)$ we have for each x in F by 6.5.13

$$\psi(x) = \psi(u_n x u_n^*) \to \phi_\lambda(x)\psi(1),$$

whence $\psi = \psi(1)\phi_\lambda$.

6.5.15. THEOREM. *The von Neumann algebras* $\mathcal{M}_\lambda = \pi_\lambda(F)''$, $0 < \lambda < \frac{1}{2}$, *are factors of type III.*

Proof. We know from 6.5.8 that each $\mathcal{M}_\lambda$ is a factor. Suppose that τ is a normal, non-zero (hence faithful), semi-finite trace on $\mathcal{M}_\lambda$. There is then by 5.3.11 a unique positive operator h on H_λ, affiliated with $\mathcal{M}_\lambda$, such that $\phi_\lambda(x) = \tau(hx)$ for all x in $\mathcal{M}_\lambda$. Since

$$\begin{aligned}\tau(\pi_\lambda(u_t^*)h\pi_\lambda(u_t)x) &= \tau(h\pi_\lambda(u_t)x\pi_\lambda(u_t^*))\\ &= \phi_\lambda(\pi_\lambda(u_t)x\pi_\lambda(u_t^*)) = \phi_\lambda(x),\end{aligned}$$

for all x in $\mathcal{M}_\lambda$, and h is unique, we see that $\pi_\lambda(u_t^*)h\pi_\lambda(u_t) = h$ for all t in Π. For each $\varepsilon > 0$ put $\psi_\varepsilon = \tau(h(\varepsilon + h)^{-1}.)$ on $\mathcal{M}_\lambda$. Since $h(\varepsilon + h)^{-1}$ commutes with all $\pi_\lambda(u_t)$, $t \in \Pi$ we see that

$$\psi_\varepsilon(\pi_\lambda(u_t x u_t^*)) = \psi_\varepsilon(\pi_\lambda(x))$$

for all x in F, whence $\psi_\varepsilon = \psi_\varepsilon(1)\phi_\lambda$ by 6.5.14. Choose x in $(\mathcal{M}_\lambda)_+$ such that $\tau(x) < \infty$ and $\phi_\lambda(x) > 0$. This is possible since τ is semi-finite. Then

$$\psi_\varepsilon(1)\phi_\lambda(x) = \psi_\varepsilon(x) = \tau(h(\varepsilon + h)^{-1}x) \nearrow \tau(x).$$

It follows that $\psi_\varepsilon(1) \nearrow \alpha < \infty$ as $\varepsilon \searrow 0$. Thus for any x in $(\mathcal{M}_\lambda)_+$

$$\alpha\phi_\lambda(x) = \operatorname{Lim}\psi_\varepsilon(1)\phi_\lambda(x) = \operatorname{Lim}\tau(h(\varepsilon + h)^{-1}x) = \tau(x),$$

whence $\tau = \alpha\phi_\lambda$. Since ϕ_λ is not a trace for $\lambda \neq \frac{1}{2}$ we have reached a contradiction. Thus $\mathcal{M}_\lambda$ is of type III.

6.5.16. Notes and remarks. The theory of product states was developed by Powers [211]. However, 6.5.6 and 6.5.7 appear already in Glimm's paper [97, pp. 584–586]. Powers established 6.5.8 in the strong form that a state ϕ of A_∞ is factorial if and only if for each x in A_∞ there is some matrix algebra $M_m \subset A_\infty$, such that $\|\phi(xy) - \phi(x)\phi(y)\| < \|y\|$ for all y in the relative commutant of M_m in A_∞. Theorem 6.5.15 was proved by Glimm [97], using earlier work by Pukanszky [213]. Powers sharpened this result considerably by showing that the von Neumann algebras $\mathscr{M}_\lambda$, $0 < \lambda < \frac{1}{2}$, are pairwise non-isomorphic, see 8.15.13. The method of proving 6.5.15 used here (via 6.5.11–6.5.14) was used by Størmer [244]. We return to the idea (asymptotic abelianness) in Sections 7.12 and 7.13.

6.6. Quasi-matrix systems

6.6.1. Let $\{m(n) \mid n \in \boldsymbol{N}\}$ be a sequence of natural numbers greater than one. A pair of sequences $\{e_n \mid n \in \boldsymbol{N}\}$, $\{v_{ni} \mid 2 \leqslant i \leqslant m(n)\}$ in a C^*-algebra A is called a *quasi-matrix system of rank* $\{m(n)\}$ if it satisfies the following conditions for each n in $\boldsymbol{N}$:

(a) $e_n \geqslant 0$, $\|e_n\| = \|v_{ni}\| = 1$ for $2 \leqslant i \leqslant m(n)$;

(b) $v_{ni}^* v_{ni} e_n = e_n$ for $2 \leqslant i \leqslant m(n)$;

(c) $v_{ni}^* v_{nj} = 0$ for $i \neq j$;

(d) $v_{ni} v_{nj} = 0$ for all i, j;

(e) $e_n e_{n+1} = e_{n+1}$, $e_n v_{n+1,i} = v_{n+1,i}$, $e_n v_{n+1,i}^* = v_{n+1,i}^*$ for $2 \leqslant i \leqslant m(n+1)$.

We say that the pair is a *matrix system of rank* $\{m(n)\}$ if furthermore it satisfies

(f) $v_{ni}^* v_{ni} = e_n$ for $2 \leqslant i \leqslant m(n)$;

(g) $e_{n+1} + \sum_{i=1}^{m(n+1)} v_{n+1,i} v_{n+1,i}^* = e_n$.

In that case the e_n's are projections by (f) and (b), so that by (f) the v_{ni}'s are partial isometries with e_n as initial projection. Condition (c) shows that all final projections of the v_{ni}'s are pairwise orthogonal and by (d) they are also orthogonal to e_n. It follows that the C^*-algebra A_n generated by $\{e_n, v_{n2}, \ldots, v_{nm(n)}\}$ is isomorphic to the algebra $\boldsymbol{M}_{m(n)}$ of $m(n) \times m(n)$-matrices. In a suitable coordinate system e_n is represented as the matrix (α_{rs}) with $\alpha_{rs} = 0$ unless $r = s = 1$, and $\alpha_{11} = 1$; while v_{ni} is represented as the matrix (α_{rs}) with $\alpha_{rs} = 0$ unless $s = 1$ and $\alpha_{r1} = \delta_{ir}$. From (g) we see that the unit for A_{n+1} is e_n, so that the C^*-algebra generated by $A_n \cup A_{n+1}$ contains A_{n+1} as a hereditary C^*-subalgebra and is isomorphic to $A_n \otimes A_{n+1}$. A straightforward inductive argument shows that the C^*-subalgebra of A generated by a matrix system of rank $\{m(n)\}$ is isomorphic to the Glimm algebra of rank $\{m(n)\}$.

6.6.2. LEMMA. *Let $\{e_n\}, \{v_{ni}\}$ be a quasi-matrix system of rank $\{m(n)\}$ in a C^*-algebra. Then*

(i) $e_n v_{mi} = 0$ *for $m \geqslant n$ and all i;*

(ii) $v_{mi}v_{nj} = 0$ *for* $m \geqslant n$ *and all* i,j;

(iii) $v_{mi}^*v_{nj} = 0$ *for* $m \neq n$ *and all* i,j.

Proof. (i) From condition (d) we see that $v_{ni}^*v_{ni}$ and $v_{ni}v_{ni}^*$ are orthogonal. Since by (e) and (b) $e_m \leqslant e_n \leqslant v_{ni}^*v_{ni}$ we conclude that e_m and $v_{ni}v_{ni}^*$ are orthogonal, whence $e_mv_{ni} = 0$ for $m \geqslant n$.

(ii) By (d) we may assume $m > n$, whence by (e) and (i)

$$(v_{mi}v_{nj})^*(v_{mi}v_{nj}) = v_{nj}^*(v_{mi}^*v_{mi})v_{nj} = v_{nj}^*(v_{mi}^*v_{mi})e_{m-1}v_{nj} = 0.$$

(iii) Replacing if necessary $v_{mi}^*v_{nj}$ by its adjoint we may assume $m > n$, whence by (e) and (i)

$$(v_{mi}^*v_{nj})^*(v_{mi}^*v_{nj}) = v_{nj}^*(v_{mi}v_{mi}^*)v_{nj} = v_{nj}^*(v_{mi}v_{mi}^*)e_{m-1}v_{nj} = 0.$$

6.6.3. Let $\{e_n\}$, $\{v_{ni}\}$ be a quasi-matrix system in a C^*-algebra. For each n in $\mathbf{N}$ let X_n be the subset of elements α in $\mathbf{N}^n$ for which $1 \leqslant \alpha(k) \leqslant m(k)$ for each $k \leqslant n$. For later use let X_n^k denote the subset of elements α in X_n for which $\alpha(j) = 1$ for $k \leqslant j \leqslant n$. With the convention $v_{n1} = 1$ define for each α in X_n

$$u_\alpha = v_{1\alpha(1)}v_{2\alpha(2)}\cdots v_{n\alpha(n)}e_n^{1/2}.$$

We claim that the elements u_α, $\alpha \in X_n$ satisfy the relations:

$$(*) \qquad u_\alpha^*u_\beta = 0 \qquad \text{if } \alpha \neq \beta,$$

$$(**) \qquad u_\alpha^*u_\alpha = e_n.$$

To prove this note that if α, $\beta \in X_n$ and $\alpha(1) = \beta(1)$ then $v_{1\alpha(1)}^*v_{1\beta(1)}$ is a unit for all v_{ki}, $2 \leqslant k \leqslant n$, and a unit for all e_k, $1 \leqslant k \leqslant n$ by (b) and (e). Consequently

$$\begin{aligned} u_\alpha^*u_\beta &= e_n^{1/2}v_{n\alpha(n)}^*\cdots v_{1\alpha(1)}^*v_{1\beta(1)}\cdots v_{n\beta(n)}e_n^{1/2} \\ &= e_n^{1/2}v_{n\alpha(n)}^*\cdots v_{2\alpha(2)}^*v_{2\beta(2)}\cdots v_{n\beta(n)}e_n^{1/2}. \end{aligned}$$

If $\alpha(1) \neq \beta(1)$, say $\beta(1) = i \neq 1$, then by (c) $v_{1j}^*v_{1i} = 0$ for $j \neq i$ unless $j = 1$ $v_{kj}^*v_{1i} = 0$ for all i, j if $k \neq 1$ unless $j = 1$ by 6.6.2. (iii) and $e_n^{1/2}v_{1i} = 0$ by 6.6.2 (i), and thus we conclude that $u_\alpha^*u_\beta = 0$. Repeating this argument with $\alpha(2)$, $\beta(2)$ instead of $\alpha(1)$, $\beta(1)$, inserting the result in the equation above and continuing in this fashion we arrive at $(*)$ and $(**)$.

6.6.4. LEMMA. *Let* $\{e_n\}$, $\{v_{ni}\}$ *be a quasi-matrix system of rank* $\{m(n)\}$ *in a* C^*-*algebra* A, *and for each* n *let* p_n *be the spectral projection of* e_n *in* A'' *corresponding to the eigenvalue* 1. *With* u_α, $\alpha \in X_n$ *and* X_n^k *as in 6.6.3 define*

$$q_n = \sum_{\alpha \in X_n} u_\alpha p_n u_\alpha^*, \quad q_{nk} = \sum_{\alpha \in X_n^k} u_\alpha p_n u_\alpha^*.$$

Then the elements q_n, q_{nk}, $1 \leqslant k \leqslant n$ are projections in A'' *satisfying the relations* (*for* $2 \leqslant i \leqslant m(k)$):

(i) $v_{ki}q_n = v_{ki}q_{nk}$;

(ii) $q_n v_{ki} = v_{ki}q_{nk}$;

(iii) $v_{ki}^* v_{ki} q_n = q_{nk}$;

(iv) $e_k q_n = q_{nk}$.

Proof. By (*) and (**) in 6.6.3 we have

$$q_n^2 = \sum_{\alpha,\beta \in X_n} u_\beta p_n u_\beta^* u_\alpha p_n u_\alpha^* = \sum_{\alpha \in X_n} u_\alpha p_n e_n p_n u_\alpha^* = q_n,$$

and similarly $q_{nk}^2 = q_{nk}$.

(i) Since $v_{ki}v_{mj} = 0$ for $k \geqslant m$ by 6.6.2. (ii), we have $v_{ki}u_\alpha = 0$ unless $\alpha \in X_n^k$; whence $v_{ki}q_n = v_{ki}q_{nk}$.

(ii) Since $v_{mj}^* v_{ki} = 0$ unless $m = n$ and $j = i$ by 6.6.2. (iii) and (c), we have $u_\beta^* v_{ki} = 0$ unless $u_\beta = v_{ki}u_\alpha$ for some α in X_n^k. Therefore

$$q_n v_{ki} = \left(\sum_{\alpha \in X_n^k} v_{ki} u_\alpha p_n u_\alpha^* v_{ki}^* \right) v_{ki} = v_{ki} q_{nk},$$

since $v_{ki}^* v_{ki}$ is a unit for v_{nj}^* and for p_n if $n > k$ by (b) and (e).

(iii) By (i) we get $v_{ki}^* v_{ki} q_n = v_{ki}^* v_{ki} q_{nk}$. However, by (b) and (e) $v_{ki}^* v_{ki}$ is a unit for v_{mj} and for p_n if $m > k$, so that $v_{ki}^* v_{ki} q_n = q_{nk}$.

(iv) Follows from (iii) by multiplying on the left with e_k and noting that $v_{ki}^* v_{ki}$ is a unit for e_k by (b) and that $e_k q_{nk} = q_{nk}$ by (e).

6.6.5. PROPOSITION. *Let* $\{e_n\}$, $\{v_{ni}\}$ *be a quasi-matrix system of rank* $\{m(n)\}$ *in a* C^**-algebra* A *and take* q_n *as in* 6.6.4. *Then* $\{q_n\}$ *is a decreasing sequence of closed projections in* A'' *with a limit* $q \neq 0$. *Moreover,* q *commutes with the* C^**-algebra generated by* $\{e_n\} \cup \{v_{ni}\}$, *and the pair* $\{e_n q\}$, $\{v_{ni} q\}$ *is a matrix system in* A'' *of rank* $\{m(n)\}$.

Proof. By 6.6.4. (iv) we have

$$q_{n+1,n} = e_n q_{n+1} = p_n q_{n+1} \leqslant p_n.$$

Consequently

$$q_{n+1} = \sum_{\alpha \in X_n} u_\alpha q_{n+1,n} u_\alpha^* \leqslant \sum_{\alpha \in X_n} u_\alpha p_n u_\alpha^* = q_n.$$

The sequence $\{q_n\}$ is therefore decreasing and has a limit projection q in A''. Recall from 3.11.10 that a projection q in A'' is closed if there is an increasing net in A_+ converging strongly to $1 - q$, or, equivalently, if q is an upper semi-continuous affine function on the quasi-state space Q of A (see 3.11.5). Since each p_n is a closed projection, all $u_\alpha p_n u_\alpha^*$, $\alpha \in X_n$ are closed; and thus q_n, being the sum of these, is closed. By (e) we have $p_{n+1} \leqslant e_n \leqslant p_n$ for each n so that $\{p_n\}$ is represented on Q as a decreasing sequence of upper semi-continuous functions of norm 1. It follows by a standard compactness argument that the limit p of $\{p_n\}$ has norm 1. For fixed $k \leqslant n$ we have $e_k q_n = q_{nk} \geqslant p_n$ by 6.6.4. (iv), and it follows that $e_k q \geqslant p$, whence $\|e_k q\| = 1$. Thus the system $\{e_n q\}$, $\{v_{ni} q\}$ satisfies condition (a) in 6.6.1, in particular $q \neq 0$.

It is immediate from 6.6.4 that each q_n commutes with the C^*-algebra generated by $\{e_k\} \cup \{v_{ki}\}$, $k \leqslant n$. Therefore q commutes with the C^*-algebra generated by $\{e_k\} \cup \{v_{ki}\}$. From this we deduce that the system $\{e_n q\}$, $\{v_{ni} q\}$ satisfies (b), (c), (d) and (e), since the original system does so. Finally, condition (f) is contained in 6.6.4. (iii) and (iv), and (g) follows from 6.6.4 (iv) and (f) since

$$
\begin{aligned}
e_n q &= e_n q_{n+1} q = q_{n+1,n} q \\
&= \left(p_{n+1} + \sum_{i=2}^{m(n+1)} v_{n+1,i} p_{n+1} v_{n+1,i}^* \right) q \\
&= e_{n+1} q + \sum_{i=2}^{m(n+1)} v_{n+1,i} q v_{n+1,i}^* .
\end{aligned}
$$

6.6.6. *Notes and remarks.* Quasi-matrix systems were introduced by Glimm [97]. The simplified treatment given here is borrowed from Lance, see [155]. The main idea is to break Glimm's frightening argument from [97] into a mildly boring exercise in algebra (Section 6.6), followed by some ingenious constructions from spectral theory (6.7.1 and 6.7.2). Glimm (and Lance) treats only the Fermion algebra. The generalization to arbitrary Glimm algebras from [206] is mainly a question of having the right notation.

6.7. Antiliminary algebras

6.7.1. LEMMA. *If A is an antiliminary C^*-algebra and if $e', x \in A_+$ with $\|e'\| = \|x\| = 1$ and $e'x = x$, then for each m in $\mathbf{N}$ there exist $e, v_2, v_3, \ldots v_m$ and x' in A satisfying the conditions:*

(a′) $e \geqslant 0$, $\|e\| = \|v_i\| = 1$ *for* $2 \leqslant i \leqslant m$;

(b′) $v_i^* v_i e = e$ *for* $2 \leqslant i \leqslant m$;

(c′) $v_i^* v_j = 0$ *for* $i \neq j$;

(d′) $v_i v_j = 0$ *for all* i, j;

(e′) $e'e = e$, $e'v_i = v_i$, $e'v_i^* = v_i^*$ for $2 \leqslant i \leqslant m$;

(∗) $x' \geqslant 0$, $\|x'\| = 1$, $ex' = x'$.

Proof. Choose positive continuous functions f, g, h on $[0, 1]$ such that

$$f(0) = g(0) = h(0) = 0, \qquad f(1) = g(1) = h(1) = 1,$$

$$\|f\| = \|g\| = 1, \qquad g(t)f(t) = f(t) \quad \text{and} \quad th(t)^2 = g(t)$$

for all t in $[0, 1]$.

Since $f(x) \neq 0$ and A is antiliminary there is by 6.2.8 an irreducible representation (π, H) of A such that $\pi(f(x))$ is not a compact operator. As $g(x)$ is a unit for $f(x)$, the eigenspace of $\pi(g(x))$ corresponding to the eigenvalue 1 is infinite-dimensional and contains therefore m orthogonal unit vectors $\xi_1, \ldots, \xi_m$. By Kadison's transitivity theorem (2.7.5) there is an element a in A_+ such that $\pi(a)\xi_i = i\xi_i$ for $1 \leqslant i \leqslant m$ and elements $u_2, \ldots, u_m$ in the unit ball of A such that $\pi(u_i)\xi_1 = \xi_i$ for $2 \leqslant i \leqslant m$.

Let $f_1, \ldots, f_m$ be positive continuous functions on $\mathbf{R}$ with norm 1 such that $f_i(i) = 1$ and $f_i f_j = 0$ for $i \neq j$, and put

$$x_i = f_i(g(x)ag(x))u_i f_1(g(x)ag(x))$$

for $2 \leqslant i \leqslant m$. It is evident that $x_i x_j = 0$ for all i, j and $x_i^* x_j = 0$ for $i \neq j$. But $\pi(x_i)\xi_1 = \xi_i$ so that $\|x_i\| = 1$ for all i.

Define $y_2 = x_2^* x_2$; $v_2 = x_2 h(y_2)$ and then inductively

$$y_{i+1} = f(y_i)x_{i+1}^* x_{i+1} f(y_i); \qquad v_{i+1} = x_{i+1} f(y_i)h(y_{i+1}),$$

for $2 \leqslant i \leqslant m$. Finally put

$$e = g(f(y_m)), \qquad x' = f(f(y_m)).$$

Since $\|y_i\| \leqslant 1$, the condition $\|\pi(y_i)\xi_1\| = 1$ is equivalent with $\pi(y_i)\xi_1 = \xi_1$, and is easily verified by induction. It follows that $\pi(v_i)\xi_1 = \xi_i$ and $\pi(e)\xi_1 = \pi(x')\xi_1 = \xi_1$ and thus the elements $v_2, \ldots, v_m, e, x'$ satisfy conditions (a') and $(*)$. Since e' is a unit for x, it is a unit for the hereditary C^*-subalgebra generated by x, which contains all the elements $v_2, \ldots, v_m, e, x'$; so (e') is also satisfied. By induction we see that $y_i x_j = 0$ for all i, j, and consequently

$$v_i v_j = x_i f(y_{i-1})h(y_i)x_j f(y_{j-1})h(y_j) = 0.$$

Clearly $v_i^* v_j = 0$ since $x_i^* x_j = 0$ so that (d') and (c') are also verified.

From the way the functions f, g, h were chosen we have

$$v_i^* v_i = h(y_i)f(y_{i-1})x_i^* x_i f(y_{i-1})h(y_i) = h(y_i)^2 y_i = g(y_i).$$

Since $gf = f$ this implies that

$$\begin{aligned}(v_i^* v_i)(v_{i+1}^* v_{i+1}) &= g(y_i)g(y_{i+1}) \\ &= g(y_i)g(f(y_i)x_{i+1}^* x_{i+1} f(y_i)) \\ &= g(f(y_i)x_{i+1}^* x_{i+1} f(y_i)) = g(y_{i+1}) = v_{i+1}^* v_{i+1}\end{aligned}$$

for $2 \leqslant i \leqslant m$. Similarly

$$v_m^* v_m e = g(y_m) g(f(y_m)) = g(f(y_m)) = e.$$

Since each $v_i^* v_i$ is a unit for the next and $v_m^* v_m$ is a unit for e we have established (b′) and the proof is complete.

6.7.2. LEMMA. *If A is an antiliminary C^*-algebra and $\{a_n\}$ is a sequence in A_{sa}, then for each sequence $\{m(n)\}$ in $\boldsymbol{N}\backslash\{1\}$ there is a quasi-matrix system $\{e_n\}$, $\{v_{ni}\}$ of rank $\{m(n)\}$ in A and a sequence $\{\tilde{a}_n\}$ in A_{sa} such that each $\tilde{a}_n$ belongs to the $*$-algebra generated by the e_k, v_{ki} with $k \leqslant n$ and*

$$(**) \qquad \left\| \left(\sum_{\alpha \in X_n} u_\alpha u_\alpha^* \right) (a_n - \tilde{a}_n) \left(\sum_{\alpha \in X_n} u_\alpha u_\alpha^* \right) \right\| \leqslant 1/n$$

for each n, with u_α, $\alpha \in X_n$ as in 6.6.3.

Proof. Suppose that we have already constructed $\{e_k\}$, $\{v_{ki}\}$ and $\{\tilde{a}_k\}$ satisfying conditions (a)–(e) in 6.6.1 and $(**)$ for all k in $\boldsymbol{N}$ with $k < n$, in such a way that each e_k satisfies $e_k x_k = x_k$ for some x_k in A_+ with $\|x_k\| = 1$.

If $n = 1$ take arbitrary elements e', x in A_+ satisfying $\|e'\| = \|x\| = 1$, $e'x = x$. If $n > 1$ take e', x to be e_{n-1}, x_{n-1}. By 6.7.1 there are elements $v_2, \ldots, v_{m(n)}, e, x'$ such that if we define $v_{ni} = v_i$, $e_n = e$ and $x_n = x'$ we have carried the induction one step further, except for condition $(**)$.

Since $\|x_n\| = 1$ there is a pure state ϕ of A such that $\phi(x_n) = 1$ (cf. 4.3.10). Regarding ϕ as a normal state on A'' its support p is closed minimal projection in A''. There is therefore a decreasing sequence $\{y_k\}$ in the unit ball of A_+ such that $y_k \searrow p$. Since $p \leqslant x_n$ we have $x_n y_k x_n \searrow p$. Thus with $y = \Sigma\, 2^{-k} x_n y_k x_n$ we see that p is the spectral projection of y corresponding to the eigenvalue 1. Using spectral theory there is therefore a decreasing sequence $\{z_k\}$ in the unit ball of A_+ such that $z_k \searrow p$ and $z_{k+1} z_k = z_{k+1}$ for all k. For later use note that $e_n z_k = z_k$ for all k, since the z_k's are functions of y and e_n is a unit for x_n.

We claim that the sequence with elements

$$(***) \qquad \sum_{\alpha, \beta \in X_n} u_\beta z_k (u_\beta^* a_n u_\alpha - \phi(u_\beta^* a_n u_\alpha)) z_k u_\alpha^*$$

is norm convergent to zero. It suffices to verify that for each α and β we have $\|z_k(b - \phi(b)) z_k\| \to 0$ where $b = u_\beta^* a_n u_\alpha$. We know that $\{z_k(b - \phi(b)) z_k\}$ is σ-weakly convergent to $p(b - \phi(b))p$, which is equal to zero, since p is a minimal projection supporting ϕ (so that $pbp = \phi(b)p$). But as the set $\{z_k(b - \phi(b)) z_k\}$ contains zero as a weak limit point, its convex hull contains zero as a limit point in norm. Thus for each $\varepsilon > 0$ there is a convex combination $\Sigma\, t_i = 1$ with

$$\left\| \sum t_i z_{k_i} (b - \phi(b)) z_{k_i} \right\| < \varepsilon.$$

It follows that whenever k is greater than all k_i we have

$$\|z_k(b - \phi(b))z_k\| \leqslant \|z_k\| \left\| \sum t_i z_{k_i}(b - \phi(b))z_{k_i} \right\| \|z_k\| < \varepsilon,$$

which establishes the claim.

Choose k such that the element in (***) has norm less than $1/n$. Now replace e_n and x_n by $\tilde{e}_n = z_k$ and $\tilde{x}_n = z_{k+1}$. Then $\tilde{e}_n\tilde{x}_n = \tilde{x}_n$ and $\|\tilde{e}_n\| = \|\tilde{x}_n\| = 1$, and since $e_n\tilde{e}_n = \tilde{e}_n$ the new system $\{e_1, e_2, \ldots, \tilde{e}_n\}$, $\{v_{ni} \mid 2 \leqslant i \leqslant m(n)\}$ will again satisfy the conditions (a)–(e) in 6.6.1. Denote by $\tilde{u}_\alpha$, $\alpha \in X_n$ the elements arising from the new system using 6.6.3, and note the simple relation $\tilde{u}_\alpha = u_\alpha \tilde{e}_n^{1/2}$ between the old and the new elements. Define

$$\tilde{a}_n = \sum_{\alpha, \beta \in X_n} \phi(u_\beta^* a_n u_\alpha)\tilde{u}_\beta \tilde{u}_\alpha^*.$$

Then $\tilde{a}_n \in A_{\mathrm{sa}}$ and belongs to the *-algebra generated by the new system. Moreover, from our choice of $\tilde{e}_n$ we have

$$\left\| \left(\sum \tilde{u}_\beta \tilde{u}_\beta^* \right) (a_n - \tilde{a}_n) \left(\sum \tilde{u}_\alpha \tilde{u}_\alpha^* \right) \right\| = \left\| \sum u_\beta \tilde{e}_n (u_\beta^* a_n u_\alpha - \phi(u_\beta^* a_n u_\alpha))\tilde{e}_n u_\alpha^* \right\| < 1/n.$$

The proof can now be completed by induction.

6.7.3. THEOREM. *For each separable C^*-algebra A which is not of type I and every Glimm algebra A_∞ there is a C^*-subalgebra B of A and a closed projection q in A'', commuting with B, such that $qAq = qB$ and qB is isomorphic to A_∞.*

Proof. It suffices to prove the theorem for some non-zero quotient of A, so by factoring out the largest ideal of A which is of type I (see 6.2.7) we may assume that A is antiliminary.

Suppose that A_∞ is of rank $\{m(n)\}$ (cf. 6.4.2) and let $\{a_n\}$ be dense in A_{sa}. There is a quasi-matrix system $\{e_n\}$, $\{v_{ni}\}$ of rank $\{m(n)\}$ in A satisfying 6.7.2. Take B as the C^*-subalgebra of A generated by $\{e_n\} \cup \{v_{ni}\}$. By 6.6.5 there is a closed projection q in A'' commuting with B such that $\{e_n q\}$, $\{v_{ni} q\}$ is a matrix system of rank $\{m(n)\}$ in A''. As we saw in 6.6.1, the C^*-algebra generated by $\{e_n q\} \cup \{v_{ni} q\}$ is isomorphic to A_∞, and clearly equals qB. Since

$$q \leqslant q_n \leqslant \sum_{\alpha \in X_n} u_\alpha u_\alpha^*,$$

it follows from (**) in 6.7.2 that $\|q(a_n - \tilde{a}_n)q\| \leqslant 1/n$ for each n. As $\{a_n\}$ is dense in A_{sa}, qB is norm closed, and the map $a \to qaq$ is continuous and self-adjoint we conclude that $qAq = qB$.

6.7.4. COROLLARY. *For each C^*-algebra A which is not of type I and every Glimm algebra A_∞ there is a C^*-subalgebra B of A and a surjective morphism of B onto A_∞.*

Proof. The separability of A was only used to obtain (**) in 6.7.2 for a dense set.

6.7.5. Notes and remarks. Theorem 6.7.3, with $A_\infty = \boldsymbol{F}$, is from Glimm [97]. Its importance was recognized by Sakai, who pointed out that 6.7.4 holds in the absence of separability (see 4.6.8 of [231]). The more precise formulation given here, involving an arbitrary Glimm algebra and using the closed projection q, is due to the author, see [206].

6.8. Glimm's theorem

6.8.1. PROPOSITION. *Let A be a separable C^*-algebra which is not of type I, and let A_∞ be a Glimm algebra. If $\mathscr{B}$ and $\mathscr{B}_\infty$ denote the enveloping Borel *-algebras for A and A_∞, respectively, then there is a closed projection q in $\mathscr{B}$ such that $q\mathscr{B}q = \mathscr{B}_\infty$.*

Proof. Let B and q be as in 6.7.3. Since $B \subset A$ we have $\mathscr{B}(B) \subset \mathscr{B}$ (cf. 3.7.9). The projection q is closed and central relative to B and by 3.11.10 there is therefore a unique closed ideal I of B such that $I = (1-q)\mathscr{B}(B) \cap B$, and $B/I = qB = A_\infty$. It follows from 4.6.2 that $\mathscr{B}(B) = \mathscr{B}(I) \oplus \mathscr{B}(B/I)$, whence

$$q\mathscr{B}q = q\mathscr{B}(B) = \mathscr{B}_\infty.$$

6.8.2. PROPOSITION. *For each separable C^*-algebra A which is not of type I and every Glimm algebra A_∞ there is a D-Borel isomorphism of $\hat{A}_\infty$ onto a D-Borel subset of $\hat{A}$.*

Proof. Let $\mathscr{C}$ and $\mathscr{C}_\infty$ be the centres of $\mathscr{B}$ and $\mathscr{B}_\infty$, respectively. Then by 6.8.1 $q\mathscr{C} = \mathscr{C}_\infty$. With $c(q)$ as the central cover of q we have $c(q) \in \mathscr{C}$ by 4.5.8, and the map $z \to qz$ is an isomorphism of $c(q)\mathscr{C}$ onto $q\mathscr{C}$. It follows that $c(q)\mathscr{C}$ is isomorphic to $\mathscr{C}_\infty$. Since points in $\hat{A}$ and $\hat{A}_\infty$ correspond to minimal projections in $\mathscr{C}$ and $\mathscr{C}_\infty$ we see that by the very definition of the D-Borel structure (4.7.1), $\hat{A}_\infty$ is isomorphic to a D-Borel subset of $\hat{A}$.

6.8.3. COROLLARY. *Any two Glimm algebras have D-Borel isomorphic spectra.*

Proof. Apply 4.6.4 to the centres of the enveloping Borel *-algebras of the two Glimm algebras.

6.8.4. PROPOSITION. *Consider the set $G = \{0, 1\}^\infty$ as a compact group and let G_0 denote the subgroup of elements whose coordinates are all zero except for a finite number. For each separable C^*-algebra A which is not of type I there is an injective Borel map of G/G_0 in its quotient Borel structure onto an M-Borel subset of $\hat{A}$.*

Proof. Let $\boldsymbol{F}$ be the Fermion algebra and $\mathscr{B}_\infty$ its enveloping Borel *-algebra. By 6.8.1 there is a closed projection q in $\mathscr{B}$ such that $\mathscr{B}_\infty = q\mathscr{B}q$. Since q is closed, the set

$$Q_q = \{\phi \in Q \mid \phi(1-q) = 0\}$$

is a weak* closed face of the quasi-state space Q of A. We define a bijection $\phi \to \bar{\phi}$ of the quasi-state space Q_∞ of $\boldsymbol{F}$ onto Q_q by setting $\bar{\phi}(x) = \phi(qxq)$ for each ϕ in Q_∞ and x in $\mathscr{B}$. Clearly the map $\phi \to \bar{\phi}$ is an affine homeomorphism; and since Q_q is a face of Q the map $\phi \to \bar{\phi}$ takes pure states to pure states.

From 6.5.5 we see that there is an injection of G into $P(\boldsymbol{F})$ and it is easy to check that it is weak* continuous and thus a homeomorphism. Combining this with the results above we have a homeomorphism $\Phi: G \to E$, where E is a compact set of $P(A)$. Let T be the image of E in $\hat{A}$ under the canonical map $\phi \to \hat{\phi}$ from $P(A)$ onto $\hat{A}$. We claim that

$$\hat{\Phi}(s) = \hat{\Phi}(t) \Leftrightarrow s - t \in G_0.$$

In view of 6.5.6 it suffices to show that the representations associated with two pure states ϕ and ψ in Q_q are equivalent if and only if their restrictions to $q\mathscr{B}q$ are equivalent. But this is obvious since irreducible representations are either equivalent or disjoint by 3.13.3. Consequently we have a commutative diagram:

$$\begin{array}{ccc} G & \xrightarrow{\Phi} & E \subset P(A) \\ \downarrow & & \downarrow \quad \downarrow \\ G/G_0 & \xrightarrow{\Psi} & T \subset \hat{A} \end{array}$$

where Ψ is a bijection. Since the Borel structures on G/G_0 and T are both quotient Borel structures and Φ is a Borel map, Ψ is a Borel map.

Finally, to show that T is an M-Borel subset of $\hat{A}$ we must prove that the saturation $\tilde{E}$ of E in $P(A)$ with respect to unitary equivalence is a Borel set. Let $\{u_n\}$ be a dense sequence in the unitary group of $\tilde{A}$. Then

$$\tilde{E} = \bigcup_n \{\phi \in P(A) \mid \|\phi - \psi\| \leq 1, \psi(u_n^* \cdot u_n) \in E\}.$$

Since E is compact and the norm function is weak* lower semi-continuous, $\tilde{E}$ is the union of a sequence of relatively compact sets and consequently a Borel set.

6.8.5. COROLLARY. *Each separable C*-algebra A which is not of type I has an uncountable family of pairwise disjoint representations, all with the same kernel.*

Proof. The representations associated with pure states in the set E from the proof of 6.8.4 all have the same kernel, namely $A \cap (1 - c(q))\mathscr{B}$.

6.8.6. PROPOSITION. *Each separable C*-algebra A which is not of type I has factor representations both of type II and III.*

Proof. By 6.8.1 we have $\mathscr{B}_\infty = q\mathscr{B}q$, where $\mathscr{B}_\infty$ is the enveloping Borel *-algebra for the Fermion algebra and q is a closed projection in $\mathscr{B}$ $(= \mathscr{B}(A))$. Let τ be the unique σ-normal tracial state on $\mathscr{B}_\infty$ (cf. 6.4.3). There is then by 5.2.8 an extension $\tilde{\tau}$ of τ to a σ-trace on $\mathscr{B}$ which vanishes on $(1 - c(q))\mathscr{B}$. The representation (π, H) associated with $\tilde{\tau}$ is clearly of type II (but not necessarily of type II_1) and since τ is factorial on $\mathscr{B}_\infty$, $\pi(q\mathscr{B}q)$ is a factor. However, $\pi(c(q)) = 1$ so that the map $z \to \pi(q)z$ is an isomorphism of the centre of $\pi(\mathscr{B})$ onto the centre of $\pi(q\mathscr{B}q)$; and thus (π, H) is a factor representation.

To show that A has factor representations of type III take by 6.5.14 a factorial state ϕ of the Fermion algebra whose associated representation is of type III. Define $\bar{\phi}$ on $\mathscr{B}$ by $\phi(x) = \phi(qxq)$ and let (π, H) be the representation of A associated with $\bar{\phi}$. As in the first part of the proof we show that $\pi(\mathscr{B})$ is a factor. If there was a non-zero finite projection in $\pi(\mathscr{B})$ then since $\pi(c(q)) = 1$ there would also be a finite projection p with $\pi(q)p\pi(q) \neq 0$. But this contradicts the assumption that $\pi(q\mathscr{B}q) = \pi(\mathscr{B}_\infty)$ is of type III; whence $\pi(\mathscr{B})$ is of type III.

6.8.7. THEOREM. *Let A be a separable C^*-algebra. The following twelve conditions are equivalent:*

(i) *A is a C^*-algebra of type I;*

(ii) *$\mathscr{B}(A)$ is a Borel *-algebra of type I;*

(iii) *A'' is a von Neumann algebra of type I;*

(iv) *A has a composition series in which each quotient has continuous trace;*

(v) *$C(H) \subset \pi(A)$ for each irreducible representation (π, H) of A;*

(vi) *$\check{A} = \hat{A}$;*

(vii) *The T-Borel structure on $\hat{A}$ is standard;*

(viii) *The D-Borel structure on $\hat{A}$ is countably separated;*

(ix) *The M-Borel structure on $\hat{A}$ is countably separated;*

(x) *$\widehat{A} = \hat{A}$;*

(xi) *A has no factor representations of type II;*

(xii) *A has no factor representations of type III.*

Proof. We use the following scheme of implications.

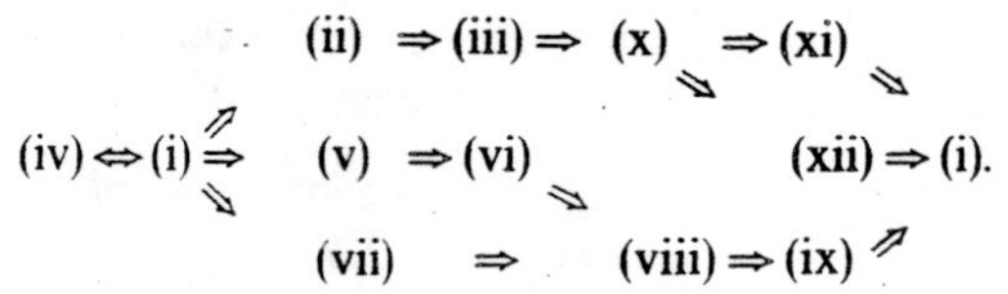

$$\begin{array}{ccccccccc}
 & & \text{(ii)} \Rightarrow \text{(iii)} \Rightarrow & \text{(x)} & \Rightarrow \text{(xi)} & & \\
 & \nearrow & & & \searrow & \searrow & \\
\text{(iv)} \Leftrightarrow \text{(i)} \Rightarrow & & \text{(v)} \Rightarrow \text{(vi)} & & & \text{(xii)} \Rightarrow \text{(i)}. & \\
 & \searrow & & \searrow & & & \\
 & & \text{(vii)} \Rightarrow & & \text{(viii)} \Rightarrow \text{(ix)} & \nearrow &
\end{array}$$

(i) ⇔ (iv) by 6.2.11, 6.1.11 and 6.2.6.
(i) ⇒ (ii) ⇒ (iii) ⇒ (x) by 6.3.4, 6.3.8, and 6.3.9, respectively
(i) ⇒ (v) ⇒ (vi) by 6.1.5.
(i) ⇒ (vii) by 6.3.2.
(vii) ⇒ (viii) ⇒ (ix) are trivial, since the Borel structures become progressively stronger by 4.7.3.

(vi) ⇒ (viii) is also easy, because (vi) implies that the T-Borel structure separates points in $\hat{A}$. Since it is always countably generated by 4.3.4 the result is immediate from 4.7.3.

(ix) ⇒ (i) by 6.8.4 because G/G_0 is not countably separated.
(x) ⇒ (xi) and (x) ⇒ (xii) are trivial.
(xi) ⇒ (i) and (xii) ⇒ (i) follow from 6.8.6.

6.8.8. Notes and remarks. The results 6.8.1–6.8.3 are taken from [206]. They lead naturally to the conjecture that any two Glimm algebras are Borel isomorphic in the sense of 4.6.8. A weaker result, that the enveloping von Neumann algebras of Glimm algebras are isomorphic, is found in [206], and is even true for a larger class of C^*-algebras, see [82]. Another result of Glimm's construction is the fact that any two Glimm algebras have isomorphic factor spectra in the Mackey–Borel structure, see [206] or [83].

Theorem 6.8.7 is the work of many hands, but the hard parts are due to Glimm. The equivalence (i) ⇔ (iv) was proved by Fell using earlier results of Kaplansky. The implication (i) ⇒ (iii) was proved by Kaplansky. The converse was established by Glimm in the stronger form that both (xi) and (xii) imply (i). Glimm also proved Mackey's conjecture from [164], that (ix) ⇔ (x). Quite surprisingly Effros [75] was able to prove Mackey's conjecture using transformation group theory and avoiding Glimm's quasi-matrix algebras.

Separability plays an important rôle in many of the arguments that go into the proof of 6.8.7. However, Sakai [228, 229] has succeeded in proving the equivalence of conditions (i), (iii), (iv), (v), (x) and (xii). The only outstanding problems now are whether (vi) ⇒ (i) and (xi) ⇒ (i). Both implications are probably true. The first is, however, unsolved even in its simplest form: if a C^*-algebra A has only one (class of) irreducible representation, is $A = \boldsymbol{C}(H)$? (Naimark). The other problem looks more hopeful now with the recent result of Anderson and Bunce [9] that the Calkin algebra, $\boldsymbol{B}(H)/\boldsymbol{C}(H)$, has a type II_∞ representation.

Chapter 7

Automorphism Groups

Until now we have studied C^*-algebras alone. Henceforth we shall assume that together with a C^*-algebra A we have a (locally compact) group G of automorphisms of A; i.e. we shall consider a triple (A, G, α) where $\alpha: G \to \mathrm{Aut}(A)$ is a (suitably continuous) homomorphism. Such a triple we call a dynamical system in analogy with the commutative case.

We briefly examine the connections between group representations and operator algebras in the first three sections, in particular we show that amenable groups are characterized by the fact that the regular representation is faithful on the group C^*-algebra. We then consider the various dynamical systems built on C^*-algebras, Borel $*$-algebras and von Neumann algebras, together with their interrelations.

The central part of the chapter deals with crossed products; certain C^*-algebras obtained from dynamical systems. If a system (A, G, α) is given, one may visualize the crossed product $G \underset{\alpha}{\times} A$ as a skew tensor product between $C^*(G)$ and A. This construction reduces to a certain extent the study of dynamical systems to ordinary C^*-algebra theory. In the converse direction it gives a whole new class of examples of C^*-algebras and von Neumann algebras. For this reason it becomes important to decide whether a given C^*-algebra is a crossed product, and we devote some effort to this problem.

The chapter concludes with an exposition of some dynamical systems of particular interest to the applications in mathematical physics: asymptotically abelian systems.

7.1. Locally compact groups

For convenience we present in this section the basic results from harmonic analysis that we shall need. At the same time the section serves to fix some notation.

Throughout the chapter G will denote a locally compact group; i.e. G is a group and a locally compact Hausdorff space, and the map $(s,t) \to s^{-1}t$ is continuous from $G \times G$ to G.

Central for the theory is the existence of a non-zero left invariant Radon measure μ_G on G; i.e. $\mu_G(sE) = \mu_G(E)$ for each Borel set E in G and each s in G. We call μ_G a *(left) Haar measure* on G. Since μ_G is unique up to scalar multiples we talk about *the* Haar measure with deliberate abuse of language. If μ_G is finite (which happens if and only if G is compact) we assume that $\mu_G(G) = 1$; and if G is discrete and infinite we assume that $\mu_G(\{e\}) = 1$ (e denoting the identity of G).

There is of course also a right Haar measure on G. This is linked with the left Haar measure by the existence of a continuous homomorphism Δ (the *modular function*) of G into the positive real numbers, such that $\mu_G(Es) = \Delta(s)\mu_G(E)$. Thus with the customary notation $\mathrm{d}s$ instead of $\mathrm{d}\mu_G(s)$ we have

$$\mathrm{d}(ts) = \mathrm{d}s, \qquad \mathrm{d}(st) = \Delta(t)\,\mathrm{d}s, \qquad \mathrm{d}(s^{-1}) = \Delta(s)^{-1}\,\mathrm{d}s.$$

7.1.2. Let $M(G)$ denote the Banach space of bounded complex Radon measures (signed measures) on G, identified with $C_0(G)^*$. For μ and ν in $M(G)$ define *convolution* and *involution* by

$$\int f(s)\,\mathrm{d}(\mu \times \nu)(s) = \int\int f(ts)\,\mathrm{d}\nu(s)\,\mathrm{d}\mu(t), \qquad f \in C_0(G)$$

$$\int f(s)\,\mathrm{d}\mu^*(s) = \overline{\int \bar{f}(s^{-1})\,\mathrm{d}\mu(s)}, \qquad f \in C_0(G).$$

(We use $\times$ for convolution instead of the usual $*$ to avoid confusion with the adjoint operation. The product measure of μ and ν on $G \times G$ will be denoted $\mu \otimes \nu$). With convolution as product $M(G)$ becomes a Banach algebra with an isometric involution; but not in general a C^*-algebra.

We denote by $L^1(G)$ the closed $*$-ideal in $M(G)$ of complex measures that are absolutely continuous with respect to Haar measure, and identify $L^1(G)$ with the (classes of) μ_G-integrable functions on G. Thus for f and g in $L^1(G)$

$$f \times g(t) = \int f(s)g(s^{-1}t)\,\mathrm{d}s = \int f(ts)g(s^{-1})\,\mathrm{d}s,$$

$$f^*(t) = \Delta(t)^{-1}\bar{f}(t^{-1}).$$

Occasionally we shall also need the transformation $f \to \tilde{f}$, where $\tilde{f}(t) = \bar{f}(t^{-1})$. This transformation is also an involution, but if G is not unimodular we may have $\tilde{f} \notin L^1(G)$ even though $f \in L^1(G)$.

For each s in G let δ_s be the point measure at s. Then for f in $L^1(G)$

$$(\delta_s \times f)(t) = f(s^{-1}t), \qquad (f \times \delta_s)(t) = \Delta(s)^{-1}f(ts^{-1}).$$

7.1.3. A *unitary representation* (u, H) of G is a homomorphism $t \to u_t$ of G into the unitary group $U(H)$ of $B(H)$ which is continuous in the weak topology on $B(H)$. Elementary calculations show that the weak, σ-weak and strong topologies coincide on $U(H)$, so that we may use either in the definition of a unitary representation. Later we shall also encounter unitary representations that are *uniformly continuous*, i.e. continuous in the norm topology on $\boldsymbol{B}(H)$.

With a view on the algebras $L^1(G)$ and $M(G)$ we define a *representation* (π, H) of a Banach algebra with involution to be an involution-preserving continuous homomorphism π into $\boldsymbol{B}(H)$.

7.1.4. PROPOSITION. *There are bijective correspondences between the sets of unitary representations of G, representations of $M(G)$ whose restrictions to $L^1(G)$ are non-degenerate, and non-degenerate representations of $L^1(G)$.*

Proof. If (u, H) is a unitary representation of G define for each μ in $M(G)$ and ξ, η in H

$$(\pi(\mu)\xi \mid \eta) = \int (u_s\xi \mid \eta)\, d\mu(s).$$

Then (π, H) is a representation of $M(G)$, and using an approximate unit for $L^1(G)$ we see that the restriction to $L^1(G)$ is non-degenerate.

Conversely, let (π, H) be a non-degenerate representation of $L^1(G)$ and choose an approximate unit $\{f_\lambda\}$ for $L^1(G)$. Since vectors of the form $\pi(f)\xi$, $f \in L^1(G)$, $\xi \in H$, are dense in H it follows that $\{\pi(f_\lambda)\}$ converges strongly to 1. We extend (π, H) to a representation $(\tilde{\pi}, H)$ of $M(G)$ by defining

$$\tilde{\pi}(\mu)(\pi(f)\xi) = \pi(\mu \times f)\xi$$

for f in $L^1(G)$ and ξ in H. Equivalently,

$$\tilde{\pi}(\mu)\xi = \operatorname{Lim}(\mu \times f_\lambda)\xi \qquad \text{(strongly)},$$

which shows that the extension is unique. The restriction of $(\tilde{\pi}, H)$ to G gives a unitary representation of G whose extension to $L^1(G)$ is precisely (π, H).

7.1.5. The *universal representation* (π_u, H_u) of $L^1(G)$ is the direct sum of all non-degenerate representations of $L^1(G)$, and the *group C*-algebra* $C^*(G)$ is the norm closure of $\pi_u(L^1(G))$ in $\boldsymbol{B}(H_u)$. The extension of (π_u, H_u) to $C^*(G)$ is thus by definition the universal representation of $C^*(G)$. Applying 7.1.4 we get an embedding of $M(G)$ into the multiplier algebra $M(C^*(G))$ of $C^*(G)$.

7.1.6. PROPOSITION. *If G is abelian and $\hat{G}$ is its dual group then $C^*(G) = C_0(\hat{G})$.*

Proof. The irreducible representations of $L^1(G)$ (hence also those of $C^*(G)$) are all one-dimensional and correspond to the points in $\hat{G}$. The topology on $\hat{G}$ is the weak* topology arising from $L^1(G)$ or $C^*(G)$, so that $\hat{G}$ is homeomorphic with the spectrum of $C^*(G)$. It follows that $C^*(G) = C_0(\hat{G})$.

7.1.7. THEOREM (Stone). *For each unitary representation (u, H) of $\boldsymbol{R}$ there is a (not necessarily bounded) self-adjoint operator h on H such that $u_t = \exp(ith)$ for each t in $\boldsymbol{R}$.*

Proof. Let $(\tilde{\pi}, H)$ be the representation of $L^1(\boldsymbol{R})$ associated with (u, H) (cf. 7.1.4) and extend it to a representation $(\tilde{\pi}, H)$ of $M(C^*(\boldsymbol{R})) = C^b(\hat{\boldsymbol{R}})$. Let h be the image under $\tilde{\pi}$ (suitably extended) of the function $s \to s$ on $\hat{\boldsymbol{R}}$. For each t in $\boldsymbol{R}$ the image in $C^b(\hat{\boldsymbol{R}})$ is the function $s \to e^{its}$, whence

$$u_t = \tilde{\pi}(e^{its}) = \exp(ith).$$

7.1.8. Returning to the general case consider the (universal) unitary representation $t \to \delta_t$ of G into $M(C^*(G))$ obtained from 7.1.5. For each functional ϕ on $C^*(G)$ we define a function Φ on G by $\Phi(t) = \phi(\delta_t)$. Since ϕ is weakly continuous on $C^*(G)''$ the function Φ is continuous and bounded by $\|\phi\|$. If $f \in L^1(G)$ then identifying $L^1(G)$ with its image in $C^*(G)$ we have

$$\phi(f) = \int \phi(\delta_t) f(t)\,dt = \int \Phi(t) f(t)\,dt$$

which shows that the map $\phi \to \Phi$ is injective. We denote by $B(G)$ the image of $C^*(G)^*$ in $C^b(G)$ under this map and let $B_+(G)$ be the generating cone obtained from the positive functionals on $C^*(G)$. The elements in $B_+(G)$ are called *positive definite functions*. The real vector space spanned by $B_+(G)$ is denoted by $B_{sa}(G)$ and corresponds to the set of self-adjoint functionals. An elementary computation shows that if Φ is the image of ϕ then $\tilde{\Phi}$ is the image of ϕ^*. Thus $\Phi \in B_{sa}(G)$ if and only if $\Phi \in B(G)$ and $\tilde{\Phi} = \Phi$. From 7.1.6 we see that if G is abelian the positive definite functions are precisely the Fourier transforms of bounded measures on $\hat{G}$, so that $B(G) = (M(\hat{G}))^{\hat{}}$.

7.1.9. PROPOSITION. *For each Φ in $C^b(G)$ the following conditions are equivalent*:

(i) *Φ is positive definite;*

(ii) $\int \Phi(s)\,d(\mu^* \times \mu)(s) \geq 0$ *for every μ in $M(G)$;*

(iii) *For each finite set $\{s_i\}$ in G the matrix (a_{ij}), with $a_{ij} = \Phi(s_i^{-1}s_j)$, is positive (definite);*

(iv) $\int \Phi(s)(f^* \times f)(s)\,ds \geq 0$ *for every f in $K(G)$.*

Proof. (i) $\Rightarrow$ (ii). Obvious.

(ii) $\Rightarrow$ (iii). Given $\{s_i\}$ take any set $\{\gamma_i\}$ of complex numbers and form the signed measure $\mu = \Sigma \gamma_i \delta_{s_i}$. The condition (ii) applied to μ gives

$$(*) \qquad \sum_{ij} \Phi(s_i^{-1}s_j)\bar{\gamma}_i\gamma_j \geq 0,$$

which implies that the matrix $(\Phi(s_i^{-1}s_j))$ is positive.

(iii) $\Rightarrow$ (iv). The function $(s, t) \to \Phi(s^{-1}t)\bar{f}(s)f(t)$ has a compact support on $G \times G$ contained in a set $C \times C$, where C is compact. Since $\mu_G(C) < \infty$ we see from Krein–Milman's theorem that $\mu_G | C$ is the weak* limit of a net $\{\mu_\lambda\}$ of measures on C with finite supports. It follows that $\mu_G \otimes \mu_G | C \times C$ is the weak* limit of the net $\{\mu_\lambda \otimes \mu_\lambda\}$. Condition (iii) in the form (*) implies that

$$\iint \Phi(s^{-1}t)\bar{f}(s)f(t)\,\mathrm{d}\mu_\lambda(s)\,\mathrm{d}\mu_\lambda(t) = \sum_{i,j} \Phi(s_i^{-1}t_j)\bar{f}(s_i)\bar{\gamma}_i f(t_j)\gamma_j \geqslant 0$$

where $\mu_\lambda = \Sigma\,\gamma_i\delta_{t_i}$ and $\gamma_i \geqslant 0$. In the weak* limit we therefore obtain

$$0 \leqslant \iint \Phi(s^{-1}t)\bar{f}(s)f(t)\,\mathrm{d}s\,\mathrm{d}t = \int \Phi(t)(f^* \times f)(t)\,\mathrm{d}t.$$

(iv) $\Rightarrow$ (i). Since $K(G)$ is dense in $L^1(G)$ the bounded functional ϕ obtained from Φ by integration will satisfy $\phi(f^* \times f) \geqslant 0$ for every f in $L^1(G)$. Then the Cauchy–Schwarz inequality holds on $L^1(G)$ with the same proof as in 3.1.3; and the Gelfand–Naimark–Segal construction can be carried out exactly as in 3.3.3 to yield a non-degenerated representation (π, H) of $L^1(G)$ with a cyclic vector ξ such that

$$\int \Phi(s)f(s)\,\mathrm{d}s = \phi(f) = (\pi(f)\xi \mid \xi)$$

for each f in $L^1(G)$. With (u, H) as the unitary representation of G corresponding to (π, H) and $\{f_\lambda\}$ an approximate unit for $L^1(G)$ we obtain

$$\Phi(t) = \operatorname{Lim} \int \Phi(s)f_\lambda(t^{-1}s)\,\mathrm{d}s = \operatorname{Lim}(\pi(\delta_t \times f_\lambda)\xi \mid \xi) = (u_t\xi \mid \xi),$$

for every t in G.

7.1.10. Proposition. *The set $B_{sa}(G)$ is a subalgebra of $C^b(G)$ and is a (real) Banach algebra in the norm inherited from $C^*(G)^*$.*

Proof. We claim that $\Phi\Psi \in B_+(G)$ for all Φ and Ψ in $B_+(G)$. By 7.1.9 it suffices to show that the pointwise product of two positive matrices (a_{ij}) and (b_{ij}) is again positive. To do so note that since $(b_{ij}) \geqslant 0$ there is a matrix (c_{ij}) such that $b_{ij} = \Sigma\, c_{ik}\bar{c}_{jk}$ for all i and j. Consequently we have for any set $\{\gamma_i\}$ of complex numbers

$$\sum (a_{ij}b_{ij})\bar{\gamma}_j\gamma_i = \sum a_{ij}(c_{ik}\bar{c}_{jk})\bar{\gamma}_j\gamma_i = \sum a_{ij}(\bar{c}_{jk}\bar{\gamma}_j)(c_{ik}\gamma_i) \geqslant 0$$

because $(a_{ij}) \geqslant 0$. Thus $(a_{ij}b_{ij}) \geqslant 0$ as desired.

From the first part of the proof it follows that $B_{sa}(G)$ is a subalgebra of $C^b(G)$. Since $B_{sa}(G)$ as a vector space is isomorphic to $C^*(G)^*_{sa}$, it is a Banach space in the norm inherited from $C^*(G)^*$. If only remains to show that the product is submultiplicative. This is obvious for positive definite functions since they attain their norm at e. If now Φ and Ψ belong to $B_{sa}(\mathfrak{G})$, let $\Phi = \Phi_+ - \Phi_-$ and $\Psi = \Psi_+ - \Psi_-$ be their decomposition into positive definite functions corresponding to the decomposition of the functionals (see 3.2.5). Then

$$\begin{aligned} \|\Phi\Psi\| &\leqq \|\Phi_+\Psi_+\| + \|\Phi_+\Psi_-\| + \|\Phi_-\Psi_+\| + \|\Phi_-\Psi_-\| \\ &= (\|\Phi_+\| + \|\Phi_-\|)(\|\Psi_+\| + \|\Psi_-\|) = \|\Phi\|\,\|\Psi\|. \end{aligned}$$

7.1.11. PROPOSITION. *The isomorphism* $\phi \to \Phi$ *defined in 7.1.8 is a homeomorphism of the state space of* $C^*(G)$ *in the weak* topology onto the set of positive definite functions* Φ *with* $\Phi(e) = 1$, *equipped with the topology of uniform convergence on compact subsets of* G.

Proof. If $\{\Phi_\lambda\}$ is a net in $B_+(G)$ converging to a function Φ uniformly on compact subsets of G then $\Phi \in B_+(G)$ by 7.1.9. Let ϕ_λ and ϕ denote the positive functionals on $C^*(G)$ associated with Φ_λ and Φ, respectively. For every f in $L^1(G)$ we have

$$\phi_\lambda(f) = \int f(s)\Phi_\lambda(s)\,ds \to \int f(s)\Phi(s)\,ds = \phi(f).$$

Thus $\phi_\lambda \to \phi$ weakly on a dense set of $C^*(G)$. Since the net $\{\phi_\lambda\}$ is bounded ($\|\phi_\lambda\| = \Phi_\lambda(e)$) we conclude that $\phi_\lambda \to \phi$ weak* in $C^*(G)^*$.

Conversely, assume that $\{\phi_\lambda\}$ is a weak* convergent net of states of $C^*(G)$ with limit ϕ where $\|\phi\| = 1$, and denote by Φ_λ and Φ the corresponding positive definite functions. Given $\varepsilon > 0$ there is an f in $K(G)_+$ with $\|f\|_1 = 1$ such that $\phi((\delta_e - f)^* \times (\delta_e - f)) < \varepsilon^2$. Since $\phi_\lambda \to \phi$ weak* and $\|\phi_\lambda\| = \|\phi\| = 1$ $(= \phi_\lambda(\delta_e))$ there is a λ_0 such that

$$\phi_\lambda((\delta_e - f)^* \times (\delta_e - f)) < \varepsilon^2$$

for all $\lambda > \lambda_0$. Thus for each t in G we have

$$\begin{aligned}|\Phi(t) - (\Phi \times \tilde{f})(t)|^2 &= |\Phi(t) - \int \Phi(s)\tilde{f}(s^{-1}t)\,ds|^2 \\ &= |\Phi(t) - \int \Phi(s)f(t^{-1}s)\,ds|^2 = |\phi(\delta_t) - \phi(\delta_t \times f)|^2 \\ &\leqslant 1 \cdot \phi((\delta_e - f)^* \times (\delta_e - f)) < \varepsilon^2.\end{aligned}$$

Similarly

$$|\Phi_\lambda(t) - (\Phi_\lambda \times \tilde{f})(t)| < \varepsilon.$$

Since

$$(\Phi_\lambda \times \tilde{f})(t) - (\Phi \times \tilde{f})(t) = (\phi_\lambda - \phi)(\delta_t \times f),$$

and $\delta_t \times f$ depends continuously on t in $C^*(G)$, we can by a standard argument for every compact set C find $\lambda_1 > \lambda_0$ such that

$$|(\Phi_\lambda \times \tilde{f})(t) - (\Phi \times \tilde{f})(t)| < \varepsilon$$

for all t in C. Combining this with our results above we obtain $|\Phi_\lambda(t) - \Phi(t)| < 3\varepsilon$ for all t in C.

7.1.12. Notes and remarks. Standard references on group theory are Loomis [162] (for appetizers) and Hewitt and Ross [115, 116] (for a full meal). A short but concise survey of the representation theory is found in Dixmier's C^*-book [65]. Proposition 7.1.4 (and 7.1.9–7.1.11) explains the close connection between the representation theory for groups and the C^*-algebra theory. The information lost in passing from G to $C^*(G)$ is compensated by having at one's disposal the smoothly functioning decomposition theory for C^*-algebras. Significantly enough, the atomic representation of C^*-algebras (4.3.7) was first established for group C^*-algebras by Gelfand and Raikov [93]. An early account of the theory of positive definite functions is found in Godement's paper [101].

7.2. The regular representation

7.2.1. Let $L^2(G)$ be the Hilbert space of (classes of) square integrable functions with respect to Haar measure. It is well known that for each μ in $M(G)$ and f in $L^2(G)$ we can define the convolution $\mu \times f$ in $L^2(G)$. It is straightforward to check that the map

$$\lambda: M(G) \to B(L^2(G))$$

given by $\lambda(\mu)f = \mu \times f$ is a representation of $M(G)$ called the *(left) regular representation*. The regular representation is faithful and its restriction to G (identifying each s in G with δ_s in $M(G)$) is the unitary representation $s \to \lambda_s$ of G on $L^2(G)$ given by

$$(\lambda_s f)(t) = f(s^{-1}t), \qquad f \in L^2(G)$$

The *reduced group C^*-algebra* is the norm closure of $\lambda(L^1(G))$ in $B(L^2(G))$ and is denoted by $C_r^*(G)$. Since each non-degenerate representation of $L^1(G)$ extends to a representation of $C^*(G)$ (by 7.1.5) we have $C_r^*(G) = \lambda(C^*(G))$. Even though λ is injective on $L^1(G)$ it is not in general an isomorphism of $C^*(G)$. Since $C_r^*(G)$ is a good deal more accessible than $C^*(G)$ we devote the next section to a discussion of those groups for which $C^*(G)$ and $C_r^*(G)$ are equal.

The *group von Neumann algebra* is the weak closure of $\lambda(L^1(G))$ in $B(L^2(G))$ and is denoted by $\mathcal{M}(G)$. Thus $\mathcal{M}(G) = C_r^*(G)''$. It follows from 7.1.4 and 7.1.5 that $\mathcal{M}(G)$ is also the von Neumann algebra generated by $\lambda(M(G))$ or by $\lambda(G)$, irrespectively.

7.2.2. The set of functionals on $C^*(G)$ of the form

$$x \to \phi(x) = \sum (\lambda(x)f_n \mid g_n),$$

where $\{f_n\} \subset L^2(G)$, $\{g_n\} \subset L^2(G)$, $\|\Sigma f_n\|_2^2 < \infty$ and $\Sigma\|g_n\|_2^2 < \infty$ is a norm closed subspace of $C^*(G)$, and can be identified with $\mathcal{M}(G)_*$. We denote by $A(G)$ the subspace of $B(G)$ consisting of functions Φ whose associated functionals belong to $\mathcal{M}(G)_*$. Thus $\Phi \in A(G)$ if $\Phi = \Sigma f_n \times \tilde{g}_n$, with $\{f_n\}$ and $\{g_n\}$ as above. Note that the series is uniformly convergent since $\|f_n \times \tilde{g}_n\|_\infty \leqslant \|f_n\|_2 \|g_n\|_2$.

$A(G)$ is a closed subspace of $B(G)$ in the norm inherited from $C^*(G)^*$, and is spanned by its positive cone, denoted by $A_+(G)$. From the above we see that $\Phi \in A_+(G)$ if and only if there is a sequence $\{f_n\}$ in $L^2(G)$ with $\Sigma \|f_n\|_2^2 < \infty$ such that $\Phi = \Sigma f_n \times \tilde{f}_n$. Since $K(G)$ is dense in $L^2(G)$ we see that each Φ in $A_+(G)$ can be approximated in the norm inherited from $C^*(G)^*$ by finite sums of positive definite functions of the form $f \times \tilde{f}$, $f \in K(G)$.

7.2.3. Define a map j on the functions on G by

$$(jf)(s) = f^*(s)\Delta(s)^{1/2} = \bar{f}(s^{-1})\Delta(s)^{-1/2}.$$

Elementary computations show that j is an involution, and that

$$(jf \mid jg) = (g \mid f); \qquad j(f \times g) = (jg) \times (jf)$$

whenever the operations make sense. In particular j is an isometry on $L^2(G)$.

For each function Φ on G such that $j\Phi \in L^1(G)$ we define an operator $\rho(\Phi)$ on $L^2(G)$ by

$$\rho(\Phi)f = j(\lambda(j\Phi))jf = j(j\Phi \times jf) = f \times \Phi.$$

The map $\Phi \to \rho(\Phi)$ (the right regular representation) is defined on $jL^1(G)$ which contains $K(G)$, and is a linear anti-homomorphism of $jL^1(G)$ into $\mathcal{M}(G)'$. If f, g belong to $K(G)$ we have

$$(\rho(\Phi)f \mid g) = (f \times \Phi \mid g = (f \mid g \times \tilde{\Phi}) = (f \mid \rho(\tilde{\Phi})g).$$

It follows that $\rho(\tilde{\Phi}) = \rho(\Phi)^*$ and from 7.1.9 (iv) we see that if $\Phi \in C^b(G)$ then $\rho(\Phi) \geq 0$ if and only if Φ is positive definite.

It will be convenient to define $\lambda(f)$ and ρ) also in some cases where f or jf does not belong to $L^1(G)$. We say that a function f on G is *left* (respectively *right*) *bounded* if there is a bounded operator denoted by $\lambda(f)$ (respectively $\rho(f)$) on $L^2(G)$ such that $\lambda(f)g = f \times g$ (respectively $\rho(f)g = g \times f$) for every g in $K(G)$. Note that if f is left bounded and Φ is right bounded then $\lambda(f)$ and $\rho(\Phi)$ commutes with every operator $\rho(g)$ and $\lambda(g)$, $g \in K(G)$, respectively.

7.2.4. LEMMA. *If $\Phi \in B_+(G) \cap K(G)$ there is a right bounded element ξ in $L^2(G)$ such that $\Phi = \xi \times \tilde{\xi}$.*

Proof. Choose an approximate unit $\{f_\lambda\}$ for $L^1(G)$ contained in $K(G)_+$. From 7.2.3 we know that $\rho(\Phi) \geq 0$, so that we may define $\xi_\lambda = \rho(\Phi)^{1/2} f_\lambda$ in $L^2(G)$. Now

$$\|\xi_\lambda - \xi_\mu\|_2^2 = (\rho(\Phi)(f_\lambda - f_\mu) \mid f_\lambda - f_\mu) = ((f_\lambda - f_\mu) \times \Phi \times (\tilde{f}_\lambda - \tilde{f}_\mu))(e),$$

which tends to zero as $\lambda, \mu \to \infty$. There is therefore a vector ξ in $L^2(G)$ such that $\xi_\lambda \to \xi$. If $f \in K(G)$ then since $\rho(\Phi) \in \mathcal{M}(G)'$

$$f \times \xi = \operatorname{Lim} \lambda(f)\rho(\Phi)^{1/2} f_\lambda = \operatorname{Lim} \rho(\Phi)^{1/2}\lambda(f) f_\lambda = \rho(\Phi)^{1/2} f,$$

which proves that ξ is right bounded with $\rho(\xi) = \rho(\Phi)^{1/2}$. Moreover, for each t in G

$$(\xi \times \tilde{\xi})(t) = (\xi \,|\, \lambda_t \xi) = \mathrm{Lim}(\rho(\Phi)^{1/2} f_\lambda \,|\, \lambda_t \rho(\Phi)^{1/2} f_\lambda)$$
$$= \mathrm{Lim}(f_\lambda \times \Phi \,|\, \lambda_t f_\lambda) = \Phi(t),$$

as desired.

7.2.5. PROPOSITION. *The set* $A_+(G)$ *is the closure (in* $C^*(G)^*$*) of* $B_+(G) \cap K(G)$.

Proof. From 7.2.4 we see that $B_+(G) \cap K(G) \subset A_+(G)$, and in 7.2.2 we showed that each element in $A_+(G)$ can be approximated by elements in $B_+(G) \cap K(G)$, so that $B_+(G) \cap K(G)$ is dense in $A_+(G)$.

7.2.6. LEMMA. *If G is separable there is a sequence* $\{\Phi_n\}$ *in* $A_+(G) \cap K(G)$ *such that* $\lambda(\Phi_n) \to 1$, *strongly, and* $\rho(\Phi_n) \nearrow 1$.

Proof. Choose an approximate unit $\{f_m\}$ for $L^1(G)$ contained in $K(G)_+$. Since the supports of the f_m's shrink towards $\{e\}$, we may assume that $\|jf_m\|_1 < 1$ and $\|\tilde{f}_m\| < 1$ for all m, whence

$$\|\rho(f_m \times \tilde{f}_m)\| \leqslant \|j(f_m \times \tilde{f}_m)\|_1 < 1.$$

Define $\Phi_1 = f_1 \times \tilde{f}_1$. Suppose that for all $k \leqslant n$ we have found Φ_k and elements f_{m_k}, where $m_k \geqslant k$, such that

(i) $\Phi_k \in A_+(G) \cap K(G)$;

(ii) $\rho(\Phi_{k-1}) \leqslant \rho(\Phi_k)$;

(iii) $\|\rho(\Phi_k)\| < 1$;

(iv) $\|\Phi_k - f_{m_k} \times \tilde{f}_{m_k}\|_1 < 1/k$.

For every ε and m we define

$$(*) \qquad \Phi = \Phi_n + (1-\varepsilon) f_m \times (\delta_e - \Phi_n) \times \tilde{f}_m.$$

Then $\Phi \in A_+(G) \cap K(G)$, since $\rho(\delta_e - \Phi_n) \geqslant 0$ by (iii), and moreover $\rho(\Phi_n) \leqslant \rho(\Phi)$. Since

$$\|\Phi - f_m \times \tilde{f}_m\|_1 = \|\Phi_n - \varepsilon f_m \times \tilde{f}_m - (1-\varepsilon) f_m \times \Phi_n \times \tilde{f}_m\|_1$$
$$\leqslant \varepsilon(\|\Phi_n\|_1 + 1) + (1-\varepsilon)\|\Phi_n - f_m \times \Phi_n \times \tilde{f}_m\|_1$$

we can arrange that $\|\Phi - f_m \times \tilde{f}_m\|_1 < 1/(n+1)$, taking ε sufficiently small and m sufficiently large. Moreover, for fixed $\varepsilon < 1 - \|\rho(\Phi_n)\|$ we have

$$\|\rho(\Phi)\| \leqslant \varepsilon\|\rho(\Phi_n)\| + (1-\varepsilon)\|\rho(\Phi_n + f_m \times \tilde{f}_m - f_m \times \Phi_n \times \tilde{f}_m)\|$$
$$\leqslant \varepsilon(1-\varepsilon) + (1-\varepsilon)(1 + \|\rho(\Phi_n - f_m \times \Phi_n \times \tilde{f}_m)\|).$$

Taking m sufficiently large we have $\|\rho(\Phi_n - f_m \times \Phi_n \times \tilde{f}_m)\| \leqslant \varepsilon^2$ whence

$$\|\rho(\Phi)\| \leqslant \varepsilon(1-\varepsilon) + (1-\varepsilon)(1+\varepsilon^2) < 1.$$

Thus if we take $\Phi_{n+1} = \Phi$ and $f_{m_{n+1}} = f_m$ the conditions (i)–(iv) are satisfied, and a sequence $\{\Phi_n\}$ satisfying these conditions can therefore be constructed by induction.

It is clear from (iv) that $\lambda(\Phi_n) \to 1$, strongly; and it follows from (ii) and (iii) that $\{\rho(\Phi_n)\}$ is an increasing sequence and therefore strongly convergent to some operator x with $0 \leqslant x \leqslant 1$. However, from (∗) we see that x must satisfy the equation

$$x = x + (1 - x),$$

since $\varepsilon \to 0$ and $\rho(f_m) \to 1$; whence $x = 1$.

7.2.7. THEOREM. *If G is separable there is a faithful σ-weight ϕ_e on $\mathcal{M}(G)$ such that $\phi_e(x^*x) < \infty$ if and only if there is a left bounded element f in $L^2(G)$ with $\lambda(f) = x$; and in this case $\phi_e(x^*x) = \|f\|_2^2$. Moreover, the representation associated with ϕ_e is spatially equivalent to the regular representation.*

Proof. Take $\{\Phi_n\}$ in $K(G) \cap A_+(G)$ satisfying 7.2.6 and put $\Phi_0 = 0$. Since $\Phi_n - \Phi_{n-1} \in K(G) \cap A_+(G)$ there is by 7.2.4 a sequence $\{\xi_n\}$ of right bounded elements in $L^2(G)$ such that $\Phi_n - \Phi_{n-1} = \xi_n \times \tilde{\xi}_n$ for each n. For every x in $\mathcal{M}(G)_+$ define

$$\phi_e(x) = \sum (x\xi_n \,|\, \xi_n).$$

From the definition it is obvious that ϕ_e is a σ-normal weight on $\mathcal{M}(G)$. Moreover, if $f \in L^1(G) \cap L^2(G)$ then

$$\begin{aligned}\phi_e(\lambda(f^* \times f)) &= \sum (f^* \times f \times \xi_n \,|\, \xi_n) \\ &= \sum (f \times \xi_n \times \tilde{\xi}_n \,|\, f) = \sum (f \times (\Phi_n - \Phi_{n-1}) \,|\, f) \\ &= \operatorname{Lim}(f \times \Phi_n \,|\, f) = \operatorname{Lim}(\rho(\Phi_n) f \,|\, f) = \|f\|_2^2 < \infty,\end{aligned}$$

which shows that ϕ_e is a σ-weight on $\mathcal{M}(G)$, since $\lambda(K(G)) \subset \mathcal{M}(G)_2^{\phi_e}$

Suppose that $x \in \mathcal{M}(G)_2^{\phi_e}$. For each g in $K(G)$ we have

$$|\phi_e(x^*\lambda(g))|^2 \leqslant \phi_e(x^*x)\phi_e(\lambda(g^* \times g)) = \phi_e(x^*x)\|g\|_2^2.$$

It follows that the map $g \to \phi_e(x^*\lambda(g))$ extends to a bounded functional on $L^2(G)$, and there is then a unique f in $L^2(G)$ such that $\phi_e(x^*\lambda(g)) = (g \,|\, f)$ for all g in $K(G)$. If g, h belong to $K(G)$ we therefore have

$$\begin{aligned}(h|f\times g) &= (h\times\tilde{g}|f) = \phi_e(x^*\lambda(h\times\tilde{g}))\\ &= \sum(x^*\lambda(h\times\tilde{g})\xi_n|\xi_n) = \sum(x^*(h\times\tilde{g}\times\xi_n)|\xi_n)\\ &= \sum(x^*\rho(\tilde{g}\times\xi_n)h|\xi_n) = \sum(x^*h|\rho(\tilde{g}\times\xi_n)^*\xi_n)\\ &= \sum(x^*h|\rho(\tilde{\xi}_n\times g)\xi_n) = \sum(x^*h|\xi_n\times\tilde{\xi}_n\times g)\\ &= \mathrm{Lim}(x^*h|\Phi_n\times g) = \mathrm{Lim}(x^*h|\lambda(\Phi_n)g) = (x^*h|g).\end{aligned}$$

Consequently $f\times g = x(g)$ for all g in $K(G)$ such that f is left bounded with $\lambda(f) = x$. Therefore

$$\begin{aligned}\phi_e(x^*x) &= \sum\|x\xi_n\|_2^2 = \sum\|f\times\xi_n\|_2^2 = \sum(f\times\xi_n\times\tilde{\xi}_n|f)\\ &= \mathrm{Lim}(f\times\Phi_n|f) = \|f\|_2^2.\end{aligned}$$

Reading the equations above backwards we see that if $x = \lambda(f)$ for some left bounded element f in $L^2(G)$ then $\phi_e(x^*x) = \|f\|_2^2 < \infty$.

Let (π_e, H_e) denote the normal representation of $\mathcal{M}(G)$ associated wtih ϕ_e. From what we proved above, there is a linear isometry u from $\mathcal{M}(G)_2^{\phi_e}$ into $L^2(G)$ given by $u\xi_x = f$, where $\lambda(f) = x$. Since $\lambda(K(G)) \subset \mathcal{M}(G)_2^{\phi_e}$ we see that u extends to an isometry of H_e onto $L^2(G)$. If $x\in\mathcal{M}(G)_2^{\phi_e}$ with $x = \lambda(f)$, $f\in L^2(G)$ then yf is left bounded for every y in $\mathcal{M}(G)$, and $\lambda(yf) = yx$. Consequently

$$u\pi_e(y)\xi_x = u\xi_{yx} = yf = yu\xi_x.$$

It follows that $\pi_e(y) = u^*yu$ for every y in $\mathcal{M}(G)$, whence π_e is spatially equivalent to the regular representation.

7.2.8 PROPOSITION. *The σ-weight ϕ_e defined in 7.2.7 satisfies $\phi_e(\lambda_t\cdot\lambda_t^*) = \Delta(t)\phi_e$ for each t in G. In particular, ϕ_e is a trace if and only if G is unimodular.*

Proof. Suppose that $f\in L^2(G)$ and that f is left bounded with $\lambda(f) = x$, $x\in\mathcal{M}(G)_2^{\phi_e}$. Define g on G by $g(s) = \Delta(t)f(st)$. It is elementary to compute that $g\in L^2(G)$ with $\|g\|_2^2 = \Delta(t)\|f\|_2^2$. Now, for each h in $K(G)$ we have

$$g\times h(s) = \int\Delta(t)f(rt)h(r^{-1}s)\,dr = \int f(r)h(tr^{-1}s)\,dr = (f\times\lambda_t^*h)(s).$$

Consequently

$$g\times h = f\times\lambda_t^*h = x\lambda_t^*(h),$$

which shows that g is left bounded with $\lambda(g) = x\lambda_t^*$. From 7.2.7 we conclude that

$$\phi_e(\lambda_t x^* x \lambda_t^*) = \|g\|_2^2 = \Delta(t)\|f\|_2^2 = \Delta(t)\phi_e(x^* x).$$

Since this is true for all x in $\mathcal{M}(G)_2^{\phi_e}$ and since $\lambda_t \mathcal{M}(G)\lambda_t^* = \mathcal{M}(G)$ we see that $\phi_e(\lambda_t \cdot \lambda_t^*) = \Delta(t)\phi_e$.

If ϕ_e is a trace then $\phi_e(\lambda_t \cdot \lambda_t^*) = \phi_e$, whence $\Delta(t) = 1$ for all t, and G is unimodular. Conversely, if G is unimodular then $f^* = jf$ for any f in $L^2(G)$. If therefore $x \in \mathcal{M}(G)_2^{\phi_e}$ with $x = \lambda(f)$, $f \in L^2(G)$ then since f^* is left bounded with $\lambda(f^*) = x^*$ we have

$$\phi_e(x^* x) = \|f\|_2^2 = \|jf\|_2^2 = \phi_e(xx^*),$$

which proves that ϕ_e is a trace.

7.2.9. Denote by $C_e(G)$ the linear subspace of functions spanned by elements $f^* \times g$, where f and g are left bounded elements of $L^2(G)$. From the formula

$$f^* \times g(t) = (g \mid f(\cdot t)) = \Delta(t)^{-1/2}(jf \mid \lambda_t(jg)),$$

easily verified by computation, we see that $C_e(G) \subset C_0(G)$. It follows from 7.2.7 that

$$\mathcal{M}(G)^{\phi_e} = \{\lambda(h) \mid h \in C_e(G)\}.$$

Moreover, if $x = \lambda(h)$, and $h = \Sigma f_n^* \times g_n$, where the f_n's and g_n's are left bounded elements of $L^2(G)$ then

$$\phi_e(x) = \sum (g_n \mid f_n) = h(e).$$

Thus on $C_e(G)$ the functional ϕ_e is just evaluation at e, whence the symbol ϕ_e. It is, however, not practical to start from this simple definition and extend ϕ_e to a σ-weight on $\mathcal{M}(G)$.

7.2.10. Notes and remarks. This section is intended as a shortcut to 7.2.7, which normally is a rather inaccessible result. Several discussions with Haagerup have eliminated several errors.

The extra structure in group C^*-algebras has invited many generalizations. An involutive algebra $\mathcal{A}$ which is at the same time a pre-Hilbert space is called a *left Hilbert algebra* if the following conditions are satisfied:

(i) $\mathcal{A}^2$ is total in the completed Hilbert space H $(=\bar{\mathcal{A}})$;

(ii) each map $L_x: y \to xy$ extends to a bounded operator on H;

(iii) the map $S: x \to x^*$ is closable in H;

(iv) $(xy \mid z) = (y \mid x^* z)$.

If, moreover, the involution satisfies the condition

(v) $(x \mid y) = (y^* \mid x^*)$,

which is equivalent to S being a (conjugate linear) unitary operator, we say that $\mathcal{A}$ is a *Hilbert algebra*. If we finally demand

(vi) $\mathcal{A} = H$,

we say that $\mathcal{A}$ is an *H^*-algebra*. The notions were introduced (of course in reverse order of time) by Tomita [259], Dixmier [56] and Ambrose [8].

The motivating example of a left Hilbert algebra is the set $L^1(G) \cap L^2(G)$. If G is unimodular we obtain in this way a Hilbert algebra. If G is compact then $L^2(G)$ is an H^*-algebra and therefore, by the general theory for H^*-algebras, a direct sum of its minimal ideals; each of which is isomorphic to the algebra of Hilbert–Schmidt operators on some Hilbert space.

Tomita's observation (later formalized by Combes in [37]) was that if ϕ is a faithful σ-weight on a von Neumann algebra $\mathcal{M}$ then the set $\mathcal{M}_2^\phi \cap (\mathcal{M}_2^\phi)^*$ is a left Hilbert algebra. Moreover, every left Hilbert algebra arises in this manner. We give a simple approach to the Tomita–Takesaki theory in section 8.13, avoiding the left Hilbert algebras.

Let F_2 denote the free group on two generators (the canonical bad apple in the next section, cf. 7.3.5). Very little is known about $C_r^*(F_2)$. Powers showed that it is simple, and clearly it is separable and has a faithful finite trace (viz. ϕ_e, since F_2 is unimodular, cf. 7.2.8). But at the time of writing it is not known whether $C_r^*(F_2)$ has any non-trivial projections, see [3]. If not, it will be the first example of a simple C^*-algebra without non-trivial projections, thus answering a problem posed by Dixmier in 1967.

7.3. Amenable groups

7.3.1. We shall need only one result (7.3.9) from the theory of amenable groups, unfortunately a rather deep result. Since the theory is not (yet) a standard part of harmonic analysis, we give a self-contained exposition in this section. First some definitions and explanatory remarks.

7.3.2. If f is a function on a locally compact group G and $s \in G$ we denote by $\lambda_s f$ the function $t \to f(s^{-1}t)$. We say that f is *left uniformly continuous* if $\|\lambda_s f - f\|_\infty \to 0$ as $s \to e$, and denote by $UC_l^b(G)$ the set of left uniformly continuous elements of $C^b(G)$. Similarly we denote by $UC_r^b(G)$ the set of functions f in $C^b(G)$ such that $\check{f} \in UC_l^b(G)$, and write

$$UC^b(G) = UC_l^b(G) \cap UC_r^b(G).$$

We then have an ascending chain of commutative C^*-algebras in the $\|\cdot\|_\infty$ norm

$$UC^b(G) \subset UC_l^b(G) \subset C^b(G) \subset L^\infty(G),$$

and it is easily verified that each of these algebras are invariant under left translation by elements in G. Moreover, we have

$$L^1(G) \times L^\infty(G) \subset UC_l^b(G), \qquad L^\infty(G) \times L^1(G)^\sim \subset UC_r^b(G).$$

7.3.3. If A is a C^*-algebra in $L^\infty(G)$, invariant under left translation, we say that a state m of A is a *left invariant mean* if $m(\lambda_s f) = m(f)$ for every f in A. We say that G is *amenable* if any of the conditions in the next proposition are fulfilled.

7.3.4. PROPOSITION. *The following conditions on a locally compact group G are equivalent*:

(i) *There is a left invariant mean on $UC^b(G)$*;

(ii) *There is a left invariant mean on $C^b(G)$*;

(iii) *There is a left invariant mean on $L^\infty(G)$*;

(iv) *There is a state m on $L^\infty(G)$ such that*

$$m(\mu \times f) = \mu(G)m(f)$$

for each μ in $M(G)$ and f in $L^\infty(G)$.

Proof. It is trivial that (iv) $\Rightarrow$ (iii) $\Rightarrow$ (ii) $\Rightarrow$ (i). We must prove that (i) $\Rightarrow$ (iv). Assume therefore that m is a left invariant mean on $UC^b(G)$. We claim that $m(g \times f) = \int g(s)\,ds\, m(f)$ for each g in $L^1(G)$ and f in $UC^b(G)$. To see this note that the functional $g \to m(g \times f)$ is bounded on $L^1(G)$ (since $\|g \times f\| \leq \|g\|_1 \|f\|_\infty$) and invariant under left translation. Since Haar measure is the unique functional satisfying this condition there is a scalar α depending only on f such that $m(g \times f) = \int g(s)\,ds\,\alpha$. Take an approximate unit $\{g_i\}$ for $L^1(G)$. Then

$$\|g_i \times f - f\|_\infty \to 0 \qquad \text{as } \lambda \to \infty$$

because $f \in UC_l^b(G)$, whence

$$\alpha = m(g_i \times f) \to m(f)$$

as desired.

Now fix an element g in $L^1(G)$ with $g \geq 0$ and $\|g\|_1 = 1$ and define $\bar{m}$ on $L^\infty(G)$ by

$$\bar{m}(f) = m(g \times f \times \tilde{g}) \qquad \forall f \in L^\infty(G).$$

Note that $g \times f \times \tilde{g} \in UC^b(G)$ so that the definition is meaningful. Clearly $\bar{m}$ is a positive functional on $L^\infty(G)$ and since $g \times 1 = 1 = 1 \times \tilde{g}$ we see that $\bar{m}$ is a state. With $\{g_i\}$ as an approximate unit for $L^1(G)$ we have for any μ in $M(G)$

and f in $L^\infty(G)$

$$\bar{m}(\mu \times f) = m(g \times \mu \times f \times \tilde{g})$$

$$= \operatorname*{Lim}_{ij} m(g \times \mu \times g_i \times g_j \times f \times \tilde{g}) = \operatorname*{Lim}_{ij} m(\mu \times g_i \times g_j \times f \times \tilde{g})$$

$$= \operatorname*{Lim}_{ij} \int \mu \times g_i(s)\,ds\, m(g_j \times f \times \tilde{g}) = \mu(G) \operatorname*{Lim}_{j} m(g \times g_j \times f \times \tilde{g})$$

$$= \mu(G) m(g \times f \times \tilde{g}) = \mu(G)\bar{m}(f),$$

and the proof is complete.

7.3.5. Every abelian group is amenable and every compact group is amenable (with Haar measure as the unique invariant mean). Every closed subgroup of an amenable group is amenable. In the converse direction, if H is a closed normal subgroup of G such that H and G/H are amenable, then G is amenable. Moreover, if G is the union of an increasing net of amenable groups then it is itself amenable.

The free group on two generators is not amenable and it is an outstanding problem whether a discrete group fails to be amenable only if it contains the free group on two generators as a subgroup. Proofs of these facts can be found in Greenleaf [103].

7.3.6. The set of functions f in $L^1(G)$ such that $f \geqslant 0$ and $\|f\|_1 = 1$ will play an important rôle in the following. We denote it by $S(G)$ and note that it can be identified with the set of normal states on $L^\infty(G)$ and that it is a semi-group under convolution.

7.3.7. PROPOSITION. *The following conditions on G are equivalent:*

(i) *G is amenable;*

(ii) *There is a net $\{g_i\}$ in $S(G)$ such that $h \times g_i - g_i \to 0$ weak* in $(L^\infty(G))^*$ for each h in $S(G)$;*

(iii) *There is a net $\{g_i\}$ in $S(G)$ such that $\|h \times g_i - g_i\|_1 \to 0$ for each h in $S(G)$;*

(iv) *For each compact set C and $\varepsilon > 0$ there is a g in $S(G)$ such that $\|\lambda_s g - g\|_1 < \varepsilon$ for every s in C.*

Proof. (i) ⇒ (ii). Take a state m on $L^\infty(G)$ satisfying 7.3.4 (iv). Since the set of normal states of $L^\infty(G)$ is weak* dense in the set of all states (being convex and separating) there is a net $\{g_i\}$ in $S(G)$ such that $(f \mid g_i) \to m(f)$ for every f in $L^\infty(G)$. Fix h in $S(G)$. Then

$$(f \mid h \times g_i) = (h^* \times f \mid g_i) \to m(h^* \times f) = m(f).$$

Consequently the net $\{h \times g_i - g_i\}$ is weak* convergent to zero in $(L^\infty(G))^*$.

(ii) $\Rightarrow$ (iii). For each finite set $\{h_1, \ldots, h_n\}$ in $S(G)$ and $\varepsilon > 0$ let X be the n-fold direct sum of $L^1(G)$. The set in X with elements

$$\bigoplus_k (h_k \times g - g), \qquad g \in S(G)$$

is convex, and by assumption it contains zero as a weak limit point. It follows from the Hahn–Banach theorem that the set contains zero as a limit point in norm. There is therefore a g in $S(G)$ such that

$$\operatorname*{Sup}_k \|h_k g - g\|_1 < \varepsilon,$$

from which the existence of a net $\{g_i\}$ satisfying (iii) is immediate.

(iii) $\Rightarrow$ (iv). Fix an element h in $S(G)$ and take a net $\{g_i\}$ in $S(G)$ satisfying (iii). Then for each s in G

$$\|\lambda_s(h \times g_i) - h \times g_i\|_1 \leqslant \|\lambda_s(h) \times g_i - g_i\| + \|h \times g_i - g_i\| \to 0.$$

Since $\lambda_s(h)$ depends continuously on s in $L^1(G)$ we can by a standard argument for each compact set C and $\varepsilon > 0$ find g_i such that

$$\|\lambda_s(h) \times g_i - g_i\|_1 + \|h \times g_i - g_i\|_1 < \varepsilon$$

for all s in C. Taking $g = h \times g_i$ we obtain $\|\lambda_s g - g\|_1 < \varepsilon$ for all s in C.

(iv) $\Rightarrow$ (i). Let I be the net of pairs (C, ε) consisting of a compact set C and $\varepsilon > 0$. For each i in I choose g_i in $S(G)$ satisfying (iv), and consider the net $\{m_i\}$ of states on $C^b(G)$ given by $m_i(f) = \int f(t) g_i(t)\, dt$. For each s in C we have

$$\begin{aligned} |m_i(\lambda_{s^{-1}} f - f)| &= |\textstyle\int (f(st) - f(t)) g_i(t)\, dt| \\ &= |\textstyle\int f(t)(g_i(s^{-1}t) - g_i(t))\, dt| \leqslant \|f\|_\infty \varepsilon. \end{aligned}$$

Thus every weak* limit point of $\{m_i\}$ in $C^b(G)^*$ will be a left invariant mean on G.

7.3.8. PROPOSITION. *The group G is amenable if and only if there is a net $\{f_i\}$ in the unit sphere of $L^2(G)$ such that $\{f_i \times \tilde{f}_i\}$ converges to 1 uniformly on compact subsets of G.*

Proof. If $g \in S(G)$ put $f = g^{1/2}$ (pointwise square root). Then $1 = \|f\|_2^2 = (f \times \tilde{f})(e)$. Moreover, for each s in G

$$\begin{aligned} |1 - (f \times \tilde{f})(s)|^2 &= |(f \times \tilde{f})(e) - (f \times \tilde{f})(s)|^2 \\ &= |\textstyle\int f(t)(\tilde{f}(t^{-1}) - \tilde{f}(t^{-1}s))\, dt|^2 = |\textstyle\int f(t)(\bar{f}(t) - \bar{f}(s^{-1}t))\, dt|^2 \\ &= |(f | f - \lambda_s f)|^2 \leqslant 1\, \|f - \lambda_s f\|_2^2 = \textstyle\int |g^{1/2}(t) - g^{1/2}(s^{-1}t)|^2\, dt \\ &\leqslant \textstyle\int |g(t) - g(s^{-1}t)|\, dt = \|g - \lambda_s g\|_1 . \end{aligned}$$

Conversely, if f belongs to the unit sphere of $L^2(G)$ put $g = |f|^2$. Then $g \in S(G)$. Moreover, for each s in G

$$\begin{aligned}\|g - \lambda_s g\|_1 &= \int \big| |f(t)|^2 - |f(s^{-1}t)|^2 \big| dt \\ &= \int |f(t) + f(s^{-1}t)| \, |f(t) - f(s^{-1}t)| dt = |((f + \lambda_s f) | (f - \lambda_s f))| \\ &\leq 2 \|f - \lambda_s f\|_2 = 2(2 - (f|\lambda_s f) - (\lambda_s f | f))^{1/2} \\ &\leq 2\sqrt{2}|1 - (\lambda_s f | f)|^{1/2} = 2\sqrt{2}|1 - (f \times \tilde{f})(s)|^{1/2}.\end{aligned}$$

Combining these two inequalities we obtain

$$|1 - (f \times \tilde{f})(s)|^2 \leq \|g - \lambda_s g\|_1 \leq 2\sqrt{2}|1 - (f \times \tilde{f})(s)|^{1/2};$$

and the result follows immediately from 7.3.7 (iv).

7.3.9. THEOREM. *The following conditions are equivalent:*

(i) *G is amenable;*

(ii) *The regular representation is faithful on $C^*(G)$.*

Proof. If G is amenable there is by 7.3.8 a net $\{\Phi_i\}$ in $A_+(G)$ with $\Phi_i(e) = 1$ such that $\{\Phi_i\}$ converges to 1 uniformly on compact subsets of G. Since $K(G) \cap A_+(G)$ is dense in $A_+(G)$ in the norm topology inherited from $C^*(G)^*$ by 7.2.5, it is *a fortiori* dense in the uniform topology on $C_0(G)$. We may therefore assume that $\{\Phi_i\} \subset K(G)$. For any Φ in $B_+(G)$ with $\Phi(e) = 1$ we have $\Phi\Phi_i \in K(G) \cap B_+(G)$ by 7.1.10, and $\Phi\Phi_i \to \Phi$ uniformly on compact subsets of G. It follows from 7.1.11 that $\mathcal{M}(G)_*$ is weak* dense in $C^*(G)^*$ and thus the regular representation has zero kernel in $C^*(G)$.

Conversely, if the regular representation if faithful on $C^*(G)$ then $\mathcal{M}(G)_*$ is weak* dense in $C^*(G)$, whence $A_+(G)$ is dense in $B_+(G)$ in the topology of uniform convergence on compact subsets of G by 7.1.11. The function Φ_1 given by $\Phi_1(t) = 1$ for all t belongs to $B_+(G)$ (and corresponds to the trivial representation of G). Consequently there is a net $\{\Phi_i\}$ in $A_+(G)$ with $\Phi_i(e) = 1$ such that $\Phi_i \to 1$ $(=\Phi_1)$ uniformly on compact subsets of G. Moreover, by 7.2.5 we may assume that $\{\Phi_i\} \subset K(G)$. For each i there is by 7.2.4 a vector ξ_i in $L^2(G)$ such that $\Phi_i = \xi_i \times \tilde{\xi}_i$, whence G is amenable by 7.3.8.

7.3.10. Notes and remarks. Invariant means on discrete groups were considered by von Neumann [172]. Since then a number of mathematicians (notably Følner and Reiter) have worked on the subject. However, a unified treatment, including the equivalences 7.3.4, 7.3.7 and 7.3.8 first appeared in Greenleaf's book [103]. Theorem 7.3.9 is due to Hulanicki [118]. Many of the results on amenable groups can be generalized to homogeneous spaces; see the notes by Eymard [88].

7.4. Dynamical systems

7.4.1. A *C^*-dynamical system* (or just a *dynamical system*) is a triple (A, G, α) consisting of a C^*-algebra A, a locally compact group G and a continuous homomorphism α of G into the group $\mathrm{Aut}(A)$ of automorphisms (i.e. $*$-automorphisms) of A equipped with the topology of pointwise convergence. This means that for each x in A the function $\alpha(x): G \to A$ defined by $t \to \alpha_t(x)$ is continuous. We are mainly interested in the case where both G and A are separable, and refer to this as a *separable dynamical* system.

7.4.2. If $\mathcal{M}$ is a von Neumann algebra we consider the topology of pointwise weak convergence on $\mathrm{Aut}(\mathcal{M})$. This is equal to the topologies of pointwise σ-weak and pointwise strong convergence, because the three weak topologies coincide on the unitary group $U(\mathcal{M})$ of $\mathcal{M}$, and $U(\mathcal{M})$ is stable under $\mathrm{Aut}(\mathcal{M})$ and generates $\mathcal{M}$ linearly. A *W^*-dynamical system* is a triple $(\mathcal{M}, G, \alpha)$ in which $\alpha: G \to \mathrm{Aut}(\mathcal{M})$ is continuous in this topology, which means that each function $\alpha(x): G \to \mathcal{M}$ is σ-weakly continuous.

7.4.3. LEMMA. *If $(\mathcal{M}, G, \alpha)$ is a W^*-dynamical system there is for each μ in $M(G)$ and x in $\mathcal{M}$ a unique element $\alpha_\mu(x)$ in $\mathcal{M}$ such that*

$$(*) \qquad \phi(\alpha_\mu(x)) = \int \phi(\alpha_t(x))\, d\mu(t)$$

for each normal functional ϕ on $\mathcal{M}$.

Proof. The equation $(*)$ defines a bounded functional on the pre-dual $\mathcal{M}_*$ of $\mathcal{M}$. Since $(\mathcal{M}_*)^* = \mathcal{M}$ we see that $\alpha_\mu(x) \in \mathcal{M}$ (cf. A3, Appendix).

7.4.4. LEMMA. *If (A, G, α) is a C^*-dynamical system there is for each μ in $M(G)$ and x in A a unique element $\alpha_\mu(x)$ in A such that*

$$(*) \qquad \phi(\alpha_\mu(x)) = \int \phi(\alpha_t(x))\, d\mu(t)$$

for each ϕ in A^*

Proof. This follows directly from A3 (Appendix).

7.4.5. PROPOSITION. *Suppose that (A, G, α) is a C^*-dynamical system. If α_t'' denotes the double transpose of α_t then the map $t \to \alpha_t''$ is a homomorphism of G into $\mathrm{Aut}(A'')$. Let $\mathcal{U}(A)$ denote the class of universally measurable elements in A''_{sa} (see 4.3.11). For each x in $\mathcal{U}(A)$ and each state ϕ the function $t \to \phi(\alpha_t''(x))$ is universally measurable on G. Moreover, for each positive μ in $M(G)$ there is a unique element $\alpha_\mu''(x)$ in $\mathcal{U}(A)$ such that*

$$(*) \qquad \phi(\alpha_\mu''(x)) = \int \phi(\alpha_t''(x))\, d\mu(t)$$

for each state ϕ of A.

Proof. Since A is σ-weakly dense in A'' there is no difficulty in verifying that each α''_t is an automorphism of A'' and that the map $t \to \alpha''_t$ is a homomorphism. The problems arise because this map has no recognizable regularity features—topologically or measure theoretically—on all of A''.

Take a positive μ in $M(G)$. The transposed α'_μ of the map α_μ on A defined by (∗) in 7.4.4 is a map on A^*. For each x in A_{sa} and each state ϕ of A we have

$$\alpha'_\mu(\phi)(x) = \int \phi(\alpha''_t(x)\,\mathrm{d}\mu(t).$$

The same formula is therefore true for each x in $(A_{sa})^m$ since the function $t \to \phi(\alpha''_t(x))$ is lower semi-continuous on G. Take now x in $\mathscr{U}(A)$. Given $\varepsilon > 0$ there are elements a and b in $(A_{sa})^m$ such that $-b \leqslant x \leqslant a$ and $\alpha'_\mu(\phi)(a + b) < \varepsilon$. This means that the function $t \to \phi(\alpha''_t(x))$ is sandwiched between the lower semi-continuous functions $t \to \phi(\alpha''_t(a))$ and $t \to -\phi(\alpha''_t(b))$ and that $\int\phi(\alpha''_t(a + b))\,\mathrm{d}\mu(t) < \varepsilon$. Since this is true for every positive μ in $M(G)$ it follows that the function $t \to \phi(\alpha''_t(x))$ is universally measurable on G. Moreover, the element $\alpha''_\mu(x)$ defined by (∗) is a bounded functional on A^*, and thus belongs to A''. Finally, if $a \in (A_{sa})^m$ then $\alpha''_\mu(a) \in (A_{sa})^m$ since μ is positive. If therefore ϕ and $\varepsilon > 0$ are given we take a and b as before and have

$$-\alpha''_\mu(b) \leqslant \alpha''_\mu(x) \leqslant \alpha''_\mu(a), \qquad \phi(\alpha''_\mu(a + b)) < \varepsilon,$$

which shows that $\alpha''_\mu(x) \in \mathscr{U}(A)$.

7.4.6. For any Borel ∗-algebra $\mathscr{A}$ we consider the Borel structure on $\mathrm{Aut}(\mathscr{A})$ induced by the functions $\alpha \to \phi(\alpha(x))$ on $\mathrm{Aut}(\mathscr{A})$, where $x \in \mathscr{A}$ and ϕ is a σ-normal state of $\mathscr{A}$. A *B^*-dynamical system* is a triple $(\mathscr{A}, G, \alpha)$ where $\alpha: G \to \mathrm{Aut}(\mathscr{A})$ is a Borel map satisfying the condition that for each μ in $M(G)$ and each x in $\mathscr{A}$ there is an element $\alpha_\mu(x)$ in $\mathscr{A}$ such that

$$(*) \qquad \phi(\alpha_\mu(x)) = \int \phi(\alpha_t(x))\,\mathrm{d}\mu(t)$$

for each σ-normal state ϕ of $\mathscr{A}$. Clearly $\alpha_\mu(x)$ is uniquely determined by (∗).

7.4.7. PROPOSITION. *If (A, G, α) is a C^*-dynamical system and $\mathscr{B}$ is the enveloping Borel ∗-algebra of A then there is a unique extension α'' of α such that $(\mathscr{B}, G, \alpha'')$ is a B^*-dynamical system.*

Proof. By double transposition we may extend α to a representation α'' of G into $\mathrm{Aut}(A'')$. The class of elements x in A''_{sa} for which $\alpha''_t(x) \in \mathscr{B}_{sa}$ for all t in G is monotone sequentially closed and contains A_{sa}. It therefore contains $\mathscr{B}_{sa}$ so that we may restrict α'' to a representation of G into $\mathrm{Aut}(\mathscr{B})$.

The class of elements x in $\mathscr{B}_{sa}$ for which all functions $t \to \phi(\alpha''_t(x))$, ϕ a state of A, are Borel measurable, is monotone sequentially closed and contains A_{sa}. It therefore equals $\mathscr{B}_{sa}$.

The class of elements x in $\mathscr{B}_{sa}$ for which $\alpha_\mu(x) \in \mathscr{B}_{sa}$ for all μ in $M(G)$, is monotone sequentially closed and contains A_{sa} by 7.4.4. It therefore equals $\mathscr{B}_{sa}$. Thus $(\mathscr{B}, G, \alpha'')$ is a B^*-dynamical system.

7.4.8. A *covariant representation* of a C^*-dynamical system (A, G, α) is a triple (π, u, H) where (π, H) is a representation of A, (u, H) is a unitary representation of G, and

$$\pi(\alpha_t(x)) = u_t \pi(x) u_t^*$$

for all x in A and t in G.

For W^*- or B^*-dynamical systems the useful concepts are of course normal or σ-normal covariant representations; i.e. covariant representations (π, u, H) in which (π, H) is normal or σ-normal, respectively.

7.4.9. LEMMA. *Let $(\mathscr{M}, G, \alpha)$ be a B^*-dynamical system where $\mathscr{M}$ is a von Neumann algebra on a separable Hilbert space K and G is separable. There is then a separable covariant representation (π, u, H) of $(\mathscr{M}, G, \alpha)$ such that (π, H) is faithful and normal. In particular, $(\mathscr{M}, G, \alpha)$ is necessarily a W^*-dynamical system.*

Proof. Let $H = L^2(G, K)$, i.e. the vectors in H are (classes of) square integrable functions $\xi: G \to K$ with inner product

$$(\xi \mid \eta) = \int (\xi(t) \mid \eta(t))\, dt.$$

If $f \in K(G)$ and $\xi_0 \in K$ then $f\xi_0 \in H$ and the linear combinations of such vectors lie dense in H (so that $H = L^2(G) \otimes K$). Since G is second countable and K is separable it follows that H is separable.

We define a unitary representation u of G on H by setting $(u_t\xi)(s) = \xi(t^{-1}s)$ for each t in G and ξ in H. Furthermore, since each function $t \to \alpha_t(x)\xi_0$, $x \in B$, $\xi_0 \in K$, is weakly Borel measurable and bounded we can define $\pi(x)$ on H by $(\pi(x)\xi)(t) = \alpha_{t^{-1}}(x)\xi(t)$ for each ξ in H. Elementary calculations verify that (π, H) is a representation of $\mathscr{M}$, and for each x in $\mathscr{M}$, t in G and ξ in H we have

$$(u_t\pi(x)u_t^*\xi)(s) = (\pi(x)u_t^*\xi)(t^{-1}s) = \alpha_{s^{-1}t}(x)\xi(s) = (\pi(\alpha_t(x))\xi)(s),$$

so that (π, u, H) is a covariant representation.

Take a faithful normal state ρ on $\mathscr{M}$. If $x \in \mathscr{M}_+$ and $x \neq 0$ then the Borel function $t \to \phi(\alpha_{t^{-1}}(x))$ is strictly positive. If therefore f is a strictly positive (continuous) function in $L^1(G)$ then $\alpha_{\tilde{f}}(x) \neq 0$. Take ξ_0 in K and define $f^{1/2}\xi_0$ in H. Then

$$(\pi(x)f^{1/2}\xi_0 \mid f^{1/2}\xi_0) = (\alpha_{\tilde{f}}(x)\xi_0 \mid \xi_0) \neq 0$$

for a suitable ξ_0, which proves that π is faithful. If $x_n \nearrow x$ in $\mathscr{M}_{sa}$ and $\xi \in H$ then for each t in G

$$(\alpha_{t^{-1}}(x_n)\xi(t) \mid \xi(t)) \nearrow (\alpha_{t^{-1}}(x)\xi(t) \mid \xi(t)),$$

whence $\pi(x_n) \nearrow \pi(x)$ by Lebesgue's monotone convergence theorem, so that (π, H) is also normal.

7.4.10. THEOREM. *Let (A, G, α) be a separable C^*-dynamical system and let $(\mathscr{B}, G, \alpha'')$ be its associated B^*-dynamical system (cf. 7.4.7). The following conditions on a separable representation (π, H) of A are equivalent and constitute the definition of a G-invariant representation.*

(i) *There is a separable covariant representation (ρ, u, K) of (A, G, α) such that (π, H) and (ρ, K) are equivalent.*

(ii) *The transposed action $t \to \alpha'_t$ of G on the dual of A leaves the pre-dual $\pi''(\mathscr{B})_*$ invariant (as a subset of the dual of A), and for each ϕ in $\pi''(\mathscr{B})_*$ the function $t \to \alpha'_t(\phi)$ is norm continuous.*

(iii) *The kernel of π in A is invariant under all α_t, $t \in G$, and each automorphism $\pi \circ \alpha_t \circ \pi^{-1}$ extends from $\pi(A)$ to an automorphism of $\pi''(\mathscr{B})$.*

(iv) *The kernel of π'' in $\mathscr{B}$ is invariant under all α''_t, $t \in G$.*

Proof. (i) $\Rightarrow$ (ii). The pre-duals of $\pi''(\mathscr{B})$ and $\rho''(\mathscr{B})$ are identical, regarded as subsets of the dual of A. If $\phi \in \rho''(\mathscr{B})_*$ then

$$\alpha'_t(\phi) = \phi(u_t \cdot u_t^*) \in \rho''(\mathscr{B})_*,$$

and the function $t \to \alpha'_t(\phi)$ is norm continuous because $t \to u_t$ is strongly continuous and ϕ can be approximated in norm by a linear combination of vector states.

(ii) $\Rightarrow$ (iii). If $x \in A$ and $\pi(x) = 0$ then for each t in G

$$\phi(\alpha_t(x)) = \alpha'_t(\phi)(x) = \alpha'_t(\phi)(\pi(x)) = 0$$

for every ϕ in $\pi''(\mathscr{B})_*$ since $\alpha'_t(\phi) \in \pi''(\mathscr{B})_*$. It follows that $\pi(\alpha_t(x)) = 0$. Thus the automorphism $\pi \circ \alpha_t \circ \pi^{-1}$ is well defined on $\pi(A)$ and it will extend to $\pi''(\mathscr{B})$ if and only if it is σ-weakly continuous. But for each ϕ in $\pi''(\mathscr{B})_*$ the function

$$x \to \phi(\pi \circ \alpha_t \circ \pi^{-1}(x)) = \alpha'_t(\phi)(x)$$

is clearly σ-weakly continuous on $\pi(A)$.

(iii) $\Rightarrow$ (iv). It follows from (iii) that the extension β_t of $\pi \circ \alpha_t \circ \pi^{-1}$ from $\pi(A)$ to $\pi''(\mathscr{B})$ satisfies

$$\beta_t \circ \pi'' = \pi'' \circ \alpha''_t$$

for every t in G since this holds on A and both maps are σ-normal. Consequently the kernel of π'' in $\mathscr{B}$ is G-invariant.

(iv) $\Rightarrow$ (i). By assumption $(\pi''(\mathscr{B}), G, \pi'' \circ \alpha \circ \pi''^{-1})$ is a B^*-dynamical system and the implication follows from 7.4.9.

7.4.11. Let (A, G, α) be a C^*-dynamical system. We say that lower semi-continuous weight ϕ of A is *G-invariant* if $\phi(\alpha_t(x)) = \phi(x)$ for all t in G and x in A_+. We say that ϕ is *G-quasi-invariant* if the associated representation (π_ϕ, H_ϕ) (see 5.1.3) is G-invariant.

If (A, G, α) is a separable system and $(\mathscr{B}, G, \alpha'')$ is its associated B^*-dynamical system then the restriction of α'' to the centre $\mathscr{C}$ of $\mathscr{B}$ gives a commutative B^*-dynamical system $(\mathscr{C}, G, \alpha'')$. Identifying $\mathscr{C}$ with $\mathscr{B}(\hat{A})$ we see that α'' is a representation of G as D-Borel transformations of $\hat{A}$ (a dynamical system in the ordinary, commutative, sense). If ϕ is a state of A and μ_ϕ is its associated central measure on $\hat{A}$ we see from 4.7.10 that ϕ is G-quasi-invariant if and only if μ_ϕ is quasi-invariant under the transformations of $\hat{A}$ induced by α''_t, $t \in G$. By 7.4.10 each G-quasi-invariant state ϕ is *G-continuous* in the sense that the function $t \to \alpha'_t(\phi)$ is norm continuous (it is of course always weak* continuous). The converse is false in general (G might be discrete).

We say that a lower semi-continuous weight ϕ of A is *G-covariant* if there is a unitary representation u of G on H_ϕ such that (π_ϕ, u, H_ϕ) is a covariant representation of (A, G, α). A G-covariant state is necessarily G-quasi-invariant. The preference of G-quasi-invariant states to G-covariant states stems from the fact that the former constitute a convex set (because the map $\phi \to \mu_\phi$ is affine and the quasi-invariant measures form a convex set), while the latter do not.

7.4.12. PROPOSITION. *Each G-invariant lower semi-continuous weight ϕ of a C^*-dynamical system (A, G, α) is G-covariant.*

Proof. Consider the representation (π_ϕ, H_ϕ) associated with ϕ and for each t in G define

$$u_t^\phi \xi_x = \xi_{\alpha_t(x)}, \qquad x \in A_2^\phi.$$

From the G-invariance of ϕ it follows immediately that u_t^ϕ is a well defined and densely defined isometry with dense range, and thus has a unique extension to a unitary operator, again denoted by u_t^ϕ. The homomorphism is weakly (= strongly) continuous since α is continuous and ϕ is lower semi-continuous so that

$$\|(1 - u_t^\phi)\xi_x\|^2 = \phi(2x^*x - x^*\alpha_t(x) - \alpha_t(x^*)x) \to 0,$$

as $t \to e$.

Finally, for each x in A, y in A_2^ϕ and t in G

$$u_t^\phi \pi_\phi(x) u_t^{\phi*} \xi_y = u_t^\phi \pi_\phi(x) \xi_{\alpha.t^{-1}(y)} = u_t^\phi \xi_{x\alpha.t^{-1}(y)} = \xi_{\alpha.t(x)y} = \pi_\phi(\alpha_t(x))\xi_y;$$

(with

$$\alpha.t \equiv \alpha_t; \qquad \alpha.t^{-1} \equiv \alpha_{t^{-1}}$$

in the subscripts) whence

$$u_t^\phi \pi_\phi(x) u_t^{\phi *} = \pi_\phi(\alpha_t(x))$$

so that $(\pi_\phi, u^\phi, H_\phi)$ is a covariant representation.

7.4.13. PROPOSITION. *Let (A, G, α) be a separable C^*-dynamical system. The set S_{qi} of G-quasi-invariant states of A is a convex, weak* dense subset of the state space S of A, and the norm closure of S_{qi} is the set of G-continuous states.*

Proof. From the characterization of quasi-invariant states given in 7.4.11 it is immediate that S_{qi} is convex.

To see that S_{qi} is weak* dense take ϕ in S and f in $L^1(G)$ and define $\alpha'_f(\phi)$ on A by $\alpha'_f(\phi)(x) = \phi(\alpha_f(x))$ for each x in A. If $\|f\|_1 = 1$ and $f \geqslant 0$ then $\alpha'_f(\phi) \in S$ and if $f(t) > 0$ for all t in G then for each x in $\mathscr{B}_+$ we have

$$0 = \alpha'_f(\phi)(x) = \int \phi(\alpha_t''(x)) f(t)\, dt,$$

if and only if $\phi(\alpha_t(x)) = 0$ almost everywhere; which implies that $\alpha'_f(\phi)(\alpha_s(x)) = 0$ for every s in G. In particular the set of null projections for $\alpha'_f(\phi) | \mathscr{C}$ is G-invariant whence $\alpha'_f(\phi) \in S_{qi}$. Since G is separable it is possible to choose an approximate unit $\{f_n\}$ in $L_1(G)$ consisting of non-negative functions, and thus for each ϕ in S and x in A

$$\alpha'_{f_n}(\phi)(x) = \int \phi(\alpha_t(x)) f_n(t)\, dt \to \phi(x),$$

which shows that S_{qi} is weak* dense in S.

At the same time we see that if ϕ were G-continuous then

$$\|\alpha'_{f_n}(\phi) - \phi\| \to 0$$

so that the norm closure of S_{qi} is the set of G-continuous states.

7.4.14. Notes and remarks. Ergodic theory is the study of commutative dynamical systems, either in the C^*-sense (a group of homeomorphisms of a locally compact space) or in the W^*-sense (a group of measure-preserving transformations on a measure space (T, μ)). A standard reference is Jacobs [119]. Dynamical systems for which $G = \boldsymbol{R}$ are particularly important for the applications in physics, and have given name to all other cases.

In quantum physics the observables are best described as non-commuting operators on a Hilbert space, and in some models it is assumed that the observables form (the self-adjoint part of) a C^*-algebra. Time evolution and/or spacial translation of the observables is then described by a (non-commutative) C^*-dynamical system. For this reason a number of mathematical physicists have worked with C^*-dynamical systems and a large part of the theory presented in this and the next chapter arose in close connection with the applications.

Theorem 7.4.10 is a reformulation of a result by Borchers [21, 22]. The results 7.4.11–7.4.13 are taken from the paper by Guichardet and Kastler [109]. There it is also shown that each quasi-invariant (resp. covariant) state ϕ can be disintegrated in centrally ergodic, quasi-invariant (resp. covariant) states, cf. 4.8.7. (A state ψ is centrally ergodic if α'' gives rise to an ergodic transformation group on $(\hat{A}, \mu_\psi)$.) The survey by Guichardet [108] treats W^*-dynamical systems in great detail.

7.5. B*-dynamical systems

7.5.1. LEMMA. *For each B^*-dynamical system $(\mathscr{A}, G, \alpha)$ the set $\mathscr{A}^c$ of elements x in $\mathscr{A}$ for which the function $t \to \alpha_t(x)$ is norm continuous is a G-invariant C^*-subalgebra of $\mathscr{A}$ generated by the elements $\alpha_f(x)$, $x \in \mathscr{A}$, $f \in L^1(G)$. If $\mathscr{M}$ is a von Neumann algebra and $(\mathscr{M}, G, \alpha)$ is a W^*-dynamical system then $\mathscr{M}^c$ is weakly dense in $\mathscr{M}$.*

Proof. It is easy to verify that the set $\mathscr{A}^c$ is a G-invariant C^*-subalgebra of $\mathscr{A}$. If $x \in \mathscr{A}$ and $f \in L^1(G)$ then for each t in G

$$\alpha_t \alpha_f(x) = \int \alpha_{ts}(x) f(s)\, ds = \int \alpha_s(x) f(t^{-1}s)\, ds.$$

We see that $\|\alpha_t \alpha_f(x) - \alpha_f(x)\| \to 0$ as $t \to e$ so that $\alpha_f(x) \in \mathscr{A}^c$. On the other hand, if $x \in \mathscr{A}^c$ and $\{f_\lambda\}$ is an approximate unit for $L^1(G)$ then

$$\|\alpha_{f_\lambda}(x) - x\| \leqslant \int \|\alpha_s(x) - x\| f_\lambda(s)\, ds \to 0,$$

from which it follows that $\mathscr{A}^c$ is generated by elements of the form $\alpha_f(x)$.

Finally, if $(\mathscr{M}, G, \alpha)$ is a W^*-dynamical system then for each normal state ϕ of $\mathscr{M}$ and each x in $\mathscr{M}$ we have

$$|\phi(x - \alpha_{f_\lambda}(x))| \leqslant \int |\phi(x - \alpha_s(x))| f_\lambda(s)\, ds \to 0,$$

whence $\mathscr{M}^c$ is σ-weakly (=weakly) dense in $\mathscr{M}$.

7.5.2. LEMMA. *For $i = 1, 2$ let $(\mathscr{A}_i, G, \alpha^i)$ be B^*-dynamical systems. If $\pi: \mathscr{A}_1 \to \mathscr{A}_2$ is a surjective, σ-normal, G-invariant morphism then $\pi(\mathscr{A}_1^c) = \mathscr{A}_2^c$.*

Proof. Since $\alpha^2 \circ \pi = \pi \circ \alpha^1$ we see that $\pi(\mathscr{A}_1^c) \subset \mathscr{A}_2^c$. Conversely, if $x \in \mathscr{A}_1$ such that $\pi(x) \in \mathscr{A}_2^c$ then with $\{f_\lambda\}$ as an approximate unit for $L^1(G)$ we have $\alpha^1_{f_\lambda}(x) \in \mathscr{A}_1^c$ by 7.5.1 and

$$\pi(\alpha^1_{f_\lambda}(x)) = \alpha^2_{f_\lambda}(\pi(x)) \to \pi(x) \qquad \text{(norm);}$$

whence $\pi(\mathscr{A}_1^c) = \mathscr{A}_2^c$.

7.5.3. LEMMA. *For $i = 1, 2, 3$ let $(\mathscr{A}_i, G, \alpha^i)$ be B^*-dynamical systems and let $\pi_i: \mathscr{A}_i \to \mathscr{A}_3$, $i = 1, 2$, be surjective σ-normal, G-invariant morphisms. If $(\mathscr{A}_1, G, \alpha^1)$ is the enveloping system for a separable C^*-dynamical system*

(A, G, α) (see 7.4.7) there is a σ-normal G-invariant morphism $\lambda: \mathscr{A}_1 \to \mathscr{A}_2$ such that $\pi_2 \circ \lambda = \lambda \circ \pi_1$. If λ' is another such morphism there is a G-invariant projection p_2 in $\mathscr{A}_2$ with $\pi_2(1 - p_2) = 0$ such that $p_2\lambda(x) = p_2\lambda'(x)$ for all x in $\mathscr{A}_1$; so that λ is 'essentially unique'.

Proof. Since $\pi_2(\mathscr{A}_2^c) = \pi_1(\mathscr{A}_1^c) \supset \pi_1(A)$ by 7.5.2 we can find a separable C^*-subalgebra B_0 of $\mathscr{A}_2^c$ such that $\pi_2(B_0) \supset \pi_1(A)$. Let G_0 be a countable, dense subgroup of G and take B as the (separable) C^*-algebra generated by $\bigcup \alpha_s^2(B_0)$, $s \in G_0$. Since $\mathscr{A}_2^c$ is G-invariant we have $B \subset \mathscr{A}_2^c$. But then B is G-invariant since it is closed and G_0-invariant, so that (B, G, α^2) is a separable C^*-dynamical system.

Let p be the strong limit in $\mathscr{A}_2$ of a countable approximate unit for the separable C^*-algebra $B \cap \ker \pi_2$. Since both B and $\ker \pi_2$ are G-invariant sets, p is G-invariant. Moreover, $\pi_2(p) = 0$ and $\pi_2 | (1 - p)B(1 - p)$ is an isomorphism. For each x in A we can therefore define $\lambda_0(x)$ as the unique element in $(1 - p)B(1 - p)$ such that $\pi_2(\lambda_0(x)) = \pi_1(x)$. For each t in G we have

$$\pi_2(\lambda_0(\alpha_t^1(x))) = \pi_1(\alpha_t^1(x)) = \alpha_t^3(\pi_1(x)) = \alpha_t^3(\pi_2(\lambda_0(x))) = \pi_2(\alpha_t^2(\lambda_0(x))),$$

whence $\lambda_0 \circ \alpha_t^1 = \alpha_t^2 \circ \lambda_0$, so that λ_0 is a G-invariant morphism. By 4.5.9 there is a unique extension of λ_0 to a σ-normal morphism $\lambda: \mathscr{A}_1 \to \mathscr{A}_2$. We have $\lambda \circ \alpha^1 = \alpha^2 \circ \lambda$ and $\pi_2 \circ \lambda = \pi_1$ since this is true for the generating algebra A.

Suppose that $\lambda': \mathscr{A}_1 \to \mathscr{A}_2$ was another σ-normal morphism such that $\lambda' \circ \alpha^1 = \alpha^2 \circ \lambda'$ and $\pi_2 \circ \lambda' = \pi_1$. Take a dense G_0-invariant sequence $\{x_n\}$ in A_{sa} and for each n let $[\lambda(x_n) - \lambda'(x_n)]$ be the range projection of $\lambda(x_n) - \lambda'(x_n)$. Put $1 - p_2 = \vee [\lambda(x_n) - \lambda'(x_n)]$. Then $\pi_2(1 - p_2) = 0$ since $\ker \pi_2$ is a monotone sequentially closed ideal in $\mathscr{A}_2$, and $p_2\lambda(x_n) = p_2\lambda'(x_n)$ for all n. However, the set

$$\{x \in (\mathscr{A}_1)_{sa} \mid p_2\lambda(x) = p_2\lambda'(x)\}$$

is monotone sequentially closed and contains $\{x_n\}$. It therefore equals $(\mathscr{A}_1)_{sa}$. Finally, to show that p_2 is G-invariant note that $1 - p_2$ is G_0-invariant and is the limit of an increasing sequence of elements from $\mathscr{A}_2^c$. Thus for each σ-normal state ϕ of $\mathscr{A}_2$ the function $t \to \phi(\alpha_t^2(1 - p_2))$ is lower semi-continuous. Since the function is constant on G_0 we conclude that it is constant on G, whence $1 - p_2$ is G-invariant.

7.5.4. PROPOSITION. (cf. 4.6.6). *For $i = 1, 2, 3$ let $(\mathscr{A}_i, G, \alpha^i)$ be a B^*-dynamical systems and let $\pi_i: \mathscr{A}_i \to \mathscr{A}_3$, $i = 1, 2$ be surjective, σ-normal, G-invariant morphisms. If $(\mathscr{A}_i, G, \alpha^i)$ are the enveloping systems for separable C^*-dynamical systems (A_i, G, α^i), $i = 1, 2$, then for each $i = 1, 2$ there is a central, G-invariant projection p_i with $\pi_i(1 - p_i) = 0$ and a G-invariant isomorphism $\lambda: p_1\mathscr{A}_1 \to p_2\mathscr{A}_2$ such that $\pi_2 \circ \lambda = \pi_1$.*

Proof. From 7.5.3 we obtain σ-normal, G-invariant morphisms $\lambda_1: \mathscr{A}_1 \to \mathscr{A}_2$ and $\lambda_2: \mathscr{A}_2 \to \mathscr{A}_1$ such that $\pi_2 \circ \lambda_1 = \pi_1$ and $\pi_1 \circ \lambda_2 = \pi_2$. Set $\rho = \lambda_2 \circ \lambda_1$ and note that $\pi_1 \circ \rho = \pi_1$. Thus ρ is 'essentially' equal to the identity map so by 7.5.3 there is a G-invariant projection y_1 in $\mathscr{A}_1$ with $\pi_1(y_1) = 0$ such that $(1 - y_1)(x - \rho(x)) = 0$ for all x in $\mathscr{A}_1$. For any projection p the central cover of p can be calculated as

$$c(p) = \bigvee u_n^* p u_n$$

where $\{u_n\}$ is any dense sequence in the unitary group of A_1 (see 2.6.3). It follows that the central cover of a G-invariant projection in $\mathscr{A}_1$ is again G-invariant. Therefore the proof of 4.6.6 applies verbatim to cover also the more complicated situation involving the group G.

7.5.5. THEOREM. *Let $(\mathscr{M}, G, \alpha)$ be a W^*-dynamical system on a separable Hilbert space H and assume that also G is separable. There is a separable C^*-dynamical system (A_1, G, α^1) and a G-invariant representation (π_1, H) of A_1 such that $\mathscr{M} = \pi_1(A_1)''$. If (A_2, G, α^2) is another dynamical system with a similar G-invariant representation (π_2, H) then with $(\mathscr{B}_i, G, \alpha^i)$, $i = 1, 2$, as the B^*-dynamical systems associated with (A_i, G, α^i) there are central, G-invariant projections p_i in $\mathscr{B}_i$ with $\pi_i(1 - p_i) = 0$ and a G-invariant isomorphism $\lambda: p_1\mathscr{B}_1 \to p_2\mathscr{B}_2$ such that $\pi_1''(x) = \pi_2''(\lambda(x))$ for all x in $p_1\mathscr{B}_1$.*

Proof. It follows from 7.5.1 and the separability of H that there is a separable C^*-algebra A_0 in $\mathscr{M}^c$ that is weakly dense in $\mathscr{M}$. As in the proof of 7.5.3 we then replace A_0 by the separable C^*-algebra A_1 generated $\cup\, \alpha_s(A_0)$, where s runs through a countable dense subgroup of G. We find as before that A_1 is G-invariant, and contained in $\mathscr{M}^c$ so that with $\alpha^1 = \alpha | A_1$ we have a separable C^*-dynamical system (A_1, G, α^1). Take π_1 as the identity map of A. Then (π_1, H) is a G-invariant representation and $\pi_1(A_1)'' = \mathscr{M}$.

The essential uniqueness of the system (A_1, G, α^1) follows directly from 7.5.4.

7.5.6. PROPOSITION. *Let (π, H) be a separable, σ-normal representation of a standard Borel *-algebra $\mathscr{A}$. If $(\pi(\mathscr{A}), G, \alpha)$ is a W^*-dynamical system in which G is separable and $t \to \alpha_t$ is uniformly continuous there is a B^*-dynamical system $(\mathscr{A}, G, \bar{\alpha})$ such that $t \to \bar{\alpha}_t$ is uniformly continuous and $\alpha \circ \pi = \pi \circ \bar{\alpha}$.*

Proof. By 7.5.5 there is a separable C^*-dynamical system (A, G, α^1) and a G-invariant representation (π_1, H) such that $\pi_1(A)'' = \pi(\mathscr{A})$. However, since the representation $t \to \alpha_t^1$ is uniformly continuous (i.e. $\|\iota - \alpha_t^1\| \to 0$ as $t \to e$) the same is true for the double transposed action (again denoted by α^1) of G on the enveloping Borel *-algebra $\mathscr{B}$ of A. If therefore q_0 is a projection in $\mathscr{B}$ with $\pi_1''(q_0) = 0$ let G_0 be a dense countable subgroup of G and define

$$q = \bigvee \alpha_t^1(q_0), \qquad t \in G_0.$$

Then $\pi_1''(q) = 0$ and $\alpha_t(q) = q$ for every t in G.

By 4.6.6 there are central projections p and p_1 in $\mathscr{A}$ and $\mathscr{B}$, respectively, and an isomorphism $\lambda : p\mathscr{A} \to p_1\mathscr{B}$ *such that* $\pi_1'' \circ \lambda = \pi$ and

$$\pi(1 - p) = \pi_1''(1 - p_1) = 0.$$

From the remark above we may assume that p_1 is G-invariant. Define $\bar{\alpha}_t$ on $\mathscr{A}$ by

$$\bar{\alpha}_t(x) = (1 - p)x + \lambda^{-1}(\alpha_t^1(\lambda(px))$$

for every x in $\mathscr{A}$. Then $t \to \bar{\alpha}_t$ is a uniformly continuous representation of G in $\mathrm{Aut}(\mathscr{A})$and

$$\pi(\bar{\alpha}_t(x)) = \pi(\lambda^{-1}(\alpha_t^1(\lambda(px)))) = \alpha_t(\pi(x))$$

for every x in $\mathscr{A}$, as desired.

7.5.7. THEOREM. *Let A be a separable C^*-algebra with enveloping Borel $*$-algebra $\mathscr{B}$. Two separable disjoint representations (π_1, H_1) and (π_2, H_2) of A will generate isomorphic von Neumann algebras $\pi_1(A)''$ and $\pi_2(A)''$ if and only if there is an automorphism α of $\mathscr{B}$ such that $(\pi_2'' \circ \alpha, H_2)$ and (π_1, H_1) are equivalent.*

Proof. If $\alpha \in \mathrm{Aut}(\mathscr{B})$ then

$$\pi_2(\alpha(A))'' = \pi_2''(\alpha(\mathscr{B})) = \pi_2''(\mathscr{B}).$$

Thus any representation equivalent with $(\pi_2'' \circ \alpha, H_2)$ generates a von Neumann algebra isomorphic with $\pi_2(A)''$.

Conversely, let $\rho : \pi_1(A)'' \to \pi_2(A)''$ be an isomorphism. If

$$(\pi, H) = (\pi_1 \oplus \pi_2, H_1 \oplus H_2)$$

then by 3.8.11 we have $\pi(A)'' = \pi_1(A)'' \oplus \pi_2(A)''$. Define an automorphism β of $\pi(A)''$ by

$$\beta(x \oplus y) = \rho^{-1}(y) \oplus \rho(x); \qquad x \in \pi_1(A)'', \qquad y \in \pi_2(A)''.$$

Since the group generated by β is a representation of $\boldsymbol{Z}$ in $\mathrm{Aut}(\pi(A)'')$ we may apply 7.5.6 to obtain an automorphism α of $\mathscr{B}$ such that

$$\begin{aligned}\pi_1''(\alpha(x)) \oplus \pi_2''(\alpha(x)) &= \beta(\pi_1''(x) \oplus \pi_2''(x)) \\ &= \rho^{-1}(\pi_2''(x)) \oplus \rho(\pi_1''(x))\end{aligned}$$

for every x in $\mathscr{B}$. It follows that

$$\pi_2'' \circ \alpha = \rho \circ \pi_1'',$$

so that $(\pi_2'' \circ \alpha, H_2)$ and (π_1'', H) are equivalent.

7.5.8. COROLLARY. *The automorphism group of a standard Borel *-algebra $\mathscr{A}$ acts transitively on the set of σ-normal factor representations of $\mathscr{A}$.*

7.5.9. *Notes and remarks.* This section is taken from the author's paper [201]. Theorem 7.5.5 is a non-commutative generalization of Mackey's result from [165] on point realization of transformation groups. In the commutative case any σ-weakly continuous automorphism group can be lifted to a group of Borel transformations, using the theory of standard Borel spaces. Thus it may well be that 7.5.6 is valid also when α is just σ-weakly continuous. For a Glimm algebra A, Powers [211] proved a very much stronger version of 7.5.7: if the von Neumann algebras $\pi_1(A)''$ and $\pi_2(A)''$ are isomorphic, there is an automorphism α of A such that (π_1, H_1) and $(\pi_2 \circ \alpha, H_2)$ are equivalent. Later this result was generalized by Bratteli [25] to faithful representations of an arbitrary AF-algebra (cf. 6.4.8).

7.6. Crossed products

7.6.1. Let (A, G, α) be a C^*-dynamical system. We define involution and convolution on the linear space $K(G, A)$ of continuous functions from G to A with compact supports by

$$y^*(t) = \Delta(t)^{-1}\alpha_t(y(t^{-1})^*)$$

$$(y \times z)(t) = \int y(s)\alpha_s(z(s^{-1}t))\,ds,$$

for all y, z in $K(G, A)$. Straightforward computations show that $K(G, A)$ becomes a *-algebra with convolution as product. For each y in $K(G, A)$ define

$$\|y\|_1 = \int \|y(t)\|\,dt.$$

Then $K(G, A)$ is a normed algebra with an isometric involution, and we denote by $L^1(G, A)$ its completion.

For each x in A and f in $L^1(G)$ we denote by $x \otimes f$ the element in $L^1(G, A)$ such that $(x \otimes f)(t) = xf(t)$. Note that the linear span of elements of the form $x \otimes f$, $x \in A$, $f \in K(G)$, is dense in $L^1(G, A)$.

7.6.2. As in 3.12.1 we define left, right and double centralizers on the Banach algebra $L^1(G, A)$; the only difference being that we *assume* continuity of the maps involved. For each x in $M(A)$ and μ in $M(G)$ with compact support, define linear maps $L(x, \mu)$ and $R(x, \mu)$ on $K(G, A)$

$$(L(x, \mu)y)(t) = x \int \alpha_s(y(s^{-1}t))\,d\mu(s),$$

$$(R(x, \mu)y)(t) = \int y(ts^{-1})\alpha_{ts^{-1}}(x)\Delta(s)^{-1}\,d\mu(s).$$

It is immediate that $L(x, \mu)$ and $R(x, \mu)$ are bounded (by $\|x\|\,\|\mu\|$) and thus extend by continuity to linear operators on $L^1(G, A)$. Having done this we see that the restriction that μ has compact support is no longer necessary.

7.6.3. LEMMA. *For each x in $M(A)$ and μ in $M(G)$ the operator $L(x, \mu)$ (respectively $R(x, \mu)$) is a left (respectively right) centralizer of $L^1(G, A)$, and the pair $(R(x, \mu), L(x, \mu))$ is a double centralizer.*

Proof. Take y and z in $K(G, A)$ and assume that μ has compact support. Then

$$\begin{aligned} L(x,\mu)(y \times z)(t) &= x \int \alpha_s((y \times z)(s^{-1}t))\, d\mu(s) \\ &= x \iint \alpha_s(y(r)\alpha_r(z(r^{-1}s^{-1}t)))\, d\mu(s)\, dr \\ &= x \iint \alpha_s(y(s^{-1}r))\, d\mu(s)\alpha_r(z(r^{-1}t))\, dr \\ &= (L(x,\mu)y \times z)(t). \end{aligned}$$

Furthermore,

$$\begin{aligned} R(x,\mu)(y \times z)(t) &= \int (y \times z)(ts^{-1})\alpha_{ts^{-1}}(x)\Delta(s^{-1})\, d\mu(s) \\ &= \iint y(r)\alpha_r(z(r^{-1}ts^{-1}))\alpha_{ts^{-1}}(x)\Delta(s^{-1})\, d\mu(s)\, dr \\ &= \int y(r)\alpha_r(\int z(r^{-1}ts^{-1})\alpha_{r^{-1}ts^{-1}}(x)\Delta(s^{-1})\, d\mu(s))\, dr \\ &= (y \times R(x,\mu)z)(t). \end{aligned}$$

Thus $L(x, \mu)$ and $R(x, \mu)$ are centralizers. To show that the pair $(R(x, \mu), L(x, \mu))$ is a double centralizer we compute

$$\begin{aligned} (R(x,\mu)y \times z)(t) &= \int R(x,\mu)y(s)\alpha_s(z(s^{-1}t))\, ds \\ &= \iint y(sr^{-1})\alpha_{sr^{-1}}(x)\Delta(r^{-1})(\alpha_s(z(s^{-1}t)))d\mu(r)\, ds \\ &= \iint y(s)\alpha_s(x)\alpha_{sr}(z(r^{-1}s^{-1}t))\, d\mu(r)\, ds \\ &= (y \times L(x,\mu)z)(t). \end{aligned}$$

7.6.4. PROPOSITION. *If (π, u, H) is a covariant representation of (A, G, α) there is a non-degenerate representation $(\pi \times u, H)$ of $L^1(G, A)$ such that*

$$(*) \qquad (\pi \times u)(y) = \int \pi(y(t))u_t\, dt$$

for every y in $K(G, A)$. Moreover, the correspondence $(\pi, u, H) \to (\pi \times u, H)$ is a bijection onto the set of non-degenerate representations of $L^1(G, A)$.

Proof. Take y in $K(G, A)$ and define $(\pi \times u)(y)$ on H by $(*)$. Then

$$\begin{aligned} (\pi \times u)(y^*) &= \int \pi(\Delta(t)^{-1}\alpha_t(y(t^{-1})^*)u_t\, dt \\ &= \int \Delta(t)^{-1}u_t\pi(y(t^{-1})^*)\, dt = \int u_t^*\pi(y(t)^*)\, dt = ((\pi \times u)(y))^*; \end{aligned}$$

$$\|(\pi \times u)(y)\| \leqslant \int \|\pi(y(t))u_t\|\, dt = \|y\|_1;$$

$$
\begin{aligned}
(\pi \times u)(y \times z) &= \int \pi \int y(s)\alpha_s(z(s^{-1}t))\,ds\,u_t\,dt \\
&= \iint \pi(y(s))u_s\pi(z(s^{-1}t))u_s^* u_t\,ds\,dt \\
&= \iint \pi(y(s))u_s\pi(z(t))u_t\,ds\,dt = (\pi \times u)(y)(\pi \times u)(z);
\end{aligned}
$$

which show that $(\pi \times u, H)$ extends to a representation of $L^1(G, A)$.

Conversely, if (ρ, H) is a non-degenerate representation of $L^1(G, A)$ take an approximate unit $\{y_\lambda\}$ for $L^1(G, A)$ contained in $K(G, A)$. One such may be obtained by putting $y_\lambda = x_\lambda \otimes f_\lambda$ where $\{x_\lambda\}$ and $\{f_\lambda\}$ are approximate units for A and for $L^1(G)$, respectively. Since (ρ, H) is non-degenerate $\{\rho(y_\lambda)\}$ converges strongly to 1. For each x in $\tilde{A}$ and t in G we define

$$
\begin{aligned}
\pi(x) &= \operatorname{Lim} \rho(L(x, \delta_e)y_\lambda) = \operatorname{Lim} \rho(R(x, \delta_e)y_\lambda); \\
u_t &= \operatorname{Lim} \rho(L(1, \delta_t)y_\lambda) = \operatorname{Lim} \rho(R(1, \delta_t)y_\lambda);
\end{aligned}
$$

where the limits are taken in the weak topology. The existence of these limits follows as in the proof of 3.12.3. Since $L(xy, \delta_e) = L(x, \delta_e)\, L(y, \delta_e)$ we see that the map $x \to \pi(x)$ is a homomorphism of $\tilde{A}$ which is *-preserving by the first statement in 7.6.3. Thus (π, H) is a representation of $\tilde{A}$. Since $L(1, \delta_{st}) = L(1, \delta_s)\, L(1, \delta_t)$ and $L(1, \delta_e) = 1$ we see that the map $t \to u_t$ is a unitary representation of G, the continuity following from the facts that the map $t \to L(1, \delta_t)y$ is continuous in the norm $\|\cdot\|_1$ for every y in $L^1(G, A)$, and that vectors of the form $\rho(y)\xi$, $\xi \in H$, are dense in H. Finally we have

$$
\begin{aligned}
u_t\pi(x)u_t^*\rho(y) &= \rho(L(1, \delta_t)(L(x, \delta_e)(L(1, \delta_{t^{-1}}))) \\
&= \rho(L(\alpha_t(x), \delta_e)y) = \pi(\alpha_t(x))\rho(y),
\end{aligned}
$$

for every y in $K(G, A)$. Taking $y = y_\lambda$ and passing to the limit we see that (π, u, H) is a covariant representation of (A, G, α).

To show that the correspondence is bijective take the covariant representation (π, u, H) obtained from a non-degenerate representation (ρ, H) of $L^1(G, A)$ and for y, z in $K(G, A)$ note that

$$
(L(y(t), \delta_e)(L(1, \delta_t)z))(s) = y(t)\alpha_t(z(t^{-1}s))
$$

for each t in G. Therefore

$$
\begin{aligned}
\int \pi(y(t))u_t\,dt\,\rho(z) &= \int \rho(L(y(t), \delta_e)(L(1, \delta_t)z))\,dt \\
&= \int \rho(y(t)\alpha_t(z(t^{-1}s)))\,dt \\
&= \rho(y \times z) = \rho(y)\rho(z).
\end{aligned}
$$

Taking $z = y_\lambda$ and passing to the limit we obtain

$$
\int \pi(y(t))u_t\,dt = \rho(y),
$$

whence $\rho = \pi \times u$. Conversely, if $\rho = \pi \times u$ take $y_\lambda = x_\lambda \otimes f_\lambda$ as before and note that for each x in A and t in G we have

$$(L(x, \delta_t)y_\lambda)(s) = x\alpha_t(x_\lambda)f_\lambda(t^{-1}s).$$

Consequently,

$$\operatorname{Lim} \rho(L(x, \delta_e)y_\lambda) = \operatorname{Lim} \int \pi(xx_\lambda)u_s f_\lambda(s)\, ds = \pi(x);$$

$$\operatorname{Lim} \rho(L(1, \delta_t)y_\lambda) = \operatorname{Lim} \int \pi(\alpha_t(x_\lambda))u_s f_\lambda(t^{-1}s)\, ds = u_t.$$

This completes the proof.

7.6.5. The *universal representation* (π_u, H_u) of $L^1(G, A)$ is the direct sum of all non-degenerate representations of $L^1(G, A)$ and the *crossed product* (or the *covariance algebra*) of the C^*-dynamical system (A, G, α) is the norm closure of $\pi_u(L^1(G, A))$ in $\boldsymbol{B}(H_u)$, denoted by $G \underset{\alpha}{\times} A$. Thus by definition the identity map of $G \underset{\alpha}{\times} A$ into $\boldsymbol{B}(H_u)$ is the universal representation of $G \underset{\alpha}{\times} A$.

7.6.6. THEOREM. *For each C^*-dynamical system (A, G, α) there is a covariant representation (π, u, H) such that*

$$G \underset{\alpha}{\times} A \subset C^*(\pi(A) \cup u_G) \subset M(G \underset{\alpha}{\times} A);$$

and for any other covariant representation (π', u', H') there is a unique representation (ρ, H') of $G \underset{\alpha}{\times} A$ such that $\pi' = \rho \circ \pi$ and $u' = \rho \circ u$.

Proof. The existence and universal properties of (π, u, H) follows from the definition of $G \underset{\alpha}{\times} A$ together with 7.6.4. The inclusion $\pi(A) \cup u_G \subset M(G \underset{\alpha}{\times} A)$ follows from the construction of the correspondence between representations of $G \underset{\alpha}{\times} A$ and covariant representations of (A, G, α) in 7.6.4. To show that $G \underset{\alpha}{\times} A$ is contained in the C^*-algebra generated by $\pi(A) \cup u_G$ take y in $K(G, A)$. For each $\varepsilon > 0$ there is a compact set C in G containing the support of y, a finite set $\{x_n\}$ in A and functions $\{f_n\}$ in $K(G)$ with supports in C such that

$$\|y(t) - \sum f_n(t)x_n\| < \varepsilon \mu_G(C)^{-1}$$

for all t in G. The elements $\int u_t f_n(t)\, dt$ belong to $C^*(u_G)$, whence

$$z = \sum \int \pi(x_n)u_t f_n(t)\, dt \in C^*(\pi(A) \cup u_G).$$

Since $y = \int \pi(y(t))u_t\, dt$ by (*) in 7.6.4 and $\|y - z\|_1 < \varepsilon$ we conclude that $y \in C^*(\pi(A) \cup u_G)$.

7.6.7. We identify A with its image in $M(G \underset{\alpha}{\times} A)$ via the map $x \to L(x, \delta_e)$ cf. 7.6.2 and let $t \to u_t$ be the unitary representation of G into $M(G \underset{\alpha}{\times} A)$ described in 7.6.6. For each functional ϕ on $G \underset{\alpha}{\times} A$ define a function

$\Phi: G \to A^*$ by

$$\Phi(t)(x) = \phi(xu_t), \qquad t \in G, x \in A.$$

The set of such functions is denoted by $B(G \underset{\alpha}{\times} A)$ and the elements in the generating cone $B_+(G \underset{\alpha}{\times} A)$ arising from positive functionals are said to be *positive definite* (with respect to α).

From the Cauchy–Schwarz inequality it follows that each Φ in $B_+(G \underset{\alpha}{\times} A)$ is continuous in the norm topology on A^* and bounded by $\|\Phi(e)\|$. From (*) in 7.6.4 we see that

$$\phi(y) = \int \Phi(t)(y(t))\,dt$$

for each y in $K(G, A)$, so that the correspondence $\phi \to \Phi$ is a bijection from $(G \underset{\alpha}{\times} A)^*$ onto $B(G \underset{\alpha}{\times} A)$.

7.6.8. Proposition. *The following conditions on a bounded, norm continuous function* $\Phi: G \to A^*$ *are equivalent*:

(i) Φ *is positive definite*;

(ii) $\Sigma_{ij}\Phi(s_i^{-1}s_j)(\alpha_{s_i^{-1}}(x_i^* x_j)) \geqslant 0$ *for all finite sets* $\{s_i\}$ *in* G *and* $\{x_i\}$ *in* A;

(iii) $\Sigma_{nm}\int \Phi(t)(y_n^* \alpha_t(y_m))(f_n^* \times f_m)(t)\,dt \geqslant 0$ *for all finite sets* $\{f_n\}$ *in* $K(G)$ *and* $\{y_n\}$ *in* A;

(iv) $\int \Phi(t)(y^* \times y)(t))\,dt \geqslant 0$ *for each* y *in* $K(G, A)$.

Proof. (i) $\Rightarrow$ (ii). Given $\{s_i\}$ and $\{x_i\}$ define $x = \Sigma x_i u_{s_i}$ in $M(G \underset{\alpha}{\times} A)$. If ϕ is the functional on $G \underset{\alpha}{\times} A$ corresponding to Φ we have

$$0 \leqslant \phi(x^*x) = \phi\left(\sum_{ij} \alpha_{s_i^{-1}}(x_i^* x_j) u_{s_i^{-1}s_j}\right)$$
$$= \sum_{ij} \Phi(s_i^{-1}s_j)(\alpha_{s_i^{-1}}(x_i^* x_j)).$$

(ii) $\Rightarrow$ (iii). Given $\{f_n\}$ in $K(G)$ and $\{y_n\}$ in A take a finite set $\{s_i\}$ in G and apply condition (ii) to the sets $\{s_{in}\}$ and $\{x_{in}\}$ where $s_{in} = s_i$ and $x_{in} = \alpha_{s_i}(y_n)f_n(s_i)$. Thus

$$0 \leqslant \sum_{ijnm} \Phi(s_i^{-1}s_j)(y_n^* \alpha_{s_i^{-1}s_j}(y_m))\bar{f}_n(s_i)f_m(s_j).$$

It follows that

$$0 \leqslant \sum_{nm} \iint \Phi(s^{-1}t)(y_n^* \alpha_{s^{-1}t}(y_m))\bar{f}_n(s)f_m(t)\,ds\,dt$$
$$= \sum_{nm} \iint \Phi(t)(y_n^* \alpha_t(y_m))f_n^*(s^{-1})\Delta(s)^{-1}f_m(st)\,ds\,dt$$
$$= \sum_{nm} \int \Phi(t)(y_n^* \alpha_t(y_m))(f_n^* \times f_m)(t)\,dt.$$

(iii) ⇒ (iv). Each y in $K(G, A)$ can be approximated (in $L^1(G, A)$) by elements of the form $y(t) = \Sigma f_n(t)\alpha_t(y_n)$. However, for such an element $y^*(t) = \Sigma f_n^*(t) y_n^*$ and

$$y^* \times y(t) = \sum_{nm} y_n^* \alpha_t(y_m) f_n^* \times f_m(t).$$

Condition (iii) tells us that $\int \Phi(t)(y^* \times y(t))\,dt \geqslant 0$ and the result follows by continuity.

(iv) ⇒ (i). For each y in $K(G, A)$ define

$$\phi(y) = \int \Phi(t)(y(t))\,dt.$$

From condition (iv) we see that ϕ is a positive functional on $K(G, A)$ and as in the proof of 7.1.9 this implies that ϕ has a unique extension to a positive functional on $G \underset{\alpha}{\times} A$.

7.6.9. COROLLARY. *If* $\Phi \in B_+(G \underset{\alpha}{\times} A)$ *and* $\Psi \in B_+(G)$ *then*

$$\Psi \cdot \Phi \in B_+(G \underset{\alpha}{\times} A).$$

Proof. Condition (ii) in 7.6.8 is equivalent to the statement that each matrix (a_{ij}) with

$$a_{ij} = \Phi(s_i^{-1} s_j)(\alpha_{s_i^{-1}}(x_i^* x_j))$$

is positive. Since the pointwise product of positive matrices is again positive (see the proof of 7.1.10) we conclude that $\Psi \cdot \Phi$ satisfies 7.6.8 (ii).

7.6.10 PROPOSITION. *Let* $S(G \underset{\alpha}{\times} A)$ *be the state space of* $G \underset{\alpha}{\times} A$ *and* $B_+^1(G \underset{\alpha}{\times} A)$ *the set of positive definite functions* $\Phi: G \to A^*$ *with* $\|\Phi(e)\| = 1$. *There is an affine homeomorphism of* $S(G \underset{\alpha}{\times} A)$ *in the weak* topology onto* $B_+^1(G \underset{\alpha}{\times} A)$ *equipped with the topology of weak* convergence uniformly on compact subsets of* G.

Proof. It is clear that the correspondence $\phi \to \Phi$ defined in 7.6.7 is an affine isomorphism of $S(G \underset{\alpha}{\times} A)$ onto $B_+^1(G \underset{\alpha}{\times} A)$.

Let $\{\phi_\lambda\}$ be a net in $S(G \underset{\alpha}{\times} A)$ that is weak* convergent to ϕ in $S(G \underset{\alpha}{\times} A)$, and let $\{\Phi_\lambda\}$ and Φ be the corresponding elements in $B_+^1(G \underset{\alpha}{\times} A)$. Combining 7.6.8 (iii) with 7.1.9 (iv) we see that for each y in $\tilde{A}$ the functions

$$\Psi_\lambda : t \to \Phi_\lambda(t)(y^* \alpha_t(y)), \qquad \Psi : t \to \Phi(t)(y^* \alpha_t(y))$$

are positive definite (in the ordinary sense) and that the corresponding net of positive functionals $\{\psi_\lambda\}$ is weak* convergent to ψ in $C^*(G)^*$. By 7.1.11 this implies that $\{\Psi_\lambda\}$ converges to Ψ uniformly on compact sets. Using the polarization identity

$$4x^*\alpha_t(y) = \sum_{k=0}^{3} \mathrm{i}^k(y + \mathrm{i}^k x)^*\alpha_t(y + \mathrm{i}^k x)$$

we see that $\{\Phi_\lambda(t)(x^*\alpha_t(y))\}$ converges to $\Phi(t)(x^*\alpha_t(y))$ uniformly on compact sets for all x and y in $\tilde{A}$. With $y = 1$ we obtain the desired result.

Conversely, if $\Phi_\lambda(t) \to \Phi(t)$ weak*, uniformly on compact subsets of G, then for each element y in $G \times_\alpha A$ of the form $y = f \otimes x$, $f \in K(G)$, $x \in A$ we have

$$\phi_\lambda(y) = \int \Phi_\lambda(t)(x) f(t)\,\mathrm{d}t \to \int \Phi(t)(x) f(t)\,\mathrm{d}t = \phi(y).$$

Since finite sums of such elements form a dense set in $G \times_\alpha A$ we conclude that $\phi_\lambda \to \phi$ weak*.

7.6.11. Notes and remarks. Crossed products of C^*-algebras with discrete groups were introduced by Turumaru [271]. Later Zeller-Meyer carried out a penetrating analysis of this case in his thesis [277]. Crossed products of abelian C^*-algebras (but arbitrary locally compact groups), i.e. transformation group C^*-algebras were introduced by Glimm [98], and studied in great detail by Effros and Hahn [77]. General crossed products were defined by Doplicher, Kastler and Robinson in [70], where also 7.6.4 was proved. Theorem 7.6.6 is essentially just a reformulation of 7.6.4. The theory of vector-valued positive definite functions runs parallel to the ordinary theory, cf. 7.1.8–7.1.11. We shall use it in section 7.7 exactly as the ordinary theory was used in section 7.3.

7.7. Regular representations of crossed products

7.7.1. Given a C^*-dynamical system (A, G, α) and a representation (π, H) of A define a covariant representation $(\tilde{\pi}, \lambda, L^2(G, H))$ by

$$(\tilde{\pi}(x)\xi)(t) = \pi(\alpha_{t^{-1}}(x))\xi(t); \qquad (\lambda_s\xi)(t) = \xi(s^{-1}t)$$

for every x in A, s in G and ξ in $L^2(G, H)$ (cf. 7.4.9). The *regular representation* of $G \times_\alpha A$ induced by (π, H) is the representation $(\tilde{\pi} \times \lambda, L^2(G, H))$ (cf. 7.6.6). From (*) in 7.6.4 we see that for each y in $K(G, A)$ and ξ in $L^2(G, H)$ we have

$$\begin{aligned}(((\tilde{\pi} \times \lambda)y)\xi)(t) &= \int (\tilde{\pi}(y(s))\lambda_s\xi)(t)\,\mathrm{d}s \\ &= \int \pi(\alpha_{t^{-1}}(y(s)))\xi(s^{-1}t)\,\mathrm{d}s.\end{aligned}$$

7.7.2. LEMMA. *If $\{\xi_i\} \subset H$, $\{f_j\} \subset L^2(G)$ such that $\{\xi_i\}$ is cyclic for (π, H) (i.e. $\Sigma\,\pi(A)\xi_i$ is dense in H) and $\{f_j\}$ is cyclic for $(\lambda, L^2(G))$, then the set $\{f_j \otimes \xi_i\}$ is cyclic for $\tilde{\pi} \times \lambda$ in $L^2(G, H)$.*

Proof. Suppose that $\xi \in L^2(G,H)$ such that

$$((\tilde{\pi} \times \lambda)(G \underset{\alpha}{\times} A) f_j \otimes \xi_i | \xi) = 0$$

for all i and j. This means that for each x in A and f in $K(G)$ we have

$$\begin{aligned} 0 &= ((\tilde{\pi} \times \lambda)(x \otimes f) f_j \otimes \xi_i \,|\, \xi) \\ &= \iint (\pi(\alpha_{t^{-1}}(x)) f(s) f_j(s^{-1}t)\xi_i \,|\, \xi(t))\, ds\, dt \\ &= \int (\pi(\alpha_{t^{-1}}(x))\xi_i \,|\, \xi(t))(f \times f_j)(t)\, dt. \end{aligned}$$

Since $K(G,H)$ is dense in $L^2(G,H)$ we can extend Lusin's theorem to the vector-valued case. Thus for each compact set E in G and $\varepsilon > 0$ there is a compact subset F of E such that $\mu_G(E \backslash F) < \varepsilon$ and $\xi | F \in C(F,H)$. From the density of the subspace spanned by $\{f \times f_j\}$, $f \in K(G)$, we see that the continuous function $t \to (\pi(\alpha_{t^{-1}}(x))\xi_i \,|\, \xi(t))$, $t \in F$, is identically zero for all x in A and all ξ_i. From the cyclicity of $\{\xi_i\}$ it follows that $\xi(t) = 0$ for all t in F. Since E and ε were arbitrary, we conclude that $\xi = 0$, whence $\{f_j \otimes \xi_i\}$ is cyclic for $\tilde{\pi} \times \lambda$.

7.7.3. PROPOSITION. *Let (A, G, α) be a C^*-dynamical system. If ϕ is a state of A with cyclic representation $(\pi_\phi, H_\phi, \xi_\phi)$ then for each z in $K(G,A)$ such that $\phi((z^* \times z)(e)) = 1$ there is a vector state $\tilde{\phi}_z$ of $G \underset{\alpha}{\times} A$ in $(\tilde{\pi}_\phi \times \lambda, L^2(G, H_\phi))$ such that*

$$\tilde{\phi}_z(y) = \phi((z^* \times y \times z)(e)), \qquad y \in K(G,A).$$

Moreover, each vector state in $(\tilde{\pi} \times \lambda, L^2(G, H_\phi))$ can be approximated in norm by states of the form $\tilde{\phi}_z$.

Proof. If $z \in C_0(G,A)$ such that the function $t \to \|z(t)\|$ belongs to $L^2(G)$ define ξ_z in $L^2(G, H_\phi)$ by $\xi_z(t) = \pi_\phi(\alpha_{t^{-1}}(z(t)))\xi_\phi$. If $y \in K(G,A)$ we have

$$\begin{aligned} ((\tilde{\pi}_\phi \times \lambda)(y^* \times y)\xi_z \,|\, \xi_z) &= \|(\tilde{\pi}_\phi \times \lambda)(y)\xi_z\|^2 \\ &= \int \left\| \int \pi_\phi(\alpha_{t^{-1}}(y(s)))\xi_z(s^{-1}t)\, ds \right\|^2 dt \\ &= \int \left\| \int \pi_\phi(\alpha_{t^{-1}}(y(s))\alpha_{t^{-1}s}(z(s^{-1}t)))\xi_\phi\, ds \right\|^2 dt \\ &= \int \|\pi_\phi(\alpha_{t^{-1}}(y \times z)(t))\xi_\phi\|^2\, dt \\ &= \int \phi(\alpha_{t^{-1}}((y \times z)(t)^* (y \times z)(t)))\, dt \\ &= \int \phi((y \times z)^*(t^{-1})\Delta(t^{-1})\alpha_{t^{-1}}((y \times z)(t))\, dt \\ &= \int \phi((y \times z)^*(t)\alpha_t((y \times z)(t^{-1})))\, dt \\ &= \phi((y \times z)^* \times (y \times z)(e)) = \tilde{\phi}_z(y^* \times y). \end{aligned}$$

It follows that $\tilde{\phi}_z$ is a vector state in $(\tilde{\pi}_\phi \times \lambda, L^2(G, H_\phi))$.

To show that every vector state in $(\tilde{\pi}_\phi \times \lambda, L^2(G, H_\phi))$ can be approximated by states of the form $\tilde{\phi}_z$ it suffices to prove that the subspace $\{\xi_z \mid z \in K(G, A)\}$ is dense in $L^2(G, H)$. A short computation shows that for each f in $K(G)$ we have

$$(\tilde{\pi}_\phi \times \lambda)(z) f \otimes \xi_\phi = \xi_{z \times f},$$

where $z \times f(t) = \int z(s) f(s^{-1}t)\, ds$. Since the subset $\{f \otimes \xi_\phi \mid f \in K(G)\}$ is cyclic for $(\tilde{\pi}_\phi \times \lambda, L^2(G, H_\phi))$ by 7.7.2, the proof is complete.

7.7.4. Let (A, G, α) be a C^*-dynamical system and let (π_u, H_u) denote the universal representation of A. The *reduced crossed product* of G and A is the C^*-algebra $(\tilde{\pi}_u \times \lambda)(G \underset{\alpha}{\times} A)$ denoted by $G \underset{\alpha r}{\times} A$.

The set of functionals ψ on $G \underset{\alpha}{\times} A$ of the form

$$\psi(y) = \sum ((\tilde{\pi}_u \times \lambda) y \xi_n \mid \eta_n)$$

where $\{\xi_n\} \subset L^2(G, H_u)$, $\{\eta_n\} \subset L^2(G, H_u)$, $\Sigma \|\xi_n\|^2 < \infty$ and $\Sigma \|\eta_n\|^2 < \infty$, is a norm closed subspace of $(G \underset{\alpha}{\times} A)^*$ and can be identified with $((G \underset{\alpha r}{\times} A)'')_*$, ($''$ denoting double commutant in $\boldsymbol{B}(L^2(G, H_u))$). We denote by $A(G \underset{\alpha}{\times} A)$ the subspace of $B(G \underset{\alpha}{\times} A)$ consisting of those A^*-valued functions whose associated functionals belong to this set (cf. 7.2.2).

Since (π_u, H_u) is the direct sum of cyclic representations of A we see from 7.7.3 that $A_+(G \underset{\alpha}{\times} A)$ is the closure (in the norm inherited from $(G \underset{\alpha}{\times} A)^*$) of finite sums of functions $\Phi = \Phi(\phi, z)$ such that for each x in A and t in G

$$\begin{aligned}
\Phi(t)(x) &= \tilde{\phi}_z(x u_t) = (\tilde{\pi}_\phi(x) \lambda_t \xi_z \mid \xi_z) \\
&= \int (\pi_\phi(\alpha_{s^{-1}}(x) \alpha_{s^{-1}t}(z(t^{-1}s)))\, \xi_\phi \mid \pi_\phi(\alpha_{s^{-1}}(z(s))) \xi_\phi)\, ds \\
&= \int \phi(\alpha_{s^{-1}}(z(s)^* x \alpha_t z(t^{-1}s)))\, ds \\
&= \int \phi(\alpha_s(z(s^{-1})^* \Delta(s)^{-1}\, x \alpha_t z(t^{-1}s^{-1})))\, ds \\
&= \phi(\textstyle\int z^*(s) \alpha_s(x) \alpha_{st}(z((st)^{-1}))\, ds,
\end{aligned}$$

where ϕ is a state of A and $z \in K(G, A)$.

7.7.5. THEOREM. *If (A, G, α) is a C^*-dynamical system and (π, H) is a faithful representation of A then $(\tilde{\pi} \times \lambda, L^2(G, H))$ is a faithful representation of $G \underset{\alpha r}{\times} A$.*

Proof. Let E denote the set of states of A contained in (π, H) and let F denote the set of functions in $A_+^1(G \underset{\alpha}{\times} A)$ whose associated states are contained in $(\tilde{\pi} \times \lambda, L^2(G, H))$. Since (π, H) is faithful on A we know that E is weak* dense in $S(A)$. To show that $(\tilde{\pi} \times \lambda, L^2(G, H))$ is faithful on $G \underset{\alpha r}{\times} A$ we must show that the set of states contained in $(\tilde{\pi} \times \lambda, L^2(G, H))$ is weak* dense in the set of states contained in $(\tilde{\pi}_u \times \lambda, L^2(G, H_u))$. By 7.6.10 this is equivalent to showing that F is dense in $A^1_+(G \underset{\alpha}{\times} A)$ in the topology of weak* convergence uniformly on compact subsets of G.

By 7.7.4 it suffices to approximate functions of the form

$$\Phi(t)(x) = \phi(\textstyle\int z^*(s)\alpha_s(x)\alpha_{st}(z(t^{-1}s^{-1}))\,ds)$$

where $\phi \in S(A)$ and $z \in K(G,A)$ with $\phi((z^* \times z)(e)) = 1$. Take a net $\{\phi_i\}$ in E which is weak* convergent to ϕ and let $\{\Phi_i\}$ denote the corresponding net of elements in $A_+(G \underset{\alpha}{\times} A)$. ($z$ remains fixed). Then $\Phi_i(t) \to \Phi(t)$ weak*, uniformly on compact subsets of G. Moreover, each Φ_i belongs to F by 7.7.3, because the representation $(\pi_{\phi_i}, H_{\phi_i})$ is equivalent to a subrepresentation of (π, H).

7.7.6. LEMMA. *Each element Φ in $B^1_+(G \underset{\alpha}{\times} A)$ with compact support belongs to $A^1_+(G \underset{\alpha}{\times} A)$.*

Proof. Let ϕ denote the state of $G \underset{\alpha}{\times} A$ corresponding to Φ and put $\psi = \Phi(e)$. From the definition of Φ (see 7.6.7) and the Cauchy–Schwarz inequality we have for each x in A

$$|\Phi(t)(\alpha_t(x))|^2 = |\phi(\alpha_t(x)u_t)|^2 = |\phi(u_t x)|^2 \leqslant \phi(x^*x)$$
$$= \Phi(e)(x^*x) = \psi(x^*x).$$

Passing to the representation $(\pi_\psi, H_\psi, \xi_\psi)$ of A it follows that for each t there is a unique vector ξ_t in H_ψ such that

$$\Phi(t)(\alpha_t(x)) = (\pi_\psi(x)\xi_\psi \mid \xi_t).$$

The function $\xi: t \to \xi_t$ has compact support and is weak* continuous. It therefore belongs to $L^2(G, H_\psi)$.

Take f_j in $K(G)$ and for each y in $K(G,A)$ compute

$$((\tilde{\pi}_\psi \times \lambda)(y) f_j \otimes \xi_\psi \mid \xi) = \textstyle\int(\int \pi_\psi(\alpha_{t^{-1}}(y(s)))f_j(s^{-1}t)\xi_\psi\,ds \mid \xi_t)\,dt$$
$$= \textstyle\int(\pi_\psi(\alpha_{t^{-1}}((y \times f_j)(t)))\xi_\psi \mid \xi_t)\,dt$$
$$= \textstyle\int \Phi(t)((y \times f_j)(t))\,dt = \phi(y \times f_j).$$

Thus the functional $\phi(\cdot f_j)$ is contained in $(\tilde{\pi}_u \times \lambda; L^2(G, H_u))$. However, taking $\{f_j\}$ as an approximate unit for $L^1(G)$ we have $y \times f_j \to y$ strongly, whence $\phi(\cdot f_j) \to \phi$ in norm. It follows that $\Phi \in A(G \underset{\alpha}{\times} A)$.

7.7.7. THEOREM. *If (A, G, α) is a C^*-dynamical system with G amenable then $G \underset{\alpha}{\times} A$ is equal to $G \underset{\alpha r}{\times} A$.*

Proof. We must show that $A_+(G \underset{\alpha}{\times} A)$ is dense in $B_+(G \underset{\alpha}{\times} A)$. Take Φ in $B_+(G \underset{\alpha}{\times} A)$. Since G is amenable there is a net $\{\Psi_i\}$ in $B_+(G) \cap K(G)$ such that $\Psi_i(t) \to 1$, uniformly on compact subsets of G (see 7.3.8). From 7.6.9 we have $\Phi \cdot \Psi_i \in B_+(G \underset{\alpha}{\times} A)$ and from 7.7.6, $\Phi \cdot \Psi_i \in A_+(G \underset{\alpha}{\times} A)$. Since $\Phi \cdot \Psi_i \to \Phi$ weak* uniformly on compact subsets of G, we are done.

7.7.8. COROLLARY. *If (A, G, α) is a (separable) C^*-dynamical system with G amenable and if (π, H) is a faithful (separable) representation of A then $(\tilde{\pi} \times \lambda, L^2(G, H))$ is a faithful (separable) representation of $G \underset{\alpha}{\times} A$.*

7.7.9. PROPOSITION. *Let (A, G, α) be a C^*-dynamical system and assume that B is a G-invariant C^*-subalgebra of A. There is then a natural injection of $G \underset{\alpha r}{\times} B$ as a C^*-subalgebra of $G \underset{\alpha r}{\times} A$; and if $B \neq A$ then $G \underset{\alpha r}{\times} B \neq G \underset{\alpha r}{\times} A$.*

Proof. It is clear that the inclusion $K(G, B) \subset K(G, A)$ gives an isometry $\iota: L^1(G, B) \to L^1(G, A)$. Thus ι is a representation of $L^1(G, B)$ into $G \underset{\alpha}{\times} A$, and therefore extends to a morphism $\iota: G \underset{\alpha}{\times} B \to G \underset{\alpha}{\times} A$. Since each representation of B extends to a representation of A (maybe on a larger space, see 4.1.8) we know that each regular representation of $G \underset{\alpha}{\times} B$ extends to a regular representation of $G \underset{\alpha}{\times} A$. Consequently ι gives an isometric morphism of $G \underset{\alpha r}{\times} B$ into $G \underset{\alpha r}{\times} A$.

If $B \neq A$ there is a non-zero functional on A that annihilates B. Thus for a suitable (faithful) representation (π, H) of A there are unit vectors ξ_0, η_0 in H such that $(\pi(x)\xi_0 \mid \eta_0) = 0$ for all x in B. Take f, g in $L^2(G)$ with $\|f\|_2 = \|g\|_2 = 1$ and define $\xi = f \otimes \xi_0$, $\eta = g \otimes \eta_0$ in $L^2(G, H)$. If $y \in K(G, A)$ then

$$((\tilde{\pi} \times \lambda)(y)\xi \mid \eta) = \iint (\pi(\alpha_{t^{-1}}(y(s)))\xi_0 \mid \eta_0) f(s^{-1}t)\bar{g}(t)\, ds\, dt.$$

Thus $((\tilde{\pi} \times \lambda)(y)\xi \mid \eta) = 0$ for every y in $K(G, B)$, whence also for y in $G \underset{\alpha r}{\times} B$. However, there is a y in $K(G, A)$ such that $|(\pi(y(e))\xi_0 \mid \eta_0) > 1$. Choosing f and g positive with small supports we find that $((\tilde{\pi} \times \lambda)(y)\xi \mid \eta) \neq 0$. Consequently $G \underset{\alpha r}{\times} B \neq G \underset{\alpha r}{\times} A$.

7.7.10. Given a C^*-algebra A and a natural number n we define $A \otimes \boldsymbol{M}_n$ as the C^*-algebra of A-valued $n \times n$-matrices. The norm on $A \otimes \boldsymbol{M}_n$ can be computed in many ways; the simplest is probably to represent A as operators on some Hilbert space H and embed $A \otimes \boldsymbol{M}_n$ in $\boldsymbol{B}(H \otimes \boldsymbol{C}^n)$. If K is an infinite dimensional Hilbert space there is for every n-dimensional subspace K_n of K a natural embedding ι of $A \otimes \boldsymbol{M}_n$ into $\boldsymbol{B}(H \otimes K)$. Moreover, if $K_n \subset K_m$ then $\iota(A \otimes \boldsymbol{M}_n) \subset \iota(A \otimes \boldsymbol{M}_m)$, so that we have an inductively ordered system of C^*-algebras. The set $\cup \iota(A \otimes \boldsymbol{M}_n)$ satisfies all axioms for a C^*-algebra except that of completeness. We denote by $A \otimes \boldsymbol{C}(K)$ the completion and note that it can be regarded as a C^*-subalgebra of $\boldsymbol{B}(H \otimes K)$. In contrast to the Glimm algebras (see 6.4), algebras of the form $A \otimes \boldsymbol{C}(K)$ have no units if $\dim(K) = \infty$.

7.7.11. PROPOSITION. *Let (A, G, α) be a C^*-dynamical system and let $\tilde{\lambda}$ denote the right regular representation of G on $L^2(G)$, i.e. $(\tilde{\lambda}_t\xi)(s) = \xi(st)\Delta(t)^{1/2}$,*

$\xi \in L^2(G)$. *There is then a* C^**-dynamical system* $(A \otimes C(L^2(G)), G, \alpha \otimes \mathrm{Ad}\tilde{\lambda})$ *such that*

$$(\alpha \otimes \mathrm{Ad}\,\tilde{\lambda})_t(x \otimes y) = \alpha_t(x) \otimes \tilde{\lambda}_t y \tilde{\lambda}_t^*$$

for every x *in* A, y *in* $C(L^2(G))$ *and* t *in* G.

Proof. Choose a faithful covariant representation (π, u, H) of (A, G, α). Then define a unitary representation w of G on $H \otimes L^2(G)$ by

$$w_t(\xi \otimes \eta) = u_t\xi \otimes \tilde{\lambda}_t\eta, \qquad \xi \in H, \eta \in L^2(G).$$

Identifying A and $\pi(A)$ we may consider $A \otimes C(L^2(G))$ as a subalgebra of $B(H \otimes L^2(G))$. If $x \in A$ and $y \in C(L^2(G))$ then

$$w_t(x \otimes y)w_t^* = u_t x u_t^* \otimes \tilde{\lambda}_t y \tilde{\lambda}_t^* = \alpha_t(x) \otimes \tilde{\lambda}_t y \tilde{\lambda}_t^*.$$

It follows that

$$w_t(A \otimes C(L^2(G)))w_t^* \subset A \otimes C(L^2(G))$$

and furthermore we see that if y is finite dimensional then the function $t \to \alpha_t(x) \otimes \tilde{\lambda}_t y \tilde{\lambda}_t^*$ is norm continuous since $t \to \tilde{\lambda}_t$ is strongly continuous. The elements z in $A \otimes C(L^2(G))$ for which the function $t \to w_t z w_t^*$ is norm continuous form a C^*-algebra (cf. 7.5.1). Since this algebra contains all elements $x \otimes y$, where $x \in A$ and y is finite dimensional, it equals $A \otimes C(L^2(G))$. It follows that if we define

$$(\alpha \otimes \mathrm{Ad}\,\tilde{\lambda})_t(z) = w_t z w_t^*, \qquad z \in A \otimes C(L^2(G)),$$

then $(A \otimes C(L^2(G)), G, \alpha \otimes \mathrm{Ad}\,\tilde{\lambda})$ is a C^*-dynamical system as desired.

7.7.12. THEOREM. *Given a* C^**-dynamical system* (A, G, α) *there exists two* C^**-dynamical systems* $(C_0(G, A), G, \rho)$ *and* $(C_0(G, A), G, \gamma)$ *where* $(\rho_s(y))(t) = y(ts)$ *and* $(\gamma_s(y))(t) = \alpha_s(y(s^{-1}t))$ *for each* y *in* $C_0(G, A)$. *The two representations* ρ *and* γ *commute and each* ρ_s *extends to an automorphism of* $G \times_{\gamma r} C_0(G, A)$ *so that we obtain a* C^**-dynamical system* $(G \times_{\gamma r} C_0(G, A), G, \rho)$. *This system is covariantly isomorphic to the system* $(A \otimes C(L^2(G)), G, \alpha \otimes \mathrm{Ad}\,\tilde{\lambda})$ *defined in 7.7.11.*

Proof. It is straightforward to verify that the definitions

$$(\rho_s(y))(t) = y(ts), \qquad (\gamma_s(y))(t) = \alpha_s(y(s^{-1}t)), \qquad y \in C_0(G, A),$$

give C^*-dynamical systems $(C_0(G, A), G, \rho)$ and $(C_0(G, A), G, \gamma)$. Evidently the two representations ρ and γ commute.

Take now a faithful representation of A on some Hilbert space H and define a faithful representation $(\pi, L^2(G, H))$ of $C_0(G, A)$ by

$$(\pi(y)\xi)(t) = y(t)\xi(t), \qquad y \in C_0(G, A), \qquad \xi \in L^2(G, H).$$

It follows from 7.7.5 that $(\tilde{\pi} \times \lambda, L^2(G \times G, H))$ is a faithful representation of $G \times_{\gamma r} C_0(G, A)$. Moreover, regarding each z in $K(G \times G, A)$ as an element of $K(G, C_0(G, A))$ we have for every ξ in $L^2(G \times G, H)$

$$(((\tilde{\pi} \times \lambda)z)\xi)(s,t) = \left(\int \pi(\gamma_{s^{-1}}(z(r,\cdot)))\xi(r^{-1}s,\cdot)\,dr\right)(t)$$
$$= \int \alpha_{s^{-1}}(z(r,st))\xi(r^{-1}s,t)\,dr.$$

Define a unitary operator on $L^2(G \times G, H)$ by

$$(w\xi)(s,t) = \xi(st,t)\Delta(t)^{1/2}, \qquad \xi \in L^2(G \times G, H).$$

Using the formula above we get

$$(w^*((\tilde{\pi} \times \lambda)z)w\xi)(s,t) = (((\tilde{\pi} \times \lambda)z)w\xi)(st^{-1},t)\Delta(t)^{-1/2}$$
$$= \int \alpha_{ts^{-1}}(z(r,s))((w\xi)(r^{-1}st^{-1},t))\,dr\,\Delta(t)^{-1/2}$$
$$= \int \alpha_{ts^{-1}}(z(r,s))\xi(r^{-1}s,t)\,dr.$$

Take a function z in $K(G \times G, A)$ of the form

$$(*) \qquad z(r,s) = \alpha_s(x)f(r^{-1}s)g(s)\Delta(r^{-1}s),$$

where $x \in A$ and $f, g \in K(G)$. Inserting this in the expression above we get

$$(w^*((\tilde{\pi} \times \lambda)z)w\xi)(s,t) = \alpha_t(x)g(s)\int f(r)\xi(r,t)\,dr.$$

Thus if we define a faithful representation $(\check{\pi}, L^2(G,H))$ of A by $(\check{\pi}(x)\xi)(t) = \alpha_t(x)\xi(t)$ then we may realize $\check{\pi}(A) \otimes C(L^2(G))$ as operators on $L^2(G \times G, H)$, and taking z as in $(*)$ we see that

$$w^*((\tilde{\pi} \times \lambda)z)w = \check{\pi}(x) \otimes v_{fg},$$

where v_{fg} is the one-dimensional operator on $L^2(G)$ such that $v_{fg}\eta = (\eta \mid \bar{f})g$, $\eta \in L^2(G)$.

Since operators of the form $v_{fg}, f, g \in K(G)$ generate $C(L^2(G))$ we see that

$$w^*((\tilde{\pi} \times \lambda)(G \times_{\gamma r} C_0(G, A)))w \supset A \otimes C(L^2(G)).$$

On the other hand, each function in $K(G \times G, A)$ can be uniformly approximated by finite sums of functions of the form $(s,t) \to \alpha_t(x)f(s)g(t)$, where $x \in A$, f, $g \in K(G)$, and therefore (applying the homeomorphism $(s,t) \to (s^{-1}t, t)$ of $G \times G$) also by functions of the form $(*)$. Since $K(G \times G, A)$ generates $G \times_{\gamma r} C_0(G, A)$ it follows that we have found a spatial isomorphism between $G \times_{\gamma r} C_0(G, A)$ and $A \otimes C(L^2(G))$.

Consider the natural extension of ρ from $C_0(G, A)$ to $K(G, C_0(G, A))$ such that $\rho_t(z)(r,s) = z(r, st)$ for each z in $K(G, C_0(G, A))$. Then $t \to \rho_t$ is a representation of G as *-automorphisms of the dense *-subalgebra

$K(G, C_0(G, A))$ of $G \underset{\gamma r}{\times} C_0(G, A)$. Moreover, with z as in (*) we have

$$\rho_p(z)(r, s) = \alpha_{sp}(x) f(r^{-1}sp) g(sp) \Delta(r^{-1}sp),$$

whence

$$\begin{aligned}(w^*((\tilde{\pi} \times \lambda)\rho_p(z))w\xi)(s, t) &= \alpha_{tp}(x) g(sp) \Delta(p) \int f(rp) \xi(r, t)\, dr \\ &= \alpha_{tp}(x) (\tilde{\lambda}_p g)(s) \int (\tilde{\lambda}_p f)(r) \xi(r, t)\, dr \\ &= ((\tilde{\pi}(\alpha_p(x)) \otimes \tilde{\lambda}_p v_{fg} \tilde{\lambda}_p^*)\xi)(s, t).\end{aligned}$$

Thus, identifying $\tilde{\pi}(A)$ and A, we have

$$w^*((\tilde{\pi} \times \lambda)\rho_p(z))w = (\alpha \otimes \operatorname{Ad} \tilde{\lambda})_p(x \otimes v_{fg}).$$

It follows that each ρ_p extend to an automorphism of $G \underset{\gamma r}{\times} C_0(G, A)$ such that we obtain a C^*-dynamical system $(G \underset{\gamma r}{\times} C_0(G, A), G, \rho)$ covariantly isomorphic to the system $(A \otimes C(L^2(G)), G, \alpha \otimes \operatorname{Ad} \tilde{\lambda})$.

7.7.13. Notes and remarks. The reduced crossed product was defined by Zeller–Meyer for discrete groups in [277] and generalized by Takai in [254]. The same distribution of credits hold for theorem 7.7.7. Proposition 7.7.9 is due to Landstad [159] and Theorem 7.7.12 is a generalization of a theorem of Takai from [254]. It serves as a major step toward the duality theorem 7.9.3.

7.8. Crossed products with abelian groups

7.8.1. In this section G will denote an abelian locally compact group and $\hat{G}$ its dual group, and we will use the additive notation for the group operations. If $t \in G$ and $\sigma \in \hat{G}$ we denote by (t, σ) the evaluation of the character σ at t, and we define the (inverse) Fourier transform of a measure μ in $M(G)$ by $\hat{\mu}(\sigma) = \int (t, \sigma)\, d\mu(t)$. Note that $\hat{\mu} \in C^b(\hat{G})$ and that $\hat{f} \in C_0(G)$ if $f \in L^1(G)$.

As a reminder that G is now abelian we shall refer to a system (A, G, α) as a C^*-dynamical *a-system*. Likewise we talk about B^*- and W^*-dynamical *a*-systems.

7.8.2. Let G (and $\hat{G}$) be given. We say that a C^*-algebra B is a *G-product* if:

(*) There is a homomorphism $t \to \lambda_t$ of G into the unitary group of $M(B)$ such that each function $t \to \lambda_t y$, $y \in B$, is continuous from G to B;

(**) There is a homomorphism $\sigma \to \hat{\alpha}_\sigma$ of $\hat{G}$ into $\operatorname{Aut}(B)$ such that $(B, \hat{G}, \hat{\alpha})$ is a C^*-dynamical a-system, and

$$\hat{\alpha}_\sigma(\lambda_t) = (t, \sigma)\lambda_t \qquad \forall t \in G, \sigma \in G.$$

Given a G-product B we say that an element x in $M(B)$ satisfies *Landstad's conditions if*:

(i) $\hat{\alpha}_\sigma(x) = x$ for all σ in $\hat{G}$;

(ii) $x\lambda_f \in B$ and $\lambda_f x \in B$ for every f in $L^1(G)$;

(iii) The map $t \to \lambda_t x \lambda_{-t}$ is continuous on G.

7.8.3. PROPOSITION. *Let (A, G, α) be a C^*-dynamical a-system and for each y in $K(G, A)$ and σ in $\hat{G}$ define $(\hat{\alpha}_\sigma(y))(t) = (t, \sigma)y(t)$. Then the $\hat{\alpha}_\sigma$'s extend to automorphisms such that $(G \times_\alpha A, \hat{G}, \hat{\alpha})$ becomes a C^*-dynamical a-system. Moreover, $G \times_\alpha A$ is a G-product and each element in A satisfies Landstad's conditions. We say that $(G \times_\alpha A, \hat{G}, \hat{\alpha})$ is the dual system of (A, G, α).*

Proof. As G is abelian, it is amenable. Thus by 7.7.7 we may consider $G \times_\alpha A$ as a subalgebra of $\boldsymbol{B}(L^2(G, H))$, where H is the universal Hilbert space for A (so that $A \subset \boldsymbol{B}(H)$). The unitary representation $t \to \lambda_t$ of G on $L^2(G, H)$ (see 7.7.1) is by 7.6.3 a homomorphism of G into $M(G \times_\alpha A)$ such that each function $t \to \lambda_t y$, $y \in G \times_\alpha A$, is norm continuous. Moreover, again from 7.6.3, we have an injection of A into $M(G \times_\alpha A)$. Note that if $\xi \in L^2(G, H)$ then

$$(*) \qquad (\lambda_s \xi)(t) = \xi(t - s); \qquad (x\xi)(t) = \alpha_{-t}(x)\xi(t)$$

for all s in G and x in A.

Define a unitary representation u of $\hat{G}$ on $L^2(G, H)$ by

$$(**) \qquad (u_\sigma \xi)(t) = (t, \sigma)\xi(t).$$

For each y in $K(G, A)$ we have

$$\begin{aligned}(u_\sigma y u_{-\sigma}\xi)(t) &= (t, \sigma)(y u_{-\sigma}\xi)(t) \\ &= (t, \sigma)\int \alpha_{-t}(y(s))((u_{-\sigma}\xi)(t - s))\,ds \\ &= \int (s, \sigma)\alpha_{-t}(y(s))\xi(t - s)\,ds = ((\sigma y)\xi)(t),\end{aligned}$$

where $(\sigma y)(t) = (t, \sigma)y(t)$. It follows immediately from this that we can define $\hat{\alpha}_\sigma$ in $\mathrm{Aut}(G \times_\alpha A)$ by $\hat{\alpha}_\sigma(y) = u_\sigma y u_{-\sigma}$, $y \in G \times_\alpha A$, and that $(G \times_\alpha A, \hat{G}, \hat{\alpha})$ becomes a C^*-dynamical a-system. Moreover, from $(*)$ and $(**)$ we see that

$$\hat{\alpha}_\sigma(\lambda_s) = (s, \sigma)\lambda_s; \qquad \hat{\alpha}_\sigma(x) = x$$

for all s in G and x in A. We have verified that $G \times_\alpha A$ is a G-product and that the elements in A satisfy the first and of course also the third of Landstad's conditions (as $\lambda_t x \lambda_{-t} = \alpha_t(x)$).

A straightforward computation shows that if $x \in A$ and $f \in K(G)$ then $x\lambda_f = L(x, f)$, with L as in 7.6.2. Moreover, in this case $L(x, f) \in K(G, A)$; in fact $L(x, f) = x \otimes f$. It follows that $x\lambda_f \in G \times_\alpha A$ for all f in $L^1(G)$. Since $\lambda_f x = (x^* \lambda_{f*})^* \in G \times_\alpha A$ we have verified also the second condition in 7.8.2.

7.8.4. Let B be a G-product. We say that an element y in $M(B)_+$ is *$\hat{\alpha}$-integrable* if there is an element $I(y)$ in $M(B)_+$ such that

$$\phi(I(y)) = \int \phi(\hat{\alpha}_\sigma(y))\,d\sigma$$

for all ϕ in $M(B)^*$. If y is not positive we say that it is $\hat{\alpha}$-integrable if it can be written as a linear combination of positive, $\hat{\alpha}$-integrable elements.

If $0 \leqslant x \leqslant y$ and y is $\hat{\alpha}$-integrable then both functions

$$\phi \to \int \phi(\hat{\alpha}_\sigma(x))\,d\sigma, \qquad \phi \to \int \phi(\hat{\alpha}_\sigma(y-x))\,d\sigma$$

are bounded and lower semi-continuous on the state space of $M(B)$. Since y is integrable their sum is continuous and they are therefore both continuous, i.e. both correspond to elements in $M(B)$ by 3.10.3. Thus x is integrable. It follows that the $\hat{\alpha}$-integrable elements form a hereditary $*$-subalgebra in $M(B)$.

The map I is an 'operator-valued weight' i.e. it is positive and linear; and it maps the $\hat{\alpha}$-integrable elements into the set of $\hat{G}$-invariant elements of $M(B)$.

7.8.5. LEMMA. *Let B be a G-product. If x, y are elements in $M(B)$ such that x^*x and y^*y are $\hat{\alpha}$-integrable then y^*x is $\hat{\alpha}$-integrable and*

$$\|I(y^*x)\|^2 \leqslant \|I(y^*y)\|\,\|I(x^*x)\|.$$

Proof. It follows from the polarization identity that y^*x is $\hat{\alpha}$-integrable (cf. 5.1.2). Realizing $M(B)$ as operators on some Hilbert space we have, for any pair of vectors ξ, η, that

$$\begin{aligned}
|(I(y^*x)\xi \mid \eta)| &= |\int (\hat{\alpha}_\sigma(x)\xi \mid \hat{\alpha}_\sigma(y)\eta)\,d\sigma| \\
&\leqslant \int \|\hat{\alpha}_\sigma(x)\xi\|\,\|\hat{\alpha}_\sigma(y)\eta\|\,d\sigma \\
&\leqslant \left(\int \|\hat{\alpha}_\sigma(x)\xi\|^2\,d\sigma \int \|\hat{\alpha}_\sigma(y)\eta\|^2\,d\sigma\right)^{1/2} \\
&= (I(x^*x)\xi \mid \xi)^{1/2}(I(y^*y)\eta \mid \eta)^{1/2}
\end{aligned}$$

from which the lemma follows.

7.8.6. LEMMA. *Let B be a G-product. Each element of the form $\lambda_f^* x \lambda_g$, $x \in M(B)$, $f, g \in L^1(G) \cap L^2(G)$ is $\hat{\alpha}$-integrable. Moreover, the function $t \to I(\lambda_f^* x \lambda_g \lambda_t)$ is norm continuous. Finally, if $x \in B$ then $\|\lambda_f^* x \lambda_f - x\| \to 0$ when f ranges over an approximate unit for $L^1(G)$ contained in $L^1(G) \cap L^2(G)$.*

Proof. For the first statement it suffices by 7.8.4 and 7.8.5 to show that $\lambda_f^*\lambda_f$ is integrable when $f \in L^1(G) \cap L^2(G)$. To this end we note that the representation $\lambda: L^1(G) \to M(B)$ extends by continuity to a morphism $\hat{\lambda}: C_0(\hat{G}) \to M(B)$ such that $\hat{\lambda}(\hat{f}) = \lambda_f$. Then it extends further to a morphism between the multiplier algebras. However, $M(C_0(\hat{G})) = C^b(\hat{G})$ and $M(M(B)) = M(B)$, so that $\hat{\lambda}: C^b(\hat{G}) \to M(B)$. If $f \in L^1(G)$ then

$$\hat{\lambda}(\delta_{-\sigma} \times \hat{f}) = \lambda_{\sigma f} = \int \lambda_t(t, \sigma) f(t)\,dt = \hat{\alpha}_\sigma(\lambda_f) = \hat{\alpha}_\sigma(\hat{\lambda}(\hat{f})),$$

whence by continuity and uniqueness of the extension $\hat{\lambda}(\delta_{-\sigma} \times g) = \hat{\alpha}_\sigma(\hat{\lambda}(g))$ for all g in $C^b(\hat{G})$ and σ in $\hat{G}$. Applying this to the case at hand, where $f \in L^1(G) \cap L^2(G)$ we get

$$\hat{\alpha}_\sigma(\lambda_f^* \lambda_f) = \hat{\alpha}_\sigma(\hat{\lambda}(|\hat{f}|^2)) = \hat{\lambda}(\delta_{-\sigma} \times |\hat{f}|^2);$$

whence

$$I(\lambda_f^* \lambda_f) = \hat{\lambda}(\int |\hat{f}|^2(\sigma)\, d\sigma) = (f^* \times f)(0) = \|f\|_2^2.$$

Using the above in conjunction with 7.8.5 we get

$$\|I(\lambda_f^* x \lambda_g(\lambda_s - \lambda_t))\|^2 \leqslant \|I(\lambda_f^* x^* x \lambda_f)\| \, \|I(\lambda_s - \lambda_t)^* \lambda_g^* \lambda_g(\lambda_s - \lambda_t))\|$$
$$\leqslant \|x\|^2 \, \|f\|_2^2 \|g \times (\delta_s - \delta_t)\|_2^2,$$

from which it follows that the function $t \to I(\lambda_f^* x \lambda_g \lambda_t)$ is continuous.

Finally, if $f \geqslant 0$ and $\int f \, dt = 1$ then

$$\|\lambda_f^* x \lambda_f - x\| = \|\int\int (\lambda_t x \lambda_s - x) f^*(t) f(s)\, dt\, ds\|$$
$$\leqslant \int \|\lambda_t x - x\| f^*(t)\, dt + \int \|x\lambda_s - x\| f(s)\, ds.$$

If $x \in B$ the functions $t \to \lambda_t x$ and $s \to x\lambda_s$ are norm continuous by 7.8.2, whence $\|\lambda_f^* x \lambda_f - x\| \to 0$ when f ranges over an approximate unit for $L^1(G)$.

7.8.7. LEMMA. *Let B be a G-product and take x in B of the form $\lambda_f^* y \lambda_f, y \in B$, $f \in L^1(G) \cap L^2(G)$. Then $I(x)$ satisfies Landstad's conditions (7.8.2) and x is the norm limit of elements of the form*

$$\int I(x\lambda_{-t})\lambda_t f_i(t)\, dt$$

where $\{f_i\} \subset L^1(G)$, $\{\hat{f}_i\} \subset L^1(\hat{G})$ and $\{\hat{f}_i\}$ is an approximate unit for $L^1(\hat{G})$.

Proof. If $f_i \in L^1(G)$ and $\hat{f}_i \in L^1(\hat{G})$ then

$$\int I(x\lambda_{-t})\lambda_t f_i(t)\, dt = \int\int \hat{\alpha}_\sigma(x)\overline{(t, \sigma)} f_i(t)\, dt\, d\sigma$$
$$= \int \hat{\alpha}_\sigma(x) \hat{f}_i(-\sigma)\, d\sigma \in B,$$

since $(B, \hat{G}, \hat{\alpha})$ is a C^*-dynamical system. Moreover, if $\{\hat{f}_i\}$ is an approximate unit for $L^1(\hat{G})$ we have

$$\int I(x\lambda_{-t})\lambda_t f_i(t)\, dt \to x \qquad \text{(in norm)}.$$

Now consider $I(x)$. The first condition in 7.8.2 is evidently satisfied. To prove the second take g in $L^1(G)$ and for $\varepsilon > 0$ choose by 7.8.6 a neighbourhood E of 0 such that

$$\|I(x\lambda_t) - I(x)\| < \varepsilon, \qquad \|g - \delta_t \times g\|_1 < \varepsilon$$

for all t in E. Take $f \geqslant 0$ with support in E and $\int f(t)\,dt = 1$, such that $f \in L^1(\hat{G})$. Then as we saw in the first part of the proof

$$\int I(x\lambda_{-t})\lambda_t f(t)\,dt \in B.$$

Moreover,

$$\begin{aligned}
&\|I(x)\lambda_g - \int I(x\lambda_{-t})\lambda_t f(t)\,dt\lambda_g\| \\
&\quad \leqslant \|I(x)(\lambda_g - \lambda_f\lambda_g)\| + \|\int (I(x) - I(x\lambda_{-t}))\lambda_t f(t)\,dt\lambda_g\| \\
&\quad \leqslant \|I(x)\|\,\|g - f \times g\|_1 + \int \|I(x) - I(x\lambda_{-t})\|\,f(t)\,dt\,\|g\|_1 \\
&\quad \leqslant (\|I(x)\| + \|g\|_1)\varepsilon.
\end{aligned}$$

Since ε was arbitrary, $I(x)\lambda_g \in B$.

Finally,

$$\lambda_t I(x)\lambda_{-t} = I(\lambda_t x\lambda_{-t}),$$

which by 7.8.6 depends norm continuously on t.

7.8.8. THEOREM. *A C^*-algebra B is a G-product (cf. 7.8.2) for a given abelian group G if and only if there is a C^*-dynamical a-system (A, G, α) such that $B = G \times_\alpha A$. This system is unique (up to covariant isomorphism) and A consists of the elements in $M(B)$ that satisfy Landstad's conditions, whereas $\alpha_t = \lambda_t \cdot \lambda_{-t}$, $t \in G$.*

Proof. As we saw in 7.8.3 the conditions (*) and (**) in 7.8.2 are necessary for B to be a crossed product. Suppose now that B is a G-product and denote by A the set of elements in $M(B)$ that satisfy Landstad's conditions. Evidently A is a C^*-subalgebra of $M(B)$ and taking $\alpha_t = \lambda_t \cdot \lambda_{-t}$, $t \in G$, we see from condition (iii) that (A, G, α) is a C^*-dynamical a-system (as the two first conditions are clearly G-invariant).

Let H denote the universal Hilbert space for $M(B)$ and consider $M(B)$ as a subalgebra of $B(H)$. Then the regular representation of $G \times_\alpha A$ on $L^2(G, H)$ is faithful because the universal Hilbert space for A is a subspace of H (since $A \subset M(B)$). Since $(B, \hat{G}, \hat{\alpha})$ is also a C^*-dynamical system we may consider the faithful representation π of $M(B)$ on $L^2(\hat{G}, H)$ where

$$(\pi(x)\eta)(\sigma) = \hat{\alpha}_{-\sigma}(x)\eta(\sigma), \qquad \eta \in L^2(\hat{G}, H).$$

Define an isometry u of $L^2(G, H)$ onto $L^2(\hat{G}, H)$ by

$$(u\xi)(\sigma) = \int \lambda_t \xi(t)\overline{(t, \sigma)}\,dt; \qquad (u^*\eta)(t) = \lambda_{-t}\int \eta(\sigma)(t, \sigma)\,d\sigma,$$

where the integrals converge in the L^2-sense. Take an element in $K(G, A)$ of the form $x \otimes f$, i.e. $(x \otimes f)(t) = xf(t)$, where $x \in A$, $f \in K(G)$ and compute

$$\begin{aligned}((x \otimes f)u^*\eta)(t) &= \int \alpha_{-t}(x) f(s) u^*\eta(t-s)\,ds \\ &= \iint \alpha_{-t}(x) f(s) \lambda_{s-t}\eta(\sigma)\check{}(t-s,\sigma)\,d\sigma\,ds \\ &= \iint \lambda_{-t} x \lambda_s f(s)(s,-\sigma)\eta(\sigma)(t,\sigma)\,ds\,ds \\ &= \int \lambda_{-t}\hat{\alpha}_{-\sigma}(x\lambda_f)\eta(\sigma)(t,\sigma)\,d\sigma \\ &= \int \lambda_{-t}(\pi(x\lambda_f)\eta)(\sigma)(t,\sigma)\,d\sigma = (u^*\pi(x\lambda_f)\eta)(t).\end{aligned}$$

It follows that $u(x \otimes f)u^* = \pi(x\lambda_f)$, and since $x\lambda_f \in B$ (condition (ii)) and elements of the form $x \otimes f$ span a dense set in $G \times_\alpha A$ we conclude that

$$u(G \times_\alpha A)u^* \subset \pi(B).$$

For the converse inclusion take an $\hat{\alpha}$-integrable element x in B of the form $\lambda_g^* y \lambda_g$, where $g \in L^1(G) \cap L^2(G)$, $y \in B$. Then for each t in G and f in $K(G)$ we have from the first part of the proof

$$u(I(x\lambda_{-t}) \otimes (\delta_t \times f))u^* = \pi(I(x\lambda_{-t})\lambda_t\lambda_f).$$

Suppose now that $f_i \in L^1(G)$ and $\hat{f}_i \in L^1(\hat{G})$. Then by 7.8.7 we can define

$$x_i = \int I(x\lambda_{-t})\lambda_t f_i(t)\,dt \in B.$$

On the other hand, the function $t \to I(x\lambda_{-t}) \otimes (\delta_t \times f)$ is norm continuous from G into $G \times_\alpha A$ by 7.8.6; whence

$$u^*\pi(x_i\lambda_f)u = \int (I(x\lambda_{-t}) \otimes (\delta_t \times f)) f_i(t)\,dt \in G \times_\alpha A.$$

Since $x_i \to x$ if $\{\hat{f}_i\}$ is an approximate unit for $L^1(\hat{G})$, and elements of the form $x = \lambda_g^* y \lambda_g$ are dense in B by 7.8.6, we conclude that $\pi(B\lambda_f) \subset u(G \times_\alpha A)u^*$. However, $t \to x\lambda_t$ is norm continuous for each x in B and thus $\|x - x\lambda_f\| \to 0$ when f ranges over an approximate unit for $L^1(G)$, so that finally $\pi(B) \subset u(G \times_\alpha A)u^*$.

To prove uniqueness take an C^*-dynamical a-system (A, G, α) and let A_1 denote the C^*-subalgebra of elements in $M(G \times_\alpha A)$ that satisfy Landstad's conditions. Then $A \subset A_1$ by 7.8.3 (i.e. we have a covariant injection of A into A_1). Since $G \times_\alpha A = G \times_\alpha A_1$ from what we proved above, we conclude from 7.7.9 that $A = A_1$. This completes the proof.

7.8.9. PROPOSITION. *Let (A, G, α) be a C^*-dynamical a-system and define $\lambda: G \to M(G \times_\alpha A)$ and $\hat{\alpha}: \hat{G} \to \mathrm{Aut}(G \times_\alpha A)$ as in 7.8.3. Consider the natural embedding of $M(A)$ into $M(G \times_\alpha A)$ (cf. 7.6.2). An element x in $M(G \times_\alpha A)$ belongs to $M(A)$ if and only if:*

(i) $\hat{\alpha}_\sigma(x) = x, \quad \forall \sigma \in \hat{G}$;

(ii) *The map $t \to \lambda_t x \lambda_{-t} y$ is (norm) continuous on G for each y in A.*

Proof. The conditions are evidently satisfied for each x in $M(A)$ (Note that (ii) is equivalent to the condition that $t \to \alpha_t(xy)$ is continuous).

Assume now that x is an $\hat{\alpha}$-invariant element in $M(G \times_\alpha A)$, and take f, g in $L^1(G) \cap L^2(G)$ such that $\hat{g} \in L^1(G)$. Then $\lambda_f^* x \lambda_g$ is $\hat{\alpha}$-integrable by 7.8.6 and

$$\begin{aligned} I(\lambda_f^* x \lambda_g) &= \iiint \lambda_t x \lambda_s(t+s, \sigma) f^*(t) g(s)\, ds\, dt\, d\sigma \\ &= \iiint \lambda_t x \lambda_{-t} \hat{\alpha}_\sigma(\lambda_s) g(s-t) f^*(t)\, ds\, d\sigma\, dt. \end{aligned}$$

Using the extension $\hat{\lambda}$ of λ from $L^1(G)$ to $C^b(\hat{G})$ defined in the proof of 7.8.6 we get

$$\begin{aligned} \iint \hat{\alpha}_\sigma(\lambda_s) g(s-t)\, ds\, d\sigma &= \iint \hat{\alpha}_\sigma(\lambda_s)(\delta_t \times g)(s)\, ds\, d\sigma \\ &= \int \hat{\alpha}_\sigma(\hat{\lambda}(\hat{\delta}_t \cdot \hat{g}))\, d\sigma = \int \hat{\lambda}(\delta_{-\sigma} \times \hat{\delta}_t \cdot \hat{g})\, d\sigma \\ &= \hat{\lambda}(\int (\delta_{-\sigma} \times \hat{\delta}_t \cdot \hat{g})\, d\sigma) = \hat{\lambda}((\hat{\delta}_t \cdot \hat{g})\hat{}\,(0)) \\ &= (\delta_t \times g)(0) = g(-t). \end{aligned}$$

Inserting this in the equation above we obtain

$$\begin{aligned} I(\lambda_f^* x \lambda_g) &= \int \lambda_t x \lambda_{-t} g(-t) \bar{f}(-t)\, dt \\ &= \int \alpha_{-t}(x) \bar{f}(t) g(t)\, dt. \end{aligned}$$

If x satisfies both (i) and (ii) and $y \in A$ then xy is $\hat{G}$-invariant, whence

$$I(\lambda_f^* x y \lambda_g) = \int \alpha_{-t}(xy) \bar{f}(t) g(t)\, dt.$$

As $y \in A$ we have $y\lambda_g \in G \times_\alpha A$ by (iii) in 7.8.2, whence $I(\lambda_f^* x y \lambda_g) \in A$ by 7.8.7. Thus $\int \alpha_{-t}(xy) \bar{f}(t) g(t)\, dt \in A$ for all sufficiently regular f and g; and since by assumption the function $t \to \alpha_{-t}(xy)$ is norm continuous we obtain $xy \in A$ taking f and g in an approximate unit for $L^1(G)$. The proof that $yx \in A$ is similar, and it follows that $x \in M(A)$, as desired.

7.8.10. Notes and remarks. This section is a simplified version, taken from [186], of the main results in Landstad's thesis [159], which characterize C^*- and W^*-crossed products with an arbitrary (locally compact) group.

7.9. Duality theory for crossed products

7.9.1. Given a C^*-algebra A and an abelian group G with dual group $\hat{G}$, define a C^*-dynamical system $(A, \hat{G}, \iota)$ by taking $\iota_\sigma(x) = x$ for all σ in $\hat{G}$ and x in A. If $A \subset \mathbf{B}(H)$ consider the regular representation of $\hat{G} \times_\iota A$ on $L^2(\hat{G}, H)$, where

$$(x\eta)(\tau) = \int x(\sigma)\eta(\tau - \sigma)\, d\sigma, \qquad x \in K(\hat{G}, A), \qquad \eta \in L^2(\hat{G}, H).$$

Define a faithful representation of $C_0(G, A)$ on $L^2(G, H)$ by

$$(y\xi)(t) = y(t)\xi(t), \qquad y \in C_0(G, A), \qquad \xi \in L^2(G, H).$$

Moreover, for each x in $K(\hat{G}, A)$ define $\hat{x}$ in $C_0(G, A)$ by $\hat{x}(t) = \int x(\sigma)(t,\sigma)\,d\sigma$. Now let u be the isometry of $L^2(G,H)$ onto $L^2(\hat{G},H)$ where

$$(u\xi)(\tau) = \int \xi(t)\overline{(t,\tau)}\,dt, \qquad (u^*\eta)(t) = \int \eta(\tau)(t,\tau)\,d\tau,$$

and the integrals converge in the L^2-sense. Then for each x in $K(\hat{G}, A)$ we have

$$\begin{aligned}(xu\xi)(\tau) &= \int x(\sigma)(u\xi)(\tau-\sigma)\,d\sigma \\ &= \iint x(\sigma)\xi(t)(t,\sigma-\tau)\,dt\,d\sigma \\ &= \int \hat{x}(t)\xi(t)\overline{(t,\tau)}\,dt = (u\hat{x}\xi)(\tau).\end{aligned}$$

Consequently $u^*xu = \hat{x}$. Since $\hat{G} \times_\iota A$ is the closure of $K(\hat{G}, A)$ in $\boldsymbol{B}(L^2(\hat{G},H))$ it follows that $\hat{G} \times_\iota A$ is isomorphic to $C_0(G,A)$.

7.9.2. LEMMA. *Given a C^*-dynamical a-system (A, G, α) consider the dual system $(G \times_\alpha A, \hat{G}, \hat{\alpha})$ defined in 7.8.3 and the trivial system $(A, \hat{G}, \iota)$ defined in 7.9.1. There exists a C^*-dynamical a-system $(\hat{G} \times_\iota A, G, \beta)$ such that $(\beta_t(x))(\tau) = \overline{(t,\tau)}\alpha_t(x(\tau))$ for each x in $K(\hat{G}, A)$. Moreover, the C^*-algebras $G \times_\beta \hat{G} \times_\iota A$ and $\hat{G} \times_{\hat{\alpha}} G \times_\alpha A$ are isomorphic.*

Proof. For each t in G we define β_t on $K(\hat{G}, A)$ by

$$(\beta_t(x))(\tau) = \overline{(t,\tau)}\alpha_t(x(\tau)),$$

and we define γ_t on $C_0(G,A)$ by

$$(\gamma_t(y))(s) = \alpha_t(y(s-t)).$$

It is easy to verify that γ_t is an automorphism of $C_0(G,A)$ and that $(C_0(G,A), G, \gamma)$ becomes a C^*-dynamical a-system. However, with $u: L^2(G,H) \to L^2(\hat{G},H)$ as in 7.9.1 a short computation gives

$$(\beta_t(x))^\wedge = \gamma_t(\hat{x}), \qquad x \in K(\hat{G}, A).$$

It follows that each β_t extends to an automorphism of $\hat{G} \times_\iota A$ and that $(\hat{G} \times_\iota A, G, \beta)$ becomes a C^*-dynamical a-system covariantly isomorphic to $(C_0(G,A), G, \gamma)$.

The regular representation of $G \times_\beta (\hat{G} \times_\iota A)$ induced from the universal representation of A on H acts on $L^2(G \times \hat{G}, H) (= L^2(G, L^2(\hat{G}, H)))$ and for each x in $K(G \times \hat{G}, A)$ (regarded as an element of $G \times_\beta (\hat{G} \times_\iota A)$ and each ξ in $L^2(G \times \hat{G}, H)$ we have

$$\begin{aligned}(x\xi)(t,\tau) &= \int \beta_{-t}(x(s,\cdot))\xi(t-s,\cdot)(\tau)\,ds \\ &= \iint \beta_{-t}\iota_{-\tau}(x(s,\sigma))\xi(t-s,\tau-\sigma)\,d\sigma\,ds \\ &= \iint \alpha_{-t}(x(s,\sigma))\xi(t-s,\tau-\sigma)(t,\sigma)\,d\sigma\,ds.\end{aligned}$$

Similarly, the regular representation of $\hat{G} \times_{\hat{\alpha}} (G \times_{\alpha} A)$ acts on $L^2(\hat{G} \times G, H)$ and for each y in $K(\hat{G} \times G, A)$ and η in $L^2(\hat{G} \times G, H)$ we have

$$\begin{aligned}(y\eta)(\tau, t) &= \int \hat{\alpha}_{-\tau}(y(\sigma, \cdot))\eta(\tau - \sigma, \cdot)(t)\, d\sigma \\ &= \iint \hat{\alpha}_{-\tau}\alpha_{-t}(y(\sigma, s))\eta(\tau - \sigma, t - s)\, ds\, d\sigma \\ &= \iint \alpha_{-t}(y(\sigma, s))\eta(\tau - \sigma, t - s)\overline{(s, \tau)}\, ds\, d\sigma.\end{aligned}$$

Define an isomorphism $\Phi: K(G \times \hat{G}, A) \to K(\hat{G} \times G, A)$ by

$$(\Phi x)(\tau, t) = (t, \tau)x(t, \tau); \qquad (\Phi^{-1}y)(t, \tau) = \overline{(t, \tau)}y(\tau, t).$$

Furthermore, define an isometry $w: L^2(G \times \hat{G}, H) \to L^2(\hat{G} \times G, H)$ by

$$(w\xi)(\tau, t) = \overline{(t, \tau)}\xi(t, \tau); \qquad (w^*\eta)(t, \tau) = (t, \tau)\eta(\tau, t).$$

Then for each x in $K(G \times \hat{G}, A)$ and ξ in $L^2(G \times \hat{G}, H)$ we have

$$\begin{aligned}(\Phi(x)w\xi)(\tau, t) &= \iint \alpha_{-t}(\Phi(x)(\sigma, s))w\xi(\tau - \sigma, t - s)\overline{(s, \tau)}\, ds\, d\sigma \\ &= \iint \alpha_{-t}(x(s, \sigma))\xi(t - s, \tau - \sigma)(s, \sigma)\overline{(t - s, \tau - \sigma)}\,\overline{(s, \tau)}\, ds\, d\sigma \\ &= \overline{(t, \tau)}\iint \alpha_{-t}(x(s, \sigma))\xi(t - s, \tau - \sigma)(t, \sigma)\, ds\, d\sigma = (wx\xi)(\tau, t).\end{aligned}$$

It follows that $w^*\Phi(x)w = x$ and since $K(\hat{G} \times G, A)$ and $K(G \times \hat{G}, A)$ are dense in $\hat{G} \times_{\hat{\alpha}} (G \times_{\alpha} A)$ and $G \times_{\beta} (\hat{G} \times_{\iota} A)$, respectively, the two C^*-algebras are isomorphic.

7.9.3. THEOREM. *Let (A, G, α) be a C^*-dynamical a-system and define the dual system $(G \times_{\alpha} A, G, \hat{\alpha})$ as in 7.8.3. Then the double dual system $(\hat{G} \times_{\hat{\alpha}} G \times_{\alpha} A, G, \hat{\hat{\alpha}})$ is covariantly isomorphic to the system $(A \otimes C(L^2(G)), G, \alpha \otimes \operatorname{Ad} \tilde{\lambda})$ defined in 7.7.11.*

Proof. From 7.7.12 we know that $(A \otimes C(L^2(G)), G, \alpha \otimes \operatorname{Ad} \tilde{\lambda})$ is covariantly isomorphic to $(G \times_{\gamma} C_0(G, A), G, \rho)$: where $\rho_t(z)(r, s) = z(r, s + t)$ for each z in $K(G, C_0(G, A))$, and γ is as in the proof of 7.9.2. As we saw there, the system $(C_0(G, A), G, \gamma)$ is covariantly isomorphic to $(\hat{G} \times_{\iota} A, G, \beta)$, so that $G \times_{\gamma} C_0(G, A)$ is isomorphic to $G \times_{\beta} \hat{G} \times_{\iota} A$ which by 7.9.2 is isomorphic to $G \times_{\hat{\alpha}} G \times_{\alpha} A$.

Now take a function z in $K(G, C_0(G, A))$ of the form $z(s, t) = xf(s)\hat{g}(t)$ where $x \in A, f \in K(G)$, and $g \in K(\hat{G})$. Regarding z as an element of $G \times_{\gamma} C_0(G, A)$ its image in $G \times_{\beta} \hat{G} \times_{\iota} A$ is the function $\check{z}$ in $K(G \times \hat{G}, A)$ where $\check{z}(s, \sigma) = xf(x)g(\sigma)$ (cf. 7.9.1). Furthermore, the image of $\check{z}$ in $\hat{G} \times_{\hat{\alpha}} G \times_{\alpha} A$ is the function $\Phi(\check{z})$ in $K(\hat{G} \times_{\hat{\alpha}} G, A)$ (cf. the proof of 7.9.2), where

$$\Phi(\check{z})(\sigma, s) = (s, \sigma)\check{z}(s, \sigma) = xf(s)g(\sigma)(s, \sigma).$$

We have $\rho_r(z)(s, t) = xf(s)\hat{g}(t + r)$, whence $\rho_r(z)^{\vee}(s, \sigma) = xf(s)g(\sigma)(r, \sigma)$

and consequently

$$\Phi(\rho_r(z)^{\check{}})(\sigma, s) = xf(s)g(\sigma)(s + r, \sigma) = (\Phi(\check{z})(\sigma, s))(r, \sigma).$$

But if $y \in K(\hat{G} \times G, A)$ then regarding y as an element of $\hat{G} \underset{\hat{\alpha}}{\times} G \underset{\alpha}{\times} A$ we have by 7.8.3

$$\hat{\hat{\alpha}}_r(y)(\sigma, s) = (r, \sigma)y(\sigma, s).$$

It follows that $\hat{\hat{\alpha}}_r(\Phi(\check{z}) = \Phi((\rho_r(z))^{\check{}})$, and since functions of the form z generate $G \underset{\gamma}{\times} C_0(G, A)$ we conclude that the systems $(G \times C_0(G, A), G, \rho)$ and $(\hat{G} \underset{\hat{\alpha}}{\times} G \underset{\alpha}{\times} A, G, \hat{\hat{\alpha}})$ are covariantly isomorphic, completing the proof.

7.9.4. We have a few applications of the duality theorem above. More will follow in Chapter 8.

If (A, G, α) is a C^*-dynamical system we say that A is *G-simple* if it has no non-trivial closed G-invariant ideals. We say that A is *G-prime* if any two non-zero (closed) G-invariant ideals of A have a non-zero intersection. This is equivalent with the demand that whenever $\alpha_G(x)A\alpha_G(y) = 0$ then $x = 0$ or $y = 0$. Equivalently it means that every non-zero closed G-invariant ideal in A is essential (cf. 3.12.7 and 3.13.7).

7.9.5. LEMMA. *Let (A, G, α) be a C^*-dynamical a-system and $(G \underset{\alpha}{\times} A, \hat{G}, \hat{\alpha})$ its dual system, and consider $G \underset{\alpha}{\times} A$ as a G-product (cf. 7.8.2). If J is a $\hat{G}$-invariant $*$-ideal in $G \underset{\alpha}{\times} A$ the set N consisting of elements in $M(G \underset{\alpha}{\times} A)$ of the form $I(y)$, where y is an $\hat{\alpha}$-integrable element in J, is a G-invariant $*$-ideal in A.*

Proof. By 7.8.8 we may identify A with the set of elements in $M(G \underset{\alpha}{\times} A)$ that satisfy Landstad's conditions. Since I is a linear $*$-preserving map we see from 7.8.4 that N is a $*$-subspace of A. If $x \in N$, say $x = I(y)$ with y in J, then

$$\alpha_t(x) = \lambda_t I(y)\lambda_{-t} = I(\lambda_t y \lambda_{-t}) \in N$$

since J is an ideal in $M(G \underset{\alpha}{\times} A)$. Furthermore, if a and b are elements in A then

$$axb = aI(y)b = I(ayb) \in N,$$

since elements in A are $\hat{G}$-invariant. Thus N is a G-invariant $*$-ideal in A.

7.9.6. PROPOSITION. *Let (A, G, α) be a C^*-dynamical a-system and $(G \underset{\alpha}{\times} A, \hat{G}, \hat{\alpha})$ its dual system. Then A is G-simple if and only if $G \underset{\alpha}{\times} A$ is $\hat{G}$-simple.*

Proof. Suppose that J is a non-trivial, closed $\hat{G}$-invariant ideal in $G \underset{\alpha}{\times} A$, and choose a state ϕ of $G \underset{\alpha}{\times} A$ that annihilates J. Consider the unique extension of ϕ to a state of $M(G \underset{\alpha}{\times} A)$ and identify A with its image in $M(G \underset{\alpha}{\times} A)$. Since $u_\lambda x \to x$ for each x in $G \underset{\alpha}{\times} A$ when $\{u_\lambda\}$ is an approximate unit for A, we do not have $\phi(A) = 0$. However, if x is an $\hat{\alpha}$-integrable element in J then

$$\phi(I(x)) = \int \phi(\hat{\alpha}_\sigma(x))\, d\sigma = 0$$

since J is $\hat{G}$-invariant. It follows that the G-invariant *-ideal N from 7.9.5 is not dense in A. On the other hand, $N \neq 0$, since it contains every element of the form $I(\lambda_f^* y \lambda_f)$, where $y \in J$ and $f \in L^1(G) \cap L^2(G)$ by 7.8.6. It follows that the closure of N is a non-trivial G-invariant ideal in A. Thus if A is G-simple, $G \times_\alpha A$ is $\hat{G}$-simple.

Conversely, if $G \times_\alpha A$ is $\hat{G}$-simple we conclude from 7.9.3 that $A \otimes C(L^2(G))$ is G-simple. But then A must be G-simple as well, for if J was a non-trivial G-invariant ideal in A then $J \otimes C(L^2(G))$ would be a non-trivial G-invariant ideal in $A \otimes C(L^2(G))$.

7.9.7. PROPOSITION. *Let (A, G, α) be a C^*-dynamical system and $(G \times_\alpha A, \hat{G}, \hat{\alpha})$ its dual system. Then A is G-prime if and only if $G \times_\alpha A$ is $\hat{G}$-prime.*

Proof. Suppose that J_1 and J_2 are non-zero, closed $\hat{G}$-invariant ideals in $G \times_\alpha A$ with $J_1 \cap J_2 = 0$. Let N_1 and N_2 be the G-invariant *-ideals corresponding to J_1 and J_2 obtained by 7.9.5. As we saw in the proof of 7.9.6 we have $N_1 \neq 0$ and $N_2 \neq 0$. However, if $x_i \in N_i$, $i = 1, 2$, say $x_i = I(y_i)$ with y_i in J_i, then for each state ϕ of $G \times_\alpha A$ we have

$$\phi(x_1 x_2) = \iint \phi(\hat{\alpha}_\sigma(y_1)\hat{\alpha}_\tau(y_2))\, d\sigma\, d\tau = 0$$

since J_1 and J_2 are $\hat{G}$-invariant. Regarding A as a subset of $M(G \times_\alpha A)$ this implies that $N_1 N_2 = 0$, so that A is not G-prime. Thus if A is G-prime, $G \times_\alpha A$ is $\hat{G}$-prime.

Conversely, if $G \times_\alpha A$ is $\hat{G}$-prime we conclude from 7.9.3 that $A \otimes C(L^2(G))$ is G-prime. But then A is G-prime as well, for if J_1 and J_2 were non-zero orthogonal G-invariant ideals in A then $J_1 \otimes C(L^2(G))$ and $J_2 \otimes C(L^2(G))$ would be non-zero orthogonal G-invariant ideals in $A \otimes C(L^2(G))$.

7.9.8. Notes and remarks. Theorem 7.9.3 is due to Takai, see [254]. The two results 7.9.6 and 7.9.7 are taken from [186].

Following Brown [27] we say that two (separable) C^*-algebras A and B are *stably isomorphic* if $A \otimes C(H)$ is isomorphic to $B \otimes C(H)$, (H a separable Hilbert space). It was shown in [27] that if B is a hereditary C^*-subalgebra of A, not contained in any proper closed ideal of A, then A and B are stably isomorphic. Note that within stable isomorphism 7.9.3 states that the two systems (A, G, α) and $(A \times_\alpha G, \hat{G}, \hat{\alpha})$ are the dual of each other. Another relation among C^*-algebras was developed by Rieffel [215] under the name Morita equivalence. It is less intuitive, but in specific cases easier to check than stable isomorphism. In [28] it was proved by Brown *et al* that the two notions coincide for separable C^*-algebras, thus leading to the belief that an important classification invariant has been found. In this connection a result from [206] should perhaps be mentioned: for any two Glimm algebras A and B the

C^*-algebras $A \otimes C(H)$ and $B \otimes C(H)$ are Borel isomorphic, which means (cf. 4.6.8) that $\mathscr{B}(A) \otimes B(H)$ and $\mathscr{B}(B) \otimes B(H)$ are isomorphic.

7.10. W^*-crossed products with abelian groups.

7.10.1. Let $(\mathcal{M}, G, \alpha)$ be a W^*-dynamical system (cf. 7.4.2). Assume that G is separable and that $\mathcal{M} \subset B(H)$, where H is separable. By 7.5.5 there is an essentially unique separable C^*-dynamical system (A, G, α) and a G-invariant representation (π, H) of A such that $\mathcal{M} = \pi(A)''$. We define the *W^*-crossed product* of G and $\mathcal{M}$ as the von Neumann algebra

$$((\tilde{\pi} \times \lambda)(G \times_\alpha A))'' \subset B(L^2(G, H))$$

(cf. 7.7.1), and denote it by $G \times_\alpha \mathcal{M}$. If α is uniformly continuous, in particular if G is discrete, $(\mathcal{M}, G, \alpha)$ may also be regarded as a C^*-dynamical system, so that the symbol $G \times_\alpha \mathcal{M}$ has two meanings. It will be clear from the context which one is intended. We see from 7.7.1 that the W^*-crossed product $G \times_\alpha \mathcal{M}$ is the von Neumann algebra on $L^2(G, H)$ generated by operators of the form $\iota(x), x \in \mathcal{M}$ and $\lambda_t, t \in G$, where

$$(\iota(x)\xi)(s) = \alpha_{s^{-1}}(x)\xi(s), \qquad (\lambda_t\xi)(s) = \xi(t^{-1}s)$$

for every ξ in $L^2(G, H)$. Equivalently, $G \times_\alpha \mathcal{M}$ is the weak closure of the $*$-algebra of operators y in $K(G, \mathcal{M})$ (σ-weakly continuous bounded functions with compact supports) where

$$(y\xi)(s) = ((\int \iota(y(t))\lambda_t \, dt)\xi)(s) = \int \alpha_{s^{-1}}(y(t))\xi(t^{-1}s) \, dt.$$

7.10.2. In the rest of this section we specialize to abelian groups, i.e. to W^*-dynamical a-systems. So now G is a (separable) locally compact abelian group and $\hat{G}$ denotes its dual group.

A von Neumann algebra $\mathcal{N}$ is said to be a *G-product* if:

(*) There is a unitary representation $t \to \lambda_t$ of G such that $\lambda_G \subset \mathcal{N}$;

(**) There is a homomorphism $\sigma \to \hat{\alpha}_\sigma$ of $\hat{G}$ into $\mathrm{Aut}(\mathcal{N})$ such that $(\mathcal{N}, \hat{G}, \hat{\alpha})$ is a W^*-dynamical a-system with

$$\hat{\alpha}_\sigma(\lambda_t) = (t, \sigma)\lambda_t, \qquad \forall t \in G, \quad \sigma \in \hat{G}.$$

7.10.3. PROPOSITION. *Let $(\mathcal{M}, G, \alpha)$ be a W^*-dynamical a-system and for each y in $K(G, \mathcal{M})$ and σ in $\hat{G}$ define $(\hat{\alpha}_\sigma(y))(t) = (t, \sigma)y(t)$. Then the $\hat{\alpha}_\sigma$'s extend to automorphisms such that $(G \times_\alpha \mathcal{M}, \hat{G}, \hat{\alpha})$ becomes a W^*-dynamical a-system. Moreover, $G \times_\alpha \mathcal{M}$ is a G-product and each element $\iota(x), x \in \mathcal{M}$, is fixed under $\hat{G}$. We say that $(G \times_\alpha \mathcal{M}, \hat{G}, \hat{\alpha})$ is the dual system of $(\mathcal{M}, G, \alpha)$.*

Proof. Choose a C^*-dynamical system (A, G, α) and a G-invariant representation (π, H) such that $\pi(A)'' = \mathscr{M}$. From the proof 7.8.3 we see that there is a unitary representation $\sigma \to u_\sigma$ of $\hat{G}$ on $L^2(G, H)$ such that

$$u_\sigma(\tilde{\pi} \times \lambda)(y)u_{-\sigma} = (\tilde{\pi} \times \lambda)(\hat{\alpha}_\sigma(y))$$

for every y in $K(G, A)$. Since $(\tilde{\pi} \times \lambda)(K(G, A))$ is σ-weakly dense in $G \underset{\alpha}{\times} \mathscr{M}$, the result follows from 7.8.3.

7.10.4. THEOREM. *A von Neumann algebra $\mathscr{N}$ is a G-product for a given abelian group G if and only if there is a W*-dynamical system $(\mathscr{M}, G, \alpha)$ such that $\mathscr{N} = G \underset{\alpha}{\times} \mathscr{M}$. This system is unique (up to covariant isomorphism) and $\iota(\mathscr{M})$ is the fixed-point algebra for $\hat{G}$ in $\mathscr{N}$, whereas $\iota \circ \alpha_t = \lambda_t \iota(\cdot)\lambda_{-t}, t \in G$.*

Proof. It follows from 7.10.3 that the conditions (*) and (**) in 7.10.2 are necessary for $\mathscr{N}$ to be a crossed product.

Conversely, assume that $\mathscr{N}$ is a G-product and denote by $\mathscr{N}^c$ the σ-weakly dense C^*-subalgebra of $\mathscr{N}$ consisting of elements x for which the function $\sigma \to \hat{\alpha}_\sigma(x)$ is norm continuous (cf. 7.5.1). Then $(\mathscr{N}^c, \hat{G}, \hat{\alpha})$ is a C^*-dynamical system. Moreover, if $t \in G$ and $x \in \mathscr{N}^c$ then $\lambda_t x \in \mathscr{N}^c$ because of condition (**) in 7.10.2, so that $\lambda_G \subset M(\mathscr{N}^c)$. Let B be the C^*-subalgebra of $\mathscr{N}^c$ consisting of elements x for which the functions $t \to \lambda_t x$ and $t \to x\lambda_t$ are norm continuous. This algebra contains all elements of the form $\lambda_f y \lambda_g$, where $y \in \mathscr{N}^c$, f, $g \in L^1(G)$, and is therefore σ-weakly dense in $\mathscr{N}^c$, whence also in $\mathscr{N}$. Clearly $\lambda_G \subset M(B)$ and B is invariant under $\hat{G}$, so that B is a G-product in the C^*-algebraic sense (see 7.8.2). By 7.8.8, $B = G \underset{\alpha}{\times} A$ where A is the set of elements in $M(B)$ that satisfy Landstad's conditions and $\alpha_t = \lambda_t \cdot \lambda_{-t}$.

Assume that $\mathscr{N} \subset B(H)$. The regular representation of B on $L^2(\hat{G}, H)$ extends to an isomorphism of $\mathscr{N}$ and as we saw in the proof of 7.8.8 there is a spatial isomorphism of B on $L^2(\hat{G}, H)$ onto $G \underset{\alpha}{\times} A$ on $L^2(G, H)$. Since B is dense in $\mathscr{N}$ it follows that $\mathscr{N}$ is isomorphic to $(G \underset{\alpha}{\times} A)'' = G \underset{\alpha}{\times} \mathscr{M}$, where $\mathscr{M}$ denotes the σ-weak closure of A in $B(L^2(G, H))$. But then $\mathscr{M}$ is also the σ-weak closure of A in $\mathscr{N}$ (on H).

Let $\mathscr{N}^{\hat{G}}$ denote the fixed-point algebra for $\hat{G}$ in $\mathscr{N}$. Clearly $\mathscr{M} \subset \mathscr{N}^{\hat{G}}$, as $A \subset \mathscr{N}^{\hat{G}}$ by 7.8.2. To prove the reverse inclusion define the operator-valued weight I on $\mathscr{N}$ as in 7.8.4. It follows as in the proof of 7.8.6 that for each f in $L^1(G) \cap L^2(G)$ the map $x \to I(\lambda_f^* x \lambda_f)$ is a positive linear (bounded) map of $\mathscr{N}$ into $\mathscr{N}^{\hat{G}}$. Furthermore we see from Lebesgue's monotone convergence theorem that the map is normal. Finally we show, as in the proof of 7.8.9, that if $x \in \mathscr{N}^{\hat{G}}$ and $f \in L^1(\hat{G})$ then

$$I(\lambda_f^* x \lambda_f) = \int \alpha_{-t}(x)|f(t)|^2 \, dt.$$

Now take x in $\mathscr{N}^{\hat{G}}$ and choose a net $\{x_i\}$ in B which is σ-weakly convergent to

x. Then

$$I(\lambda_f^* x_i \lambda_f) \to \alpha_g(x) \qquad (\sigma\text{-weakly})$$

where $g(t) = |f(-t)|^2$. By 7.8.7 we have

$I(\lambda_f^* x_i \lambda_f) \in A$ for all i, whence $\alpha_g(x) \in \mathcal{M}$. Since $(\mathcal{M}, G, \alpha)$ is a W^*-dynamical system, we have $\alpha_g(x) \to x$, σ-weakly, as f ranges over an approximate unit for $L^1(G)$, whence $x \in \mathcal{M}$, i.e. $\mathcal{M} = \mathcal{N}^{\hat{G}}$.

To prove uniqueness take a W^*-dynamical a-system $(\mathcal{M}, G, \alpha)$ and put $\mathcal{N} = G \times_\alpha \mathcal{M}$. As we saw in the proof of 7.8.3 there is a C^*-dynamical a-system (A, G, α) and a G-invariant representation (π, H) of A such that $\mathcal{M} = \pi(A)''$ and $\mathcal{N} = ((\tilde{\pi} \times \lambda)(G \times_\alpha A))''$. By 7.8.8 we can identify A with the set of elements in $M(G \times_\alpha A)$ that satisfy Landstad's conditions. But the argument above shows that $\tilde{\pi}(A)'' = \mathcal{N}^{\hat{G}}$ and since $\iota(\mathcal{M}) = \tilde{\pi}(A)''$ we conclude that $\mathcal{M}$ is uniquely determined from $\mathcal{N}$

7.10.5. Given a von Neumann algebra $\mathcal{M}$ on a Hilbert space H, and another Hilbert space K, we denote by $\mathcal{M} \otimes \boldsymbol{B}(K)$ the smallest von Neumann algebra on $H \otimes K$ that contains all operators $x \otimes y$ with x in $\mathcal{M}$ and y in $\boldsymbol{B}(K)$. Each operator x on $H \otimes K$ has a matrix representation $x = (x_{ij})$ with x_{ij} in $\boldsymbol{B}(H)$ and $1 \leq i \leq \dim(K)$, $1 \leq j \leq \dim(K)$. It is easy to verify that $\mathcal{M} \otimes \boldsymbol{B}(K)$ consists of those operators x for which $x_{ij} \in \mathcal{M}$ for all i,j. Thus if $\rho: \boldsymbol{B}(H) \to \boldsymbol{B}(H \otimes K)$ is the amplification map defined in 2.1.4, it follows that

$$\mathcal{M} \otimes \boldsymbol{B}(K) = (\rho(\mathcal{M}'))'.$$

Note that if A is a σ-weakly dense C^*-subalgebra of $\mathcal{M}$ then

$$(A \otimes \boldsymbol{C}(K))'' = (\rho(\mathcal{M}'))' = \mathcal{M} \otimes \boldsymbol{B}(H),$$

with $A \otimes \boldsymbol{C}(K)$ as defined in 7.7.10.

7.10.6. THEOREM. *Let $(\mathcal{M}, G, \alpha)$ be a W^*-dynamical a-system and define the dual system $(G \times_\alpha \mathcal{M}, \hat{G}, \hat{\alpha})$ as in 7.10.3. Then the double dual system $(\hat{G} \times_{\hat{\alpha}} G \times_\alpha \mathcal{M}, G, \hat{\hat{\alpha}})$ is covariantly isomorphic to the W^*-dynamical system $(\mathcal{M} \otimes \boldsymbol{B}(L^2(G)), G, \alpha \otimes \mathrm{Ad}\,\tilde{\lambda})$, where $\tilde{\lambda}$ is the inverse regular representation of G on $L^2(G)$ (i.e. $(\tilde{\lambda}_t \xi)(s) = \xi(s + t)$) and*

$$(\alpha \otimes \mathrm{Ad}\,\tilde{\lambda})_t(x \otimes y) = \alpha_t(x) \otimes \tilde{\lambda}_t y \tilde{\lambda}_{-t}$$

for each x in $\mathcal{M}$ and y in $\boldsymbol{B}(L^2(G))$.

Proof. Choose a C^*-dynamical system (A, G, α) and a G-invariant representation (π, H) of A such that $\mathcal{M} = \pi(A)''$. Then $G \times_\alpha \mathcal{M} = ((\tilde{\pi} \times \lambda)(G \times_\alpha A))''$ and $(\tilde{\pi} \times \lambda, L^2(G, H))$ is a $\hat{G}$-invariant representation of $G \times_\alpha A$. Thus with $(\hat{\lambda}, L^2(\hat{G}))$ as the regular representation of $\hat{G}$ we have

$$\hat{G} \underset{\hat{\alpha}}{\times} G \underset{\alpha}{\times} \mathcal{M} = (((\tilde{\pi} \times \lambda)^{\sim} \times \tilde{\lambda})(\hat{G} \underset{\hat{\alpha}}{\times} G \underset{\alpha}{\times} A))''.$$

From the proofs of 7.7.12 and 7.9.3 we see that $\hat{G} \underset{\hat{\alpha}}{\times} G \underset{\alpha}{\times} A$ on $L^2(\hat{G} \times G, H)$ is spatially isomorphic to $\tilde{\pi}(A) \otimes \boldsymbol{C}(L^2(G))$ on $L^2(G \times G, H)$, where $(\tilde{\pi}, L^2(G, H))$ is the representation of A such that

$$(\tilde{\pi}(x)\xi)(t) = \pi(\alpha_t(x))\xi(t), \qquad x \in A, \qquad \xi \in L^2(G, H).$$

Consequently $\hat{G} \underset{\hat{\alpha}}{\times} G \underset{\alpha}{\times} \mathcal{M}$ is isomorphic to $\tilde{\pi}(A)'' \otimes \boldsymbol{B}(L^2(G))$ which in turn is isomorphic to $\mathcal{M} \otimes \boldsymbol{B}(L^2(G))$. Furthermore, since $((\tilde{\pi} \times \lambda)^{\sim} \times \tilde{\lambda}, L^2(\hat{G} \times G, H))$ and $(\tilde{\pi} \otimes \iota, L^2(G \times G, H))$ are G-invariant representations of $G \underset{\hat{\alpha}}{\times} G \underset{\alpha}{\times} A$ and $A \otimes \boldsymbol{C}(L^2(G))$, respectively, we conclude from 7.9.3 that the two W^*-dynamical systems

$$(G \underset{\hat{\alpha}}{\times} G \underset{\alpha}{\times} \mathcal{M}, G, \hat{\hat{\alpha}}) \qquad \text{and} \qquad (\mathcal{M} \otimes \boldsymbol{B}(L^2(G)), G, \alpha \otimes \operatorname{Ad}\tilde{\lambda})$$

are covariantly isomorphic.

7.10.7. Let H be a separable Hilbert space of dimension n $(1 \leqslant n \leqslant \infty)$ and choose a discrete abelian group G with cardinality n. Denote by $\hat{G}$ the (compact) dual group of G. Choose an orthonormal basis $\{\xi_s \mid s \in G\}$ for H and define unitary representations λ and u for G and $\hat{G}$ by

$$\lambda_t(\Sigma \gamma_s \xi_s) = \Sigma \gamma_s \xi_{s-t}, \qquad t \in G,$$

$$u_\sigma(\Sigma \gamma_s \xi_s) = \Sigma \gamma_s \overline{(s, \sigma)} \xi_s, \quad \sigma \in \hat{G}.$$

Define $\hat{\alpha}: \hat{G} \to \operatorname{Aut}(\boldsymbol{B}(H))$ by

$$\hat{\alpha}_\sigma(x) = u_\sigma x u_{-\sigma}, \qquad x \in \boldsymbol{B}(H), \qquad \sigma \in \hat{G}.$$

Since $\sigma \to u_\sigma$ is strongly continuous it follows that $\sigma \to \hat{\alpha}_\sigma(x)$ is norm continuous for each finite dimensional operator x. From this it is easy to see that $(\boldsymbol{C}(H), \hat{G}, \hat{\alpha})$ is a C^*-dynamical system whereas $(\boldsymbol{B}(H), \hat{G}, \hat{\alpha})$ is a W^*-dynamical system. Furthermore, each function $t \to \lambda_t x, x \in \boldsymbol{C}(H)$ is norm continuous. Finally we see that

$$\hat{\alpha}_\sigma(\lambda_t) = u_\sigma \lambda_t u_{-\sigma} = (t, \sigma)\lambda_t$$

so that both $\boldsymbol{C}(H)$ and $\boldsymbol{B}(H)$ are G-products; the former as a C^*-algebra, the latter as a von Neumann algebra.

Denote by $\mathfrak{A}$ the von Neumann algebra of operators that are diagonalizable with respect to $\{\xi_s\}$. Thus $\mathfrak{A}$ is commutative, and isomorphic to $l^\infty(G)$. Put $A = \boldsymbol{C}(H) \cap \mathfrak{A}$ and note that A isomorphic to $c_0(G)$. It is straightforward to prove that $\mathfrak{A}$ coincides with the set of fixed-points of $\hat{G}$ in $\boldsymbol{B}(H)$. Furthermore.

A coincides with the elements in $\boldsymbol{B}(H)$ that satisfy Landstad's conditions (cf. 7.8.2). It follows from 7.8.8 and 7.9.4 that with $\alpha_t = \lambda_t \cdot \lambda_{-t}$ we have

$$C(H) = G \underset{\alpha}{\times} A; \qquad \boldsymbol{B}(H) = G \underset{\alpha}{\times} \mathfrak{A}.$$

This result can also be obtained from 7.7.12, taking $A = \boldsymbol{C}$.

7.10.8. We shall give yet another application of the notion of G-products; this time to Glimm algebras and the von Neumann algebras they generate.

As in 6.4.2 let $\{m(n) \mid n \in \boldsymbol{N}\}$ be a sequence of natural numbers greater than one, and for each n let $\boldsymbol{Z}_n$ denote the cyclic group of order n $(=\hat{\boldsymbol{Z}}_n)$. From 7.10.7 there are unitary representations λ^n and u^n of $\boldsymbol{Z}_{m(n)}$ and $\hat{\boldsymbol{Z}}_{m(n)}$, respectively, into $M_{m(n)}$ and we may put $\hat{\alpha}^n_\sigma = u^n_\sigma \cdot u^n_{-\sigma}$, $\sigma \in \boldsymbol{Z}_{m(n)}$ to obtain a representation $\hat{\alpha}^n : \hat{\boldsymbol{Z}}_{m(n)} \to \mathrm{Aut}(M_{m(n)})$ such that $\boldsymbol{M}_{m(n)}$ is a $\boldsymbol{Z}_{m(n)}$-product.

Define

$$G = \bigoplus_{n=1}^{\infty} \boldsymbol{Z}_{m(n)} \qquad \text{(direct sum)},$$

$$\hat{G} = \prod_{n=1}^{\infty} \hat{\boldsymbol{Z}}_{m(n)} \qquad \text{(direct product)},$$

so that G and $\hat{G}$ is a dual pair of groups with G discrete and $\hat{G}$ compact. As in 6.4.2 let

$$\boldsymbol{M}_{m(n)!} = \boldsymbol{M}_{m(1)} \otimes \boldsymbol{M}_{m(2)} \otimes \cdots \otimes \boldsymbol{M}_{m(n)}.$$

Put $\boldsymbol{M}_\infty = \cup \boldsymbol{M}_{m(n)}$ and denote by A_∞ the completion of $\boldsymbol{M}_\infty$. Each t in G is a finite sum $t = t_1 + t_2 + \cdots + t_n$ and we define λ_t in $M_{m(n)!}$ by

$$\lambda_t = \lambda^1_{t_1} \otimes \lambda^2_{t_2} \otimes \cdots \otimes \lambda^n_{t_n}.$$

It is easy to verify that λ is a unitary representation of G into $\boldsymbol{M}_\infty$, hence into A_∞. Each element in $\boldsymbol{M}_\infty$ is a linear span of elements of the form

$$x = x_1 \otimes x_2 \otimes \cdots \otimes x_n;$$

so for each σ in $\hat{G}$ we can define

$$\begin{aligned} \hat{\alpha}_\sigma(x) &= \hat{\alpha}^1_{\sigma_1}(x_1) \otimes \hat{\alpha}^2_{\sigma_2}(x_2) \otimes \cdots \otimes \hat{\alpha}^n_{\sigma_n}(x_n) \\ &= (u^1_{\sigma_1} \otimes u^2_{\sigma_2} \otimes \cdots \otimes u^n_{\sigma_n}) x (u^1_{-\sigma_1} \otimes u^2_{-\sigma_2} \otimes \cdots \otimes u^n_{-\sigma_n}). \end{aligned}$$

It is not difficult to show that each $\hat{\alpha}_\sigma$ extends to an automorphism of A_∞ and that $(A_\infty, \hat{G}, \hat{\alpha})$ becomes a C^*-dynamical system. Evidently

$$\begin{aligned} \hat{\alpha}_\sigma(\lambda_t) &= \hat{\alpha}^1_{\sigma_1}(\lambda^1_{t_1}) \otimes \hat{\alpha}^2_{\sigma_2}(\lambda^2_{t_2}) \otimes \cdots \otimes \hat{\alpha}^n_{\sigma_n}(\lambda^n_{t_n}) \\ &= \prod_k (t_k, \sigma_k)(\lambda^1_{t_1} \otimes \lambda^2_{t_2} \otimes \cdots \otimes \lambda^n_{t_n}) = (t, \sigma)\lambda_t, \end{aligned}$$

so that A_∞ is a G-product.

To find the elements x in A_∞ that satisfy Landstad's conditions (7.8.2) note that (iii) is vacuous (G is discrete), (ii) merely says that $x \in A_\infty$ (since $\delta_e \in L^1(G)$); so that (i) is the only restraint. Let $C_{m(n)}$ denote the set of fixed-points under $\hat{G}$ in $M_{m(n)}$ and note from 7.10.7 that $C_{m(n)} = C(\hat{\mathbf{Z}}_{m(n)})$. Put

$$C_{m(n)!} = C_{m(1)} \otimes C_{m(2)} \otimes \cdots \otimes C_{m(n)}$$

(regarded as a subalgebra of $M_{m(n)!}$) and further let $C_\infty = \cup C_{m(n)!}$, and $C = (C_\infty)^-$. We have $C_{m(n)!} = C(\hat{\mathbf{Z}}_{m(1)} \times \hat{\mathbf{Z}}_{m(2)} \times \cdots \times \hat{\mathbf{Z}}_{m(n)})$; whence $C = C(\hat{G})$. An easy induction argument shows that $C_{m(n)!}$ is the fixed-points of $\hat{G}$ in $M_{m(n)!}$. It follows that each element in C is fixed under $\hat{G}$. Conversely, assume that x is a fixed-point for $\hat{G}$ in A_∞. Given $\varepsilon > 0$ there is a y in some $M_{m(n)!}$ with $\|x - y\| \leq \varepsilon$. Define

$$z = \int_{\hat{G}} \hat{\alpha}_\sigma(y)\, d\sigma$$

and note that $z \in M_{m(n)!}$, $\hat{\alpha}_\sigma(z) = z$ for all σ in $\hat{G}$, and $\|x - z\| \leq \varepsilon$ (cf. the proof of 6.5.4). Thus $z \in C_{m(n)!}$ whence in the limit $x \in C$. Consequently $C = C(\hat{G})$ is the set of elements in A_∞ that satisfy Landstad's conditions. With $\alpha_t = \lambda_t \cdot \lambda_{-t}$ on C we note that the transposed action α'_t of α_t on $\hat{G}$ is the translation $\sigma \to \sigma - t$ (regarding G as a subset of $\hat{G}$). From 7.8.8

$$A_\infty = G \times_\alpha C(\hat{G}).$$

Let ϕ_Λ be the product state of A_∞ determined by a sequence $\{\Lambda^n\}$ of convex combinations, each of length $m(n)$ (cf. 6.5.2). The restriction of ϕ_Λ to $M_{m(n)}$ is the state

$$x \to \sum \Lambda_i^n x_{ii}, \qquad x \in M_{m(n)};$$

and since $C_{m(n)}$ is the diagonal operators we see that the restriction of ϕ_Λ to $C_{m(n)}$ corresponds to a probability measure μ_n on $\hat{\mathbf{Z}}_{m(n)}$. Consequently, the restriction of ϕ_Λ to C corresponds to the product measure $\mu = \otimes \mu_n$ on $\hat{G}$. Assuming that $\Lambda_i^n > 0$ for all i $(1 \leq i \leq m(n))$ and all n (i.e. excluding the pure product states, cf. 6.5.5) the measure μ is quasi-invariant under the action of G on $\hat{G}$. Therefore the automorphisms α_t, $t \in G$, extend to automorphisms of $L^\infty_\mu(\hat{G})$ and we obtain a W^*-dynamical system $(L^\infty_\mu(\hat{G}), G, \alpha)$.

Consider the representation (π_Λ, H_Λ) associated with ϕ_Λ, and put $\mathcal{M}_\Lambda = \pi_\Lambda(A)''$. Note that $\mathcal{M}_\Lambda$ is a factor by 6.5.8. For each n we have a representation $\hat{\alpha}^n : \mathbf{Z}_{m(n)} \to \mathrm{Aut}(\mathcal{M}_\Lambda)$ induced by the unitary representation $u^n : \mathbf{Z}_{m(n)} \to A_\infty$ mentioned above. It follows that there is a representation $\hat{\alpha} : \hat{G} \to \mathrm{Aut}(\mathcal{M}_\Lambda)$ such that $(\mathcal{M}_\Lambda, \hat{G}, \hat{\alpha})$ is a W^*-dynamical system. Since $\pi_\Lambda(\lambda_G) \subset \mathcal{M}_\Lambda$ we see that $\mathcal{M}_\Lambda$ is a G-product. Now $\pi_\Lambda(C(\hat{G}))''$ is isomorphic to $L^\infty_\mu(\hat{G})$ and in the same manner as above we show that $\pi_\Lambda(C(\hat{G}))''$ is the set of fixed-points for $\hat{G}$ in $\mathcal{M}_\Lambda$. It follows from 7.10.4 that

$$\mathcal{M}_\Lambda = G \times_\alpha L^\infty_\mu(\hat{G}).$$

7.10.9. Notes and remarks. In full generality W^*-crossed products first appeared in Takesaki's paper [263], where 7.10.6 was proved. Although we have chosen to present the von Neumann algebra results as corollaries ('localized versions') of the corresponding C^*-algebra results it must be remembered that the historical development was quite the opposite. Thus 7.10.6 came first and actually made Takesaki conjecture 7.9.3. Theorem 7.10.4 is due to Landstad, see [159] or [186]. The example in 7.10.9 appeared in [258]. To understand it, Takesaki had to develop the duality theory.

In contrast to the C^*-algebra case the notion of 'stable isomorphism' (cf. 7.9.8) between von Neumann algebras is completely understood. If $\mathscr{M}$ is a von Neumann algebra on a separable Hilbert space H then $\mathscr{M}$ is isomorphic to $\mathscr{M} \otimes \boldsymbol{B}(H)$ if and only if $\mathscr{M}$ is properly infinite (no non-zero finite central projections, cf. 5.4.1), see p. 298 of [59]. In particular any type III von Neumann algebra is stable, so that 7.10.6 is an honest duality theorem.

7.11. W^*-crossed products with discrete groups

7.11.1. Let $(\mathscr{M}, G, \alpha)$ be a W^*-dynamical system. Assume that $\mathscr{M} \subset \boldsymbol{B}(H)$, where H is separable Hilbert space, and assume moreover that G is discrete (and countable). In this case the W^*-crossed product $G \underset{\alpha}{\times} \mathscr{M}$ is relatively easy to handle; indeed, this construction furnished the earliest examples of exotic von Neumann algebras (i.e. algebras of type II and III).

7.11.2. Since G is discrete the Hilbert space $L^2(G, H)$ is the direct sum of card(G) copies of H and for each t in G we let p_t denote the projection of $L^2(G, H)$ onto the t'th summand. If $y \in K(G, \mathscr{M})$ we have $y = \Sigma \iota(y(s))\lambda_s$ so that $p_s y p_t = p_s \iota(y(st^{-1}))\lambda_{st^{-1}} p_t$. Since the map $y \to p_s y p_t$ is σ-weakly continuous and $K(G, \mathscr{M})$ is σ-weakly dense in $G \underset{\alpha}{\times} \mathscr{M}$ we can associate to each y in $G \underset{\alpha}{\times} \mathscr{M}$ a bounded function $\check{y}: G \to \mathscr{M}$ such that

$$(*) \qquad y = \sum \iota(\check{y}(s))\lambda_s \qquad \text{(strong convergence).}$$

We do of course not assert that every such function defines an element. An equivalent formulation of the formula $(*)$ says that each element y in G has a matrix $(y_{s,t})$ with entries in $\mathscr{M}$ such that

$$(**) \qquad y_{s,t} = \alpha_{s^{-1}}(\check{y}(st^{-1})).$$

Note that if $x \in \mathscr{M}$, then $(\iota(x))\check{}(s) = 0$ if $s \neq e$ and $(\iota(x)\check{}(e) = x$. Moreover, if $t \in G$ then $(\lambda_t)\check{}(s) = 0$ if $s \neq t$ and $(\lambda_t)\check{}(t) = 1$. Finally, if y, z are elements of $G \underset{\alpha}{\times} \mathscr{M}$ then the function corresponding to yz is the convolution of the two functions (cf. 7.6.1):

$$(yz)\check{}(s) = \sum_t \check{y}(t)\alpha_t(\check{z}(t^{-1}s)) \qquad \text{(strong convergence).}$$

7.11.3. LEMMA. *For each y in $G \underset{\alpha}{\times} \mathcal{M}$ define $\pi(y) = \check{y}(e)$. The map $\pi: G \underset{\alpha}{\times} \mathcal{M} \to \mathcal{M}$ is a faithful normal positive linear map satisfying the conditions:*

(i) $\pi(y^*y) = \Sigma_s \alpha_{s^{-1}}(\check{y}(s)^*\check{y}(s))$; $\qquad \pi(yy^*) = \Sigma_s \check{y}(s)\check{y}(s)^*$;

(ii) $\pi(\lambda_t y \lambda_t^*) = \alpha_t(\pi(y))$;

(iii) $\pi(\iota(x_1)y\iota(x_2)) = x_1\pi(y)x_2$; $x_1, x_2 \in \mathcal{M}$.

Proof. We may identify $\pi(y)$ with $p_e y p_e$ (cf. 7.11.2). From this it is clear that π is a normal positive linear map. Thus by continuity it suffices to prove (i), (ii) and (iii) for elements y in $K(G, \mathcal{M})$, which is a straightforward calculation. Finally, (i) shows that π is faithful.

7.11.4. LEMMA. *If x is a central fixed-point under G in $\mathcal{M}$ then $\iota(x)$ belongs to the centre of $G \underset{\alpha}{\times} \mathcal{M}$. Conversely, if y is central in $G \underset{\alpha}{\times} \mathcal{M}$ then $\pi(y)$ is a central fixed-point under G in $\mathcal{M}$.*

Proof. Since $\iota(x)$ commutes with both $\iota(\mathcal{M})$ and λ_G by 7.11.3 it belongs to the centre of $G \underset{\alpha}{\times} \mathcal{M}$. Conversely, by 7.11.3 (iii) we have for each x in $\mathcal{M}$:

$$x\pi(y) = \pi(\iota(x)y) = \pi(y\iota(x)) = \pi(y)x.$$

Moreover, by 7.11.3 (ii)

$$\alpha_t(\pi(y)) = \pi(\lambda_t y \lambda_t^*) = \pi(y).$$

7.11.5. We define a relation $\underset{G}{\sim}$ in $\mathcal{M}_+$ be setting $x \underset{G}{\sim} y$ if there is a countable set $\{u_{nt} \mid n \in \mathbf{N}, t \in G\}$ in $\mathcal{M}$ such that

$$x = \sum \alpha_{t^{-1}}(u_{nt}^* u_{nt}); \qquad y = \sum u_{nt} u_{nt}^*.$$

With $\approx$ defined as in 5.2.6 for Borel $*$-algebras we have the following lemma.

7.11.6. LEMMA. *If y, z belong to $(G \underset{\alpha}{\times} \mathcal{M})_+$ and $y \approx z$ then $\pi(y) \underset{G}{\sim} \pi(z)$. If x, y belong to $\mathcal{M}_+$ then $x \underset{G}{\sim} y$ if and only if $\iota(x) \approx \iota(y)$. In particular, $\underset{G}{\sim}$ is a countably additive equivalence relation in $\mathcal{M}_+$.*

Proof. Suppose that $y \approx z$ so that $y = \Sigma u_n^* u_n$ and $z = \Sigma u_n u_n^*$ for some sequence $\{u_n\}$ in $G \underset{\alpha}{\times} \mathcal{M}$. Applying π to these equations and using 7.11.3 (i) we get $\pi(y) \underset{G}{\sim} \pi(z)$.

Conversely, take x and y in $\mathcal{M}_+$. If $\iota(x) \approx \iota(y)$ then $x \underset{G}{\sim} y$ from the preceding, since $\pi \circ \iota$ is the identity map on $\mathcal{M}$ by 7.11.3 (iii). But if

$$x = \sum \alpha_{t^{-1}}(u_{nt}^* u_{nt}), \qquad y = \sum u_{nt} u_{nt}^*$$

we see from 7.11.3 (ii) that

$$\iota(x) = \sum \lambda_{t^{-1}} \iota(u_{nt}^* u_{nt}) \lambda_t = \sum (\iota(u_{nt})\lambda_t)^* (\iota(u_{nt})\lambda_t,$$

$$\iota(y) = \sum \iota(u_{nt} u_{nt}^*) = \sum (\iota(u_{nt})\lambda_t)(\iota(u_{nt})\lambda_t)^*,$$

whence $\iota(x) \approx \iota(y)$.

Since $\gtrsim$ is a countably additive equivalence relation in $(G \times_\alpha \mathcal{M})_+$ the same is true for $\underset{G}{\sim}$ in $\mathcal{M}_+$.

7.11.7. PROPOSITION. *A projection p in $\mathcal{M}$ is $\underset{G}{\sim}$-finite (i.e. $0 \leqslant x \leqslant p$, $x \underset{G}{\sim} p$ implies $x = p$) if and only if $\iota(p)$ is a finite projection in $G \times_\alpha \mathcal{M}$.*

Proof. If p is $\underset{G}{\sim}$-finite and q is a projection in $G \times_\alpha \mathcal{M}$ such that $q \leqslant \iota(p)$ and $q \sim \iota(p)$ then $\pi(q) \leqslant p$ and by 7.11.6 $\pi(q) \underset{G}{\sim} p$, whence $\pi(q) = p$ by assumption. Thus $\pi(\iota(p) - q) = 0$ and since π is faithful $\iota(p) = q$, so that $\iota(p)$ is finite by 5.4.10.

Conversely, assume that $\iota(p)$ is a finite projection and that $0 \leqslant x \leqslant p$ with $x \underset{G}{\sim} p$. Then $\iota(x) \leqslant \iota(p)$ and $\iota(x) \gtrsim \iota(p)$ by 7.11.6. By 5.4.6 there is a faithful normal semi-finite trace ϕ on $c(\iota(p))(G \times_\alpha \mathcal{M})$ with $\phi(\iota(p)) < \infty$. Since $\phi(\iota(x)) = \phi(\iota(p))$ this implies that $\iota(x) = \iota(p)$, i.e. $x = p$; so that p is $\underset{G}{\sim}$-finite.

7.11.8. COROLLARY. *The von Neumann algebra $G \times_\alpha \mathcal{M}$ is finite if and only if there is a faithful, normal, G-invariant finite trace on $\mathcal{M}$.*

7.11.9. Let α be an automorphism of a von Neumann algebra $\mathscr{Z}$. We say that α acts *freely* if there is a set of projections $\{p_i\}$ in $\mathscr{Z}$ with $\Sigma\, p_i = 1$ such that $p_i \alpha(p_i) = 0$ for each i. If $\mathscr{Z}$ is commutative this is equivalent to the demand that for each non-zero projection p in $\mathscr{Z}$ there exists a non-zero projection $q \leqslant p$ with $q\alpha(q) = 0$.

Now let $(\mathcal{M}, G, \alpha)$ be a W^*-dynamical system as before and denote by $\mathscr{Z}$ the centre of $\mathcal{M}$. We say that G acts *centrally freely* on $\mathcal{M}$ if each α_t, $t \neq e$, acts freely on $\mathscr{Z}$.

7.11.10. LEMMA. *Let $(\mathcal{M}, G, \alpha)$ be a W^*-dynamical system and assume that G acts centrally freely. Then*

$$\iota(\mathscr{Z})' \cap (G \times_\alpha \mathcal{M}) = \iota(\mathcal{M}); \qquad \iota(\mathcal{M})' \cap (G \times_\alpha \mathcal{M}) = \iota(\mathscr{Z}).$$

Proof. Take y in $G \times_\alpha \mathcal{M}$ and x in $\mathcal{M}$. Then

$$(\iota(x)y)\check{}(s) = \sum \iota(x)\check{}(t)\alpha_t(\check{y}(t^{-1}s)) = x\check{y}(s);$$

$$(y\iota(x))\check{}(s) = \sum \check{y}(t)\alpha_t(\iota(x)\check{}(t^{-1}s)) = \check{y}(s)\alpha_s(x)$$

(cf. 7.11.2). If now $y \in \iota(\mathscr{Z})'$ then $x\check{y}(s) = \alpha_s(x)\check{y}(s)$ for every x in $\mathscr{Z}$. If $s \neq e$ there is by assumption a set of projections $\{p_i\}$ in $\mathscr{Z}$ with $\Sigma\, p_i = 1$ such that $p_i \alpha_s(p_i) = 0$. Consequently

$$p_i \check{y}(s) = p_i \alpha_s(p_i)\check{y}(s) = 0$$

for each i, whence $\check{y}(s) = 0$. This implies that $y = \iota(\check{y}(e)) \in \iota(\mathcal{M})$.

If, moreover, $y \in \iota(\mathcal{M})'$ then from the above we infer that $x\check{y}(e) = \check{y}(e)x$ for every x in $\mathcal{M}$, so that $\check{y}(e) \in \mathscr{Z}$; i.e. $y \in \iota(\mathscr{Z})$.

7.11.11. THEOREM. *Let $(\mathcal{M}, G, \alpha)$ be a W^*-dynamical system where G is discrete.*

If $G \underset{\alpha}{\times} \mathcal{M}$ is a factor then G is ergodic on the centre of $\mathcal{M}$. The converse holds if G acts centrally freely on $\mathcal{M}$.

Proof. If $G \underset{\alpha}{\times} \mathcal{M}$ is a factor each central fixed-point in $\mathcal{M}$ is a scalar multiple of 1 by 7.11.4.

Assume now that G acts centrally freely on $\mathcal{M}$ and take y in the centre of $G \underset{\alpha}{\times} \mathcal{M}$. Then $y \in \iota(\mathcal{M})'$ so that $y = \iota(x)$ by 7.11.10, where $x \in \mathcal{Z}$. Moreover, by 7.11.4 $x\,(=\pi(y))$ is a fixed point for G. Thus if G is ergodic on $\mathcal{Z}$ then y is a scalar multiple of 1.

7.11.12. LEMMA. *Let $\mathcal{U}$ be a group of unitaries in a von Neumann algebra $\mathcal{M}$. If ϕ is a faithful normal semi-finite trace on $\mathcal{M}$ then for each x in $\mathcal{M}_2^\phi$ we have $\mathcal{K} \cap \mathcal{U}' \neq \emptyset$, where $\mathcal{K}$ denotes the σ-weak convex closure of the set $\{u^*xu \mid u \in \mathcal{U}\}$.*

Proof. We may assume that $\|x\| \leqslant 1$ and that $\phi(x^*x) \leqslant 1$. The set $\mathcal{K}$ is convex and σ-weakly compact in the unit ball of $\mathcal{M}$. Moreover, since $\phi(y) = \Sigma(\pi_\phi(y)\xi_n \mid \xi_n)$ where $\{\xi_n\} \subset H_\phi$ (cf. 5.1.5) the map $y \to \|\xi_y\|$ (ξ_y denoting the image of y in H_ϕ) is σ-weakly lower semi-continuous on $\mathcal{M}_2^\phi$. It follows that ξ_y belongs to the unit ball of H_ϕ for each y in $\mathcal{K}$, in particular $\mathcal{K} \subset \mathcal{M}_2^\phi$. Since $\mathcal{K}$ is compact there exists an element y in $\mathcal{K}$ for which $\|\xi_y\|$ is minimal. As $\mathcal{K}$ is convex and the norm on H_ϕ is strictly convex, the element y is unique. Since $\mathcal{U}$ is a group it follows easily that $u^*\mathcal{K}u = \mathcal{K}$ for each u in $\mathcal{U}$. But

$$\|\xi_{u^*yu}\|^2 = \phi(u^*y^*uu^*yu) = \phi(y^*y) = \|\xi_y\|^2$$

so the uniqueness of y implies that $u^*yu = y$, i.e. $y \in \mathcal{U}'$.

7.11.13. THEOREM. *Let $(\mathcal{M}, G, \alpha)$ be a W^*-dynamical system where G is discrete and acts centrally freely on $\mathcal{M}$. Then $\phi = \phi \circ \iota \circ \pi$ for every faithful normal semi-finite trace ϕ on $G \underset{\alpha}{\times} \mathcal{M}$.*

Proof. Since $(G \underset{\alpha}{\times} \mathcal{M})^\phi$ is a σ-weakly dense ideal in $G \underset{\alpha}{\times} \mathcal{M}$ and $\phi \circ \iota \circ \pi$ is normal it suffices to show that $\phi(y) = \phi(\iota(\pi(y)))$ for each y in $(G \underset{\alpha}{\times} \mathcal{M})_+^\phi$. Let $\mathcal{U}$ denote the unitary group in $\iota(\mathcal{Z})$, where $\mathcal{Z}$ is the centre of $\mathcal{M}$. By 7.11.12 there is an element z in the σ-weak convex closure of the set $\{uyu^* \mid u \in \mathcal{U}\}$ such that z commutes with $\iota(\mathcal{Z})$. By 7.11.10 we have $z = \iota(x)$ for some x in $\mathcal{M}_+$. Since $\pi(uyu^*) = \pi(y)$ by 7.11.3 (iii) for every u in $\mathcal{U}$ we conclude that $\pi(y) = \pi(z) = x$. On the other hand we also have $\phi(uyu^*) = \phi(y)$ for each u in $\mathcal{U}$ so that $\phi(z) \leqslant \phi(y)$. It follows that

$$\phi(\iota(\pi(y))) = \phi(\iota(x)) = \phi(z) \leqslant \phi(y).$$

This proves that $\phi \circ \iota$ is a faithful normal semi-finite and G-invariant trace on $\mathcal{M}$. Consequently $\phi \circ \iota \circ \pi$ is a faithful normal semi-finite trace on $G \underset{\alpha}{\times} \mathcal{M}$ and since $\phi \circ \iota \circ \pi \leqslant \phi$ we get $\phi \circ \iota \circ \pi = \phi(h.)$ by 5.3.4 for some central element h in $G \underset{\alpha}{\times} \mathcal{M}$ with $0 \leqslant h \leqslant 1$. However, since h is central $h = \iota(k)$

where k is a fixed-point under G in $\mathcal{M}$ (cf. the proof of 7.11.11). For each x in $\mathcal{M}_+$ such that $\iota(x) \in (G \underset{\alpha}{\times} \mathcal{M})^\phi$ we have

$$\phi(\iota(x)) = \phi(\iota(\pi(\iota(x)))) = \phi(\iota(k)\iota(x)) = \phi(\iota(kx)),$$

whence $(1 - k)x = 0$. Since $\iota(\mathcal{M}_+) \cap (G \underset{\alpha}{\times} \mathcal{M})^\phi$ contains 1 as a strong limit point we conclude that $k = 1$, i.e. $\phi = \phi \circ \iota \circ \pi$.

7.11.14. COROLLARY. *If $(\mathcal{M}, G, \alpha)$ is a W^*-dynamical system where G is discrete and acts centrally freely on $\mathcal{M}$ then $G \underset{\alpha}{\times} \mathcal{M}$ is semi-finite if and only if there is a faithful, normal, semi-finite and G-invariant trace on $\mathcal{M}$.*

7.11.15. The preceding results apply in particular to the case where $\mathcal{M} = L^\infty_\mu(T)$ for some σ-finite Borel measure μ on a standard Borel space T. Suppose that α' is a representation of the discrete group G as Borel isomorphisms of T. If μ is a quasi-invariant measure (i.e. $\mu(N) = 0$ implies $\mu(\alpha'_t(N)) = 0$ for all t in G) then by transposition we obtain a W^*-dynamical system $(L^\infty_\mu(T), G, \alpha)$.

7.11.16. THEOREM. *Let $(L^\infty_\mu(T), G, \alpha)$ be a W^*-dynamical system as described above, and assume that G acts freely and ergodicly on T. Assume furthermore that μ is diffuse (i.e. has no atoms). If μ is G-invariant then $G \underset{\alpha}{\times} L^\infty_\mu(T)$ is a factor of type II_1 (respectively II_∞) provided that $\mu(T) < \infty$ (resp. $\mu(T) = \infty$). If μ is not G-invariant, but is invariant under a subgroup G_0 of G which is ergodic on T, then $G \underset{\alpha}{\times} L^\infty_\mu(T)$ is a factor of type III.*

Proof. It follows from 7.11.11 that $G \underset{\alpha}{\times} L^\infty_\mu(T)$ is a factor. Define $\phi(x) = \int x \, d\mu$ for each x in $L^\infty_\mu(T)_+$. Then ϕ is a faithful, normal, semi-finite trace on $L^\infty_\mu(T)$. If μ is G-invariant then so is ϕ, whence $\phi \circ \pi$ is a trace on $G \underset{\alpha}{\times} L^\infty_\mu(T)$. Since all traces on a factor are proportional we see that $G \underset{\alpha}{\times} L^\infty_\mu(T)$ is finite or infinite according to whether $\mu(T) < \infty$ or $\mu(T) = \infty$. Since μ is diffuse we can find a decreasing sequence of Borel sets $\{T_n\}$ in T with $0 < \mu(T_n) < \infty$ such that $\mu(T_n) \searrow 0$. If p_n denotes the projection corresponding to T_n then $\phi(p_n) \searrow 0$ but $0 < \phi(p_n) < \infty$. This shows that $G \underset{\alpha}{\times} L^\infty_\mu(T)$ is not of type I (cf. 5.5.8) and proves the first half of the theorem.

Assume now that ϕ is invariant under a subgroup G_0 of G which is ergodic on T. If $G \underset{\alpha}{\times} L^\infty_\mu(T)$ is semi-finite there is by 7.11.14 a faithful, normal, semi-finite G-invariant trace ψ on $L^\infty_\mu(T)$. Since $\phi + \psi$ is semi-finite by 5.3.6, there is by 5.3.4 a unique element h in $L^\infty_\mu(T)$ with $0 \leqslant h \leqslant 1$ such that $\phi = (\phi + \psi)(h\cdot)$. Since both ϕ and ψ are G_0-invariant and h is unique it follows that h is a fixed-point under G_0. As G_0 is ergodic $h = \lambda 1$. But then $\phi = (1 - \lambda)^{-1}\lambda\psi$ (since $0 < \lambda < 1$) so that ϕ is G-invariant. This proves that $G \underset{\alpha}{\times} L^\infty_\mu(T)$ is of type III whenever μ is G_0—but not G-invariant, as desired.

7.11.17. To see that 7.11.16 is not vacuously true take T to be the real line $\boldsymbol{R}$ or the circle $\boldsymbol{T}$, both equipped with Lebesgue measure μ. Let G be the group of translation/rotations of T by rational numbers (modulo 2π if $T = \boldsymbol{T}$). It is well known that G is ergodic on T. If $t \neq 0$ there is a neighbourhood E of 0 such that $(E + t) \cap E = \emptyset$. If therefore $S \subset T$ with $\mu(S) > 0$ then $\mu(S \cap (E + s)) > 0$ for some s in G whence

$$(S \cap (E + s) + t) \cap (S \cap (E + s)) = \emptyset.$$

It follows that G acts freely on T so that $G \times_{\alpha} L^{\infty}_{\mu}(T)$ is a factor of type II_{∞} or II_1 according to whether $T = \boldsymbol{R}$ or $T = \boldsymbol{T}$.

As an example of a type III factor take again $T = \boldsymbol{R}$ with Lebesgue measure μ, but let G be the group of affine isomorphisms $s \to as + b$, $a > 0$, with rational coefficients a and b. As before we show that G acts freely on T (If $b = 0$ choose a neighbourhood F of 1 such that $aF \cap F = \emptyset$). Clearly μ is invariant under the subgroup G_0 of pure translations $(a = 1)$, which is ergodic on T; but μ is not G-invariant so that $G \times_{\alpha} L^{\infty}(T)$ is of type III.

7.11.18. Notes and remarks. The material in this section is largely taken from Zeller–Meyer's paper [277]. However, 7.11.5–7.11.8 are borrowed from [209]. Lemma 7.11.12 is due to Dixmier, see [59]. Theorem 7.11.16 is von Neumann's classical result from [176].

Following Zeller–Meyer we say that a discrete group G in a C^*-dynamical system (A, G, α) acts centrally freely if G acts centrally freely on A'' in its bi-transposed action α''. When A is separable this is equivalent to the condition that G acts freely as a transformation group on the factor spectrum $\hat{A}$ of A (i.e. the stabilizer of each point in $\hat{A}$ is trivial). With centrally free action a C^*-algebraic version of 7.11.11 is valid: $G \times_{\alpha r} A$ is simple if and only if A is G-simple; cf. 4.20 of [277]. For abelian groups we shall obtain sharper results on the ideals in a crossed product in section 8.11.

7.12. Large groups of automorphisms

7.12.1. In this section we shall study the G-invariant functionals of a C^*-algebra A, assuming that G has a representation in $\mathrm{Aut}(A)$ of a certain type (Large). The topology of G will play no rôle in this, so that although we frame the results for an arbitrary C^*-dynamical system (A, G, α) one may assume that G is discrete.

We adopt the convention that for any set M (vectors, functionals, operators etc.) on which G acts as transformations, the symbol M^G will denote the set of fixed-point in M under the action of G.

7.12.2. LEMMA. *Let G be a group of unitaries on a Hilbert space H and let p denote the projection of H onto the subspace H^G of G-invariant vectors. For each*

ξ in H, $p\xi$ is the unique vector in H such that for each $\varepsilon > 0$ and x in Conv(G) *there is a y in* Conv(G) *such that $\|zyx\xi - p\xi\| < \varepsilon$ for every z in* Conv(G).

Proof. Put $H_0 = (1 - p)H$ and note that both H_0 and H^G are G-invariant subspaces of H. Fix ξ_0 in H_0 and let K denote the norm closure of $\{x\xi_0 \mid x \in \text{Conv}(G)\}$. There is then a unique vector η in K nearest to zero, i.e. with smallest norm. Since G is a group K is G-invariant, whence $u\eta \in K$ for every u in G. As $\|u\eta\| = \|\eta\|$ we conclude from the uniqueness of η that $u\eta = \eta$, i.e. $\eta \in H^G$. But $\eta \in H_0$ as well, so that $\eta = 0$. It follows that for each ξ_0 in H_0 and $\varepsilon > 0$ there is a y in Conv(G) such that $\|y\xi_0\| < \varepsilon$; whence also $\|zy\xi_0\| < \varepsilon$ for all z in Conv(G).

Given ξ in H we have $\xi = \xi_0 + p\xi$, with ξ_0 in H_0. Applying the argument above to $x\xi_0$ we find y such that $\|zyx\xi_0\| < \varepsilon$ for all z, whence

$$\|zyx\xi - p\xi\| = \|zyx\xi_0\| < \varepsilon.$$

7.12.3. LEMMA. *Let G, H and p be as in 7.12.2. Then p belongs to the strong closure of* Conv(G).

Proof. Given $\xi_1, \ldots, \xi_n$ in H and $\varepsilon > 0$ we argue by induction. Suppose that for each natural number $i < k$ we have chosen x_i in Conv(G) such that

$$(*) \qquad \|x_i x_{i-1} \ldots x_1 \xi_j - p\xi_j\| < \varepsilon, \qquad \forall j \leqslant i.$$

(Note that for $k = 1$ we have chosen no x_i.) By 7.12.2 we can find x_k in Conv(G) such that

$$\|x_k x_{k-1} \ldots x_1 \xi_k - p\xi_k\| < \varepsilon$$

and since $p\xi_j \in H^G$ we see from $(*)$ that

$$\|x_k x_{k-1} \ldots x_1 \xi_j - p\xi_j\| < \varepsilon$$

for all $j \leqslant i$. By induction we find an element $x = x_n \ldots x_1$ such that $\|x\xi_j - p\xi_j\| < \varepsilon$ for all $j \leqslant n$. Since Conv(G) is a multiplicative semi-group, $x \in \text{Conv}(G)$ and x approximates p strongly.

7.12.4. PROPOSITION. *Let $(\mathcal{M}, G, \alpha)$ be a W^*-dynamical system and (π, u, H) a faithful, normal, covariant representation. Let p denote the projection of H on the subspace H^{u_G} and assume that no non-trivial projection in the centre of $\pi(\mathcal{M}^G)$ majorizes p. There exists then a unique faithful normal positive linear map $\Phi: \mathcal{M} \to \mathcal{M}^G$ such that*

$$\Phi(xyz) = x\Phi(y)z, \qquad y \in \mathcal{M}, \qquad x, z \in \mathcal{M}^G.$$

If $x \in \mathcal{M}$ then $\Phi(x)$ is the unique element in

$$\mathcal{M}^G \cap (\text{Conv}(\alpha_G(x)))^{-w}.$$

Moreover, the transposed map $\Phi_*:(\mathcal{M}^G)_* \to \mathcal{M}_*$ *takes* $(\mathcal{M}^G)_*$ *isometrically onto the set* $(\mathcal{M}_*)^G$ *of G-invariant functionals in* $\mathcal{M}_*$.

Proof. We may identify $\mathcal{M}$ with $\pi(\mathcal{M})$. By 7.12.3 there is a net $\{a_\lambda\}$ in $\mathrm{Conv}(u_G)$ with elements of the form $a_\lambda = \Sigma \gamma_t u_t$, such that $a_\lambda \to p$ strongly. Given x in $\mathcal{M}$ let y be a weak limit point of the bounded net (indexed by λ) with elements $\Sigma \gamma_t \alpha_t(x)$. Then, since $u_t p = p$ for all t,

$$pxp = \mathrm{Lim}\, a_\lambda xp = \mathrm{Lim}\, \Sigma \gamma_t u_t xp = \mathrm{Lim}\, \Sigma \gamma_t \alpha_t(x) u_t p = yp.$$

Likewise $pxp = py$. We claim that $yp = 0$ implies $y = 0$ for each y in $\mathcal{M}$. Indeed, $yp = 0$ implies $qp = 0$ where $q = \vee [\alpha_t(y)]$. Clearly, $q \in \mathcal{M}^G$. However, $p \in (\mathcal{M}^G)'$ by 7.12.3 and thus $ep = 0$, where $e = \vee wqw^*$ and w ranges over the unitary group in $\mathcal{M}^G$. Now e belongs to the centre of $\mathcal{M}^G$ and $p \leqslant 1 - e$. By assumption this implies that $e = 0$, whence $y = 0$, and the claim is established. Consequently we may define $\Phi(x)$ as the unique element in $\mathcal{M}$ such that $pxp = \Phi(x)p$. We saw above that $\Phi(x)p = p\Phi(x)$, whence

$$\alpha_t(\Phi(x))p = u_t\Phi(x)u_t^*p = u_t\Phi(x)p = u_tp\Phi(x) = \Phi(x)p.$$

From the uniqueness of $\Phi(x)$ it follows that $\Phi(x) \in \mathcal{M}^G$. It is easy to check that $\Phi: \mathcal{M} \to \mathcal{M}^G$ is a faithful normal positive linear map, and since $p \in (\mathcal{M}^G)'$ we have for all y in $\mathcal{M}$ and x, z in $\mathcal{M}^G$

$$\Phi(xyz)p = pxyzp = xpypz = x\Phi(y)zp,$$

whence $\Phi(xyz) = x\Phi(y)z$. Clearly $\Phi(x) \in \mathcal{M}^G \cap \mathrm{Conv}(\alpha_G(x))^{-w}$. However, Φ is G-invariant so if $y \in \mathrm{Conv}(\alpha_G(x))$ then $\Phi(y) = \Phi(x)$. The same is therefore true for any weak limit point y, since Φ is normal. If furthermore $y \in \mathcal{M}^G$ we conclude that $y = \Phi(y) = \Phi(x)$, so that $\Phi(x)$ is the unique element in $\mathcal{M}^G \cap \mathrm{Conv}(\alpha_G(x))^{-w}$.

Clearly Φ_* is an isometry of $(\mathcal{M}^G)_*$ into $(\mathcal{M}_*)^G$. But if $\phi \in (\mathcal{M}_*)^G$ then $\phi(x) = \phi(\Phi(x))$ for every x in $\mathcal{M}$ whence $\phi = \Phi_*(\phi \mid \mathcal{M}^G)$. This shows that Φ_* is surjective and completes the proof.

7.12.5. Let (A, G, α) be a C^*-dynamical system and denote by S^G (resp. Q^G) the set of G-invariant states (resp. quasi-states) of A. We say that G is represented as a *large group* of automorphisms of A if

$$\pi_\phi(\mathrm{Conv}(\alpha_G(x)))^{-w} \cap \pi_\phi(A)' \neq \varnothing$$

for every x in A and ϕ in S^G. The virtue of this definition lies in the fact that it is implied by several other more transparent conditions.

7.12.6. THEOREM. *Let* (A, G, α) *be a* C^**-dynamical system where* G *is a large group of automorphisms. For each* ϕ *in* S^G *consider the cyclic covariant representation* $(\pi_\phi, u^\phi, H_\phi, \xi_\phi)$ *(cf. 7.4.12), and put* $\mathcal{M} = \pi_\phi(A)''$,

$\mathscr{Z} = \mathscr{M} \cap \mathscr{M}'$. *There is a unique normal, positive linear, G-invariant map* $\Phi: \mathscr{M} \to \mathscr{Z}^G$ *such that*

$$\Phi(xyz) = x\Phi(y)z, \qquad y \in \mathscr{M}, \qquad x, z \in \mathscr{Z}^G.$$

If $x \in A$ *then* $\Phi(\pi_\phi(x))$ *is the unique element in*

$$\pi_\phi(\mathrm{Conv}(\alpha_G(x)))^{-w} \cap \mathscr{Z}^G.$$

Moreover, if $\psi \in (\mathscr{M}_*)^G$ *then* $\psi = \psi \cdot \Phi$; *and if* $\psi \in Q^G$ *with* $\psi \leq \phi$ *there is a unique h in* $\mathscr{Z}^G$ *with* $0 \leq h \leq 1$ *such that* $\psi(x) = (\pi_\phi(x)\xi_\phi \mid h\xi_\phi)$ *for all x in A; and the map* $\psi \to h$ *is an order isomorphism between the set of such functionals and* $(\mathscr{Z}^G)_+^1$.

Proof. Let p be the projection on H_ϕ^{uG} and q the projection on the closure of $\mathscr{Z}^G\xi_\phi$. Both p and q belong to $(\mathscr{Z}^G)'$ and since $\xi_\phi \in H_\phi^{uG}$ we have $q \leq p$. If $x \in \mathscr{M}$ then $qxq \in (\mathscr{Z}^G q)'$ and since $\mathscr{Z}^G q$ is a commutative von Neumann algebra (on qH) with a cyclic vector (viz. ξ_ϕ), it is maximal commutative by 2.8.3; whence $qxq = yq$ for some y in $\mathscr{Z}^G$. Moreover, this y is unique; for if $yq = 0$ then $y\mathscr{M}\xi_\phi = \mathscr{M}yq\xi_\phi = 0$, whence $y = 0$. Define $\Phi(x) = y$. It is straightforward to see that Φ is a normal, positive linear, G-invariant map of $\mathscr{M}$ onto $\mathscr{Z}^G$ such that $\Phi(xyz) = x\Phi(y)z$ whenever $y \in \mathscr{M}$ and $x, z \in \mathscr{Z}^G$.

If e is a projection in $\mathscr{Z}^G$ orthogonal to p then $eq = 0$, whence $e = 0$ from the preceding. Thus the conditions of 7.12.4 are satisfied with $\mathscr{M}$ replaced by $\mathscr{Z}$ and we have a map $\Phi_1: \mathscr{Z} \to \mathscr{Z}^G$ such that $pzp = \Phi_1(z)p$ and

$$\{\Phi_1(z)\} = (\mathrm{Conv}(\alpha_G(z)))^{-w} \cap \mathscr{Z}^G.$$

Note that if $z \in \mathscr{Z}$ then

$$\Phi(z)q = qzq = q(pzp)q = q\Phi_1(z)pq = \Phi_1(z)q,$$

whence $\Phi(z) = \Phi_1(z)$.

Take x in A. By assumption there exists an element z in

$$\pi_\phi(\mathrm{Conv}(\alpha_G(x)))^{-w} \cap \mathscr{M}' = \pi_\phi(\mathrm{Conv}(\alpha_G(x)))^{-w} \cap \mathscr{Z}.$$

Since Φ is normal and G-invariant this implies that $\Phi(\pi_\phi(x)) = \Phi(z) = \Phi_1(z)$. Consequently

$$\Phi(\pi_\phi(x)) \in \mathrm{Conv}(\alpha_G(z))^{-w} \cap \mathscr{Z}^G \subset \mathrm{Conv}(\alpha_G(\pi_\phi(x)))^{-w} \cap \mathscr{Z}^G.$$

Moreover, $\Phi(\pi_\phi(x))$ is the only element in the latter set; for if

$$y \in \mathrm{Conv}(\alpha_G(\pi_\phi(x)))^{-w} \cap \mathscr{Z}^G \qquad \text{then } y = \Phi(y) = \Phi(x).$$

If $\psi \in (\mathscr{M}_*)^G$ then $\psi(\pi_\phi(x)) = \psi(y)$ for every y in $\mathrm{Conv}(\pi_\phi(\alpha_G(x)))^{-w}$; in particular $\psi(\pi_\phi(x)) = \psi(\Phi(\pi_\phi(x)))$. Since $\pi_\phi(A)$ is dense in $\mathscr{M}$ we conclude that $\psi = \psi \circ \Phi$.

If $\psi \in Q^G$ and $\psi \leqslant \phi$ then $\psi(x) = (\pi_\phi(x)\xi_\phi \mid a\xi_\phi)$ for some unique a in $\mathcal{M}'$ and all x in A by 3.3.5. Since $u_t \mathcal{M} u_t^* = \mathcal{M}$ we also have $u_t \mathcal{M}' u_t^* = \mathcal{M}'$ for all t in G and the uniqueness of a implies that $a \in (\mathcal{M}')^G$. Identifying ψ with its image in $\mathcal{M}_*$ we now see that $\psi \in (\mathcal{M}_*)^G$ whence $\psi = \psi \circ \Phi$ from the preceding. Since $\mathscr{Z}^G$ is commutative there is by the Radon–Nikodym theorem a unique h in $\mathscr{Z}^G$ with $0 \leqslant h \leqslant 1$ such that $\psi = \phi(h\cdot)$ on $\mathscr{Z}^G$. But then,

$$\phi(hx) = \phi(\Phi(hx)) = \phi(h\Phi(x)) = \psi(\Phi(x)) = \psi(x)$$

for every x in $\mathcal{M}$. Clearly the map $\psi \to h$ is an order isomorphism from the set of functionals ψ in Q^G with $\psi \leqslant \phi$ onto $(\mathscr{Z}^G)_+^1$.

7.12.7. COROLLARY. *If (A, G, α) is a C^*-dynamical system where G is a large group of automorphisms then $(A_{sa}^*)^G$ is a vector lattice.*

7.12.8. PROPOSITION. *Let (A, G, α) be a C^*-dynamical system where G is a large group of automorphisms. If $\phi \in S^G$ then with notations as in 7.12.6 the following conditions are equivalent:*

(i) ϕ is an extreme point of S^G;

(ii) $(\pi_\phi(A) \cup u_G^\phi)'' = B(H_\phi)$, i.e. $(\pi_\phi(A)')^G = \mathbf{C}1$;

(iii) $\mathscr{Z}^G = \mathbf{C}1$;

(iv) $S^G \cap \mathcal{M}_* = \{\phi\}$;

(v) $(H_\phi)^G = \mathbf{C}\xi_\phi$;

(vi) $\phi(\Phi(x)y) = \phi(x)\phi(y), \qquad x, y \in \mathcal{M}$.

Proof. (i) $\Rightarrow$ (ii). If $a \in (\mathcal{M}')^G$ and $0 \leqslant a \leqslant 1$, then, defining $\psi(x) = (\pi_\phi(x)\xi_\phi \mid a\xi_\phi)$, we have $0 \leqslant \psi \leqslant \phi$ and $\psi = \lambda\phi$ if and only if $a = \lambda 1$. If therefore ϕ is extreme, $(\mathcal{M}')^G = \mathbf{C}1$, whence

$$B(H_\phi) = ((\mathcal{M}')^G)' = (\mathcal{M}' \cap (u_G^\phi)')' = (\pi_\phi(A) \cup u_G^\phi)''.$$

(ii) $\Rightarrow$ (iii) is obvious since $\mathscr{Z}^G \subset (\mathcal{M}')^G$.

(iii) $\Rightarrow$ (i) and (iii) $\Rightarrow$ (iv) follow from 7.12.6.

(iv) $\Rightarrow$ (v) and (v) $\Rightarrow$ (iii) $\Rightarrow$ (vi) are obvious.

(vi) $\Rightarrow$ (i) follows from 7.12.6; for if $0 \leqslant \psi \leqslant \phi$ and $\psi \in Q^G$ then $\psi = \phi(h\cdot)$ with $h \in \mathscr{Z}^G$, whence $\psi = \phi(h)\phi$.

7.12.9. COROLLARY. *Two G-invariant states whose representations are equivalent are equal if one of them is extremal (and G is large).*

Proof. Apply 7.12.8 (iv).

7.12.10. *Notes and remarks.* This and the next section are lifted from Størmer's survey [246]. Proposition 7.12.4 is an ergodic type result first used by Kovacs and Szücs [151] and appearing in the form above in [71]. Large groups of automorphisms were defined by Stormer in [242] where also 7.12.6 and 7.12.8

were proved. Note that the group of inner automorphisms of a C^*-algebra is large, so that 7.12.7 provides a neat proof of Thoma's result from [264], that the set of tracial states on a C^*-algebra form a simplex. Another proof of this result, which also works for infinite, densely defined traces is found in [193] and uses the Riesz decomposition property from 1.4.10.

7.13. Asymptotically abelian systems

7.13.1. Let (A, G, α) be a C^*-dynamical system. We say that the system is *asymptotically abelian* (respectively *weakly asymptotically abelian*) if there is a net Λ in G such that

$$\|x\alpha_t(y) - \alpha_t(y)x\| \to 0 \quad \text{as } t \to \infty \text{ in } \Lambda$$

$$(\phi(x\alpha_t(y) - \alpha_t(y)x) \to 0 \quad \text{as } t \to \infty \text{ in } \Lambda)$$

for all x, y in A (and all ϕ in A^*). Most often $\Lambda = G$ and the ordering is such that $t \to \infty$ in Λ means that $t \to \infty$ in the usual sense (i.e. t moves out of any compact set). The difference between the two notions of asymptotic Abelianess is slight: if (A, G, α) is weakly asymptotically abelian there is by Hahn–Banach's theorem for every x, y in A, $\varepsilon > 0$ and t in Λ a convex combination $z = \Sigma\varepsilon_n\alpha_{t_n}(y)$ with t_n in Λ, $t_n > t$, such that $\|xz - zx\| < \varepsilon$.

We say that a G-invariant state ϕ of A is *asymptotically multiplicative* (with respect to Λ) if

$$\phi(\alpha_t(x)y) \to \phi(x)\phi(y) \qquad \text{as } t \to \infty \text{ in } \Lambda$$

for all x, y in A. Such states appear in the applications of the theory to quantum statistical mechanics under the names of *strongly clustering* and *strongly mixing* states.

7.13.2. LEMMA. *If (A, G, α) is a weakly asymptotically abelian C^*-dynamical system then G is a large group of automorphisms.*

Proof. Take ϕ in S^G and x in A. If z denotes any weak limit point of the bounded net $\{\pi_\phi(\alpha_t(x)) \mid t \in \Lambda\}$ then evidently $z \in \pi_\phi(\text{Conv}(\alpha_G(x)))^{-w}$ Moreover, for any y in A and ψ in $(\pi_\phi(A)'')_*$

$$\psi(z\pi_\phi(y) - \pi_\phi(y)z) = \text{Lim}\,\psi(\pi_\phi(\alpha_t(x)y - y\alpha_t(x))) = 0,$$

whence $z \in \pi_\phi(A)'$. Consequently G is large.

7.13.3. PROPOSITION. *Let (A, G, α) be a weakly asymptotically abelian system and consider a G-invariant state ϕ of A with covariant cyclic representation*

$(\pi_\phi, u^\phi, H_\phi, \xi_\phi)$. The following conditions are equivalent:

(i) *ϕ is asymptotically multiplicative;*

(ii) *ϕ is an extreme point of S^G and for each x in A the net $\{\pi_\phi(\alpha_t(x)) \mid t \in \Lambda\}$ is weakly convergent to $\phi(x)1$ in $\boldsymbol{B}(H_\phi)$;*

(iii) *The net $\{u_t^\phi \mid t \in \Lambda\}$ is weakly convergent in $\boldsymbol{B}(H_\phi)$ to the one-dimensional projection on $\boldsymbol{C}\xi_\phi$.*

Proof. (i) $\Rightarrow$ (ii) If ϕ is not extreme there is by 7.12.8 (v) a unit vector η in H_ϕ^G orthogonal to ξ_ϕ. Given $\varepsilon > 0$ choose y in A such that $\|\pi_\phi(y)\xi_\phi - \eta\| < \varepsilon$. Then for each x in A^1 we have

$$\begin{aligned}|(\pi_\phi(x)\xi_\phi \mid \eta)| &= (\pi_\phi(x)u_t^*\xi_\phi \mid u_t^*\eta)| = |(\pi_\phi(\alpha_t(x)\xi_\phi \mid \eta)\\ &\leqslant |(\pi_\phi(y^*\alpha_t(x))\xi_\phi \mid \xi_\phi)| + \varepsilon = |\phi(y^*\alpha_t(x))| + \varepsilon\\ &\leqslant \operatorname*{Lim\,sup}_{t\in\Lambda} |\phi(y^*\alpha_t(x))| + \varepsilon = |\phi(y^*)\phi(x)| + \varepsilon\\ &\leqslant |(\xi_\phi \mid \pi_\phi(y)\xi_\phi)| + \varepsilon \leqslant .|(\xi_\phi \mid \eta)| + 2\varepsilon = 2\varepsilon.\end{aligned}$$

It follows that $(\pi_\phi(x)\xi_\phi \mid \eta) = 0$ for all x in A whence $\eta = 0$, a contradiction. Thus ϕ is extreme is S^G.

Take x in A and let z be a weak limit point of the bounded net $\{\pi_\phi(\alpha_t(x)) \mid t \in \Lambda\}$. As we saw in the proof of 7.13.2, z belongs to the centre of $\pi_\phi(A)''$. Let Λ_0 be a subnet for which $\pi_\phi(\alpha_t(x)) \to z$ as $t \to \infty$ in Λ_0. Then for all a, b in A

$$\begin{aligned}(z\xi_a \mid \xi_b) &= (z\pi_\phi(b^*a)\xi_\phi \mid \xi_\phi) = \operatorname*{Lim}_{\Lambda_0} \phi(\alpha_t(x)b^*a)\\ &= \phi(x)\phi(b^*a) = \phi(x)(\xi_a \mid \xi_b).\end{aligned}$$

It follows that $z = \phi(x)1$, as desired.

(ii) $\Rightarrow$ (iii). For all x, y in A we have by assumption

$$(u_t^\phi \xi_x \mid \xi_y) = (\pi_\phi(\alpha_t(x)\xi_\phi \mid \xi_y) \to \phi(x)(\xi_\phi \mid \xi_y).$$

It follows that the net $\{u_t^\phi \mid t \in \Lambda\}$ is weakly convergent to an operator w with $\|w\| \leqslant 1$ such that

$$(*) \qquad (w\xi_x \mid \xi_y) = \phi(x)\phi(y^*)$$

for all x, y in A. In particular, $(w\xi_\phi \mid \xi_\phi) = 1$, whence $w\xi_\phi = \xi_\phi$. Let e denote the projection on $\boldsymbol{C}\xi_\phi$. From $(*)$ it follows that for every t in G

$$(u_t^\phi w\xi_x \mid \xi_y) = \phi(x)\phi(\alpha_{t^{-1}}(y^*)) = \phi(x)\phi(y^*),$$

from which we conclude that $w\xi_x \in H_\phi^G$. But by 7.12.8(v), $H_\phi^G = eH_\phi$, whence

$$w\xi_x = (w\xi_x \mid \xi_\phi)\xi_\phi = \phi(x)\xi_\phi = (\xi_x \mid \xi_\phi)\xi_\phi = e\xi_x.$$

It follows that $w = e$.

(iii) $\Rightarrow$ (i) Take x, y in A, t in Λ and compute

$$\phi(y^*\alpha_t(x)) = (\pi_\phi(\alpha_t(x))\xi_\phi \mid \xi_y) = (u_t^\phi \pi_\phi(x) u_t^{\phi*}\, \xi_\phi \mid \xi_y)$$

$$= (u_t^\phi \xi_x \mid \xi_y) \underset{\Lambda}{\rightarrow} (e\xi_x \mid \xi_y) = (\xi_x \mid \xi_\phi)(\xi_\phi \mid \xi_y) = \phi(y^*)\phi(x).$$

This completes the proof.

7.13.4. COROLLARY. *Take (A, G, α) and ϕ as in 7.13.3. If ϕ is a factor state it is asymptotically multiplicative.*

Proof. Take Φ as in 7.12.6. Since $\mathscr{M} = \pi_\phi(A)''$ is a factor we have $\mathscr{Z} = \mathscr{Z}^G = \boldsymbol{C}1$ whence $\Phi(x) = \phi(x)1$ for all x in $\mathscr{M}$. By 7.12.8 (vi) ϕ is extreme in S^G and since every weak limit point of $\{\alpha_t(x) \mid t \in \Lambda\}$ belongs to $\mathscr{Z}\,(=\mathscr{Z}^G)$ it must equal $\Phi(x)$ $(=\phi(x))$, whence ϕ is asymptotically multiplicative by 7.13.3 (ii).

7.13.5. COROLLARY. *Let (A, G, α) be weakly asymptotically abelian and let ϕ be asymptotically multiplicative. If G is abelian the von Neumann algebra $(u_G^\phi)'$ has precisely one minimal projection, namely the projection on $\boldsymbol{C}\xi_\phi$.*

Proof. Let e denote the projection on $\boldsymbol{C}\xi_\phi$. Since $\xi_\phi \in H_\phi^G, e \in (u_G^\phi)'$. On the other hand, by 7.13.3 (iii) $e \in (u_G^\phi)''$.

Let p be a minimal projection in $(u_G^\phi)'$. Then the unitary representation $t \rightarrow u_t p$ is irreducible and thus a homomorphism, since G is abelian. Passing to a subnet of Λ we may assume that $u_t p \rightarrow \lambda p$ as $t \rightarrow \infty$ in Λ, where $|\lambda| = 1$. However, by 7.13.3 (iii) $u_t \rightarrow e$, whence $ep = \lambda p$. Consequently $e = p$.

7.13.6. THEOREM. *Let (A, G, α) be a C^*-dynamical system where G is a large group of automorphisms. Consider a G-invariant factor state ϕ with cyclic covariant representation $(\pi_\phi, u^\phi, H_\phi, \xi_\phi)$. Put $\mathscr{M} = \pi_\phi(A)''$ and denote by $\tilde{\phi}$ the vector state on $\boldsymbol{B}(H_\phi)$ determined by ξ_ϕ. Then*

(i) *$\mathscr{M}$ is finite if and only if $\tilde{\phi}$ is a trace on $\mathscr{M}$;*

(ii) *$\mathscr{M}$ is semi-finite, but infinite, if and only if $\tilde{\phi}$ is a trace on $\mathscr{M}'$, but not on $\mathscr{M}$;*

(iii) *$\mathscr{M}$ is of type III if and only if $\tilde{\phi}$ is not a trace on $\mathscr{M}'$.*

Proof. (iii) If $\mathscr{M}$ is of type III then so is $\mathscr{M}'$ by 5.4.3, thus $\tilde{\phi}$ is not a trace on $\mathscr{M}'$. Assume now that $\mathscr{M}$ and $\mathscr{M}'$ are semi-finite and let τ be a faithful normal semi-finite trace on $\mathscr{M}'$. Since $u_t \mathscr{M}' u_t^* = \mathscr{M}'$ and τ is unique to to scalar multiples

(5.3.7) there is a $\gamma(t) > 0$ such that $\tau(u_t \cdot u_t^*) = \gamma(t)\tau$; and the map $\gamma: G \to \boldsymbol{R}_+$ is a homomorphism. We claim that $\gamma = 1$. If not, then $\tau(u \cdot u^*) = \gamma\tau$ with $\gamma < 1$ for some unitary u in u_G^ϕ. Take x, y in $(\mathscr{M}')_+^\tau$. Then

$$\tau(yu^n xu^{-n}) \leqslant \|y\|\,\tau(u^n xu^{-n}) = \|y\|\,\gamma^n \tau(x) \to 0.$$

Since this holds for all y we conclude from 5.3.11 that $u^n xu^{-n} \to 0$ σ-weakly. However, $\tilde{\phi}(u^n xu^{-n}) = \tilde{\phi}(x)$, since $\xi_\phi \in I_\phi^G$; whence $\tilde{\phi}(x) = 0$. As ξ_ϕ is cyclic for $\mathscr{M}$ this implies that $x = 0$ for all x in $(\mathscr{M}')^\tau$, a contradiction. Consequently $\gamma = 1$ so that τ is G-invariant. By 5.3.11 there is a unique positive (possibly unbounded) operator h on H_ϕ affiliated with $\mathscr{M}'$ such that $\tilde{\phi} = \tau(h \cdot)$ on $\mathscr{M}'$. Since both $\tilde{\phi}$ and τ are G-invariant and h is unique we conclude that $u_t^\phi h u_t^{\phi *} = h$ for all t in G. But then each spectral projection q of h is G-invariant, whence $q\xi_\phi \in H_\phi^G$. By 7.12.8 (v) this implies that $q\xi_\phi = \lambda\xi_\phi$ where $\lambda = 0, 1$. Since $q \in \mathscr{M}'$ and ξ_ϕ is cyclic for $\mathscr{M}$ we see that either $q = 1$ or $q = 0$. It follows that $h = \lambda 1$ for some $\lambda > 0$ whence $\tilde{\phi} = \lambda\tau$ so that $\tilde{\phi}$ is a trace on $\mathscr{M}'$. This completes the proof of (iii).

(i) If $\tilde{\phi}$ is a trace on $\mathscr{M}$ then $\mathscr{M}$ is finite. If $\mathscr{M}$ is finite let τ be a normalized $(\tau(1) = 1)$ trace on $\mathscr{M}$. Such a trace is unique and therefore G-invariant. By 7.12.8(iv) $\tau = \phi$.

(ii) follows from (i) and (iii).

7.13.7. THEOREM. *Let (A, G, α) be a weakly asymptotically abelian system with G abelian and consider a G-invariant factor state ϕ of A with cyclic covariant representation $(\pi_\phi, u^\phi, H_\phi, \xi_\phi)$. Put $\mathscr{M} = \pi_\phi(A)''$ and denote by $\tilde{\phi}$ the vector state on $B(H_\phi)$ determined by ξ_ϕ.*

(i) *$\mathscr{M} = \boldsymbol{C}1$ if and only if ϕ is multiplicative;*

(ii) *$\mathscr{M} = B(H_\phi)$, $\dim(H_\phi) = \infty$, if and only if ϕ is a pure state but not multiplicative;*

(iii) *$\mathscr{M}$ is of type II_1 if and only if ϕ is a trace but not multiplicative;*

(iv) *$\mathscr{M}$ is of type II_∞ if and only if $\tilde{\phi}$ is a trace on $\mathscr{M}'$ but ϕ is neither pure nor a trace;*

(v) *$\mathscr{M}$ is of type III if and only if $\tilde{\phi}$ is not a trace on $\mathscr{M}'$.*

Proof. (i) is trivial and (v) follows from 7.13.6(iii). (ii) If ϕ is a pure state, but not multiplicative, then $\dim(H_\phi) = \infty$. Otherwise $\mathscr{M}$ is finite so that ϕ is a trace (7.13.6) which is impossible. Conversely, if $\mathscr{M}$ is of type I_∞ then $\mathscr{M}'$ is of type I by 5.5.11 and $\tilde{\phi}$ is a trace on $\mathscr{M}'$ by 7.13.6 so that $\mathscr{M}'$ is of type I_n, $n < \infty$. The projection on $\mathscr{M}'\xi_\phi$ is therefore finite-dimensional and belongs to $(u_G^\phi)'$ since $\mathscr{M}'$ is a G-invariant subset of $B(H_\phi)$. It then follows from 7.13.5 that $\mathscr{M}'\xi_\phi = \boldsymbol{C}\xi_\phi$; which implies that $\mathscr{M}' = \boldsymbol{C}1$ since ξ_ϕ is cyclic for $\mathscr{M}$. Consequently $\mathscr{M} = B(H_\phi)$, so that ϕ is a pure state by 3.13.2; and ϕ is not multiplicative since $\dim(H_\phi) \neq 1$. (iii) and (iv) follow from (i) and (ii) by 7.13.6.

7.13.8. As an example of an asymptotically abelian system take the Fermion algebra $\boldsymbol{F}$ defined in 6.4.2 and consider the (discrete) group Π of permutations of $\boldsymbol{N}$ which leave all but a finite number of elements fixed. As we saw in 6.5.11 there is a unitary representation $u:\Pi \to \boldsymbol{F}$ and consequently also a representation $\alpha:\Pi \to \mathrm{Aut}(\boldsymbol{F})$, viz. $\alpha_t = u_t \cdot u_t^*$. By 6.5.12 there is a sequence $\{t_n\}$ in Π such that if x, y belong to M_{2^k} *then* $\alpha_{t_n}(x)y = y\alpha_{t_n}(x)$ whenever $n \geqslant k$. Since $\boldsymbol{F}$ is the completion of the union of the subalgebras M_{2^k}, $k \in \boldsymbol{N}$, it follows that the system (F, Π, α) is asymptotically abelian.

The product states ϕ_λ $(0 \leqslant \lambda \leqslant \frac{1}{2})$ defined in 6.5.10 are Π-invariant (6.5.12), extreme in S^Π (combine 6.5.14 and 7.12.8) and asymptotically multiplicative (combine 6.5.13 and 7.13.3). In fact they are factor states by 6.5.8. It is not difficult to show that for $0 < \lambda < \frac{1}{2}$ the extension $\bar{\phi}_\lambda$ to $\pi_\lambda(\boldsymbol{F})'$ is not a trace, so that $\pi_\lambda(\boldsymbol{F})''$ is of type III by 7.13.6. However, the direct proof in 6.5.15 is simpler; for since Π is represented as inner automorphisms of $\boldsymbol{F}$ any trace on $\pi_\lambda(\boldsymbol{F})''$ is necessarily Π-invariant.

7.13.9. Notes and remarks. The notion of asymptotically abelian systems grew naturally out of the C^*-algebraic approach to quantum physics mentioned in 7.4.14. The basic idea is that if two observables are given and one of them is translated far away (in a space-like direction in quantum field theory), then they should become independent of each other, i.e. their representing operators should commute. Asymptotically abelian C^*-dynamical systems were defined by Doplicher, Kastler and Robinson [70] and independently by Ruelle [219]. Weaker notions (weak asymptotic abelianness, large groups etc.) promptly appeared, and a unifying treatment was given in [71].

Proposition 7.13.3 was proved by Størmer [244]. In weaker forms the result was obtained by Kastler and Robinson [146] and by Borchers [18]. Theorems 7.13.6 and 7.13.7 are both found in [244].

Chapter 8

Spectral Theory for Automorphism Groups

The study of C^*-dynamical systems (A, G, α) for which G is an abelian (locally compact) group may properly be thought of as harmonic analysis on operator algebras. The present chapter is devoted to such a study and includes some of the deepest results in the theory of operator algebras.

We use the first two sections for a general study of harmonic analysis on Banach spaces. The main idea is that if G has a representation $t \to \alpha_t$ as isometries on a Banach space X then we can associate to each subset Ω of the dual group Γ of G a (spectral) subspace $R^\alpha(\Omega)$ of X. If X is a Hilbert space, so that α is a unitary representation, the subspaces $R^\alpha(\Omega)$, $\Omega \subset \Gamma$, correspond to the spectral projections $\mu(\Omega)$, where μ is the spectral measure obtained from Stone's theorem, i.e. $\alpha_t = \int(t, \tau)\, d\mu(\tau)$. If X is a C^*-algebra and α is an automorphic representation there is just enough Hilbert space structure around to allow the construction of a spectral measure μ on Γ such that $\mu(\Omega)$ corresponds to the support projection of $R^\alpha(\Omega)$ in X'' for a large class of sets Ω in Γ. Taking $u_t = \int(t, \tau)\, d\mu(\tau)$ we obtain a unitary representation of G which under certain (spectrum) conditions is covariant for α, i.e.

$$\alpha_t(x) = u_t x u_{-t}, \quad x \in X.$$

The discussion of this rather complex situation is carried out in sections three to five. We then devote two sections to derivations and the automorphisms they generate. In sections eight to eleven we explore certain subsets of Γ (Borchers and Connes spectrum) which serve as obstructions for automorphisms to be inner. We conclude with four sections discussing one-parameter dynamical systems; the main new tool being the introduction of complex function theory. Very properly the book finishes in an area of the theory of operator algebras where progress is most rapid, and where a conclusive account seems hopeless at the moment.

8.1. Spectral subspaces and Arveson spectrum

8.1.1. Let X and X_* be a pair of Banach spaces in duality via a bilinear form $\langle \cdot, \cdot \rangle$. This means that $\langle x, \cdot \rangle \in (X_*)^*$ and $\langle \cdot, \xi \rangle \in X^*$ for every x in X and ξ in X_*, and moreover, that the maps $x \to \langle x, \cdot \rangle$ and $\xi \to \langle \cdot, \xi \rangle$ are isometries of X and X_* onto weak* dense subspaces of $(X_*)^*$ and X^*, respectively.

Let $\boldsymbol{B}_\sigma(X)$ and $\boldsymbol{B}_\sigma(X_*)$ denote the sets of bounded linear operators on X and X_* that are continuous in the $\sigma(X, X_*)$- and $\sigma(X_*, X)$-topologies, respectively. Note that an operator α in $\boldsymbol{B}(X)$ belongs to $\boldsymbol{B}_\sigma(X)$ if and only if it has a transposed operator α' in $\boldsymbol{B}(X_*)$ (which then necessarily belongs to $\boldsymbol{B}_\sigma(X_*)$).

A *representation* of a locally compact group G on X is a $\sigma(X, X_*)$-continuous homomorphism $t \to \alpha_t$ of G into the group of invertible isometries in $\boldsymbol{B}_\sigma(X)$. We say that α is an *integrable representation* if for each μ in $M(G)$ there is an operator α_μ in $\boldsymbol{B}_\sigma(X)$ (necessarily unique) such that

$$\langle \alpha_\mu(x), \xi \rangle = \int \langle \alpha_t(x), \xi \rangle \, d\mu(t)$$

for each x in X and ξ in X_*. It follows by direct computation that in this case the map $\mu \to \alpha_\mu$ is a norm-decreasing homomorphism of $M(G)$ into $\boldsymbol{B}_\sigma(X)$.

It is immediate from the definitions that the transposed map $t \to \alpha_t'$ of a representation on X is a representation on X_*, and that α' is integrable whenever α is.

8.1.2. LEMMA. *If X is a Banach space and $X_* = X^*$ then every homomorphism $t \to \alpha_t$ of a locally compact group G into the group of invertible isometries on X such that each function $t \to \alpha_t$, $x \in X$, is norm continuous, is an integrable representation of G on X.*

Proof. This follows directly from (A3, Appendix).

8.1.3. From now on G will denote a locally compact *abelian* group and Γ its dual group. The unit elements in G and Γ will be denoted by 0 and θ. If $t \in G$ and $\tau \in \Gamma$ we denote by (t, τ) the value of τ at t, and we write $\hat{\mu}(\tau) = \int (t, \tau) \, d\mu(t)$ for each μ in $M(G)$ (inverse Fourier transform).

We shall be especially concerned with the dense ideal $K^1(G)$ of $L^1(G)$ consisting of the functions f such that $\hat{f}$ has compact support in Γ.

Let X and X_* be as in 8.1.1 and let α be an integrable representation of G on X. For each *open* subset Ω of Γ define the *spectral R-subspace* $R^\alpha(\Omega)$ as the $\sigma(X, X_*)$-closure in X of the linear subspace of elements $\alpha_f(x)$, where $x \in X$, $f \in K^1(G)$ and supp.$\hat{f} \subset \Omega$. Similarly we define $R^{\alpha'}(\Omega)$ in X_*. For each *closed* subset Λ of Γ define the *spectral M-subspace* $M^\alpha(\Lambda)$ in X as the annihilator of $R^{\alpha'}(\Gamma \backslash \Lambda)$. Thus $x \in M^\alpha(\Lambda)$ if and only if $\langle x, \alpha_f'(\xi) \rangle = 0$ for all ξ in X_* and all f in $K^1(G)$ with supp.$\hat{f} \subset \Gamma \backslash \Lambda$; i.e. if $\alpha_f(x) = 0$ for all such f. Similarly we define $M^{\alpha'}(\Lambda)$ in X_*.

8.1.4. THEOREM. *The R- and M-spaces defined in 8.1.3 are G-invariant (as sets) and satisfy the relations:*

(i) $R^\alpha(\Omega_1) \subset R^\alpha(\Omega_2)$ *if* $\Omega_1 \subset \Omega_2$;

(ii) $M^\alpha(\Lambda_1) \subset M^\alpha(\Lambda_2)$ *if* $\Lambda_1 \subset \Lambda_2$;

(iii) $(\Sigma_i R^\alpha(\Omega_i))^{-\sigma} = R^\alpha(\cup_i \Omega_i)$;

(iv) $\cap_i M^\alpha(\Lambda_i) = M^\alpha(\cap \Lambda_i)$;

(v) $R^\alpha(\Omega) \subset M^\alpha(\Lambda)$ *if* $\Omega \subset \Lambda$;

(vi) $M^\alpha(\Lambda) \subset R^\alpha(\Omega)$ *if* $\Lambda \subset \Omega$;

(vii) $M^\alpha(\Lambda) = \cap_i R^\alpha(\Omega_i)$ *if* $\Lambda = \cap_i \Omega_i = \cap_i \overline{\Omega}_i$;

(viii) $R^\alpha(\Omega) = (\Sigma_i M^\alpha(\Lambda_i))^{-\sigma}$ *if* $\Omega = \cup_i \Lambda_i = \cup_i \mathring{\Lambda}_i$;

(ix) $R^\alpha(\varnothing) = M^\alpha(\varnothing) = \{0\}$; $R^\alpha(\Gamma) = M^\alpha(\Gamma) = X$.

Proof. If Ω is open in Γ and $f \in K^1(G)$ with $\text{supp.} f \subset \Omega$ then since

$$\text{supp.}(\mu \times f)\hat{} = \text{supp.}(\hat{\mu}\hat{f}) \subset \text{supp.} f$$

for every μ in $M(G)$ we see that $\alpha_\mu(R^\alpha(\Omega)) \subset R^\alpha(\Omega)$. In particular

$$\alpha_t(R^\alpha(\Omega)) = R^\alpha(\Omega)$$

for every t in G. Since $M^\alpha(\Lambda)$ is the annihilator of a G-invariant subspace in X_*, it is itself G-invariant.

Condition (i) is evident from the definition and (ii) follows by duality. Thus of condition (iii) we already have

$$\left(\sum_i R^\alpha(\Omega_i)\right)^{-\sigma} \subset R^\alpha\left(\bigcup_i \Omega_i\right).$$

To prove the converse inclusion take an x in $R^\alpha(\cup_i \Omega_i)$ of the form $\alpha_f(y)$, where $f \in K^1(G)$ and $\text{supp.} f \subset \cup \Omega_i$. Since supp. f is compact it is covered by a finite number of sets, say $\Omega_1, \Omega_2, \ldots, \Omega_n$. Put $\Omega' = \Omega_1 \cup \cdots \cup \Omega_{n-1}$ and choose an open set Ω'' with compact closure Λ contained in Ω_n such that $\text{supp.} f \subset \Omega' \cup \Omega''$. By 2.6.2 of [217] there is a function g in $K^1(G)$ with $\text{supp.}\, \hat{g} \subset \Omega_n$ such that $\hat{g} = 1$ on Λ. Thus $\text{supp.}(f \times g)\hat{} \subset \Omega_n$ and setting $f' = f - f \times g$ we see that

$$\text{supp.} f' \subset \text{supp.} f \backslash \Omega'' \subset \Omega'.$$

Working by induction we can therefore write $f = f_1 + \cdots + f_n$ where $f_i \in K^1(G)$ and $\text{supp.} f_i \subset \Omega_i$ for all i. But then

$$x = \alpha_f(y) \in \sum_i R^\alpha(\Omega_i).$$

We have shown that a dense subset of $R^\alpha(\cup \Omega_i)$ is contained in $(\Sigma_i R^\alpha(\Omega_i))^{-\sigma}$, and this suffices since the latter is closed. Condition (iv) follows from (iii) by duality.

To prove (v) let $x = \alpha_f(y)$, where supp. $f \subset \Omega$. For every g in $K^1(G)$ with supp. $\hat{g} \subset \Gamma\backslash\Lambda$ we have $\alpha_g(x) = \alpha_{g \times f}(y) = 0$ since $g \times f = 0$. This proves that $x \in M^\alpha(\Lambda)$. Consequently $R^\alpha(\Omega) \subset M^\alpha(\Lambda)$.

(vi): Take x in $M^\alpha(\Lambda)$ and ξ in $M^{\alpha'}(\Gamma\backslash\Omega)$ and consider the integral $\int\langle\alpha_t(x), \xi\rangle f(t)\,dt$ for f in $K^1(G)$. If supp. $f \subset \Gamma\backslash\Lambda$ the integral is zero since $x \in M^\alpha(\Lambda)$. If supp. $f \subset \Omega$ the integral is also zero since $\xi \in M^{\alpha'}(\Gamma\backslash\Omega)$. But as we saw in the proof of condition (iii), each f in $K^1(G)$ can be written as a sum of such functions. Thus the integral is zero for a dense set of functions in $L^1(G)$, whence $\langle\alpha_t(x), \xi\rangle = 0$ for all t. In particular $\langle x, \xi\rangle = 0$, whence

$$M^\alpha(\Lambda) \subset M^{\alpha'}(\Gamma\backslash\Omega)^\perp = R^\alpha(\Omega)^{\perp\perp} = R^\alpha(\Omega).$$

(vii): It follows from (vi) that $M^\alpha(\Lambda) \subset \cap_i R^\alpha(\Omega_i)$. However, from (v) and (iv) we get

$$\bigcap_i R^\alpha(\Omega_i) \subset \bigcap_i M^\alpha(\bar{\Omega}_i) = M^\alpha(\Lambda).$$

(viii) follows from (vii) by duality and (ix) is evident.

8.1.5. PROPOSITION. *Suppose that* $G = G_1 \times G_2$, *whence* $\Gamma = \Gamma_1 \times \Gamma_2$, *and let* α *be an integrable representation of* G *on* X. *If* $\alpha^i = \alpha \mid G_i$, $i = 1, 2$, *then for any open sets* Ω_i *and any closed sets* Λ_i *in* Γ_i, $i = 1, 2$, *we have*

$$R^\alpha(\Omega_1 \times \Gamma_2 \cup \Gamma_1 \times \Omega_2) = (R^{\alpha^1}(\Omega_1) + R^{\alpha^2}(\Omega_2))^{-\sigma}.$$

$$M^\alpha(\Lambda_1 \times \Lambda_2) = M^{\alpha^1}(\Lambda_1) \cap M^{\alpha^2}(\Lambda_2).$$

Proof. Take f in $K^1(G_1)$ with supp. $f \subset \Omega_1$ and take any g in $K^1(G_2)$. Then $f \otimes g \in K^1(G)$ and by computation we see that

$$\alpha_{f\otimes g}(x) = \alpha_f^1(\alpha_g^2(x)) = \alpha_g^2(\alpha_f^1(x))$$

for any x in A. Elements of the form $\alpha_{f\otimes g}(x)$ generate $R^{\alpha^1}(\Omega_1)$, for if g ranges over an approximate unit for $L^1(G_2)$ then $\alpha_{f\otimes g}(x) \to \alpha_f^1(x)$. It follows that

$$R^{\alpha^1}(\Omega_1) \subset R^\alpha(\Omega_1 \times \Gamma_2).$$

If Λ_1 is a closed subset of Γ_1 let $\{\Omega_i\}$ be the family of open sets containing Λ_1. From the result above together with 8.1.4 (vii) we obtain

$$M^{\alpha^1}(\Lambda_1) = \cap R^{\alpha^1}(\Omega_i) \subset \cap R^\alpha(\Omega_i \times \Gamma_2) = M^\alpha(\Lambda_1 \times \Gamma_2).$$

This is true with $\alpha^{1\prime}$ and α' instead of α^1 and α so taking $\Lambda_1 = \Gamma_1\backslash\Omega_1$ we obtain by duality

$$R^{\alpha^1}(\Omega_1) \supset R^\alpha(\Omega_1 \times \Gamma_2).$$

Consequently the two spaces are equal; and since the same is true with Ω_2 in place of Ω_1 we see from 8.1.4 (iii) that

$$R^\alpha(\Omega_1 \times \Gamma_2 \cup \Gamma_1 \times \Omega_2) = (R^\alpha(\Omega_1 \times \Gamma_2) + R^\alpha(\Gamma_1 \times \Omega_2))^{-\sigma}$$
$$= (R^{\alpha^1}(\Omega_1) + R^{\alpha^2}(\Omega_2))^{-\sigma}.$$

The second formula follows by duality.

8.1.6. From 8.1.4 (iv) it follows that there is a smallest closed set Λ in Γ such that $M^\alpha(\Lambda) = X$. We say that Λ is the *Arveson spectrum* of α and denote it by $\mathrm{Sp}(\alpha)$. From the definition it follows that $\Gamma\backslash\mathrm{Sp}(\alpha)$ is the largest open set such that $R^{\alpha'}(\Gamma\backslash\mathrm{Sp}(\alpha)) = 0$. Thus $R^{\alpha'}(\Omega) = X_*$ for every open set Ω containing $\mathrm{Sp}(\alpha)$ by 8.1.4 (iv), whence $M^{\alpha'}(\mathrm{Sp}(\alpha)) = X_*$ by 8.1.4 (vii). It follows that $\mathrm{Sp}(\alpha') \subset \mathrm{Sp}(\alpha)$, whence by symmetry $\mathrm{Sp}(\alpha') = \mathrm{Sp}(\alpha)$.

For a closer investigation of the Arveson spectrum we need the following lemma.

8.1.7. LEMMA. *Given σ in Γ, $\varepsilon > 0$ and a compact subset K of G there is a compact neighbourhood Λ of σ such that*

$$\|\alpha_t(x) - (t, \sigma)x\| < \varepsilon\|x\|$$

for all t in K and all x in $M^\alpha(\Lambda)$.

Proof. Let Λ_1 be a compact neighbourhood of σ and choose a function f in $K^1(G)$ such that $\hat{f} = 1$ on Λ_1. For each t in K the function f^t defined by

$$f^t(s) = f(s - t) - (t, \sigma)f(s)$$

satisfies $(f^t)\hat{}(\sigma) = 0$. By 2.6.3 of [217] there is then a function g in $K^1(G)$ such that $\hat{g} = 1$ on some neighbourhood of σ and $\|f^t \times g\|_1 < \varepsilon$. Since the map $t \to f^t$ is continuous and K is compact we only need a finite number of functions g in order to satisfy the requirement above for all t in K. There is therefore a compact neighbourhood Λ_2 of σ such that for each t in K there is a function g in $K^1(G)$ with $\hat{g} = 1$ on Λ_2 and $\|f^t \times g\|_1 < \varepsilon$.

Let Λ be a compact neighbourhood of σ contained in an open set $\Omega \subset \Lambda_1 \cap \Lambda_2$. We claim that

$$\alpha_{f \times g}(x) = x, \qquad \forall x \in M^\alpha(\Lambda).$$

If $\xi \in R^{\alpha'}(\Gamma\backslash\Lambda)$ then

$$\langle \alpha_{f \times g}(x), \xi \rangle = 0 = \langle x, \xi \rangle.$$

If $\xi \in R^{\alpha'}(\Omega)$ then

$$\langle \alpha_{f \times g}(x), \xi \rangle = \langle x, \alpha'_{f \times g}(\xi) \rangle = \langle x, \xi \rangle,$$

since $(f \times g)\hat{} = 1$ on Ω. Since $R^{\alpha'}(\Gamma\backslash\Lambda) + R^{\alpha'}(\Omega)$ is dense in X_* by 8.1.4(iii), the claim is established. It follows that for each x in $M^{\alpha}(\Lambda)$ and t in K

$$\begin{aligned}\|\alpha_t(x) - (t,\sigma)x\| &= \|\alpha_t(\alpha_{f\times g}(x)) - (t,\sigma)\alpha_{f\times g}(x)\| \\ &= \|\alpha_{f^t\times g}(x)\| \leqslant \|f^t \times g\|_1\,\|x\| < \varepsilon\,\|x\|.\end{aligned}$$

8.1.8. COROLLARY. *If $\sigma\in\Gamma$ then $x\in M^{\alpha}(\{\sigma\})$ if and only if $\alpha_t(x) = (t,\sigma)x$ for all t in G (x is an eigenvalue for α). In particular, $M^{\alpha}(\{0\})$ is the set of fixed-points in X under α.*

Proof. If $\alpha_t(x) = (t,\sigma)x$ for all t then $\alpha_f(x) = \hat{f}(\sigma)x$, whence $\alpha_f(x) = 0$ for each f in $K^1(G)$ with $\text{supp}.\hat{f}\subset\Gamma\backslash\{\sigma\}$. Thus $x\in M^{\alpha}(\{\sigma\})$. The converse follows from 8.1.7 and 8.1.4 (ii).

8.1.9. PROPOSITION. *Let α be an integrable representation of G on X. For each σ in Γ the following conditions are equivalent:*

(i) $\sigma\in\text{Sp}(\alpha)$.

(ii) *$R^{\alpha}(\Omega)\neq 0$ for every neighbourhood Ω of σ.*

(iii) *There is a net $\{x_i\}$ in the unit sphere of X such that $\|\alpha_t(x_i) - (t,\sigma)x_i\|\to 0$ uniformly on compact subsets of G.*

(iv) *For every μ in $M(G)$ we have $|\hat{\mu}(\sigma)|\leqslant\|\alpha_\mu\|$.*

(v) *For every f in $L^1(G)$ we have $|\hat{f}(\sigma)|\leqslant\|\alpha_f\|$.*

(vi) *If $f\in L^1(G)$ such that $\alpha_f = 0$ then $\hat{f}(\sigma) = 0$.*

Proof. If $R^{\alpha}(\Omega) = 0$ then $\text{Sp}(\alpha)\subset\Gamma\backslash\Omega$ so that (i)$\Rightarrow$(ii). Conversely, if $\sigma\notin\text{Sp}(\alpha)$ there is an open neighbourhood Ω of σ disjoint from $\text{Sp}(\alpha)$, whence $R^{\alpha}(\Omega) = 0$. Thus we have shown that (i)$\Leftrightarrow$(ii).

(ii)$\Rightarrow$(iii) follows from 8.1.7.

(iii)$\Rightarrow$(iv). Given μ and $\varepsilon > 0$ there is a compact set K in G such that $|\mu|(G\backslash K) < \varepsilon$. Assuming that $x_i\in X$ and $\|\alpha_t(x_i) - (t,\sigma)x_i\| < \varepsilon$ for all t in K we have

$$\begin{aligned}|\hat{\mu}(\sigma)| &= \|\hat{\mu}(\sigma)x_i\| = \|\textstyle\int(t,\sigma)x_i\,d\mu(t)\| \\ &\leqslant \|\textstyle\int(\alpha_t(x_i) - (t,\sigma)x_i)\,d\mu(t)\| + \|\textstyle\int\alpha_t(x_i)\,d\mu(t)\| \\ &\leqslant \varepsilon\,|\mu|(K) + 2\,|\mu|(G\backslash K) + \|\alpha_\mu(x_i)\| \\ &\leqslant \varepsilon\,\|\mu\| + 2\varepsilon + \|\alpha_\mu\|.\end{aligned}$$

Since ε is arbitrary $|\hat{\mu}(\sigma)|\leqslant\|\alpha_\mu\|$.

(iv)$\Rightarrow$(v) is obvious, and so is (v)$\Rightarrow$(vi).

(vi)$\Rightarrow$(ii). Given any neighbourhood Ω of σ there is an f in $K^1(G)$ with $\text{supp}.\hat{f}\subset\Omega$ such that $\hat{f}(\sigma) = 1$. By assumption this implies that $\alpha_f(x)\neq 0$ for some x in X, whence $R^{\alpha}(\Omega)\neq 0$.

8.1.10. PROPOSITION. *Let α be an integrable representation of G on X. If A is the commutative Banach algebra in $B(X)$ generated by elements of the form α_f, $f \in L^1(G)$, then the Arveson spectrum of α is homeomorphic to the Gelfand spectrum of A.*

Proof. Let $\hat{A}$ denote the spectrum of A. The dual of the homomorphism $\alpha: L^1(G) \to A$ defines a continuous injection α_* of $\hat{A}$ into Γ (because Γ is the spectrum of $L^1(G)$). Since $\hat{A}$ is locally compact α_* is a homeomorphism onto its image. From 8.1.9 (v) we see that $\sigma \in \alpha_*(\hat{A})$ if and only if $\sigma \in \mathrm{Sp}(\alpha)$.

8.1.11. COROLLARY. *Let α be an invertible isometry on a Banach space X, and consider the representation $n \to \alpha^n$ of $\boldsymbol{Z}$ on X. The Arveson spectrum of this representation is equal to the spectrum of α as an element in $B(X)$.*

Proof. Let A be the Banach algebra in $B(X)$ generated by α. The spectrum of α is the range of its Gelfand transform on $\hat{A}$ and thus, since α is a generator for A, homeomorphic to $\hat{A}$. The result now follows from 8.1.10.

8.1.12. THEOREM. *Let α be an integrable representation of G on X. The following conditions are equivalent:*

(i) $\mathrm{Sp}(\alpha)$ *is compact;*

(ii) *The representation α is uniformly continuous, i.e.*

$$\|\iota - \alpha_t\| \to 0 \qquad \text{as } t \to 0.$$

Proof. (i) $\Rightarrow$ (ii). Choose a function f in $K^1(G)$ such that $\hat{f} = 1$ on an open set Ω containing $\mathrm{Sp}(\alpha)$. Then $\alpha_f(x) = x$ for every x in $R^\alpha(\Omega)$ and since $R^\alpha(\Omega) = X$ we have $\alpha_f = 1$. But then

$$\|x - \alpha_t(x)\| \leqslant \|f - \delta_t \times f\|_1 \|x\|$$

for every x, whence $\|\iota - \alpha_t\| \to 0$ as $t \to 0$.

(ii) $\Rightarrow$ (i). If (f_λ) is an approximate unit for $L^1(G)$

$$\|\alpha_{f_\lambda}(x) - x\| \leqslant \int \|\alpha_t(x) - x\| f_\lambda(t)\,dt$$
$$\leqslant \int \|\alpha_t - \iota\| f_\lambda(t)\,dt\, \|x\|,$$

whence $\alpha_{f_\lambda} \to 1$, so that 1 belongs to the Banach algebra generated by $\alpha(L^1(G))$. But then $\mathrm{Sp}(\alpha)$ is compact by 8.1.10.

8.1.13. PROPOSITION. *Let $\pi: G_1 \to G_2$ be a continuous homomorphism between two groups G_1 and G_2 and denote by $\hat{\pi}$ the transposed homomorphism between the dual groups Γ_2 and Γ_1. If α is an integrable representation of G_2 on X then $\alpha \circ \pi$ is an integrable representation of G_1 on X and $\mathrm{Sp}(\alpha \circ \pi)$ is the closure of $\hat{\pi}(\mathrm{Sp}(\alpha))$ in Γ_1.*

Proof. Clearly $\alpha \circ \pi$ is an integrable representation. Take μ in $M(G_1)$ and denote by $\pi\mu$ its image in $M(G_2)$. Then

$$(\alpha \circ \pi)_\mu(x) = \int_{G_1} (\alpha \circ \pi)_t(x)\,d\mu(t) = \int_{G_2} \alpha_s(x)\,d\pi\mu(s) = \alpha_{\pi\mu}(x)$$

for every x in X; whence $(\alpha \circ \pi)_\mu = \alpha_{\pi\mu}$. If therefore $\sigma \in \mathrm{Sp}(\alpha)$ then by 8.1.9 (iv)

$$|\hat{\mu}(\hat{\pi}(\sigma))| = |(\pi\mu)\hat{}\,(\sigma)| \leqslant \|\alpha_{\pi\mu}\| = \|(\alpha \circ \pi)_\mu\|,$$

whence (again by 8.1.9 (iv)) $\hat{\pi}(\sigma) \in \mathrm{Sp}(\alpha \circ \pi)$.

Conversely, if τ is a point in Γ_1 not in the closure of $\hat{\pi}(\mathrm{Sp}(\alpha))$ choose disjoint open sets Ω_1 and Ω_2 such that $\tau \in \Omega_1$ and $\hat{\pi}(\mathrm{Sp}(\alpha)) \subset \Omega_2$. If $\mu \in M(G_1)$ with supp. $\hat{\mu} \subset \Omega_1$ then $(\pi\mu)\hat{}\,(\sigma) = \hat{\mu}(\hat{\pi}(\sigma)) = 0$ for every σ in $\hat{\pi}^{-1}(\Omega_2)$. It follows that for every g in $K^1(G_2)$ we have $\pi\mu \times g \in K^1(G_2)$ and

$$\mathrm{supp.}(\pi\mu \times g)\hat{} \subset \Gamma_2 \backslash \hat{\pi}^{-1}(\Omega_2) \subset \Gamma_2 \backslash \mathrm{Sp}(\alpha).$$

Since $R^\alpha(\Gamma_2 \backslash \mathrm{Sp}(\alpha)) = 0$ this implies that

$$(\alpha \circ \pi)_\mu(\alpha_g(x)) = \alpha_{\pi\mu \times g}(x) = 0$$

for all x in X and g in $K^1(G)$. Thus $(\alpha \circ \pi)_\mu = 0$. Applying this with $\mu = f\mu_G$ where $f \in K^1(G_1)$ and supp. $\hat{f} \subset \Omega_1$ we conclude that $R^{\alpha \circ \pi}(\Omega_1) = 0$, whence $\tau \notin \mathrm{Sp}(\alpha \circ \pi)$.

8.1.14. COROLLARY. *Let α be an integrable representation of G on X. For each t in G the spectrum of α_t as an element of $B(X)$ is the closure of the set*

$$\{(t, \sigma) \mid \sigma \in \mathrm{Sp}(\alpha)\}.$$

Proof. Consider the homomorphism $\pi: \mathbf{Z} \to G$ given by $\pi(n) = nt$ and apply 8.1.13 and 8.1.11.

8.1.15. Notes and remarks. Spectral subspaces were introduced by Godement [100] and a brief account of their properties can be found on p. 545 of [116]. A systematic study of spectral subspaces and their applications to dynamical systems was presented by Arveson [13]. Arveson treated mainly W^*-dynamical systems, but as shown by Olesen [183] the theory functions equally well (or better) for C^*-dynamical systems. The presentation of the theory given here is the result of numerous consultations with Olesen.

The definition of integrable representations in 8.1.1 is slightly more general than the Assumption 1.1 in Arveson's paper. More importantly, it describes what is needed and is obviously fulfilled in all cases we shall encounter. Proposition 8.1.9 is due to Connes, cf. [38], and theorem 8.1.12 was proved by Olesen [181].

8.2 Composition of representations

8.2.1. Assume now that we have two pairs (X, X_*) and (Y, Y_*) of Banach spaces in duality. Denote by $\boldsymbol{B}_\sigma(X, Y)$ the subspace of $\boldsymbol{B}(X, Y)$ consisting of operators that are $\sigma(X, X_*)$- $\sigma(Y, Y_*)$-continuous. This is a closed subspace of $\boldsymbol{B}(X, Y)$; for if z is the norm limit of a sequence (z_n) in $\boldsymbol{B}_\sigma(X, Y)$ then for each η in Y_* we have $\|z'(\eta) - z'_n(\eta)\| \to 0$. whence $z'(\eta) \in X_*$, so that $x \to \langle z(x), \eta\rangle$ is continuous on X, whence $z \in \boldsymbol{B}_\sigma(X, Y)$. Let $\boldsymbol{B}_\sigma(X, Y)_*$ denote the norm closure in $\boldsymbol{B}_\sigma(X, Y)^*$ of the linear span of functionals of the form $x \otimes \eta$, $x \in X$, $\eta \in Y_*$, where

$$\langle z, x \otimes \eta\rangle = \langle z(x), \eta\rangle.$$

Clearly $\boldsymbol{B}_\sigma(X, Y)_*$ is weak* dense in $\boldsymbol{B}_\sigma(X, Y)^*$, so that the spaces $\boldsymbol{B}_\sigma(X, Y)$ and $\boldsymbol{B}_\sigma(X, Y)_*$ are in duality.

8.2.2. Let α and β be representations of G on X and Y, respectively, and consider the conditions:

(i) $X_* = X^*$ and each function $t \to \alpha_t(x)$ is norm continuous;

(i′) $X = (X_*)^*$ and each function $t \to \alpha'_t(\xi)$ is norm continuous;

(ii) $Y_* = Y^*$ and each function $t \to \beta_t(y)$ is norm continuous;

(ii′) $Y = (Y_*)^*$ and each function $t \to \beta'_t(\eta)$ is norm continuous.

8.2.3. PROPOSITION. *Let α and β be integrable representations of G on X and Y, respectively. Assume either of the combinations of the conditions in 8.2.2 hold:* (i) *and* (ii); (i′) *and* (ii′); *and* (i) *and* (ii′). *Define $\Phi_t(z) = \beta_t z \alpha_{-t}$ for every z in $\boldsymbol{B}_\sigma(X, Y)$. Then we have an integrable representation $t \to \Phi_t$ of G on $\boldsymbol{B}_\sigma(X, Y)$.*

Proof. Clearly $\Phi_t(z) \in \boldsymbol{B}_\sigma(X, Y)$ for every z in $\boldsymbol{B}_\sigma(X, Y)$ and

$$\Phi'_t(x \otimes \eta) = \alpha_{-t}(x) \otimes \beta'_t(\eta).$$

Take z in $\boldsymbol{B}_\sigma(X, Y)$, x in X and η in Y_*. Then

$$\langle z(\alpha_{-s}(x) - \alpha_{-t}(x)), \beta'_s(\eta) - \beta'_t(\eta)\rangle$$
$$(*) \qquad = \langle \Phi_t(z), x \otimes \eta\rangle + \langle \Phi_s(z), x \otimes \eta\rangle - \langle z\alpha_{-s}(x), \beta'_t(\eta)\rangle$$
$$- \langle z\alpha_{-t}(x), \beta'_s(\eta)\rangle$$

for all s, t in G. The left-hand side is bounded by

$$(**) \qquad \|z\| \, \|\alpha_{-s}(x) - \alpha_{-t}(x)\| \, \|\beta'_s(\eta) - \beta'_t(\eta)\|$$

which tend to zero as $s \to t$, for any choice of our assumptions. On the right hand side of $(*)$ we have a constant (for fixed t), an undecided term, and two continuous functions of s, both converging to $\langle \Phi_t(z), x \otimes \eta\rangle$ as $s \to t$.

follows that $\Phi_s(z) \to \Phi_t(z)$ weakly on $\boldsymbol{B}_\sigma(X, Y)_*$ as $s \to t$, so that Φ is a representation of G on $\boldsymbol{B}_\sigma(X, Y)$. Let μ be a probability measure on G with compact support K. For fixed x and η, and $\varepsilon > 0$ there is by (*) and (**) for each s in K a neighbourhood E of s such that

$$|\langle z, \Phi'_t(x \otimes \eta)\rangle + \langle z, \Phi'_s(x \otimes \eta)\rangle - \langle z\alpha_{-s}(x), \beta'_t(\eta)\rangle - \langle z\alpha_{-t}(x), \beta'_s(\eta)\rangle| \leqslant \varepsilon \|z\|$$

for all t in E and all z in $\boldsymbol{B}_\sigma(X, Y)$. Since K is compact we can thus find a finite covering of K with disjoint sets E_n and points s_n in E_n such that with $\mu_n = \mu \mid E_n$ we have

$$|\textstyle\int \langle z, \Phi'_t(x \otimes \eta)\rangle \, d\mu(t) + \sum_n (\mu(E_n)\langle z, \Phi'_{s_n}(x \otimes \eta)\rangle$$

$$- \langle z, \alpha_{-s_n}(x) \otimes \beta'_{\mu_n}(\eta)\rangle) - \langle z, \alpha_{\tilde{\mu}_n}(x) \otimes \beta'_{s_n}(\eta)\rangle \,| \leqslant \varepsilon \|z\|,$$

for all z in $\boldsymbol{B}_\sigma(X, Y)$. Thus $\Phi'_\mu(x \otimes \eta)$ can be approximated in norm by elements in $\boldsymbol{B}_\sigma(X, Y)_*$. whence $\Phi'_\mu(x \otimes \eta) \in \boldsymbol{B}_\sigma(X, Y)_*$. By continuity $\Phi'_\mu(\zeta) \in \boldsymbol{B}_\sigma(X, Y)_*$ for every ζ in $\boldsymbol{B}_\sigma(X, Y)_*$ and every μ in $M(G)$.

To show that Φ'_μ is weakly continuous we first note that there is a bijective correspondence between elements z in $\boldsymbol{B}_\sigma(X, Y)$, and bilinear forms $b(\cdot, \cdot)$ on $X \times Y_*$ that are separately weakly continuous in each variable, given by

$$\langle z(x), \eta\rangle = b(x, \eta).$$

Indeed, if $b(\cdot, \cdot)$ is given, then for each x in X the element $b(x, \cdot)$ is a $\sigma(Y_*, Y)$-continuous functional on Y_*, whence $b(x, \cdot) \in Y$. Moreover, the bounded operator $x \to b(x, \cdot)$ belongs to $\boldsymbol{B}_\sigma(X, Y)$, since for each η in Y_* the functional $x \to b(x, \eta)$ is $\sigma(X, X_*)$-continuous.

Now assume (i) and (ii′). Then Φ' is pointwise norm continuous on functionals of the form $x \otimes \eta$, and therefore pointwise norm continuous on $\boldsymbol{B}_\sigma(X, Y)_*$. Moreover, $(\boldsymbol{B}_\sigma(X, Y)_*)^* = \boldsymbol{B}_\sigma(X, Y)$. Indeed, each element ϕ in $(\boldsymbol{B}_\sigma(X, Y)_*)^*$ determines a bilinear form $b(\cdot, \cdot)$, where $b(x, \eta) = \langle \phi, x \otimes \eta\rangle$, and $b(\cdot, \cdot)$ is separately (norm) continuous in each variable. From the preceding we conclude that $\phi \in \boldsymbol{B}_\sigma(X, Y)$. It follows from 8.1.2 that Φ' (and therefore Φ) is integrable in this case.

Assume then either (i) and (ii) or (i′) and (ii′). Take μ in $M(G)$ and z in $\boldsymbol{B}_\sigma(X, Y)$. If $\eta \in Y_*$ then the function

$$t \to \langle \Phi_t(z) \cdot, \eta\rangle = \langle \beta_t z\alpha_{-t}(\cdot), \eta\rangle = \langle (\cdot), \alpha'_{-t} z' \beta'_t(\eta)\rangle$$

from G to X_* is either weak* continuous (if (i) and (ii) are satisfied) or norm continuous (if (i′) and (ii′) are satisfied). In either case its integral exists in X_*, i.e. there exists ζ in X_* such that

$$\int \langle \Phi_t(z)x, \eta\rangle \, d\mu(t) = \langle x, \zeta\rangle$$

for all x in X. In exactly the same way we show that for fixed x in X there exists y in Y such that

$$\int\langle\Phi_t(z)x,\eta\rangle\,d\mu(t) = \langle y,\eta\rangle$$

for all η in Y_*. It follows that the bilinear form

$$b(\cdot,\cdot) = \int\langle\Phi_t(z)\cdot,\cdot\rangle\,d\mu(t)$$

on $X \times Y_*$ is separately weakly continuous in each variable. From the preceding it follows that there is an element w in $\boldsymbol{B}_\sigma(X, Y)$ such that

$$\langle w(x),\eta\rangle = b(x,\eta) = \int\langle\Phi_t(z)x,\eta\rangle\,d\mu(t).$$

But then

$$\langle w, x\otimes\eta\rangle = \langle z, \Phi'_\mu(x\otimes\eta)\rangle$$

for all x, η which shows that $w = \Phi_\mu(z)$ and thus Φ'_μ is weakly continuous also in the last two cases.

8.2.4. LEMMA. *Let α and β be integrable representations of G on X and Y, respectively, and assume that $\Phi = \beta.\alpha^{-1}$ is integrable on $\boldsymbol{B}_\sigma(X, Y)$. Then for all open sets Ω_1, Ω_2 and all closed sets Λ_1, Λ_2 in Γ we have*

$$R^\Phi(\Omega_1)R^\alpha(\Omega_2) \subset R^\beta(\Omega_1 + \Omega_2), \qquad M^\Phi(\Lambda_1)M^\alpha(\Lambda_2) \subset M^\beta((\Lambda_1 + \Lambda_2)^-).$$

Proof. Take f and g in $K^1(G)$ such that $\hat{f}$ and $\hat{g}$ have supports in Ω_1 and Ω_2, respectively. Let $\Lambda = \text{supp.}\hat{f} + \text{supp.}\hat{g}$. Then Λ is a compact set and $\Lambda \subset \Omega_1 + \Omega_2$.

Take any function h in $K^1(G)$ with $\text{supp.}\hat{h} \subset \Gamma\backslash\Lambda$. Then for z in $\boldsymbol{B}_\sigma(X, Y)$, x in X and η in Y_* consider the expression

$$\begin{aligned}\langle\Phi_f(z)\alpha_g(x), \beta'_h(\eta)\rangle &= \iiint\langle\beta_t z\alpha_{-t+s}(x), \beta'_r(\eta)\rangle f(t)g(s)h(r)\,dt\,ds\,dr\\ &= \iiint\langle\beta_r z\alpha_s(x),\eta\rangle\, h(r-t)f(t)g(s+t)\,dt\,ds\,dr\\ &= \iint\langle\beta_r z\alpha_s(x),\eta\rangle\,(h\times(f\cdot(\delta_{-s}\times g)))(r)\,ds\,dr.\end{aligned}$$

We have

$$\begin{aligned}\text{supp.}(h\times(f\cdot(\delta_{-s}\times g)))^\wedge &= \text{supp.}(\hat{h}\cdot(\hat{f}\times(s\cdot\hat{g})))\\ &\subset \text{supp.}\hat{h}\cap(\text{supp.}\hat{f} + \text{supp.}\hat{g}) = \varnothing\end{aligned}$$

whence $h \times (f\cdot(\delta_{-s} \times g)) = 0$. This shows that

$$\Phi_f(z)\alpha_g(x) \in M^\beta(\Lambda) \subset R^\beta(\Omega_1 + \Omega_2).$$

The last inclusion is independent of f and g, whence

$$R^{\Phi}(\Omega_1)R^{\alpha}(\Omega_2) \subset R^{\beta}(\Omega_1 + \Omega_2).$$

To prove the other inclusion note that by 8.1.4 (vi) we have, for any open neighbourhood Ω of θ

$$M^{\Phi}(\Lambda_1)M^{\alpha}(\Lambda_2) \subset R^{\Phi}(\Lambda_1 + \Omega)R^{\alpha}(\Lambda_2 + \Omega) \subset R^{\beta}(\Lambda_1 + \Lambda_2 + \Omega + \Omega)$$
$$\subset M^{\beta}((\Lambda_1 + \Lambda_2 + \Omega + \Omega)^{-}).$$

Taking the intersection over all Ω and using 8.1.4 (iv) we obtain the desired result.

8.2.5. THEOREM. *Let α and β be integrable representations of G on X and Y, respectively, and assume that $\Phi = \beta \cdot \alpha^{-1}$ is integrable on $\boldsymbol{B}_{\sigma}(X, Y)$. Let Λ be a closed subset of Γ which is the closure of its interior. For each τ_0 in Γ and z in $\boldsymbol{B}_{\sigma}(X, Y)$ the following conditions are equivalent:*

(i) $z \in M^{\Phi}(\tau_0 + \bigcap_{\sigma \in \Lambda} (\Lambda - \sigma))$;

(ii) $zM^{\alpha}(\tau + \Lambda) \subset M^{\beta}(\tau_0 + \tau + \Lambda)$ *for all τ in Γ.*

Proof. (i) $\Rightarrow$ (ii) is immediate from 8.2.4 because

$$\tau_0 + \bigcap_{\sigma \in \Lambda} (\Lambda - \sigma) = \{\sigma' \in \Gamma \mid \sigma' + \Lambda \subset \tau_0 + \Lambda\}.$$

(ii) $\Rightarrow$ (i). Take f and g in $K^1(G)$ such that $\hat{f}$ and $\hat{g}$ have supports in $\Gamma \backslash \Lambda$ and $\mathring{\Lambda}$ respectively. If $x \in X, \eta \in Y_*$ and $\tau \in \Gamma$ then with $f_1 = f \cdot \bar{\tau}_0 \cdot \bar{\tau}$ and $g_1 = g \cdot \bar{\tau}$ we have

$$\alpha_{g_1}(x) \in M^{\alpha}(\tau + \Lambda), \qquad \beta'_f(\eta) \in R^{\beta'}(\Gamma \backslash (\tau_0 + \tau + \Lambda)).$$

From our assumption it follows that

$$\begin{aligned} 0 &= \langle z\alpha_{g_1}(x), \beta'_{f_1}(\eta) \rangle \\ &= \iint \langle \beta_t z \alpha_s(x), \eta \rangle f(t) g(s) \overline{(t, \tau_0)} \overline{(s + t, \tau)} \, ds \, dt \\ &= \iint \langle \Phi_t(z) \alpha_s(x), \eta \rangle f(t) g(s - t) \overline{(t, \tau_0)} \overline{(s, \tau)} \, dt \, ds \\ &= \hat{h}(-\tau), \end{aligned}$$

where

$$h(s) = \int \langle \Phi_t(z) \alpha_s(x), \eta \rangle f(t) g(s - t) \overline{(t, \tau_0)} \, dt.$$

Since h is continuous and $\hat{h}$ vanishes on Γ we see that $h = 0$; in particular

$$(*) \qquad 0 = h(0) = \int \langle \Phi_t(z) x, \eta \rangle f(t) g(-t) \overline{(t, \tau_0)} \, dt.$$

Put $k(t) = \langle \Phi_t(z), x \otimes \eta \rangle$. Then the equation (*) says that $(kf\bar{\tau}_0)\hat{}$ is orthogonal to $\tilde{g}\hat{}$ $(=\bar{\hat{g}})$ in $L^2(\Gamma)$. Since this holds for all g in $K^1(G)$ with supp.$\hat{g} \subset \mathring{\Lambda}$ we conclude that supp.$(kf\bar{\tau}_0)\hat{} \subset \Gamma\backslash\mathring{\Lambda}$, i.e.

$$\text{supp.}(kf)\hat{} \subset \Gamma\backslash(\tau_0 + \mathring{\Lambda}).$$

It follows by continuity that

$$0 = (kf)\hat{}(\sigma) = \int \langle \Phi_t(z), x \otimes \eta \rangle f(t)(t, \sigma)\, dt$$

for all σ in $\tau_0 + \Lambda$. Since this holds for all f in $K^1(G)$ with supp.$f \subset \Gamma\backslash\Lambda$ and for all $x \otimes \eta$ we conclude that

$$z \in R^{\Phi'}(\Gamma\backslash(\Lambda - \sigma))^{\perp} = M^{\Phi}(\Lambda - \sigma)$$

for all σ in $\tau_0 + \Lambda$. Equivalently, $z \in M^{\Phi}(\tau_0 + (\Lambda - \sigma))$ for all σ in Λ, and the theorem follow from 8.1.4 (iv).

8.2.6. COROLLARY. *Let α and β be integrable representations of G on X and assume that $\Phi = \beta \cdot \alpha^{-1}$ is integrable on $\boldsymbol{B}_\sigma(X)$. Let Λ be a closed subset of Γ with $-\Lambda \cap \Lambda = \{\theta\}$ and assume that Λ is the closure of its interior. If*

$$M^{\alpha}(\tau + \Lambda) \subset M^{\beta}(\tau + \Lambda) \qquad \text{and} \qquad M^{\alpha}(\tau - \Lambda) \subset M^{\beta}(\tau - \Lambda)$$

for all τ in Γ then $\alpha = \beta$.

Proof. Let 1 denote the identity operator in $\boldsymbol{B}_\sigma(X)$. It follows from 8.2.5 that $1 \in M^{\Phi}(\Lambda)$ and that also $1 \in M^{\Phi}(-\Lambda)$. By 8.1.4 (iv) this implies that $1 \in M^{\Phi}(\{\theta\})$, i.e. $\beta_t 1 \alpha_{-t} = 1$ for all t by 8.1.8.

8.2.7. Notes and remarks. Proposition 8.2.3 is obviously a cheap solution. However, the result is sufficient for our purposes and the proof, while not being elegant, is quite elementary. The all-important theorem 8.2.5 is due to Arveson [13].

8.3. The minimal positive representation

8.3.1. We shall apply the techniques developed in the preceding sections to three cases.

(i) A strongly continuous unitary representation $t \to u_t$ of G on a Hilbert space H;

(ii) A continuous representation $t \to \alpha_t$ of G as automorphisms of a C^*-algebra A, i.e. a C^*-dynamical system in the sense of 7.4.1 (an a-system using the terminology from 7.8.1);

(iii) A σ-weakly continuous representation $t \to \alpha_t$ of G as automorphisms of a von Neumann algebra $\mathcal{M}$, i.e. a W^*-dynamical system (a-system).

For simplicity we shall assume throughout the chapter that G is separable, although this is probably not necessary for the validity of the results.

Note that by 7.4.3 and 7.4.4 the representations $t \to \alpha_t$ are integrable in cases (ii) and (iii), and clearly also in case (i), so that the theory of spectral subspaces can be applied.

Case (i) is simple and satisfying:

8.3.2. PROPOSITION. *If $t \to u_t$ is a unitary representation of an abelian group G on a Hilbert space H there is a unique spectral measure μ on the Borel sets of Γ with values in $B(H)$ such that*

$$(*) \qquad u_t = \int (t, \tau)\, d\mu(\tau), \qquad \forall t \in G.$$

If Ω is open and Λ is closed in Γ then

$$R^u(\Omega) = \mu(\Omega)H \qquad \textit{and} \qquad M^u(\Lambda) = \mu(\Lambda)H.$$

Proof. For each f in $L^1(G)$ define $\pi(f) = \int u_t f(t)\, dt$. Then π is a *-representation of $L^1(G)$ into $B(H)$. However, since each $\pi(f)$ is a normal operator, $\|\pi(f)\| \leq \|\hat{f}\|$. Therefore π extends by continuity to a representation of the C^*-algebra $C_0(\Gamma)$ and then by 4.5.9 and 4.5.14 to a σ-normal representation of $\mathscr{B}(\Gamma)$. Restricting π to the projections in $\mathscr{B}(\Gamma)$ we obtain a spectral measure μ on Γ satisfying $(*)$ cf. 4.7.7.

If Ω is open in Γ we can find a sequence $\{f_n\}$ in $K^1(G)$ such that $\hat{f}_n \leq \chi_\Omega$ and $\hat{f}_n \to \chi_\Omega$ pointwise on Γ. We have $\pi(f_n)H \subset R^u(\Omega)$ but also $\pi(f_n)\xi \to \xi$ for every ξ in $R^u(\Omega)$. It follows that $R^u(\Omega) = \pi(\chi_\Omega)H = \mu(\Omega)H$.

The formula $M^u(\Lambda) = \mu(\Lambda)H$ follows by duality.

8.3.3. LEMMA. *Let (A, G, α) be a C^*-dynamical a-system. Then*

(i) $R^\alpha(\Omega)^* = R^\alpha(-\Omega)$.

(ii) $M^\alpha(\Lambda)^* = M^\alpha(-\Lambda)$.

(iii) $R^\alpha(\Omega_1)R^\alpha(\Omega_2) \subset R^\alpha(\Omega_1 + \Omega_2)$.

(iv) $M^\alpha(\Lambda_1)M^\alpha(\Lambda_2) \subset M^\alpha((\Lambda_1 + \Lambda_2)^-)$.

Proof. (i) If $x \in A$ and $f \in K^1(G)$ with supp.$\hat{f} \subset \Omega$ then

$$\alpha_f(x)^* = (\int \alpha_t(x) f(t)\, dt)^* = \alpha_{\bar{f}}(x^*).$$

Since $(\bar{f})\hat{}(\tau) = (\hat{f}(-\tau))^-$ we see that $\alpha_f(x)^* \in R^\alpha(-\Omega)$.

(ii) This follows by duality or by using 8.1.4 (vii).

(iii) Let $\Phi = \alpha \cdot \alpha^{-1}$ be the representation of $B_\sigma(A)$ defined in 8.2.3 and note that Φ is integrable since the conditions (i) and (ii) in 8.2.2 hold. For each x in A define $\lambda(x)$ in $B_\sigma(A)$ by $\lambda(x)y = xy$. We claim that $\Phi_\mu(\lambda(x)) = \lambda(\alpha_\mu(x))$ for every μ in $M(G)$. Indeed, if $y \in A$ then

$$\Phi_\mu(\lambda(x))y = \int \alpha_t \lambda(x) \alpha_{-t}(y)\, d\mu(t)$$
$$= \int \alpha_t(x) y\, d\mu(t) = \alpha_\mu(x) y = \lambda(\alpha_\mu(x))y.$$

For any f in $K^1(G)$ with supp. $f \subset \Omega$ we have $\lambda(\alpha_f(x)) = \Phi_f(\lambda(x))$ for every x in A and it follows that $\lambda(R^\alpha(\Omega)) \subset R^\Phi(\Omega)$. Using 8.2.4 we conclude that

$$R^\alpha(\Omega_1)R^\alpha(\Omega_2) = \lambda(R^\alpha(\Omega_1))(R^\alpha(\Omega_2))$$
$$\subset R^\Phi(\Omega_1)(R^\alpha(\Omega_2)) \subset R^\alpha(\Omega_1 + \Omega_2).$$

(iv) follows from (iii) using 8.1.4 (vii).

8.3.4. COROLLARY. *Let (A, G, α) be a C^*-dynamical a-system. If $\sigma \in \mathrm{Sp}(\alpha)$ then also $-\sigma \in \mathrm{Sp}(\alpha)$.*

8.3.5. In an attempt to carry the results in 8.3.2 over to C^*-dynamical a-systems we make the following definition: Given (A, G, α) and a closed set Λ in Γ let $p(\Lambda)$ denote the largest projection in A'' such that $p(\Lambda)R^\alpha(\Gamma\backslash\Lambda) = 0$. Equivalently, $1 - p(\Lambda)$ is the smallest projection in A'' such that $(1 - p(\Lambda))x = x$ for every x in $R^\alpha(\Gamma\backslash\Lambda)$.

8.3.6. LEMMA. *Let (A, G, α) be a C^*-dynamical a-system. The projections $p(\Lambda)$ defined in 8.3.5 satisfy the following conditions:*

(i) *each $p(\Lambda)$ is a closed projection in A'';*

(ii) $p(\Lambda_1) \leqslant p(\Lambda_2)$ *if* $\Lambda_1 \subset \Lambda_2$;

(iii) $\wedge_i p(\Lambda_i) = p(\cap_i \Lambda_i)$;

(iv) $p(\Lambda) = 0$ *if* $\theta \notin \Lambda$;

(v) $p(\Lambda) = 1$ *if* $\mathrm{Sp}(\alpha) \subset \Lambda$;

(vi) $\alpha_t''(p(\Lambda)) = p(\Lambda)$ *for every* t;

(vii) $p(\Lambda)$ *commutes with every universally measurable element in A''_{sa} which is a fixed-point for α''_G.*

Proof. Let R denote the closed right ideal in A generated by $R^\alpha(\Gamma\backslash\Lambda)$. Then $p(\Lambda)$ is the left annihilator of R or, equivalently, $1 - p(\Lambda)$ is the support of $R \cap R^*$. Thus $1 - p(\Lambda)$ is open, whence $p(\Lambda)$ is closed (cf. 3.11.10).

(ii) is obvious from 8.1.4 (i).

(iii) follows from 8.1.4 (iii).

(iv) If $\theta \notin \Lambda$ there is an f in $K^1(G)$ with supp. $f \subset \Gamma\backslash\Lambda$ and $\hat{f}(\theta) = 1$. Let $\{u_\lambda\}$ be an approximate unit for A and take x in A. We have

$$p(\Lambda)\alpha_f(u_\lambda)x = 0$$

for every λ, since $\alpha_f(u_\lambda) \in R^\alpha(\Gamma\backslash\Lambda)$. However, $\|\alpha_t(u_\lambda)x - x\| \to 0$ for every t, uniformly on compact subsets of G, whence $\|\alpha_f(u_\lambda)x - x\| \to 0$. It follows that $p(\Lambda)x = 0$ for all x, whence $p(\Lambda) = 0$.

(v) If $\mathrm{Sp}(\alpha) \subset \Lambda$ then $R^\alpha(\Gamma\backslash\Lambda) = 0$ by 8.1.6 whence $p(\Lambda) = 1$.

(vi) Since $R^\alpha(\Gamma\backslash\Lambda)$ is a G-invariant subset by 8.1.4 its annihilator is a fixed point for α''_G.

(vii) Take any open set Ω with compact closure contained in $\Gamma\backslash\Lambda$. There is then a neighbourhood Ω_0 of θ such that $\Omega + \Omega_0 \subset \Gamma\backslash\Lambda$. Choose a function f in $K^1(G)$ with $\mathrm{supp}.f \subset \Omega_0$ and $f(\theta) = 1$. By 8.3.3 (iii) we have

$$p(\Lambda)\alpha_f(x)R^\alpha(\Omega) = 0$$

for every x in A. It follows from (*) in 7.4.5 that the same is true for any x in $\mathscr{U}(A)$ (with $\alpha_f(x)$ replaced by $\alpha''_f(x)$). If moreover x is a fixed-point for α''_G we see, again from (*) in 7.4.5, that $\alpha''_f(x) = x$. Thus $p(\Lambda)\, xR^\alpha(\Omega) = 0$ whence

$$p(\Lambda)x\,(1 - p(\Gamma\backslash\Omega)) = 0.$$

Since this is true for all relatively compact open sets Ω with $\bar{\Omega} \subset \Gamma\backslash\Lambda$ we conclude from (iii) that $p(\Lambda)x(1 - p(\Lambda)) = 0$. This gives

$$p(\Lambda)x = p(\Lambda)xp(\Lambda) = xp(\Lambda)$$

since $\mathscr{U}(A) \subset A''_{sa}$.

8.3.7. LEMMA. *Let (A, G, α) be a C*-dynamical a-system. Suppose that $\{\Lambda_n\}$ is a sequence of closed sets with $\cup\Lambda_n = \Gamma$ and that for each compact neighbourhood Λ_0 of θ and each m there is a number n such that $(\Gamma\backslash\Lambda_n) + \Lambda_0 \subset \Gamma\backslash\Lambda_m$. Then the projection $p(\infty) = \vee_n\, p(\Lambda_n)$ belongs to the centre of A''.*

Proof. Take x in A and f in $K^1(G)$ with $f(\theta) = 1$. By assumption there is for each m an n such that

$$(\Gamma\backslash\Lambda_n) + \mathrm{supp}.f \subset \Gamma\backslash\Lambda_m.$$

It follows from 8.3.3 that

$$p(\Lambda_m)\alpha_f(x)R^\alpha(\Gamma\backslash\Lambda_n) = 0$$

whence $p(\Lambda_m)\alpha_f(x)(1 - p(\Lambda_n)) = 0$. Since $p(\infty) \geqslant p(\Lambda_n)$ this implies that $p(\Lambda_m)\alpha_f(x)(1 - p(\infty)) = 0$ for every m, whence $p(\infty)\alpha_f(x)(1 - p(\infty)) = 0$. If f runs through an approximate unit for $L^1(G)$ we have $\alpha_f(x) \to x$, whence $p(\infty)x(1 - p(\infty)) = 0$; and since x is arbitrary $p(\infty)$ commutes with A.

8.3.8. A closed subset Γ_+ of a locally compact group Γ will be called a *positive cone* if it satisfies the following conditions.

(i) $\Gamma_+ + \Gamma_+ \subset \Gamma_+$.

(ii) $-\Gamma_+ \cap \Gamma_+ = \{\theta\}$.

(iii) $\Gamma_+ - \Gamma_+ = \Gamma$.

(iv) Γ_+ is the closure of its interior.

8.3.9. LEMMA. *If Γ_+ is a positive cone in a locally compact separable abelian group Γ there is a sequence $\{\tau_n\}$ in Γ_+ such that for each τ in Γ_+ and each compact set Λ_0 we have*

$$\tau - \Gamma_+ - \Lambda_0 \subset \tau_n - \Gamma_+$$

for some n. In particular, if. $\Lambda_n = \tau_n - \Gamma_+$ the sequence $\{\Lambda_n\}$ satisfies the requirement in 8.3.7.

Proof. By a compactness argument in conjunction with (iii) and (iv) we see that $\tau - \Lambda_0 \subset \cup\sigma_k - \Gamma_+$ for some finite set $\{\sigma_k\}$ in Γ_+. But if $\sigma = \Sigma\,\sigma_k$ then $\sigma_k - \Gamma_+ \subset \sigma - \Gamma_+$ for all k. Moreover, if we take $\{\tau_n\}$ as a dense sequence in Γ_+ then $\tau_n \in \sigma + \Gamma_+$ for some n, whence

$$\tau - \Gamma_+ - \Lambda_0 \subset \sigma - \Gamma_+ - \Gamma_+ \subset \tau_n - \Gamma_+.$$

If we put $\Lambda_n = \tau_n - \Gamma_+$ then from the above we have for each m an n such that $\Lambda_m - \Lambda_0 \subset \Lambda_n$. Consequently $((\Gamma \backslash \Lambda_n) + \Lambda_0) \cap \Lambda_m = \varnothing$, as desired.

8.3.10. Let (A, G, α) be a C^*-dynamical a-system. A positive cone Γ_+ in Γ is *admissible* for (A, G, α) if there is a spectral measure μ on the Borel sets in Γ with values in A'' such that

$$\mu(\tau - \Gamma_+) = p(\tau - \Gamma_+)$$

for every τ in Γ, with $p(\tau - \Gamma_+)$ as defined in 8.3.5. Since the sets $\tau - \Gamma_+$, $\tau \in \Gamma$, generate the Borel structure in Γ (they separate points and produce a countably generated structure) such a spectral measure is unique.

If Γ_+ is admissible for (A, G, α) then by 8.3.6 (iv)

$$\mu(\Gamma) = p(\infty) = \mu(\Gamma_+)$$

where $p(\infty) = \vee\, p(\tau - \Gamma_+)$; and by 8.3.7 and 8.3.9 $p(\infty)$ is a central projection in A'' and a sequential limit of closed projections. Define

$$u_t = \int (t, \tau)\, d\mu(\tau), \qquad \forall t \in G.$$

It is straightforward to verify that $t \to u_t$ is a σ-weakly continuous unitary representation of G into $A''p(\infty)$. From 8.3.2 we see that μ is uniquely determined by u. In particular,

$$\mathrm{Sp}(u) = \mathrm{supp}.\, \mu \subset \Gamma_+.$$

We say that u is the *minimal positive representation* of G associated with Γ_+. The reason for this terminology will be apparent from the following.

8.3.11. THEOREM. *Let (A, G, α) be a C*-dynamical a-system and assume that Γ_+ is an admissible positive cone in Γ for (A, G, α). Consider the minimal positive unitary representation u of G into $A''p(\infty)$ described in 8.3.10 (i.e. $u_t = \int(t,\tau)\,d\mu(\tau)$). Then*

$$u_t x u_t^* = \alpha_t(x)p(\infty), \qquad \forall t \in G, \qquad \forall x \in A.$$

Moreover, for any covariant representation (ρ, v, K) of (A, G, α) with $\mathrm{Sp}(v) \subset \Gamma_+$ we have $\rho''(p(\infty)) = 1$, and defining $w_t = v_t \rho''(u_t^*)$ we obtain a unitary representation (w, K) of G with $\mathrm{Sp}(w) \subset \Gamma_+$.

Proof. Put $H = p(\infty)H_u$ and consider the representation of A on H given by $\pi(x) = xp(\infty)$. We must show that (π, u, H) is a convariant representation for (A, G, α).

Fix τ_0 in Γ and take x in $M^\alpha(\tau_0 - \Gamma_+)$. Then $x^* \in M^\alpha(\Gamma_+ - \tau_0)$ by 8.3.3 (ii). For any compact set Λ contained in $\Gamma\backslash(\tau - \Gamma_+)$ we have

$$\Gamma_+ - \tau_0 + \Lambda \subset \Gamma\backslash(\tau - \tau_0 - \Gamma_+).$$

It follows trom 8.3.3 (iv) and 8.1.4 (viii) that

$$x^* R^\alpha(\Gamma\backslash(\tau - \Gamma_+)) \subset R^\alpha(\Gamma\backslash(\tau - \tau_0 - \Gamma_+)).$$

Consequently $p(\tau - \tau_0 - \Gamma_+)x^* R^\alpha(\Gamma\backslash\tau - \Gamma_+)A = 0$, whence

$$p(\tau - \tau_0 - \Gamma_+)x^*(1 - p(\tau - \Gamma_+)) = 0.$$

From 8.3.2 and 8.3.10 we know that $M^u(\tau - \Gamma_+) = (p(\tau - \Gamma_+))H$, so the equation above can be rephrased to say that

$$\pi(x)M^u(\tau - \tau_0 - \Gamma_+) \subset M^u(\tau - \Gamma_+).$$

Since this holds for every τ in Γ we conclude from 8.2.5 (with $\alpha = \beta = u$, $X = Y = H$ and $\Lambda = -\Gamma_+$) that $\pi(x) \in M^\beta(\tau_0 - \Gamma_+)$, where $\beta = u \cdot u^{-1}$ is a representation of G as automorphisms on $\boldsymbol{B}(H)$.

The conclusion above is valid for every x in $M^\alpha(\tau_0 - \Gamma_+)$ which shows that the map $\pi: A \to \boldsymbol{B}(H)$ satisfies

$$(*) \qquad \pi(M^\alpha(\tau_0 - \Gamma_+)) \subset M^\beta(\tau_0 - \Gamma_+),$$

for every τ_0 in Γ. It follows from 8.2.5 (but now with $(X, \alpha) = (A, \alpha)$, $(Y, \beta) = \boldsymbol{B}(H), \beta$ and $\Lambda = -\Gamma_+$) that $\pi \in M^\Phi(-\Gamma_+)$, with $\Phi = \beta \cdot \alpha^{-1}$ However, π is *-preserving so $(*)$ and 8.3.3 (ii) gives

$$\pi(M^\alpha(\tau_0 + \Gamma_+)) \subset M^\beta(\tau_0 + \Gamma_+)$$

for every τ_0, whence $\pi \in M^\Phi(\Gamma_+)$ by 8.2.5. Since $-\Gamma_+ \cap \Gamma_+ = \{\theta\}$ it follows from 8.1.4 that $\pi \in M^\Phi(\{\theta\})$, i.e. $\pi \circ \alpha = \beta \circ \pi$. But this means exactly that

$$u_t \pi(x) u_t^* = \pi(\alpha_t(x)), \qquad \forall t \in G, \qquad \forall x \in A.$$

Suppose that (ρ, v, K) is another covariant representation of (A, G, α) with $\mathrm{Sp}(v) \subset \Gamma_+$, and let ν denote the spectral measure from Γ into $B(K)$ associated with (v, K) (cf. 8.3.2). Put $\mathcal{M} = \rho(A)''$ and let $\gamma = v \cdot v^{-1}$ on $\mathcal{M}$. Then $\rho \circ \alpha = \gamma \circ \rho$; and since $\rho(A)$ is σ-weakly dense in $\mathcal{M}$ it follows that $R^{\gamma}(\Omega)$ is the σ-weak closure in $\mathcal{M}$ of $\rho(R^{\alpha}(\Omega))$, for every open set Ω in Γ. Thus $\rho''(p(\tau - \Gamma_+))$ is the left annihilator in $\mathcal{M}$ of $R^{\gamma}(\Gamma \backslash (\tau - \Gamma_+))$.

For every compact subset Λ of $\Gamma \backslash (\tau - \Gamma_+)$ we have by 8.2.4 and 8.3.2

$$M^{\gamma}(\Lambda)K = M^{\gamma}(\Lambda)M^{v}(\Gamma_+) \subset M^{v}(\Lambda + \Gamma_+) \subset R^{v}(\Gamma \backslash (\tau - \Gamma_+))$$
$$= (1 - \nu(\tau - \Gamma_+))K.$$

Since this holds for every Λ it follows that

$$\nu(\tau - \Gamma_+)R^{\gamma}(\Gamma \backslash (\tau - \Gamma_+)) = 0,$$

whence $\rho''(p(\tau - \Gamma_+)) \geqslant \nu(\tau - \Gamma_+)$ from what we proved above. This holds for every τ in Γ, so in particular

$$\rho''(p(\infty)) = \vee \rho''(p(\tau - \Gamma_+)) = 1.$$

We now consider the two unitary representations $t \to v_t$ and $t \to \rho''(u_t)$ on K. Since

$$v_t\rho''(u_s) = \gamma_t(\rho''(u_s))v_t = \rho''(\alpha_t''(u_s))v_t = \rho''(u_s)v_t,$$

we can define a unitary representation w by $w_t = v_t\rho''(u_t^*)$. We have $v_t = \int(t, \tau)\,\mathrm{d}\nu(\tau)$ and $\rho''(u_t) = \int(t, \tau)\,\mathrm{d}\mu'(\tau)$, where $\mu' = \rho'' \circ \mu$; and we proved above that $\mu'(\tau - \Gamma_+) \geqslant (\tau - \Gamma_+)$ for all τ in Γ. If $f \in L^1(G)$ then since μ' and ν commute

$$\int w_t f(t)\,\mathrm{d}t = \iiint(t, \tau - \sigma) f(t)\,\mathrm{d}t\,\mathrm{d}\nu(\tau)\,\mathrm{d}\mu'(\sigma)$$
$$= \iint \hat{f}(\tau - \sigma)\,\mathrm{d}\nu(\tau)\,\mathrm{d}\mu'(\sigma) = \iint \hat{f}(\tau)\,\mathrm{d}\nu(\tau + \sigma)\,\mathrm{d}\mu'(\sigma).$$

Fix τ and σ. If $\tau \notin \Gamma_+$ then $\tau + 2\omega \notin \Gamma_+$ for some ω in the interior of Γ_+. However, the set $\Lambda = (\sigma + \omega - \Gamma_+) \cap (\sigma - \omega + \Gamma_+)$ is a neighbourhood of σ, and

$$\nu(\tau + \Lambda) \leqslant \nu(\tau + \sigma + \omega - \Gamma_+) \leqslant \mu'(\tau + \sigma + \omega - \Gamma_+);$$

$$\mu'(\Lambda) \leqslant \mu'(\sigma - \omega + \Gamma_+).$$

Since $(\tau + \sigma + \omega - \Gamma_+) \cap (\sigma - \omega + \Gamma_+) = \varnothing$ by our assumption $(\tau + 2\omega \notin \Gamma_+)$, we see that $\nu(\tau + \Lambda)\mu'(\Lambda) = 0$. It follows that the integral $\int w_t f(t)\,\mathrm{d}t$ vanishes if $\mathrm{supp}.\hat{f} \cap \Gamma_+ = \varnothing$. Thus $\mathrm{Sp}(w) \subset \Gamma_+$, as desired.

8.3.12. Notes and remarks. Proposition 8.3.2 is Stone's theorem. Thus the theory of spectral subspaces may be viewed as an attempt to extend Stone's theorem on unitary representations to automorphic representations.

The lemmas 8.3.3 and 8.3.6 are found in [13], and 8.3.7 is an observation of Borchers from [23]. Theorem 8.3.11 is essentially due to Olesen, cf. [183].

8.4. Spectrum conditions

8.4.1. Every positive cone Γ_+ in a locally compact abelian group Γ induces a partial order (viz. $\sigma \leqslant \tau$ if $\tau - \sigma \in \Gamma_+$). A case of particular interest arises when this order is total, which happens when condition (iii) in 8.3.8 is replaced by the stronger condition

(iii′) $$-\Gamma_+ \cup \Gamma_+ = \Gamma.$$

We say in this case that Γ is an *ordered group* and that Γ_+ is a *maximal* positive cone.

8.4.2. LEMMA. *If G is an abelian group then its dual group Γ can be ordered if and only if $G = \boldsymbol{R}^n \times H$, where $0 \leqslant n \leqslant 1$ and H is a compact, connected group. Moreover, in this case there exists for each subset Λ of Γ satisfying (i) and (ii) of 8.3.8 a maximal positive cone Γ_+ containing Λ.*

Proof. If Γ ($\neq 0$) is compact and Γ_+ is a positive cone in Γ then since $(\tau_1 + \Gamma_+) \cap (\tau_2 + \Gamma_+) \supset \tau_1 + \tau_2 + \Gamma_+$ for all τ_1, τ_2 in Γ_+ there exists an element σ_0 in $\cap(\tau + \Gamma_+)$. In particular (with $\tau = 2\sigma_0$) there is some σ in Γ_+ such that $\sigma_0 = 2\sigma_0 + \sigma$. But then $\sigma_0 = \theta$ and $\sigma_0 \notin \tau + \Gamma_+$ if $\tau > \theta$, so we have a contradiction.

If Γ is ordered we see from above that it contains no non-trivial compact subgroups. Consequently G have no discrete quotient groups, i.e. G is connected. But then $G = \hat{\boldsymbol{R}}^n \times \hat{H}$. Since $\boldsymbol{R}^n$ can not be ordered for $n > 1$ (because $\boldsymbol{R}^n \backslash \{0\}$ is connected) the conditions in 8.4.2 are necessary.

For the converse we may assume that G is compact and connected, since there is no difficulty in ordering $\boldsymbol{R}$ (and no choice, either). Thus Γ is discrete. By Zorn's lemma the family of subsets (partially ordered by inclusion) containing Λ and satisfying (i) and (ii) of 8.3.8 has a maximal element Γ_+.

If $\sigma \in \Gamma$ and $\boldsymbol{N}\sigma \cap \Gamma_+ = \varnothing$, put $\Gamma'_+ = \Gamma_+ - \boldsymbol{Z}_+\sigma$. Then

$$\Gamma'_+ + \Gamma'_+ \subset \Gamma'_+$$

and if

$$-\Gamma'_+ \cap \Gamma'_+ \neq \{\theta\}$$

then

$$-(\tau_1 - n\sigma) = \tau_2 - m\sigma \neq \theta.$$

This gives $(n+m)\sigma = \tau_1 + \tau_2$, whence $n = m = 0$ and $\tau_1 = \tau_2 = \theta$, a contradiction. Since Γ_+ is maximal, $\Gamma'_+ = \Gamma_+$, i.e. $-\boldsymbol{N}\sigma \subset \Gamma_+$.

If $l\sigma \in \Gamma_+$ for some $l > 0$ put $\Gamma'_+ = \Gamma_+ + \boldsymbol{Z}_+\sigma$. Then $\Gamma'_+ + \Gamma'_+ \subset \Gamma'_+$ and if $-\Gamma'_+ \cap \Gamma'_+ \neq \{\theta\}$ then

$$-(\tau_1 + n\sigma) = \tau_2 + m\sigma \neq \theta.$$

This gives $l(n+m)\sigma \in -\Gamma_+ \cap \Gamma_+ = \{\theta\}$. Since G is connected Γ contains no elements of finite order, whence $n = m = 0$ and $\tau_1 = \tau_2 = \theta$, a contradiction. Since Γ_+ is maximal, $\Gamma'_+ = \Gamma_+$, i.e. $\boldsymbol{N}\sigma \subset \Gamma_+$.

We have shown that $-\Gamma_+ \cup \Gamma_+ = \Gamma$, as desired.

8.4.3. THEOREM. *Let $(A, \boldsymbol{R}, \alpha)$ be a C^*-dynamical system. There is then a central G-invariant projection $p(\infty)$ in A'' and a one-parameter unitary group $t \to u_t$ in $A''p(\infty)$ with $\mathrm{Sp}(u) \subset \boldsymbol{R}_+$ such that*

$$\alpha_t(x)p(\infty) = u_t x u_t^*, \qquad t \in \boldsymbol{R}, \qquad x \in A.$$

If (ρ, v, K) is any covariant representation of $(A, \boldsymbol{R}, \alpha)$ with $\mathrm{Sp}(v) \subset \boldsymbol{R}_+$ then $\rho''(p(\infty)) = 1$ so that $(\rho, \rho''(u), K)$ is covariant. Moreover, $\mathrm{Sp}(v\rho(u^*)) \subset \boldsymbol{R}_+$ (u is minimal positive).

Proof. We must show that the maximal cone $\boldsymbol{R}_+$ in $\boldsymbol{R}$ is admissible for the system $(A, \boldsymbol{R}, \alpha)$. For each τ in $\boldsymbol{R}$ consider the projection $p_\tau = p(\tau - \boldsymbol{R}_+)$ defined in 8.3.5. The function $\tau \to p_\tau$ is increasing and continuous from the right by 8.3.6 (ii) and (iii). Thus a Lebesgue–Stieltjes type argument shows that the definition $\mu(]\sigma, \tau]) = p_\tau - p_\sigma$ gives a spectral measure μ on the ring of disjoint unions of half-open intervals of $\boldsymbol{R}$. This then extends uniquely to a spectral measure μ on $\boldsymbol{R}$. We have shown that $\boldsymbol{R}_+$ is an admissible cone and the theorem now follows from 8.3.11.

8.4.4. THEOREM. *Let (A, G, α) be a C^*-dynamical a-system with G compact and connected. Any maximal cone Γ_+ in Γ such that the set*

$$(*) \qquad \mathrm{Sp}(\alpha) \cap (\tau - \Gamma_+) \cap (-\tau + \Gamma_+)$$

is finite for every τ is admissible for (A, G, α).

Proof. As in 8.4.3 we define $p_\tau = p(\tau - \Gamma_+)$ for each τ in Γ and obtain an increasing projection—valued function on Γ. Since Γ is countable (G is separable) $\mathrm{Sp}(\alpha)$ is a totally ordered countable set, and condition $(*)$ ensures that each point has an immediate successor. Thus $\mathrm{Sp}(\alpha)$ is order-isomorphic with $\boldsymbol{Z}$ or with $[-n, n] \cap \boldsymbol{Z}$.

Given τ in Γ there is, again by $(*)$, a largest τ_n in $\mathrm{Sp}(\alpha)$ such that $\tau_n \leqslant \tau$. Thus by 8.1.4 (iii)

$$\begin{aligned} R^\alpha(\{\sigma \mid \sigma > \tau_n\}) &= R^\alpha(\{\sigma \mid \sigma > \tau\} \cup \{\sigma \mid \tau \geqslant \sigma > \tau_n\}) \\ &= R^\alpha(\{\sigma \mid \sigma > \tau\}), \end{aligned}$$

since $\tau \geqslant \sigma > \tau_n$ implies $\sigma \notin \mathrm{Sp}(\alpha)$. Consequently $p_\tau = p_{\tau_n}$. Define $q_n = p_{\tau_n} - p_{\tau_{n-1}}$ so that $\{q_n\}$ is a sequence of pairwise orthogonal projections in A'', and for each subset $\Omega \subset \Gamma$ define

$$\mu(\Omega) = \sum q_k, \qquad \tau_k \in \Omega.$$

Then μ is a spectral measure on Γ and if τ_n is the largest minorant for τ in $\mathrm{Sp}(\alpha)$ then

$$\mu(\tau - \Gamma_+) = \sum_{\tau_k \leqslant \tau} q_k = p_{\tau_n} = p_\tau = p(\tau - \Gamma_+),$$

whence Γ_+ is admissible for (A, G, α).

8.4.5. LEMMA: *Let Γ be a discrete group without elements of finite order. For any semi-group Σ in Γ (i.e. $\Sigma + \Sigma \subset \Sigma$) the following conditions are equivalent:*

(i) *There exists an element σ in Γ such that $-\boldsymbol{N}\sigma \cap \Sigma = \varnothing$.*

(ii) *There exists an element σ in Γ such that*

$$-(\boldsymbol{N}\sigma + \Sigma) \cap (\boldsymbol{N}\sigma + \Sigma) = \varnothing.$$

(iii) *There exists an element σ in Γ and a maximal positive cone Γ_+ such that $\boldsymbol{N}\sigma + \Sigma \subset \Gamma_+$.*

Proof. (i) $\Rightarrow$ (ii). If $-(n\sigma + \tau_1) = m\sigma + \tau_2$ then $-(n+m)\sigma \in \Sigma$, a contradiction.

(ii) $\Rightarrow$ (iii). Define $\Lambda = (\boldsymbol{N}\sigma + \Sigma) \cup \{\theta\}$ and apply 8.4.2.

(iii) $\Rightarrow$ (i). If $\sigma = \theta$ we take any $\tau \notin \Gamma_+$ and have

$$\boldsymbol{N}\tau \cap \Sigma \subset (-\Gamma_+ \backslash \{\theta\}) \cap \Gamma_+ = \varnothing.$$

If $\sigma \neq \theta$ we claim that $-\boldsymbol{N}2\sigma \cap \Sigma = \varnothing$. For if $-2n\sigma \in \Sigma$ then by our assumption

$$\sigma = (2n+1)\sigma - 2n\sigma \in \Gamma_+, \qquad -\sigma = (2n-1)\sigma - 2n\sigma \in \Gamma_+;$$

a contradiction.

8.4.6. Let (A, G, α) be a C^*-dynamical a-system with G compact and connected. We say that a covariant representation (ρ, v, H) satisfies a *spectrum condition* if there is a σ in Γ and a maximal cone Γ_+ such that $\boldsymbol{N}\sigma + \mathrm{Sp}(v) \subset \Gamma_+$ and each set

$$\mathrm{Sp}(\alpha) \cap (\tau - \Gamma_+) \cap (-\tau + \Gamma_+)$$

is finite. A necessary condition for (ρ, v, H) to satisfy a spectrum condition is easily derived from 8.4.5.

8.4.7. PROPOSITION. *Let (A, G, α) be a C*-dynamical system with G compact and connected. For every covariant representation (ρ, v, H) of (A, G, α) satisfying a spectrum condition there is a σ-weakly continuous unitary representation $t \to u_t$ of G into $\rho(A)''$ such that (ρ, u, H) is covariant for (A, G, α).*

Proof. Replacing v_t by $w_t = (\iota, \sigma)v_t$ we have a covariant representation (ρ, w, H) with $\mathrm{Sp}(w) \subset \mathrm{Sp}(v) - \sigma$. By 8.4.6. we may therefore assume that $\mathrm{Sp}(w) \subset \Gamma_+$ for some maximal cone Γ_+ in Γ satisfying (*) in 8.4.4. The existence of the representation $t \to u_t$ now follows from 8.4.4 and 8.3.11.

8.4.8. PROPOSITION. *Let (A, G, α) be a C*-dynamical a-system. If $G = \prod_{1 \leq k \leq n} G_k$, whence $\Gamma = \prod_{1 \leq k \leq n} \Gamma_k$, and if for each k there is an admissible positive cone Γ_{k+} for the dynamical system $(A, G_k, \alpha | G_k)$, then the cone $\Gamma_+ = \prod_{1 \leq k \leq n} \Gamma_{k+}$ is admissible for (A, G, α).*

Proof. Fix k and consider the embedding of G_k into G and of Γ_k into Γ. We may then write $G = G_k \times G_k'$, $\Gamma = \Gamma_k \times \Gamma_k'$. Put $\alpha_k = \alpha | G_k$ and consider the dynamical system (A, G_k, α_k). From 8.1.5 we see that

$$R^{\alpha_k}(\Omega) = R^{\alpha}(\Omega \times \Gamma')$$

for each open set Ω in Γ_k. Consequently the projections $p_k(\tau - \Gamma_{k+})$, $\tau \in \Gamma_k$, defined in 8.3.5 relative to α_k commute with every universally measurable fixed-point for α_G'' (cf. 8.3.6 (vii)). The same is therefore true for all projections $\mu_k(\Lambda)$, $\Lambda \subset \Gamma_k$, where μ_k is the spectral 'masure on Γ_k determined by the projections $p_k(\tau - \Gamma_{k+})$, $\tau \in \Gamma_k$. Let $\mathscr{U}^G(A)$ denote the set of universally measurable fixed-points of α_G'' in A_{sa}''. From 4.5.12 we know that $\mathscr{U}^G(A)$ is monotone sequentially closed. Since $p_k(\tau - \Gamma_{k+}) \in \mathscr{U}^G(A)$ for every τ in Γ_k by 8.3.6 (i) and (vi), we conclude that

$$\mu_k(\Lambda) \in \mathscr{U}^G(A) \cap \mathscr{U}^G(A)'$$

for every Borel set Λ in Γ_k.

From the above it follows that we can define a spectral measure μ on Γ by $\mu = \otimes_{1 \leq k \leq n} \mu_k$. Put $\Gamma_+ = \prod_k \Gamma_{k+}$. Then Γ_+ is a positive cone for Γ and for each $\tau = (\tau_k)$ in Γ we have

$$\begin{aligned}\mu(\tau - \Gamma_+) &= \prod_k \mu_k(\tau_k - \Gamma_{k+}) = \prod_k p_k(\tau_k - \Gamma_{k+}) \\ &= \bigwedge_k p((\tau_k - \Gamma_{k+}) \times \Gamma_k') = p\left(\bigcap_k (\tau_k - \Gamma_{k+}) \times \Gamma'_k\right) \\ &= p(\tau \dot{-} \Gamma_+),\end{aligned}$$

by 8.3.6 (iii). It follows that Γ_+ is an admissible cone for (A, G, α).

8.4.9. LEMMA. *Let Λ be a closed, convex set in $\mathbf{R}^n$. The following conditions are equivalent:*

(i) *No straight line is contained in Λ.*

(ii) *There is a closed convex cone Π in $\mathbf{R}^n$ containing Λ and Π is not degenerate: i.e. if ω denotes the vertex of Π then $-(\Pi-\omega)\cap(\Pi-\omega) = \{0\}$.*

(iii) *There is a basis $\{\sigma_1,\dots,\sigma_n\}$ for $\mathbf{R}^n$ and a vector ω such that $\Lambda-\omega$ is contained in the simplicial cone with edges $\mathbf{R}_+\sigma_k$, $1\leq k\leq n$.*

(iv) *For every vector ω the set $-(\Lambda-\omega)\cap(\Lambda-\omega)$ is compact.*

Proof. (i)$\Rightarrow$(ii). Take a point ω outside Λ and let Π be the smallest closed cone containing Λ with vertex ω. Since Λ contains no straight line, Π is not degenerate.

(ii)$\Rightarrow$(iii). Assume that $\omega = 0$. If Π is not degenerate then its dual cone Π^* in $(\mathbf{R}^n)^*$ is generating: i.e. there is a basis $\{s_1,\dots,s_n\}$ for $(\mathbf{R}^n)^*$ contained in Π^*. Let $\{\sigma_1,\dots,\sigma_n\}$ be the dual basis for $\mathbf{R}^n$. Then the simplicial cone with edges $\mathbf{R}_+\sigma_k$, $1\leq k\leq n$, contains $(\Pi^*)^* = \Pi$, as desired.

(iii)$\Rightarrow$(iv) is immediate.

(iv)$\Rightarrow$(i). If $\omega+\mathbf{R}\sigma\subset\Lambda$, then $\mathbf{R}\sigma\subset -(\Lambda-\omega)\cap(\Lambda-\omega)$.

8.4.10. We say that a covariant representation (ρ, v, H) of a C^*-dynamical system $(A,\mathbf{R}^n,\alpha)$ satisfies a *spectrum condition* if $\mathrm{Sp}(\alpha)$ is contained in a convex set Λ satisfying 8.4.9.

If (A, G, α) is a C^*-dynamical a-system where G is connected, we say that a covariant representation (ρ, v, H) satisfies a *spectrum condition* if both $v|G_1$ and $v|G_2$ satisfy spectrum conditions (as defined above and in 8.4.6), where G_1 is the Euclidean subgroup of G and G_2 is the compact subgroup.

8.4.11. COROLLARY. *Let (A, G, α) be a C^*-dynamical a-system with G connected. For every covariant representation (ρ, v, H) of (A, G, α) satisfying a spectrum condition there is a σ-weakly continuous unitary representation $t\to u_t$ of G into $\rho(A)''$ such that (ρ, u, H) is covariant for (A, G, α).*

Proof. As in the proof of 8.4.7 we may assume that $\mathrm{Sp}(v)\subset\Gamma_+$, for some admissible positive cone Γ_+ in Γ obtained by 8.4.8. Then 8.3.11 applies and completes the proof.

8.4.12. COROLLARY. *Let $\mathcal{M}$ be a von Neumann algebra on a Hilbert space H and let (v, H) be a unitary representation of an abelian connected group G, satisfying a spectrum condition. If $v_t\mathcal{M}v_{-t} = \mathcal{M}$ for all t in G there is a unitary representation (u, H) of G such that $u_t\in\mathcal{M}$ and $v_tu_{-t}\in\mathcal{M}'$ for every t in G.*

Proof. Combine 7.5.1 and 8.4.11.

8.4.13. PROPOSITION. *Let $\mathcal{M}$ be a von Neumann algebra on a Hilbert space H and let (v, H) be a unitary representation of an abelian connected group G such that $\mathrm{Sp}(v) \subset \Gamma_+$ where Γ_+ is an admissible cone. If $v_t \mathcal{M} v_{-t} = \mathcal{M}$ for all t in G and if ζ_0 is a cyclic vector for $\mathcal{M}$ such that $v_t \xi_0 = \xi_0$ for all t then $v_G \subset \mathcal{M}$.*

Proof. Applying 8.3.11 to the system $(\mathcal{M}^c, G, \alpha)$, where $\alpha_t(x) = v_t x v_{-t}$, we obtain a unitary representation (u, H) of G with $\mathrm{Sp}(u) \subset \Gamma_+$ and $u_G \subset \mathcal{M}$ such that u and v commute. Moreover, if $w_t = v_t u_{-t}$ then (w, H) is a unitary representation of G with $\mathrm{Sp}(w) \subset \Gamma_+$ and $w_G \subset \mathcal{M}'$.

Take f in $K^1(G)$ and consider the integral

$$\int w_t \xi_0 f(t)\, dt = \int u_{-t} \xi_0 f(t)\, dt.$$

Since $\mathrm{Sp}(w) \subset \Gamma_+$ the integral is zero whenever $\mathrm{supp}.f \subset \Gamma \backslash \Gamma_+$. Thus $\xi_0 \in M^w(\Gamma_+)$. However, we also have $\mathrm{Sp}(u) \subset \Gamma_+$ so the integral is zero whenever $\mathrm{supp}.f \subset \Gamma \backslash (-\Gamma_+)$. Thus $\xi_0 \in M^w(-\Gamma_+)$. By 8.1.4 (iv) $\xi_0 \in M^w(\{\theta\})$, whence ξ_0 is a fixed-point for w by 8.1.8. It follows that for each t in G and each x in $\mathcal{M}$

$$w_t x \xi_0 = x w_t \xi_0 = x \xi_0.$$

Since ξ_0 is cyclic for $\mathcal{M}$ we conclude that $w_t = 1$ for all t, whence $u = v$ as desired.

8.4.14. Notes and remarks. Theorem 8.4.3. was found independently by Arveson and Borchers, cf. [13] and [23]. Theorem 8.4.4 appeared in incorrect form in [207]. The present version is inspired by Kraus [301] who treats this problem in detail. Proposition 8.4.8 is an observation of Olesen from [183]. The corollaries 8.4.11 and 8.4.12 were proved by Borchers in [20] and [19] for the case $G = \mathbf{R}^n$. Proposition 8.4.13 is an earlier result by Araki [10], see also [131].

8.5 Uniformly continuous representations

8.5.1. The uniformly continuous representations of an abelian group are the easiest to handle and we have detailed information about them. However, in order to bring the uniform topology effectively into play (as being different from the usual pointwise-norm topology) it is necessary to exclude discrete groups. This is best done by demanding that the group is connected. We propose to study discrete groups (i.e. single automorphisms) in 8.9.

8.5.2. THEOREM. *Let (A, G, α) be a C*-dynamical a-system with G connected, and assume that α is uniformly continuous. There exists a uniformly continuous unitary representation $t \to u_t$ of G into A'' such that*

$$u_t x u_t^* = \alpha_t(x), \qquad \forall t \in G, \qquad \forall x \in A.$$

Proof. We know from 8.1.12 that $\mathrm{Sp}(\alpha)$ is compact. Since $G = \boldsymbol{R}^n \times H$ where H is compact and connected, there is by 8.4.4 and 8.4.8 a positive cone Γ_+ in Γ which is admissible for (A, G, α). Since $\Gamma = \cup(\tau - \Gamma_+), \tau \in \Gamma_+$, we conclude that $\mathrm{Sp}(\alpha) \subset \tau_\infty - \Gamma_+$ for some τ_∞ in Γ_+, whence $p(\tau_\infty - \Gamma_+) = 1$ by 8.3.6 (v). Consequently $p(\infty) = 1$ and by 8.3.11 there is a unitary representation $t \to u_t$ of G into A'' such that $u_t x u_t^* = \alpha_t(x)$ for all t in G and x in A. Moreover, $\mathrm{Sp}(u) \subset \Gamma_+$. However, from the construction of the spectral measure μ determining u (see 8.3.10) we have $\mu(\tau_\infty - \Gamma_+) = p(\tau_\infty - \Gamma_+) = 1$; whence

$$\mathrm{Sp}(u) = \mathrm{supp}.\,\mu \subset \Gamma_+ \bigcap (\tau_\infty - \Gamma_+).$$

It follows that $\mathrm{Sp}(u \mid \boldsymbol{R}^n)$ is compact, whence $u \mid \boldsymbol{R}^n$ is uniformly continuous. For the compact subgroup H of G we argue more directly: Since $\mathrm{Sp}(\alpha \mid H)$ is finite in $\hat{H}$, the spectral measure μ arising from the projections $p(\tau - \Gamma_+)$ is finite-valued on H and thus $u \mid H$ is a finite sum of uniformly continuous functions, therefore itself uniformly continuous.

8.5.3. COROLLARY. *Let $(\mathcal{M}, G, \alpha)$ be a W^*-dynamical a-system with G connected, and assume that α is uniformly continuous. There exists a uniformly continuous unitary representation $t \to u_t$ of G into $\mathcal{M}$ such that*

$$u_t x u_t^* = \alpha_t(x), \qquad \forall t \in G, \qquad \forall x \in \mathcal{M}.$$

8.5.4. COROLLARY. *Let $(\mathscr{A}, G, \alpha)$ be a B^*-dynamical a-system where $\mathscr{A}$ is countably generated, G is connected and α is uniformly continuous. There is then a uniformly continuous, unitary representation $t \to u_t$ of G into $\mathscr{A}$ such that*

$$u_t x u_t^* = \alpha_t(x), \qquad \forall t \in G, \qquad \forall x \in \mathscr{A}.$$

Proof. There is a separable C^*-dynamical a-system (A, G, α) with enveloping system $(\mathscr{B}, G, \alpha)$, in which α is still uniformly continuous, such that $\mathscr{A} = \pi(\mathscr{B})$ for some α-invariant σ-normal morphism π of $\mathscr{B}$ (cf. 7.5). Thus it suffices to prove the result for the system $(\mathscr{B}, G, \alpha)$.

By 8.5.2 we have a representation $t \to u_t$ of G into A'' implementing α on $\mathscr{B}$. Moreover, $u_t = \int (t, \tau)\, d\mu(\tau)$ for a spectral measure μ on Γ. Since $\mu(\tau - \Gamma_+)$ belongs to $\mathscr{B}$ for all τ in Γ by 8.3.6 (i), we conclude that $\mu(\Lambda) \in \mathscr{B}$ for every Borel set Λ in Γ; whence $u_t \in \mathscr{B}$, as desired.

8.5.5. PROPOSITION. *In a dynamical system $(A, \boldsymbol{R}, \alpha)$ in which α is uniformly continuous, there is a unique element h in $((A_+)^m)^-$ such that $\exp(ith) = u_t$ for every t in $\boldsymbol{R}$ where u is the minimal positive representation of $\boldsymbol{R}$. As a consequence, if (ρ, v, H) is a covariant representation of $(A, \boldsymbol{R}, \alpha)$ such that $v_t = \exp(itk)$ for some positive operator k in $\boldsymbol{B}(H)$ then $\rho''(h) \leqslant k$.*

Proof. We know from 8.5.2 that there is a spectral measure μ on a compact subset of $\hat{\boldsymbol{R}}_+$ such that if $u_t = \int (t, \tau)\, d\mu(\tau)$, then u is the minimal positive representation of $\boldsymbol{R}$. Put $h = \int \tau \, d\mu(\tau)$. Then $h \in A''_+$, and

$$\exp(ith) = \int \exp(it\tau) \, d\mu(\tau) = u_t.$$

Given $\varepsilon > 0$, let $0 = \tau_0 < \tau_1 < \cdots < \tau_n$ be a partition of an interval $[0, \tau_n]$ in $\hat{\boldsymbol{R}}$ such that $\mu([0, \tau_n]) = 1$ and $\tau_k - \tau_{k-1} < \varepsilon$ for all k. Put

$$h_\varepsilon = \sum_{k=1}^{n} \tau_{k-1} \mu(]\tau_{k-1}, \tau_k]) = \sum_{k=1}^{n} (\tau_k - \tau_{k-1}) \mu(]\tau_k, \infty[).$$

From the first expression we see that $0 \leqslant h - h_\varepsilon \leqslant \varepsilon$. From the second we conclude that $h_\varepsilon \in (A_+)^m$, since

$$\mu(]\tau_k, \infty[) = 1 - p(\tau_k - \boldsymbol{R}_+) \in (A_+)^m$$

by 8.3.6 (i) and 8.3.10. It follows that $h \in ((A_+)^m)^-$.

If (ρ, v, H) is a covariant representation with $v_t = \exp(itk)$ for some k in $\boldsymbol{B}(H)_+$ then

$$\mathrm{Sp}(v) = \mathrm{Sp}(k) \subset \boldsymbol{R}_+.$$

It follows from 8.3.11 (last part of the proof) that

$$\rho''(\mu(\tau - \boldsymbol{R}_+)) \geqslant q[0, \tau]$$

for all $\tau \geqslant 0$, where $q[0, \tau]$ denotes the spectral projection of k corresponding to the interval $[0, \tau]$. Consequently $\rho''(h) \leqslant k$.

8.5.6. Let (A, G, α) be a C^*-dynamical a-system and assume that the dual group Γ of G is ordered by a maximal positive cone Γ_+. If α is uniformly continuous then each representation (π, H) of A is G-invariant by 8.5.2 and $\pi\alpha$ is again uniformly continuous (where $(\pi\alpha)_t(x) = \pi(\alpha_t(x))$). Actually,

$$R^{\pi\alpha}(\Omega) = \pi(R^\alpha(\Omega))$$

for every open set Ω in Γ, whence $\mathrm{Sp}(\pi\alpha) \subset \mathrm{Sp}(\alpha)$. There is therefore a largest element τ_π in the compact set $\mathrm{Sp}(\pi\alpha)$, whence $-\tau_\pi$ is the smallest element by 8.3.4. Since τ_π depends only on the equivalence class of (π, H) we can define a function $|\alpha| : \hat{A} \to \Gamma_+$ by

$$|\alpha|(\dot{\pi}) = \tau_\pi, \qquad \dot{\pi} \in \hat{A}, \qquad (\pi, H) \in \dot{\pi}.$$

8.5.7. LEMMA. *Let (A, G, α), Γ_+ and $|\alpha|$ be as in 8.5.6, and define $\tilde{\alpha}$ by $\tilde{\alpha}_t = \alpha_{-t}$. If u and $\tilde{u}$ are the minimal positive representations of G relative to (A, G, α) and $(A, G, \tilde{\alpha})$, respectively, then for each irreducible representation (π, H) of A*

$$\pi''(u_t \tilde{u}_t) = (t, |\alpha|(\dot{\pi})) 1_H, \qquad t \in G.$$

Proof. Put $\tau_0 = |\alpha|(\dot{\pi})$ and consider the unitary representation w on H given by $w_t = (t, \tau_0)\pi''(u_{-t})$. From the definition of $|\alpha|$ it follows that $\mathrm{Sp}(w) \subset \Gamma_+$.

Since (π, w, H) is covariant for $(A, G, \tilde{\alpha})$ and $\tilde{u}$ is the minimal positive representation, it follows from 8.3.11 that $\mathrm{Sp}(w\pi''(\tilde{u}^*)) \subset \Gamma_+$. However, $w_t\pi''(\tilde{u}_t^*)(=(t, \tau_0)\pi''(u_{-t}\tilde{u}_{-t}))$ is a multiple of the identity, since it commutes with $\pi(A)$, whence

$$w_t = (t, \tau_0)\pi''(u_{-t}) = (t, \tau)\pi''(\tilde{u}_t)$$

for all t in G and some τ in Γ_+. By construction $\theta \in \mathrm{Sp}(w)$, and since $\mathrm{Sp}(\tilde{u}) \subset \Gamma_+$ it follows that $\tau = \theta$. Thus $\pi''(u_t\tilde{u}_t) = (t, \tau_0)1_H$, as desired.

8.5.8. THEOREM. *Let (A, G, α) be a C^*-dynamical a-system such that α is uniformly continuous and the dual group of G is ordered. If the function $|\alpha|$ defined in 8.5.6 is continuous, then the minimal positive representation u maps G into the multiplier algebra $M(A)$ of A.*

Proof. It will be convenient to consider separately the cases $G = \boldsymbol{R}$ and G compact. If G is a product of such groups we apply 8.4.8.

If $G = \boldsymbol{R}$ the function $|\alpha|$ corresponds to a positive central element z in $M(A)$ by 4.4.8. Applying 8.5.5 to the systems $(A, \boldsymbol{R}, \alpha)$ and $(A, \boldsymbol{R}, \tilde{\alpha})$ we obtain elements h and $\tilde{h}$ in $((A_+)^m)^-$ such that $u_t = \exp(ith)$, $\tilde{u}_t = \exp(it\tilde{h})$ where u and $\tilde{u}$ are the minimal positive representations of $\boldsymbol{R}$. By 8.5.7 we have $\pi''(u_t\tilde{u}_t) = \pi''(\exp(itz))$ for every irreducible representation (π, H) of A, whence $u_t\tilde{u}_t = \exp(itz)$ by 4.3.15 for all t. Differentiating at $t = 0$ we obtain $h + \tilde{h} = z$. Since $M(A)_{sa} = (\tilde{A}_{sa})^m \cap (\tilde{A}_{sa})_m$ by 3.12.9, and $((A_{sa})^m)^- \subset (\tilde{A}_{sa})^m$ by 3.11.7, we conclude that

$$(\tilde{A}_{sa})^m \ni h = z - \tilde{h} \in (\tilde{A}_{sa})_m.$$

It follows that $h \in M(A)$ whence $u_t \in M(A)$ for all t.

If G is compact, $\hat{A}$ is the disjoint union of open and closed sets on which $|\alpha|$ is constant. These sets correspond to mutually orthogonal ideals in A, and since every closed ideal in A is G-invariant by 8.5.2 it suffices to consider each of them separately. Thus we may assume that $|\alpha|$ is constant. We have

$$\mathrm{Sp}(\alpha) = \{-\tau_n, \ldots, -\tau_1, \theta, \tau_1, \ldots, \tau_n\}$$

and thus the minimal positive representation u of G has the form $u_t = \Sigma_0^n(t, \tau_k)p_k$, where $p_k = \mu(\{\tau_k\})$ and $0 \leqslant k \leqslant n$. Similarly $\tilde{u}_t = \Sigma_0^m(t, \tilde{\tau}_j)\tilde{p}_j$, where $\tilde{u}$ is the minimal positive representation of G corresponding to $(A, G, \tilde{\alpha})$. By 8.5.7 we have $\tilde{u}_t u_t = (t, \tau_n)1$, since this equation holds for every irreducible representation. Consequently

$$\sum_0^m (t, \tilde{\tau}_j)\tilde{p}_j = \sum_0^n (t, \tau_n - \tau_k)p_k,$$

whence $m = n$, $\tilde{p}_{n-k} = p_k$ and $\tilde{\tau}_{n-k} = \tau_n - \tau_k$ for $0 \leqslant k \leqslant n$. Each projection $\Sigma_{j=0}^k p_j$ and $\Sigma_{j=0}^k \tilde{p}_j$ is closed (cf. 8.3.10 and 8.3.6 (i)); but from the equations above it follows that one is the complement of each other, and consequently open. As before we apply 3.12.9 to conclude that $p_k \in M(A)$ for $1 \leqslant k \leqslant n$, whence $u_t \in M(A)$ for all t in G.

8.5.9. COROLLARY. *Let (A, G, α) be a C*-dynamical a-system. Assume that A is simple, G is connected, and α is uniformly continuous. Then there is a uniformly continuous unitary representation u of G into $M(A)$ such that*

$$u_t x u_t^* = \alpha_t(x), \qquad \forall t \in G, \qquad x \in A.$$

Proof. Write G as a product of groups each of whose dual group can be ordered and apply 8.5.8 to each. As we saw in 8.4.8 the minimal positive representations of subgroups commute, and therefore can be combined to a representation of G.

8.5.10. COROLLARY. *In a dynamical system $(A, \boldsymbol{R}, \alpha)$ where α is uniformly continuous, the norm of the minimal positive generator h defined in 8.5.5 satisfies*

$$\mathrm{Conv}(\mathrm{Sp}(\alpha)) = [-\|h\|, \|h\|].$$

Proof. In the proof of 8.5.8 we showed that $h + \tilde{h} = z$ where $\pi''(z) = |\alpha|(\dot{\pi})1_H$ for each irreducible representation (π, H). Since $\tilde{h} \geqslant 0$ this implies that $\|h\| \leqslant \sup |\alpha|(\dot{\pi})$, whence $\|h\| \in \mathrm{Conv}(\mathrm{Sp}(\alpha))$. On the other hand, since $\pi''(h) + \pi''(\tilde{h}) = |\alpha|(\dot{\pi})1_H$ we conclude from spectral theory that $\|\pi''(h)\| = \|\pi''(\tilde{h})\| = |\alpha|(\dot{\pi})$, whence

$$\mathrm{Sp}(\alpha) \subset [-\|h\|, \|h\|].$$

8.5.11. Notes and remarks. Theorem 8.5.2 was proved by Olesen in [181] and previously by Moffat in [167] with a non-constructive proof. Proposition 8.5.5 is due to Arveson in the von Neumann algebra case (even with a version which does not require α to be uniformly continuous, see [13]). The semi-continuity of the generator h in 8.5.5 was established by Olesen and the present author in [184]. Corollary 8.5.9 was proved by Olesen in [181].

8.6. Derivations

8.6.1. A *derivation* of a C*-algebra is a linear map $\delta: A \to A$ such that

$$\delta(xy) = x\delta(y) + \delta(x)y$$

for all x and y in A. We say that δ is a **-derivation* if moreover $\delta(A_{sa}) \subset A_{sa}$. Note that if δ is a derivation then so is the operator δ^* defined by $\delta^*(x) = \delta(x^*)^*$ (cf. 3.1.1). Since

$$\delta = \tfrac{1}{2}(\delta + \delta^*) - i\tfrac{1}{2}i(\delta - \delta^*)$$

we see that every derivation is a (unique) combination of *-derivations.

If $d \in A$ we define a derivation $\operatorname{ad} d$ on A by $(\operatorname{ad} d)(x) = dx - xd$. If u is invertible in A we define an algebra automorphism $\operatorname{Ad} u$ on A by $\operatorname{Ad} u(x) = uxu^{-1}$. We say that $\operatorname{ad} d$ is an *inner derivation* and that $\operatorname{Ad} u$ is an *inner automorphism*. Note that $\operatorname{ad} d$ is a *-derivation if $d \in iA_{sa}$ and that $\operatorname{Ad} u$ is a *-automorphism if u is unitary.

Define operators $\lambda(d)$ and $\rho(d)$ on A by $\lambda(d)x = dx$ and $\rho(d)x = xd$. Then $\operatorname{ad} d = \lambda(d) - \rho(d)$ and if u is invertible $\operatorname{Ad} u = \lambda(u)\rho(u^{-1})$. Since $\lambda(d)$ and $\rho(d)$ are commuting operators on A we see from Banach algebra theory that

$$\begin{aligned}\exp(\operatorname{ad} d) &= \exp(\lambda(d) - \rho(d)) = \exp(\lambda(d))\exp(-\rho(d)) \\ &= \lambda(\exp(d))\rho(\exp(-d)) = \operatorname{Ad}(\exp(d)).\end{aligned}$$

8.6.2. LEMMA. *If δ is a *-derivation on A then $\phi(\delta(x)) = 0$ for every pair (x, ϕ) in $A_+ \times A_+^*$ such that $\phi(x) = \|x\|\,\|\phi\|$.*

Proof. If $1 \in A$ then $\delta(1) = 1\delta(1) + \delta(1)1 = 2\delta(1)$, whence $\delta(1) = 0$. If $1 \notin A$ we extend δ to a derivation $\tilde{\delta}$ of $\tilde{A}$ by defining $\tilde{\delta}(\lambda 1 + x) = \delta(x)$ for every λ in $\boldsymbol{C}$ and x in A. Thus we may assume that $1 \in A$.

It suffices to consider the case where $\|x\| = \|\phi\| = \phi(x) = 1$. Set $y = (1 - x)^{1/2}$. Since $\delta(1) = 0$ we get

$$\begin{aligned}|\phi(\delta(x))|^2 &= |\phi(\delta(y^2))|^2 = |\phi(y\delta(y) + \delta(y)y)|^2 \\ &\leqslant 4|\phi(y\delta(y))|^2 \leqslant 4\phi(y^2)\phi(\delta(y)^2) = 0,\end{aligned}$$

because $\phi(1) = \phi(x)(=1)$.

8.6.3. PROPOSITION. *Every derivation of a C*-algebra is bounded.*

Proof. It suffices to consider a *-derivation δ on A, and we may assume $1 \in A$ (cf. the proof of 8.6.2). If δ is unbounded, $\delta \mid A_{sa}$ is unbounded; and then by the closed graph theorem there is a sequence $\{x_n\}$ in A_{sa} such that $x_n \to 0$ but $\delta(x_n) \to y$, where $y \in A_{sa} \backslash \{0\}$. Multiplying with a suitable scalar we may assume that $\|y\| = 1$ and that $1 \in \operatorname{Sp}(y)$. Replacing x_n with $x_n + \|x_n\| 1$ (a harmless operation since $\delta(1) = 0$) we may assume that $\{x_n\} \subset A_+$.

For each n choose a state ϕ_n of A such that $\phi_n(y_+ + x_n) = \|y_+ + x_n\|$ (cf. 4.3.10). Then $\phi_n(\delta(y_+ + x_n)) = 0$ by 8.6.2. Since the state space is weak* compact we can find a state ϕ which is a limit point of $\{\phi_n\}$. As $x_n \to 0$ and $\delta(x_n) \to y$ in norm we conclude that

$$\phi(y_+) = \|y_+\| \qquad \text{and} \qquad \phi(\delta(y_+) + y) = 0.$$

But then $\phi(\delta(y_+)) = 0$ by 8.6.2, whence $\phi(y) = 0$. This is impossible since $\phi(y_+) = 1(=\|y_+\|)$, and ϕ is a state. Consequently δ is bounded.

8.6.4. PROPOSITION. *If δ is a *-derivation of a C*-algebra A, define $\alpha_t = \exp(t\delta)$ for each t in $\boldsymbol{R}$. Then $t \to \alpha_t$ is a uniformly continuous one-parameter group of automorphisms of A and* $\mathrm{Sp}(\delta) = \mathrm{i}\,\mathrm{Sp}(\alpha)$.

Proof. Since δ is bounded by 8.6.3 we have a convergent series expression

$$(*) \qquad \alpha_t = \sum_{n=0}^{\infty} \frac{t^n}{n!}\delta^n.$$

We have $\exp((s+t)\delta) = \exp(s\delta)\exp(t\delta)$ so that $t \to \alpha_t$ is a uniformly continuous representation of $\boldsymbol{R}$ as bounded, invertible, *-preserving operators on A.

Take x and y in A. A standard induction argument shows that δ satisfies Leibniz' formulae; i.e. for each n we have

$$(**) \qquad \delta^n(xy) = \sum_{k=0}^{n} \binom{n}{k} \delta^k(x)\delta^{n-k}(y).$$

Using $(*)$ and $(**)$ we obtain

$$\begin{aligned}(\exp\delta)(x)(\exp\delta)(y) &= \sum_{k=0}^{\infty}\sum_{m=0}^{\infty} \frac{1}{k!}\frac{1}{m!}\delta^k(x)\delta^m(y)\\ &= \sum_{n=0}^{\infty}\sum_{k=0}^{n} \frac{1}{k!}\frac{1}{(n-k)!}\delta^k(x)\delta^{n-k}(y)\\ &= \sum_{n=0}^{\infty} \frac{1}{n!}\delta^n(xy) = (\exp\delta)(xy).\end{aligned}$$

Consequently every $\alpha_t, t \in \boldsymbol{R}$, is an automorphism of A.

Let $\mathfrak{A}$ denote the Banach algebra in $\boldsymbol{B}(A)$ generated by δ and ι. Then $\mathfrak{A}$ is also generated by the family $\{\alpha_t \mid t \in \boldsymbol{R}\}$ and, since $\mathrm{Sp}(\alpha)$ is compact, $\mathfrak{A}$ is generated by the family $\{\alpha_f \mid f \in L^1(\boldsymbol{R})\}$ as well. It follows from 8.1.10 that the three sets $\mathrm{Sp}(\delta)$, $\hat{\mathfrak{A}}$ and $\mathrm{Sp}(\alpha)$ are homeomorphic and that the map $\lambda \to \mathrm{i}\lambda$ implements the homeomorphism between $\mathrm{Sp}(\alpha)$ and $\mathrm{Sp}(\delta)$.

8.6.5. THEOREM. *Let δ be a *-derivation of a C*-algebra A. Then the spectral radius of δ equals $\|\delta\|$ and there exists an element h in $((A_+)^m)^-$ such that $\delta = \mathrm{ad}(\mathrm{i}h)$ and $\|\delta\| = \|h\|$. Moreover, if k is any positive operator such that $\delta = \mathrm{ad}(\mathrm{i}k)$ then $h \leqslant k$.*

Proof. Put $\alpha_t = \exp(t\delta), t \in \boldsymbol{R}$. By 8.6.4, α is a uniformly continuous representation of $\boldsymbol{R}$ into $\mathrm{Aut}(A)$. Let u be the minimal positive representation of $\boldsymbol{R}$ (cf. 8.5.2) and use 8.5.5 to obtain h in $((A_+)^m)^-$ such that $u_t = \exp(\mathrm{i}th)$ for all t. For each x in A we have

$$\begin{aligned}\delta(x) &= \operatorname{Lim} t^{-1}(\alpha_t - \iota)(x)\\ &= \operatorname{Lim} t^{-1}(\exp(ith)x - x\exp(ith))\exp(-ith)\\ &= ihx - xih = \operatorname{ad}(ih)(x).\end{aligned}$$

Since $\operatorname{Sp}(\delta) = \mathrm{i}\operatorname{Sp}(\alpha)$ by 8.6.4 we conclude from 8.5.10 that the spectral radius of δ equals $\|h\|$. However, $\delta = \operatorname{ad}(\mathrm{i}(h - \frac{1}{2}\|h\|))$ whence

$$\|\delta(x)\| \leqslant 2\|h - \tfrac{1}{2}\|h\|\|\,\|x\| \leqslant \|h\|\,\|x\|$$

for every x in A. Consequently $\|\delta\| \leqslant \|h\|$, which proves the first half of the theorem.

If $A \subset \boldsymbol{B}(H)$ for some Hilbert space H and if $k \in \boldsymbol{B}(H)_+$ such that $\delta = \operatorname{ad}(\mathrm{i}k)$, then we define $\lambda(k)$ and $\rho(k)$ on $\boldsymbol{B}(\boldsymbol{B}(H))$ by $\lambda(k)x = kx$, $\rho(k)x = xk$. These two operators commute and $\operatorname{ad}(\mathrm{i}k) = \lambda(\mathrm{i}k) + \rho(-\mathrm{i}k)$. Consequently

$$\begin{aligned}\alpha_t &= \exp(\operatorname{ad}(\mathrm{i}tk)) = \exp(\lambda(\mathrm{i}tk) + \rho(-\mathrm{i}tk))\\ &= \exp(\lambda(\mathrm{i}tk))\exp(\rho(-\mathrm{i}tk)) = \lambda(\exp(\mathrm{i}tk))\rho(\exp(-\mathrm{i}tk)).\end{aligned}$$

It follows that the unitary group given by $v_t = \exp(\mathrm{i}tk)$ implements α, whence $h \leqslant k$ by 8.5.5.

8.6.6. COROLLARY. *Every derivation δ of a von Neumann algebra $\mathcal{M}$ is inner. Moreover, if $\delta = \delta^*$ there exists an h in $\mathcal{M}_+$ with $\|h\| = \|\delta\|$ such that $\delta = \operatorname{ad}(\mathrm{i}h)$; and $\alpha(h) = h$ for every automorphism α of $\mathcal{M}$ such that $\alpha \circ \delta = \delta \circ \alpha$.*

Proof. If $\mathcal{M} \subset \boldsymbol{B}(H)$ apply 8.6.5 to the identical representation $\pi: \mathcal{M} \to \boldsymbol{B}(H)$ and put $h = \pi''(h_0)$, where h_0 is the minimal positive generator for δ in $(\mathcal{M}_+)^m$. (In fact $h = h_0$). If $\alpha\circ\delta = \delta\circ\alpha$ then $\delta = \operatorname{ad}(\mathrm{i}\alpha(h))$, whence $h \leqslant \alpha(h)$ by 8.6.5. Replacing α by α^{-1} we conclude that $h = \alpha(h)$.

8.6.7. From the definition of a derivation δ it is evident that $\delta(I) \subset I$ for every ideal I of A. It follows that for each representation (π, H) of A we can define a derivation $\pi\delta$ on $\pi(A)$ by $\pi\delta(x) = \pi(\delta(x))$. The norm of this derivation (as a bounded operator on $\pi(A)$) depends only on the equivalence class of (π, H). We can therefore define a (bounded) function $|\delta|: \hat{A} \to \boldsymbol{R}_+$ by

$$|\delta|(\dot{\pi}) = \|\pi\delta\|, \qquad (\pi, H) \in \dot{\pi}, \qquad \dot{\pi} \in \hat{A}.$$

8.6.8. LEMMA. *Let δ be a *-derivation of a C^*-algebra A, put $\alpha_t = \exp(t\delta)$ and write $\delta = \operatorname{ad}(\mathrm{i}h)$ with h in 8.6.5. With $|\delta|$ as defined in 8.6.7 and $|\alpha|$ as in 8.5.6 we have*

$$|\alpha|(\dot{\pi}) = \|\pi(h)\| = |\delta|(\dot{\pi})$$

for every $\dot{\pi}$ *in* $\hat{A}$ *and* $(\pi, H) \in \dot{\pi}$.

Proof. It follows from the definition of h in 8.5.5 that $|\alpha|(\dot{\pi}) = \|\pi(h)\|$. Furthermore, applying 8.6.4 to the derivation $\pi\delta$ on $\pi(A)$ we see that $|\alpha|(\dot{\pi})$ equals the spectral radius of $\pi\delta$, which in turn equals $|\alpha|(\dot{\pi})$ by 8.6.5.

8.6.9. THEOREM. *Let* δ *be a *-derivation of a* C^**-algebra* A. *If the function* $|\delta|$ *defined in 8.6.7 is continuous on* $\hat{A}$ *then* δ *is inner in* $M(A)$.

Proof. This follows immediately from 8.6.8 and 8.5.8.

8.6.10. COROLLARY. *Every derivation* δ *of a simple* C^**-algebra* A *is inner in* $M(A)$. *Moreover, if* $\delta = \delta^*$ *there exists an* h *in* $M(A)_+$ *with* $\|h\| = \|\delta\|$ *such that* $\delta = \mathrm{ad}(ih)$; *and* $\alpha(h) = h$ *for every automorphism* α *of* A *such that* $\alpha \circ \delta = \delta \circ \alpha$.

8.6.11. PROPOSITION. *Every derivation of a* C^**-algebra* A *with continuous trace is inner in* $M(A)$, *provided that* $\hat{A}$ *is paracompact; a condition that is automatically fulfilled if* A *has a countable approximate unit, and, a fortiori, when* A *is separable.*

Proof. We know from 6.1.11 that $\hat{A}$ is locally compact Hausdorff space. If A has a countable approximate unit, then it has a strictly positive element (3.10.5), and consequently $\hat{A}$ is σ-compact by 4.4.4. But a σ-compact, locally compact Hausdorff space is paracompact (and normal).

Let $\pi_a = \oplus_{t \in \hat{A}} \pi_t$ denote the atomic representation of A on $H_a = \oplus_{t \in \hat{A}} H_t$ (cf. 4.3.7). For each t in $\hat{A}$ there is by 6.1.11 an abelian element x in A such that $\pi_s(x)$ is a one-demensional projection for all s in some neighbourhood E of t. Since $\hat{A}$ is paracompact there is a locally finite covering $\{E_i\}$ of $\hat{A}$ with such sets, and we denote by x_i the abelian element corresponding to E_i. Choose a partition of unity $\{f_i\}$ subordinate to the covering $\{E_i\}$ (i.e. $f_i \in C_0(E_i)_+$ and $\Sigma f_i = 1$); and denote by I_i the closed ideal of A generated by $f_i x_i$ (note that $f_i x_i \in A$ by 4.4.8). Since $\hat{x}_i = 1$ on E_i, we see that $\hat{I}_i = \{t \in \hat{A} \mid f_i(t) > 0\}$; which implies that $f_i x \in I_i$ for every x in A.

Now let δ be a derivation of A and assume (by 8.6.5) that $\delta = \mathrm{ad}\, k$ for some k in A''. Let $\delta_t = \mathrm{ad}\, \pi_t(k)$ on $B(H_t)$. Changing if necessary $\pi_t(k)$ with a scalar multiple of the identity we can arrange for each t in a given E_i that $\delta_t = \mathrm{ad}\, h_t$ where h_t is the (unique) element in $B(H_t)$ for which $\pi_t(x_i) h_t \pi_t(x_i) = 0$. Since $\|h_t\| \leqslant 2\|k\|$ for all t in E_i we can define h_i in $B(H_a)$ by $h_i = h_t$ on H_t if $t \in E_i$, $h_i = 0$ on H_t otherwise. For every x in A and t in E_i we have

$$\pi_t(\delta(x)) = \delta_t(\pi_t(x)) = \pi_t''(\mathrm{ad}\, h_i(x)),$$

which implies that $\delta = \mathrm{ad}\, h_i$ on I_i. Furthermore,

$$h_i(f_i x_i) = f_i h_i x_i^2 = f_i(h_i x_i - x_i h_i + x_i h_i) x_i = \delta(f_i x_i) \in I_i$$

since $x_i h_i x_i = 0$. Since h_i derives I_i the set

$$J_i = \{x \in I_i \mid h_i x \in I_i\}$$

is a two-sided ideal in I_i. We have just shown that $f_i x_i \in J_i$, whence $J_i = I_i$.

Define $h = \Sigma f_i h_i$. Since the covering $\{E_i\}$ is locally finite, $\pi_t''(h) \in B(H_t)$ for every t in $\hat{A}$. Moreover,

$$\|\pi_t''(h)\| \leqslant \sum f_i(t)\|\pi_t''(h_i)\| \leqslant \sum f_i(t)2\|k\| = 2\|k\|,$$

whence $h \in B(H_a)$. For every x in A we know that $f_i x \in I_i$ and that the sum $\Sigma f_i x$ is norm convergent to x (since $\check{x} \in C_0(\hat{A})$ by 4.4.4). Consequently

$$\operatorname{ad} h(x) = \Sigma \operatorname{ad} h_i(f_i x) = \Sigma f_i \delta(x) = \delta(x),$$

and

$$hx = \sum h_i(f_i x) \in A,$$

whence $h \in M(A)$ as desired.

8.6.12. LEMMA. *Let δ be a *-derivation of a separable C^*-algebra A with unit. There exists an increasing sequence $\{h_n\}$ in A_+ such that $h_n \nearrow h$, where $\delta = \operatorname{ad} ih$ as in 8.6.5, and such that*

$$\|\delta(h_n)\| \to 0 \qquad \text{and} \qquad \|\operatorname{ad} ih_n(x) - \delta(x)\| \to 0$$

for every x in A.

Proof. Since $h \in (A_+)^m$ there is an increasing sequence $\{k_n\}$ in A_+ such that $k_n \nearrow h$. Given $x_1, \ldots, x_m$ in A and $\varepsilon > 0$ let B denote the direct sum of A with itself $m+1$ times. The sequence $\{d_n\}$ in B with elements

$$(d_n)_0 = \delta(k_n), \qquad (d_n)_j = \operatorname{ad} ik_n(x_j) - \delta(x_j), \qquad 1 \leqslant j \leqslant m,$$

converges σ-weakly to zero in B''. In particular, the norm closed convex hull K of $\{d_n\}$ contains zero as a σ-weak limit point. By Hahn–Banach's theorem $0 \in K$. Consequently there is a convex combination $\Sigma \lambda_n d_n$ with norm less than ε. Taking $h_\varepsilon = \Sigma \lambda_n k_n$ this implies that

$$\|\delta(h_\varepsilon)\| < \varepsilon \qquad \text{and} \qquad \|\operatorname{ad} ih_\varepsilon(x_j) - \delta(x_j)\| < \varepsilon, \qquad 1 \leqslant j \leqslant m.$$

Let $\{x_j\}$ be a dense sequence in A. Assume that for each $m < p$ we have found a convex combination h_m of elements from $\{k_n\}$ such that

(i) $\|\delta(h_m)\| < 1/m$;

(ii) $\|\operatorname{ad} ih_m(x_j) - (x_j)\| < 1/m$ for $1 \leqslant j \leqslant m$;

(iii) $k_m \leqslant h_m$ and $h_{m-1} \leqslant h_m$.

Let i be the largest index occuring among the k_n's used in the expression for

h_{p-1}. Applying the first part of the proof to the sequence $\{k_n \mid n > \max\{i, p\}\}$, the set $x_1, \ldots, x_p$ and $\varepsilon = 1/p$ we find an element h_p satisfying (i) and (ii) for $m = p$. However h_p is a convex combination of elements each of which dominate both k_p and h_{p-1} (since they dominate every component in h_{p-1}). Consequently $k_p \leqslant h_p$ and $h_{p-1} \leqslant h_p$.

By induction (starting with $h_0 = 0$) we can construct a sequence $\{h_n\}$ in A_+ satisfying (i), (ii) and (iii). It is evident from (iii) that $h_n \nearrow h$. Moreover, $\|\delta(h_n)\| \to 0$ by (i). From (ii) we see that $\|\operatorname{ad} ih_n(x) - \delta(x)\| \to 0$ for a dense set of elements x in A. Since the sequence $\{\operatorname{ad} ih_n\}$ is bounded in $\boldsymbol{B}(A)$ this implies the convergence to zero for every x in A.

8.6.13. LEMMA. *Let a and b be operators on a Hilbert space H and consider the two analytic functions $\zeta \to \alpha_\zeta$ and $\zeta \to \beta_\zeta$ from $\boldsymbol{C}$ to $\boldsymbol{B}(\boldsymbol{B}(H))$ given by*

$$\alpha_\zeta = \exp(\operatorname{ad}(\zeta a)), \qquad \beta_\zeta = \exp(\operatorname{ad}(\zeta b)), \qquad \zeta \in \boldsymbol{C}.$$

For each x in $\boldsymbol{B}(H)$ and ζ in $\boldsymbol{C}$ we have

$$\|\alpha_\zeta(x) - \beta_\zeta(x)\| \leqslant \exp(2\|\zeta a\|)\exp(2\|\zeta b)(\|\operatorname{ad}(a-b)x\| + |\zeta|\,\|ab - ba\|\,\|x\|).$$

Proof. Put $\delta = \operatorname{ad}(a - b)$. The two functions $\zeta \to \alpha_\zeta - \beta_\zeta$ and $\zeta \to \alpha_\zeta \circ \beta_{-\zeta}$ and both analytic. Differentiating at $\zeta = 0$ we obtain

$$\zeta^{-1}(\alpha_\zeta - \beta_\zeta - 0) = \zeta^{-1}(\alpha_\zeta - \iota - (\beta_\zeta - \iota)) \to \delta;$$

$$\zeta^{-1}(\alpha_\zeta \circ \beta_{-\zeta} - \iota) = \zeta^{-1}(\alpha_\zeta - \beta_\zeta) \circ \beta_{-\zeta} \to \delta.$$

Using this we compute

$$\begin{aligned}
\alpha_\zeta - \beta_\zeta &= \int_0^1 \frac{\mathrm{d}}{\mathrm{d}s}(\alpha_{s\zeta} \circ \beta_{(1-s)\zeta})\Big|_{s=t}\,\mathrm{d}t \\
&= \int_0^1 \frac{\mathrm{d}}{\mathrm{d}s}(\alpha_{(s+t)\zeta} \circ \beta_{(1-s-t)\zeta})\Big|_{s=0}\,\mathrm{d}t \\
&= \int_0^1 \alpha_{t\zeta} \circ \frac{\mathrm{d}}{\mathrm{d}s}(\alpha_{s\zeta} \circ \beta_{-s\zeta})\Big|_{s=0} \circ \beta_{(1-t)\zeta}\,\mathrm{d}t \\
&= \int_0^1 \alpha_{t\zeta} \circ \delta \circ \beta_{(1-t)\zeta}\,\mathrm{d}t.
\end{aligned}$$

Note that by 8.6.1

$$\|\alpha_\zeta\| = \|\operatorname{Ad}(\exp(\zeta a))\| \leqslant \|\exp(\zeta a)\|\,\|\exp(-\zeta a)\| \leqslant \exp(2\|\zeta a\|).$$

Moreover,

$$\begin{aligned}
\|\delta(\exp(\zeta b))\| &= \|\textstyle\sum (1/n!)\delta((\zeta b)^n)\| \\
&\leqslant \textstyle\sum (1/n!) n \|\zeta b\|^{n-1}\, \|\delta(\zeta b)\| \\
&= \exp(\|\zeta b\|)\|\delta(\zeta b)\|.
\end{aligned}$$

Inserting this in the formula above we obtain for every x in $\boldsymbol{B}(H)$ with $\|x\| \leqslant 1$

$$
\begin{aligned}
\|\alpha_\zeta(x) - \beta_\zeta(x)\| &\leqslant \int_0^1 \|\alpha_{(1-t)\zeta}\| \, \|\delta(\beta_{t\zeta}(x))\| \, dt \\
&\leqslant \exp(2\|\zeta a\|) \int_0^1 \|\delta(\exp(t\zeta b)\, x \exp(-t\zeta b)\| \, dt \\
&\leqslant \exp(2\|\zeta a\|)\exp(2\|\zeta b\|) \int_0^1 (\|\delta(t\zeta b)\| + \|\delta(x)\| + \|\delta(-t\zeta b)\|)\, dt \\
&\leqslant \exp(2\|\zeta a\|)\exp(2\|\zeta b\|)(\|\delta(\zeta b)\| + \|\delta(x)\|) \\
&= \exp(2\|\zeta a\|)\exp(2\|\zeta b\|)(|\zeta|\,\|ab - ba\| + \|ad(a-b)x\|).
\end{aligned}
$$

8.6.14. PROPOSITION. *Let* $t \to \alpha_t$ *be a uniformly continuous one-parameter group of automorphisms of a separable C*-algebra A with unit. There exists a sequence* $\{t \to \alpha_t^n\}$ *of inner automorphism groups such that*

$$\|\alpha_t(x) - \alpha_t^n(x)\| \to 0$$

for every x in A, uniformly on compact subsets of $\boldsymbol{R}$.

Proof. We know (cf. 8.5.5) that $\alpha_t = \exp(t\delta)$ for some *-derivation δ of A. Thus 8.6.12 applies and the result follows from 8.6.13.

8.6.15. THEOREM. *Let* $\pi: B \to A$ *be a surjective morphism between separable C*-algebras B and A. If* δ *is a derivation of A there is a derivation* $\bar{\delta}$ *of B such that* $\pi \circ \bar{\delta} = \delta \circ \pi$. *Moreover, if* δ *is a *-derivation then* $\bar{\delta}$ *can be chosen as a *-derivation as well, with* $\|\bar{\delta}\| = \|\delta\|$.

Proof. We may assume that δ is a *-derivation with $\|\delta\| = 1$ and that A and B have units such that $\pi(1) = 1$. With $\delta = \operatorname{ad} ih$ as in 8.6.5 we have $\|h\| = \|\delta\|$. Applying 8.6.12 we find a sequence $\{h_n\}$ in A_+ such that $h_n \nearrow h$ and such that $\|\operatorname{ad}(h - h_n)(x_k)\| < 2^{-n}$ for all $k \leqslant n$, where $x_k = \pi(y_k)$ and $\{y_k\}$ is a dense sequence in B. By successive applications of 1.5.10 we can find an increasing sequence $\{k_n\}$ in B_+^1 such that $\pi(k_n) = h_n$ for every n. Put $h_0 = 0$ and $k_0 = 0$ and let $\{u_\lambda\}$ be an approximate unit for $\ker \pi$ which is quasi-central for B (see 3.12.14). By 1.5.4 we have for each $k \leqslant n$

$$
\begin{aligned}
\operatorname{Lim} \|(1 - u_\lambda)\operatorname{ad}(k_n - k_{n-1})(y_k)\| &= \|\pi(\operatorname{ad}(k_n - k_{n-1})(y_k))\| \\
&= \|\operatorname{ad}(h_n - h_{n-1})(x_k)\| < 2 \cdot 2^{-n}.
\end{aligned}
$$

Since $\{u_\lambda\}$ is quasi-central there is therefore a λ such that with

$$z_n = (k_n - k_{n-1})^{1/2}(1 - u_\lambda)(k_n - k_{n-1})^{1/2}$$

we have $\|(\operatorname{ad} z_n)(y_k)\| < 2^{1-n}$ for every $k \leqslant n$. Since $z_n \leqslant k_n - k_{n-1}$, the element $k = \Sigma z_n$ exists in B'' and belongs to $(B_+^1)^m$. We define the *-derivation

$\bar{\delta} = \operatorname{ad} ik$ on B''. For each y_k we have $\|\operatorname{ad} z_n(y_k)\| < 2^{-n+1}$ whenever $n \geqslant k$. It follows that

$$\bar{\delta}(y_k) = \sum \operatorname{ad} z_n(y_k) \in B,$$

and since $\{y_k\}$ is dense in B we conclude that $\bar{\delta}(B) \subset B$. Finally, for each y in B

$$\begin{aligned}\pi(\bar{\delta}(y)) &= \sum \pi(\operatorname{ad} z_n(y)) \\ &= \sum \operatorname{ad}(h_n - h_{n-1})(\pi(y)) = \delta(\pi(y));\end{aligned}$$

and the proof is completed by observing that $\|\bar{\delta}\| \leqslant \|k\| = 1$, whence $\|\bar{\delta}\| = \|\delta\|$.

8.6.16. COROLLARY. *Let $\pi : B \to A$ be a surjective morphism between separable C^*-algebras B and A. If $t \to \alpha_t$ is a uniformly continuous one-parameter group of automorphisms of A there is a uniformly continuous one-parameter group $t \to \bar{\alpha}_t$ of automorphisms of B such that $\pi \circ \bar{\alpha}_t = \alpha_t \circ \pi$ for all t. Moreover,* $\operatorname{Conv}(\operatorname{Sp}(\alpha)) = \operatorname{Conv}(\operatorname{Sp}(\bar{\alpha}))$.

8.6.17. Notes and remarks. Derivations of von Neumann algebras were studied by Kaplansky [144], who proved that if the algebra was of type I, the derivations were all inner. At the same time he conjectured 8.6.3 which was proved by Sakai [224]. The proof given here uses the dissipative character of derivations and is borrowed from Kishimoto's paper [149]. The outstanding problem whether derivations are inner on any von Neumann algebra was solved by Sakai [227] and by Kadison [131]. Both proofs were essentially non-constructive, but minimality conditions on the generators were established by Kadison, Lance and Ringrose [134]. Later it was realized by the gang of four in [2], and independently by Kadison in [133], that the existence of a minimal positive generator could be used to give a fairly simple proof of the derivation result in 8.6.6. Finally Arveson proved 8.6.6 using the spectral subspace method and Olesen extended his result to cover also derivations of AW^*-algebras, cf. [180]. From this it was just a short step to apply Arveson's theory to derivations of C^*-algebras, from which 8.6.5 followed, see [184]. Theorem 8.6.9 was proved in [2] and again in [184] with a constructive proof. It gives the natural generalization of Sakai's famous result (8.6.10) from [230] and [232].

Proposition 8.6.11 was established in [2]. Combining it with 8.6.10 we see that if a separable C^*-algebra A has the form $A_1 \oplus A_2$, where A_1 has continuous trace and A_2 is the direct sum of simple algebras (i.e. $\check{A}_2$ is discrete), then every derivation of A is inner in $M(A)$. The converse was proved by Akemann and the author in [6]. Theorem 8.6.15 is a result by the author, see [204] or [205]. A generalization to certain one-parameter automorphism groups was obtained in [185].

We have only considered bounded derivations in this section, but of course there is also a theory for unbounded derivations. The major problem there is that a (densely defined, closed) *-derivation δ need not be the infinitesimal generator of a one-parameter group of automorphisms (the converse is of course true), so extra conditions must be added to reach this conclusion. The problem is quite relevant for the applications of C^*-algebras to quantum statistical mechanics. For further information we refer to the exhaustive discussion in the books by Sakai [233] and by Bratteli and Robinson [26].

8.7. Derivable automorphisms

8.7.1. Let α be an automorphism of a C^*-algebra A, and consider the following hierarchy of conditions:

(i) α is inner (in $M(A)$ if $1 \notin A$);

(ii) $\alpha = \exp \delta$ for some *-derivation δ on A; we say in this case that α is *derivable*;

(iii) α is *universally weakly inner*, i.e. α'' is inner in A'';

(iv) α is (π, H)-*weakly inner*, i.e. (π, H) is an α-invariant representation of A and $\pi\alpha$ extends to an inner automorphism of $\pi(A)''$.

It follows from 8.6.5 that (ii) ⇒ (iii), and clearly (i) ⇒ (iii) ⇒ (iv). Moreover, if A is simple (ii) ⇒ (i) by 8.6.10. In 8.9 we shall use a localized version of (ii) to characterize inner automorphisms of von Neumann algebras, thereby characterizing automorphisms of C^*-algebras satisfying (iii) and (iv).

8.7.2. From 8.6.4 we know that the exponential of a *-derivation is an automorphism. The converse, that the logarithm of an automorphism is a *-derivation, is not a meaningful statement because the logarithm is not a well-defined function on $\boldsymbol{C}$. However, if α is an automorphism of a C^*-algebra A then its spectrum (as an element of $\boldsymbol{B}(A)$) must be a symmetric subset of the unit circle by 8.1.14. If therefore $\nu(\iota - \alpha) < 2$ (ν denotes the spectral radius in this section and ι is the identity map) then $-1 \notin \mathrm{Sp}(\alpha)$ and thus with Log as the principal branch of the logarithm (defined on $\boldsymbol{C}\backslash(-\boldsymbol{R}_+)$) we can define $\operatorname{Log}\alpha$ in $\boldsymbol{B}(A)$. However, $\operatorname{Log}\alpha$ is not necessarily a derivation. It suffices to take $A = \boldsymbol{C}^3$ and let α be a cyclic permutation. Then $\mathrm{Sp}(\alpha)$ consists of the third roots of unity, so $\nu(\iota - \alpha) = \sqrt{3}$; but $\operatorname{Log}\alpha$ is not a derivation. If it were, it would be inner, whence α would be inner. The sufficient conditions for $\operatorname{Log}\alpha$ to be a derivation are contained in the next lemma.

8.7.3. LEMMA. *If α is an automorphism of a von Neumann algebra $\mathcal{M}$ and if either $\|\iota - \alpha\| < 2$ or $\nu(\iota - \alpha) < \sqrt{3}$ then α is the identity on $\mathcal{M} \cap \mathcal{M}'$.*

Proof. Suppose that $\|\iota - \alpha\| < 2$. If p is a projection then $2p - 1$ is unitary whence

$$2 > \|2p - 1 - \alpha(2p - 1)\| = 2\|\alpha(p) - p\|.$$

If p is central then p and $\alpha(p)$ commute, and then $\|\alpha(p) - p\| < 1$ implies $\alpha(p) = p$. It follows that α fixes every element in the centre of $\mathcal{M}$.

To prove the second half note that if $\sigma \in \mathrm{Sp}(\alpha | \mathcal{M} \cap \mathcal{M}')$ then the same holds for σ^n for every $n > 1$. This follows from the equation

$$\alpha(x^{n+1}) - \sigma^{n+1}x^{n+1} = (\alpha(x^n) - \sigma^n x^n)\alpha(x) + \sigma^n x^n(\alpha(x) - \sigma x),$$

in conjunction with 8.1.9 and 8.1.11 (note also that $\|x^n\| = \|x\|^n$ for normal elements). If therefore $v(\iota - \alpha) < \sqrt{3}$ then $\iota = \alpha$ on $\mathcal{M} \cap \mathcal{M}'$.

8.7.4. LEMMA. *If α is an automorphism of $\boldsymbol{B}(H)$ then $\alpha = \mathrm{Ad}\, u$ for some unitary u on H.*

Proof. If p is a one-dimensional (i.e. minimal) projection on H then so is $\alpha(p)$. There is then a unitary v on H such that $v\alpha(p)v^* = p$. Replacing α with $(\mathrm{Ad}\, v)\circ\alpha$ (an admissible operation; for if one is inner, so is the other) we may assume that $\alpha(p) = p$.

Fix a unit vector ξ in $p(H)$. For each x in $\boldsymbol{B}(H)$ we have

$$\begin{aligned}\|\alpha(x)\xi\|^2 &= (\alpha(x^*x)\xi \mid \xi) = \|p\alpha(x^*x)p\| \\ &= \|\alpha(px^*xp)\| = \|px^*xp\| = (x^*x\xi \mid \xi) \\ &= \|x\xi\|^2.\end{aligned}$$

Thus if we define

$$u(x\xi) = \alpha(x)\xi, \qquad x \in \boldsymbol{B}(H),$$

then u is unitary on H (with $u^{-1}(x\xi) = \alpha^{-1}(x)\xi$). Furthermore, for all x, y in $\boldsymbol{B}(H)$

$$(uxu^*)y\xi = ux\alpha^{-1}(y)\xi = \alpha(x)y\xi,$$

whence $\mathrm{Ad}\, u = \alpha$.

8.7.5. LEMMA. *If α is an automorphism of $\boldsymbol{B}(H)$ with $\|\iota - \alpha\| < 2$, then $\alpha = \mathrm{Ad}\, u$ for some unitary u on H such that $\mathrm{Re}\,\sigma \geqslant (1 - \frac{1}{4}\|\iota - \alpha\|)^{1/2}$ for every σ in $\mathrm{Sp}\, u$.*

Proof. From 8.7.4 we know that $\alpha = \mathrm{Ad}\, u$. Suppose that $\sigma \in \mathrm{Conv}(\mathrm{Sp}(u))$, i.e. $\sigma = \Sigma\gamma_n\sigma_n$ where $\gamma_n > 0$, $\Sigma\gamma_n = 1$ and $\{\sigma_n\} \subset \mathrm{Sp}(u)$. Fix $\varepsilon > 0$ and choose mutually orthogonal unit vectors ξ_n such that $\|(u - \sigma_n)\xi_n\| \leqslant \varepsilon\gamma_n^{1/2}$ for every n. Put $\xi = \Sigma\gamma_n^{1/2}\xi_n$. Then $\|\xi\| = 1$ and

$$\begin{aligned}|(u\xi \mid \xi)| &= |(\textstyle\sum \gamma_n^{1/2}((u\xi_n - \sigma_n\xi_n) + \sigma_n\xi_n \mid \xi)| \\ &\leqslant \textstyle\sum \gamma_n^{1/2}\,\varepsilon\gamma_n^{1/2} + |(\sum \gamma_n^{1/2}\sigma_n\xi_n \mid \sum \gamma_m^{1/2}\xi_m)| = \varepsilon + |\sigma|.\end{aligned}$$

If therefore p denotes the projection on $\boldsymbol{C}\xi$ then

$$\begin{aligned}\tfrac{1}{2}\|\iota - \alpha\| &\geqslant \tfrac{1}{2}\|2p - 1 - \alpha(2p - 1)\| = \|p - \alpha(p)\| \\ &\geqslant ((p - upu^*)\xi \,|\, \xi) \geqslant 1 - (pu^*\xi \,|\, u^*\xi) \\ &= 1 - |(u^*\xi \,|\, \xi)|^2 \geqslant 1 - (\varepsilon + |\sigma|)^2.\end{aligned}$$

Since ε is arbitrary we have $|\sigma|^2 \geqslant 1 - \frac{1}{2}\|\iota - \alpha\|$.

Take σ_0 to be the point in Conv(Sp(u)) nearest to 0. Multiplying u with a suitable complex number of modulus 1 we may assume that $\sigma_0 \in \boldsymbol{R}_+$. It follows (from plane geometry) that $\operatorname{Re}\sigma \geqslant \sigma_0$ for every σ in Sp(u), as desired.

8.7.6. LEMMA. *If α is an automorphism of $\boldsymbol{B}(H)$ with $\nu(\iota - \alpha) < \sqrt{3}$, then $\alpha = \operatorname{Ad} u$ for some unitary u on H such that $\operatorname{Re}\sigma > \frac{1}{2}$ for every σ in* Sp(u).

Proof. We know that $\alpha = \operatorname{Ad} u$, and claim that

$$(\operatorname{Sp}(u))(\operatorname{Sp}(u^*)) \subset \operatorname{Sp}(\alpha).$$

Indeed, if σ, τ belong to Sp(u) $(\sigma \neq \tau)$ and $\varepsilon > 0$, there are orthogonal unit vectors ξ and η in H such that $\|u\xi - \sigma\xi\| < \varepsilon$ and $\|u\eta - \tau\eta\| < \varepsilon$. Let v denote the partial isometry of dimension two that exchanges ξ and η. Then $v = v^*$ and

$$\|uvu^*\xi - \tau\sigma^{-1}v\xi\| < 2\varepsilon, \qquad \|uvu^*\eta - \tau\sigma^{-1}v\eta\| < 2\varepsilon;$$

and since $\|uvu^*(1 - v^2)\| < \varepsilon$ we conclude that $\|\alpha(v) - \tau(\sigma)^{-1}v\| < 2\varepsilon$. Since ε is arbitrary $\tau\sigma^{-1} \in \operatorname{Sp}(\alpha)$ by 8.1.8.

Assuming as we may that $1 \in \operatorname{Sp}(u)$ we see from the result above that $\exp(it) \notin \operatorname{Sp}(u)$ if $2\pi/3 \leqslant t \leqslant 4\pi/3$. Let σ and τ be the points in Sp(u) farthest away from 1 in the upper and lower half plane, respectively. Then $\sigma = \exp(it)$ and $\tau = \exp(-is)$ where $0 \leqslant t < 2\pi/3$ and $0 \leqslant s < 2\pi/3$. Since also $\sigma\tau^{-1} \in \operatorname{Sp}(\alpha)$ and since $s + t < 4\pi/3$ we conclude that in fact $s + t < 2\pi/3$. But then the spectrum of u is contained in one third of the circle, and we may arrange that $\operatorname{Re}\sigma > \frac{1}{2}$ for every σ in Sp(u).

8.7.7. THEOREM. *If α is an automorphism of a C^*-algebra A such that either $\|\iota - \alpha\| < 2$ or $\nu(\iota - \alpha) < \sqrt{3}$ then $\operatorname{Log}\alpha$ is a $*$-derivation of A.*

Proof. Let α'' be the canonical extension of α to an automorphism of A'', and note that $\operatorname{Sp}(\alpha) = \operatorname{Sp}(\alpha'')$ and that $\|\iota - \alpha\| = \|\iota'' - \alpha''\|$. Thus by 8.7.3 each central element in A'' is fixed under α''. Let

$$\pi_a = \bigoplus_{\iota \in \hat{A}} \pi_\iota \qquad \text{on} \quad H_a = \bigoplus_{\iota \in \hat{A}} H_\iota$$

be the atomic representation of A (cf. 4.3.7). Then for each ι in $\hat{A}$ we have $\alpha_\iota = \pi_\iota'' \circ \alpha'' \circ \pi_\iota''^{-1}$ and α_ι is an automorphism of $\pi_\iota''(A'') = \boldsymbol{B}(H_\iota)$. Since

$\|\iota - \alpha_t\| \leqslant \|\iota - \alpha\|$ and $\mathrm{Sp}(\alpha_t) \subset \mathrm{Sp}(\alpha)$, there is by 8.7.5 and 8.7.6 a unitary u_t on H_t with spectrum in the open right half-plane such that $\alpha_t = \mathrm{Ad}\, u_t$. Put $\mathscr{M} = \pi_a(A'')$ and $u = \oplus u_t$. Then u is a unitary in $\mathscr{M}$ with spectrum in the open right half-plane and, identifying A with its image $\pi_a(A)$ in $\mathscr{M}$, we have $\alpha = \mathrm{Ad}\, u$ on A. By abuse of notation we will also write $\alpha = \mathrm{Ad}\, u$ on $\mathscr{M}$.

For every y in $\mathscr{M}$ define $\lambda(y)$ and $\rho(y)$ in $B(\mathscr{M})$ by $\lambda(y)x = yx$ and $\rho(y)x = xy$, $x \in \mathscr{M}$. Then λ and ρ are linear isometries of $\mathscr{M}$ into $B(\mathscr{M})$, and λ is an isomorphism whereas ρ is an anti-isomorphism. It follows that $\mathrm{Sp}(\lambda(y)) \subset \mathrm{Sp}(y)$ and $\mathrm{Sp}(\rho(y)) \subset \mathrm{Sp}(y)$. Note that $\rho(y)\lambda(z) = \lambda(z)\rho(y)$ for all y, z in $\mathscr{M}$.

We have $\alpha = \lambda(u)\rho(u^*)$ on $\mathscr{M}$ and $\alpha, \lambda(u)$ and $\rho(u^*)$ are commuting operators. Let $\mathfrak{A}$ be a maximal commutative subalgabra of $B(\mathscr{M})$ containing these three operators. Because of the maximality the spectra of the three operators are unchanged. Since $\mathrm{Log}(\sigma\tau) = \mathrm{Log}(\sigma) + \mathrm{Log}(\tau)$ provided that σ and τ belong to the open right half-plane we see that

$$\begin{aligned}\omega(\mathrm{Log}\,\alpha) = \mathrm{Log}\,\omega(\alpha) &= \mathrm{Log}\,\omega(\lambda(u)) + \mathrm{Log}\,\omega(\rho(u^*)) \\ &= \omega(\mathrm{Log}\,\lambda(u) + \mathrm{Log}\,\rho(u^*))\end{aligned}$$

for every character ω in the spectrum of $\mathfrak{A}$. As $\mathfrak{A}$ is semi-simple it follows that

$$\begin{aligned}\mathrm{Log}\,\alpha &= \mathrm{Log}\,\lambda(u) + \mathrm{Log}\,\rho(u^*) \\ &= \lambda(\mathrm{Log}\,u) - \rho(\mathrm{Log}\,u) = \mathrm{ad}(\mathrm{Log}\,u).\end{aligned}$$

This shows that $\mathrm{Log}\,\alpha$ is an inner *-derivation on $\mathscr{M}$. (Note that $\mathrm{Log}\,u \in i\mathscr{M}_{\mathrm{sa}}$). By Runge's theorem Log can be approximated by complex polynomials uniformly on the compact set $\mathrm{Sp}(\alpha)$. Consequently $\mathrm{Log}\,\alpha$ can be approximated uniformly in $B(\mathscr{M})$ by polynomials in α. Each of these polynomials leaves A invariant (as a set), and since A is closed in $\mathscr{M}$ we conclude that $\mathrm{Log}\,\alpha \mid A$ is a *-derivation of A.

8.7.8. COROLLARY. *Given a C^*-algebra A consider* $\mathrm{Aut}(A)$ *with the uniform topology (in which it is a topological group). Each derivable automorphism of A lies in the connected component* $\mathrm{Aut}_0(A)$ *of the identity of* $\mathrm{Aut}(A)$. *Conversely, each automorphism in* $\mathrm{Aut}_0(A)$ *is a product of derivable automorphisms.*

Proof. If $\alpha = \exp(\delta)$ then α lies on the connected orbit of the uniformly continuous group $\{\exp(t\delta) \mid t \in \boldsymbol{R}\}$, whence $\alpha \in \mathrm{Aut}_0(A)$. On the other hand, the family of products of derivable automorphisms is a subgroup G of $\mathrm{Aut}_0(A)$. Since G contains a neighbourhood of ι by 8.7.7, G is open, thus also closed; whence $G = \mathrm{Aut}_0(A)$ since the latter is connected.

8.7.9. PROPOSITION. *Let α be an automorphism of a von Neumann algebra $\mathscr{M}$. If $\|\iota - \alpha\| < 2$ or if $\nu(\iota - \alpha) < \sqrt{3}$ then $\alpha = \mathrm{Ad}\, u$ for some unitary u in $\mathscr{M}$ with*

$$\mathrm{Sp}(u) \subset \{\exp(it) \mid |t| \leqslant \mathrm{Arc\,sin}(\tfrac{1}{2}\nu(\iota - \alpha))\}.$$

Proof. Put $\delta = \operatorname{Log}\alpha$. By 8.7.7 δ is a *-derivation of $\mathscr{M}$ and $\operatorname{Sp}(\delta) = \operatorname{Log}(\operatorname{Sp}(\alpha))$. From plane geometry we see that

$$\nu(\iota - \alpha) = 2\sin(\tfrac{1}{2}\nu(\delta)).$$

By 8.6.6 $\delta = \operatorname{ad}(ih)$ for some h in $\mathscr{M}_+$ with $\|h\| = \|\delta\|$ $(= \nu(\delta)$ cf. 8.6.5). Put $v = \exp(ih)$. Then v is a unitary in $\mathscr{M}$ and $\alpha = \operatorname{Ad} v$. Moreover,

$$\operatorname{Sp}(v) \subset \{\exp(it) \mid 0 \leqslant t \leqslant \nu(\delta)\}.$$

Taking $u = \exp(-\tfrac{1}{2}\nu(\delta))v$ we obtain the desired result.

8.7.10. COROLLARY. *Let α be an automorphism of a C^*-algebra A. If $\nu(\iota - \alpha) < \sqrt{3}$ or if $\|\iota - \alpha\| < 2$ then $\|\iota - \alpha\| = \nu(\iota - \alpha)$.*

Proof. By 8.7.9 (applied to A'') we have $\alpha = \operatorname{Ad} u$, whence

$$\begin{aligned}\|\iota - \alpha\| = \|\operatorname{ad} u\| &= \operatorname*{Inf}_{\lambda} \|\operatorname{ad}(u - \lambda 1)\| \\ &\leqslant \operatorname*{Inf}_{\lambda} 2\,\|u - \lambda 1\| \\ &\leqslant 2(\tfrac{1}{2}\nu(\iota - \alpha)) = \nu(\iota - \alpha).\end{aligned}$$

8.7.11. Notes and remarks. Lance [154] proved that if A is a Glimm algebra the conditions (i) and (iii) in 8.7.1 are equivalent. Since then it has been an outstanding problem whether this is the case for every simple C^*-algebra. Theorem 8.7.7 was proved by Kadison and Ringrose [136] for the case $\|\iota - \alpha\| < 2$. Later Serre gave a short proof for $\nu(\iota - \alpha) < \sqrt{3}$, see p. 314 of [59]. That the two results are equivalent is seen from 8.7.10, which was established by Borchers [24].

8.8. Borchers and Connes spectrum

8.8.1. Let (A, G, α) be a C^*-dynamical a-system. If B is a G-invariant C^*-subalgebra of A we may consider the dynamical system $(B, G, \alpha \mid B)$. Clearly $\operatorname{Sp}(\alpha \mid B) \subset \operatorname{Sp}(\alpha)$. Since automorphisms with small spectra tend to be inner (or at least derivable cf. 8.7.7), we may regard the spectral values $(\neq 0)$ that belong to $\operatorname{Sp}(\alpha \mid B)$ for all B as the essential obstructions for α to be inner.

8.8.2. Let $\mathscr{H}^\alpha(A)$ denote the set of G-invariant, hereditary, non-zero C^*-subalgebras of A; and let $\mathscr{H}^\alpha_B(A)$ denote the subset consisting of algebras B in $\mathscr{H}^\alpha(A)$ such that the closed ideal of A generated by B is essential in A (cf. 3.12.7). Define the *Connes spectrum* of α by

$$\Gamma(\alpha) = \bigcap \operatorname{Sp}(\alpha \mid B), \qquad B \in \mathscr{H}^\alpha(A).$$

Furthermore, define the *Borchers spectrum* of α by

$$\Gamma_B(\alpha) = \bigcap \mathrm{Sp}(\alpha)B), \qquad B \in \mathscr{H}_B^\alpha(A).$$

Evidently $\Gamma(\alpha) \subset \Gamma_B(\alpha)$, and usually they are different. Both spectra are needed for the description of α and each of them have convenient structural properties which the other may lack. It is fortunate, therefore, that the two notions coincide for a large class of interesting C^*-algebras: Algebras in which every non-zero closed ideal is essential, i.e. prime algebras (cf. 3.13.7). Note also that if A is G-prime (cf. 7.9.4) then $\Gamma_B(\alpha) = \Gamma(\alpha)$.

8.8.3. If $(\mathscr{M}, G, \alpha)$ is a W^*-dynamical a-system we define

$$\Gamma(\alpha) = \bigcap \mathrm{Sp}(\alpha \mid p\mathscr{M}p)$$

where p ranges over the set of non-zero, G-invariant projections in $\mathscr{M}$ and

$$\Gamma_B(\alpha) = \bigcap \mathrm{Sp}(\alpha \mid p\mathscr{M}p),$$

where p ranges over the set of G-invariant projections in $\mathscr{M}$ with $c(p) = 1$.

If now $(\mathscr{M}, G, \alpha)$ is a W^*-dynamical a-system where α is uniformly continuous (G might be discrete), then $(\mathscr{M}, G, \alpha)$ can also be regarded as a C^*-dynamical a-system and we have apparently two definitions of both $\Gamma(\alpha)$ and $\Gamma_B(\alpha)$. However, they produce the same result. For if Ω is open in Γ and $B \in \mathscr{H}^\alpha(\mathscr{M})$ then

$$R^{\alpha|B}(\Omega)^- = R^{\alpha|B}(\Omega)$$

($^-$ denoting weak closure). Consequently $\mathrm{Sp}(\alpha \mid B) = \mathrm{Sp}(\alpha \mid \bar{B})$ by 8.1.7 (ii) and since $\bar{B} = p\mathscr{M}p)$ for some G-invariant projection p we see that $\Gamma(\alpha)$ has a unique meaning. Since furthermore the norm closed ideal generated by B is essential in $\mathscr{M}$ if and only if $c(p) = 1$ we see that also $\Gamma_B(\alpha)$ is uniquely determined.

8.8.4. PROPOSITION. *Let (A, G, α) be a C^*-dynamical a-system. If $\sigma_1 \in \Gamma(\alpha)$ and $\sigma_2 \in \mathrm{Sp}(\alpha)$ then $\sigma_1 + \sigma_2 \in \mathrm{Sp}(\alpha)$. Moreover, $\Gamma(\alpha)$ is a closed subgroup of Γ.*

Proof. If Ω is a neighbourhood of $\sigma_1 + \sigma_2$ there are neighbourhoods Ω_1 and Ω_2 of σ_1 and σ_2, respectively, such that $\Omega_1 + \Omega_2 \subset \Omega$. By assumption (cf. 8.1.8 (ii)) there is a non-zero element x_2 in $R^\alpha(\Omega_2)$. Let B denote the hereditary C^*-subalgebra of A generated by the orbit $\{\alpha_t(x_2^* x_2) \mid t \in G\}$. This implies that if x is a non-zero element of B, then $\alpha_t(x_2^* x_2)x \neq 0$ for some t in G. Since B is G-invariant there is a non-zero element x in $R^{\alpha|B}(\Omega_1)$. Thus $\alpha_t(x_2)x_1 \neq 0$ for some t in G. As $\alpha_t(x_2) \in R^\alpha(\Omega_2)$ (cf. 8.1.4) we see from 8.3.3 (iii) that $\alpha_t(x_2)x_1$ is a non-zero element of $R^\alpha(\Omega_1 + \Omega_2)$. Thus $R^\alpha(\Omega) \neq 0$, and since this holds for every Ω we have $\sigma_1 + \sigma_2 \in \mathrm{Sp}(\alpha)$.

If. $\sigma_1, \sigma_2 \in \Gamma(\alpha)$ the argument above shows that $\sigma_1 + \sigma_2 \in \mathrm{Sp}(\alpha \mid B)$ for every B in $\mathscr{H}^\alpha(A)$. Consequently $\sigma_1 + \sigma_2 \in \Gamma(\alpha)$; and $\Gamma(\alpha)$, being the intersection of symmetric, closed sets, is a closed subgroup of Γ.

8.8.5. PROPOSITION. *Let (A, G, α) be a C*-dynamical a-system. If $\sigma \in \Gamma_B(\alpha)$ then $n\sigma \in \Gamma_B(\alpha)$ for every n in* $\mathbf{Z}$.

Proof. We claim that for any neighbourhood Ω of σ, any B in $\mathscr{H}^\alpha_B(A)$ and any n there exists elements $x_1, \cdots, x_n$ in $R^\alpha(\Omega) \cap B$ such that the product $x_1 x_2 \cdots x_n$ is non-zero. Since $\sigma \in \Gamma_B(\alpha)$ this is true for $n = 1$; and we now assume that the claim has been established for some n.

Let $\{C_i\}$ be a maximal collection of algebras in $\mathscr{H}^\alpha(B)$ such that the ideals generated by the C_i's are mutually orthogonal and such that for each i there is an element x_i in $R^\alpha(\Omega)$ such that C_i is the hereditary C*-algebra generated by the orbit $\{\alpha_t(x_i^* x_i) \mid t \in G\}$. Let C be the direct sum of the C_i's. Either $C \in \mathscr{H}^\alpha_B(B)$ or we can find (using a maximality argument) a closed, G-invariant ideal I in B, orthogonal to (the ideal generated by) C such that $C + I \in \mathscr{H}^\alpha_B(B)$. In either case, $I = 0$ or $I \neq 0$, we must have $R^\alpha(\Omega) \cap I = 0$. Otherwise we contradict the maximality of the family $\{C_i\}$.

Since $C + I \in \mathscr{H}^\alpha_B(B)$ and $B \in \mathscr{H}^\alpha_B(A)$ we have $C + I \in \mathscr{H}^\alpha_B(A)$. Indeed, if J is a non-zero ideal of A then $B \cap J$ is non-zero. But it is also an ideal in B, whence $(C + I) \cap J \neq 0$, as claimed. By the induction hypothesis there are therefore elements $x_1, \ldots, x_n$ in $R^\alpha(\Omega) \cap (C + I)$ such that the product $y = x_1 x_2 \ldots x_n$ is non-zero. As $R^\alpha(\Omega) \cap I = 0$, all the x_k's belong to $R^\alpha(\Omega) \cap C$. Thus $y \in C$. But then $\alpha_t(x_i)y \neq 0$ for some t in G and some i, since $C = \oplus C_i$. Since $\alpha_t(x_i) \in R^\alpha(\Omega) \cap B$ we have established the claim above for $n + 1$. By induction it is therefore true for all n in $\mathbf{N}$.

To show that $n\sigma \in \Gamma_B(\alpha)$ given that $\sigma \in \Gamma_B(\alpha)$ we may assume that $n > 0$, since $\Gamma_B(\alpha)$ is a symmetric set. Given any neighbourhood Ω_n of $n\sigma$ choose a neighbourhood Ω of σ such that $\Omega + \cdots + \Omega \subset \Omega_n$. Given B in $\mathscr{H}^\alpha_B(A)$ we apply the first part of the proof to obtain elements $x_1, \ldots, x_n$ in $R^\alpha(\Omega) \cap B$ such that the product $y = x_1, x_2, \ldots, x_n$ is non-zero. By 8.3.3 (iii) we have

$$y \in R^\alpha(\Omega + \cdots + \Omega) \cap B \subset R^\alpha(\Omega_n) \cap B.$$

Since this is valid for all Ω_n, $n\sigma \in \mathrm{Sp}(\alpha \mid B)$. As B was arbitrary this implies that $n\sigma \in \Gamma_B(\alpha)$.

8.8.6. LEMMA. *Let (A, G, α) be a C*-dynamical a-system. If $B_0 \in \mathscr{H}^\alpha(A)$ and $B_1 \in \mathscr{H}^\alpha_B(A)$ there are, for every symmetric neighbourhood Ω of 0, algebras C_0 and C_1 in $\mathscr{H}^\alpha(A)$ with $C_0 \subset B_0$, $C_1 \subset B_1$, such that*

$$\mathrm{Sp}(\alpha \mid C_0) \subset \mathrm{Sp}(\alpha \mid C_1) + \Omega, \qquad \mathrm{Sp}(\alpha \mid C_1) \subset \mathrm{Sp}(\alpha \mid C_0) + \Omega.$$

Proof. Define

$$L_0 = \{x \in A \mid x^* x \in B_0\}; \qquad L_1 = \{x \in A \mid x^* x \in B_1\}.$$

Then L_0 and L_1 are closed, non-zero left ideals of A. We claim that

$L_0^* \cap L_1 \neq \{0\}$. Otherwise the closures of AL_0^* and L_1A would be orthogonal, non-zero ideals in A. However, $B_1 \subset L_1A$ and the ideal generated by B_1 is essential, which is a contradiction. Consequently there is a non-zero x in $L_0^* \cap L_1$.

Take Ω_1 such that $\Omega_1 - \Omega_1 \subset \Omega$. Then choose f in $K^1(G)$ with supp. $\hat{f} \subset \Omega_1$ such that the element $y = \alpha_f(x)$ is non-zero. Now

$$y \in L_0^* \cap L_1 \cap R^\alpha(\Omega_1).$$

Let C_1 be the hereditary C^*-subalgebra generated by the orbit $\{\alpha_t(y^*y) \mid t \in G\}$ and let C_0 be the hereditary C^*-subalgebra generated by $\{\alpha_t(yy^*) \mid t \in G\}$. Then C_0 and C_1 belong to $\mathscr{H}^\alpha(A)$, and $C_0 \subset B_0, C_1 \subset B_1$.

If $\sigma_0 \in \mathrm{Sp}(\alpha \mid C_0)$ there is for each neighbourhood Ω_0 of σ_0 a non-zero x_0 in $R^{\alpha|C_0}(\Omega_0)$. Then for some s and t in G the element $x_1 = \alpha_s(y^*)x_0\alpha_t(y)$ is non-zero. Otherwise

$$\alpha_s(yy^*)x_0\alpha_t(yy^*) = 0$$

for all s and t. Since $x_0 \in C_0$ and C_0 is the hereditary algebra generated by $\alpha_t(yy^*)$, this would imply that $x_0 = 0$; a contradiction. Thus $x_1 \neq 0$. Clearly $x_1 \in C_1$ and by 8.3.3

$$x_1 \in R^\alpha(-\Omega_1 + \Omega_0 + \Omega_1) \subset R^\alpha(\Omega_0 + \Omega).$$

It follows that $\mathrm{Sp}(\alpha \mid C_1) \cap (\Omega_0 + \Omega) \neq \varnothing$. Since this holds for every Ω_0 we conclude that $\sigma_0 \in \mathrm{Sp}(\alpha \mid C_1) + \Omega$, whence

$$\mathrm{Sp}(\alpha \mid C_0) \subset \mathrm{Sp}(\alpha \mid C_1) + \Omega.$$

Similarly

$$\mathrm{Sp}(\alpha \mid C_1) \subset \mathrm{Sp}(\alpha \mid C_0) + \Omega.$$

8.8.7. PROPOSITION. *Let (A, G, α) be a C^*-dynamical a-system and assume that Γ_0 is a closed subgroup of Γ such that $\mathrm{Sp}(\alpha)/\Gamma_0$ is compact (in Γ/Γ_0). For each neighbourhood Ω of θ there is a B in $\mathscr{H}^\alpha_B(A)$ such that*

$$\mathrm{Sp}(\alpha \mid B) \subset \Gamma_B(\alpha) + \Omega + \Gamma_0.$$

Proof. Choose an open neighbourhood Ω_0 with compact closure contained in Ω. If $\{\Lambda_i\}$ denotes the family of compact neighbourhoods of θ then

$$\bigcap_i \bigcap_{\mathscr{H}^\alpha_B(A)} (\mathrm{Sp}(\alpha \mid B) + \Lambda_i)/\Gamma_0 \subset (\Gamma_B(\alpha) + \Omega_0)/\Gamma_0.$$

Since $(\mathrm{Sp}(\alpha \mid B) + \Lambda_i)/\Gamma_0$ is compact and $(\Gamma_B(\alpha) + \Omega_0)/\Gamma_0$ is open it follows that

$$\bigcap_k (\mathrm{Sp}(\alpha \mid B_k) + \Lambda_k) \subset \Gamma_B(\alpha) + \Omega_0 + \Gamma_0$$

for some finite sets $\{B_k\}$ and $\{\Lambda_k\}$ in $\mathscr{H}^{\alpha}_{B}(A)$ and $\{\Lambda_i\}$, respectively. Let B_0 be an arbitrary algebra in $\mathscr{H}^{\alpha}(A)$. Applying 8.8.6 repeatedly we obtain an algebra C in $\mathscr{H}^{\alpha}(A)$ such that $C \subset B_0$ and

$$\mathrm{Sp}(\alpha \mid C) \subset \bigcap_k (\mathrm{Sp}(\alpha \mid B_k) + \Lambda_k) \subset \Gamma_B(\alpha) + \Omega_0 + \Gamma_0.$$

Let $\{C_j\}$ be a maximal family of algebras in $\mathscr{H}^{\alpha}(A)$ such that $\mathrm{Sp}(\alpha \mid C_j) \subset \Gamma_B(\alpha) + \Omega_0 + \Gamma_0$ and such that the closed ideals generated by the C_j's are mutually orthogonal. Take B to be the direct sum of the C_j's. Clearly $B \in \mathscr{H}^{\alpha}(A)$, but in fact $B \in \mathscr{H}^{\alpha}_{B}(A)$. For if I denotes the closed ideal generated by B (necessarily G-invariant), and if J was a non-zero closed ideal orthogonal to I (which could then be taken to be G-invariant), then from the first part of the proof (with $J = B_0$) there would be a C in $\mathscr{H}^{\alpha}(A)$ with

$$\mathrm{Sp}(\alpha \mid C) \subset \Gamma_B + \Omega_0 + \Gamma_0 \quad \text{and} \quad C \subset J;$$

contradicting the maximality of the family $\{C_j\}$.

It remains to show that $\mathrm{Sp}(\alpha \mid B) \subset \Gamma_B + \Omega + \Gamma_0$. But since $\Gamma_B(\alpha)/\Gamma_0$ and $\overline{\Omega}_0/\Gamma_0$ are both compact we know that $\Gamma_B(\alpha) + \overline{\Omega}_0 + \Gamma_0$ is closed in Γ. Thus for each $\sigma \notin \Gamma_B + \Omega + \Gamma_0$ there is an open neighbourhood Ω_1 of σ disjoint from $\Gamma_B(\alpha) + \Omega_0 + \Gamma_0$. Consequently $R^{\alpha}(\Omega_1) \cap C_j = \{0\}$ for all j by 8.1.8; whence $R^{\alpha}(\Omega_1) \cap B = \{0\}$, proving that $\sigma \notin \mathrm{Sp}(\alpha \mid B)$.

8.8.8. Let $(\mathscr{M}, G, \alpha)$ be a W^*-dynamical a-system. A verbatim repetition of the arguments in the proof of 8.8.4–8.8.7 show that these results are valid for $(\mathscr{M}, G, \alpha)$ with $\Gamma(\alpha)$ and $\Gamma_B(\alpha)$ as defined in 8.8.3. Another way to reach the results for W^*-dynamical systems is offered by the next proposition (8.8.10).

8.8.9. LEMMA. *Let $(\mathscr{M}, G, \alpha)$ be a W^*-dynamical a-system and let A be a G-invariant, σ-weakly dense C^*-subalgebra of $\mathscr{M}$ contained in $\mathscr{M}^c$ (cf. 7.5.1). Set $\beta = \alpha \mid A$ and consider the C^*-dynamical system (A, G, β). Then* $\mathrm{Sp}(\alpha) = \mathrm{Sp}(\beta)$, $\Gamma(\alpha) \subset \Gamma(\beta)$ *and* $\Gamma_B(\alpha) \subset \Gamma_B(\beta)$.

Proof. Let Ω be an open set in Γ and B a G-invariant hereditary C^*-subalgebra of A. Since A is dense in $\mathscr{M}$, $R^{\beta}(\Omega)$ is dense in $R^{\alpha}(\Omega)$. Moreover, the σ-weak closure of B has the form $p\mathscr{M}p$ for some G-invariant projection p in $\mathscr{M}$. We see that $R^{\beta}(\Omega) \cap B = \{0\}$ if and only if $R^{\alpha}(\Omega) \cap p\mathscr{M}p = \{0\}$. Consequently $\mathrm{Sp}(\alpha \mid p\mathscr{M}p) = \mathrm{Sp}(\beta \mid B)$. Taking $B = A$ we obtain $\mathrm{Sp}(\alpha) = \mathrm{Sp}(\beta)$; and taking the intersection over all B in $\mathscr{H}^{\beta}(A)$ we obtain $\Gamma(\alpha) \subset \Gamma(\beta)$. Finally we note that if $B \in \mathscr{H}^{\beta}_{B}(A)$ then $c(p) = 1$; so the intersection over all B in $\mathscr{H}^{\beta}_{B}(A)$ yields $\Gamma_B(\alpha) \subset \Gamma_B(\beta)$.

8.8.10. PROPOSITION. *Let $(\mathscr{M}, G, \alpha)$ be a W^*-dynamical a-system and consider the C^*-dynamical system $(\mathscr{M}^c, G, \alpha \mid \mathscr{M}^c)$. Then the Arveson, Borchers and Connes spectrum of α on $\mathscr{M}$ is equal to the corresponding spectrum for α on $\mathscr{M}^c$.*

Proof. Set $\beta = \alpha | \mathcal{M}^c$. We know from 8.8.9 that $\mathrm{Sp}(\beta | B) = \mathrm{Sp}(\alpha | p\mathcal{M}p)$ for each B in $\mathscr{H}^\beta(\mathcal{M}^c)$, where $B^{-w} = p\mathcal{M}p$. Moreover $c(p) = 1$ whenever $B \in \mathscr{H}^\beta_B(\mathcal{M}^c)$. Now observe that if p is a non-zero, G-invariant projection in $\mathcal{M}$ then $p \in \mathcal{M}^c$ so that $p\mathcal{M}^c p \in \mathscr{H}^\beta(\mathcal{M}^c)$. Moreover $p\mathcal{M}^c p \in \mathscr{H}^\beta_B(\mathcal{M}^c)$ if $c(p) = 1$. Thus when B ranges over $\mathscr{H}^\beta(\mathcal{M}^c)$ (resp. $\mathscr{H}^\beta_B(\mathcal{M}^c)$) then p ranges over all non-zero G-invariant projections in $\mathcal{M}$ (with $c(p) = 1$). Since $\mathrm{Sp}(\beta | B) = \mathrm{Sp}(\alpha | p\mathcal{M}p)$ it follows immediately that $\Gamma(\beta) = \Gamma(\alpha)$ and that $\Gamma_B(\beta) = \Gamma_B(\alpha)$.

8.8.11. *Notes and remarks.* The Connes spectrum was defined for W^*-dynamical systems as in 8.3.3 by Connes [38], and a terminology close to the Borchers spectrum was developed by Borchers [24]. The generalization of these spectra to C^*-dynamical systems was done by Olesen [182], where 8.8.4 and 8.8.7 were also established. Connes [38] showed that $\Gamma(\alpha) = \cap \mathrm{Sp}(\alpha | p\mathcal{M}p)$, where p ranges over the projections in the centre of the fixed-point algebra $\mathcal{M}^G$. A similar formula for C^*-dynamical systems has not been found (yet).

8.9. Inner automorphisms

8.9.1. LEMMA. *Let α be an automorphism of a von Neumann algebra $\mathcal{M}$ and assume that there is a partial isometry w in $\mathcal{M}$ such that with $p = w^*w$ we have $\alpha(pxp) = wxw^*$ for every x in $\mathcal{M}$. If $c(p) = 1$ there is a unique unitary u in $\mathcal{M}$ with $up = w$ such that $\alpha = \mathrm{Ad}\, u$.*

Proof. Using Zorn's lemma we take a maximal family of partial isometries $\{v_i\}$ in $\mathcal{M}$ such that $v_i^* v_i \leq p$ and such that the projections $v_i v_i^*$ are pairwise orthogonal. Since $c(p) = 1$ we have $\Sigma\, v_i v_i^* = 1$ (cf. 5.4.8).

Define $u = \Sigma\, \alpha(v_i) w v_i^*$ (strong convergence). Then $u \in \mathcal{M}$ and since $v_i^* v_j \leq p$ for all i and j

$$u^*u = \sum_{i,j} v_i w^* \alpha(v_i^* v_j) w v_j^* = \sum_{i,j} v_i v_i^* v_j v_j^* = 1.$$

Moreover, for each x in $\mathcal{M}$

$$uxu^* = \sum_{i,j} \alpha(v_i) w v_i^* x v_j w^* \alpha(v_j^*) = \sum_{i,j} \alpha(v_i v_i^* x v_j v_j^*) = \alpha(x).$$

It follows that u is unitary and that $\alpha = \mathrm{Ad}\, u$. Finally,

$$up = \sum \alpha(v_i) w v_i^* p w^* w = \sum \alpha(v_i v_i^* p) w = \alpha(p) w = w.$$

If v was another unitary in $\mathcal{M}$ with $vp = w$ and $\alpha = \mathrm{Ad}\, v$ then $u^*v \in \mathcal{M}'$ since $(\mathrm{Ad}\, u^*)(\mathrm{Ad}\, v) = \iota$. Thus u^*v is a central element and

$$u^*vp = u^*w = u^*up = p,$$

whence $(u^*v - 1)p = 0$. Since $c(p) = 1$ this implies that $u^*v = 1$, i.e. $u = v$.

8.9.2. PROPOSITION. *Let α be an automorphism of a von Neumann algebra $\mathcal{M}$ of type I. If α fixes every point in the centre of $\mathcal{M}$ then α is inner.*

Proof. Take by 5.5.3 an abelian projection p in $\mathcal{M}$ with $c(p) = 1$. Then also $\alpha(p)$ is abelian with $c(\alpha(p)) = 1$. By 5.5.2 there is a partial isometry w in $\mathcal{M}$ such that $w^*w = p$ and $ww^* = \alpha(p)$. Since p is abelian there is for each x in $\mathcal{M}$ a central element z such that $pxp = zp$. Consequently

$$\alpha(pxp) = z\alpha(p) = wzw^* = wzpw^* = wxw^*.$$

Thus 8.9.1 applies and α is inner.

8.9.3. PROPOSITION. *Let α be an automorphism of a von Neumann algebra $\mathcal{M}$. The following conditions are equivalent:*

(i) *α is inner;*

(ii) *For every $\varepsilon > 0$ there is an α-invariant projection p in $\mathcal{M}$ with $c(p) = 1$ such that $\|\iota - \alpha | p\mathcal{M}p\| < \varepsilon$;*

(iii) *$\Gamma_B(\alpha) = 0$;*

(iv) *There is an α-invariant projection p in $\mathcal{M}$ with $c(p) = 1$ such that $\|\iota - \alpha | p\mathcal{M}p\| < 2$;*

(v) *There is an α-invariant projection p in $\mathcal{M}$ with $c(p) = 1$ such that $\nu(\iota - \alpha | p\mathcal{M}p) < \sqrt{3}$.*

Proof. (i) $\Rightarrow$ (ii). If $\alpha = \operatorname{Ad} u$ for some unitary u in $\mathcal{M}$ there exists by a standard maximality argument for each $\varepsilon > 0$ a set of spectral projections $\{p_i\}$ of u with pairwise orthogonal central covers, and a corresponding set of spectral values $\{\lambda_i\}$ with $\|p_i(u - \lambda_i)\| \leqslant \frac{1}{2}\varepsilon$ for each i; such that $\Sigma c(p_i) = 1$. Put $p = \Sigma p_i$. Then $c(p) = 1$; and if $x \in p\mathcal{M}p$ then

$$\|\alpha(x) - x\| = \|ux - xu\| = \operatorname*{Sup}_i \|up_i x - xp_i u\| \leqslant \varepsilon \|x\|.$$

(ii) $\Rightarrow$ (iii) and (ii) $\Rightarrow$ (iv) are obvious; and (iii) $\Rightarrow$ (v) by 8.8.7 (cf. 8.8.8). Finally (iv) $\Rightarrow$ (i) and (v) $\Rightarrow$ (i) follow by combining 8.7.9 and 8.9.1.

8.9.4. THEOREM. *Let $(\mathcal{M}, G, \alpha)$ be a W^*-dynamical a-system and denote by G_0 the annihilator of $\Gamma_B(\alpha)$ in G. If $\alpha_t = \operatorname{Ad} u$ for some unitary u in the fixed-point algebra of G in $\mathcal{M}$ then $t \in G_0$. The converse holds if $\operatorname{Sp}(\alpha)/\Gamma(\alpha)$ is compact (in $\Gamma/\Gamma(\alpha)$).*

Proof. If $\alpha_t = \operatorname{Ad} u$ there exists for every $\varepsilon > 0$ a G-invariant projection p in $\mathcal{M}$ with $c(p) = 1$ such that $\|\iota - \alpha_t | p\mathcal{M}p\| < \varepsilon$ (cf. the implication (i) $\Rightarrow$ (ii) in 8.9.3). If $\sigma \in \Gamma_B(\alpha)$ then $(t, \sigma) \in \operatorname{Sp}(\alpha_t | p\mathcal{M}p)$ by 8.1.14 whence $|1 - (t, \sigma)| < \varepsilon$. Since ε is arbitrary $(t, \sigma) = 1$ and consequently $t \in G_0$.

Assume now that $Sp(\alpha)/\Gamma(\alpha)$ is compact and take t in G_0. Choose a neighbourhood Ω of θ such that $|1-(t,\Omega)| < \sqrt{3}$. By 8.8.7 there is a G-invariant projection p in $\mathcal{M}$ with $c(p) = 1$ such that

$$Sp(\alpha \,|\, p\mathcal{M}p) \subset \Gamma_B(\alpha) + \Omega.$$

(Note that $\Gamma_B(\alpha) + \Gamma(\alpha) \subset \Gamma_B(\alpha)$ by 8.8.4). It follows from 8.1.14 that $\nu(\iota - \alpha_t \,|\, p\mathcal{M}p) < \sqrt{3}$, whence $\alpha_t = \mathrm{Ad}\, u$ for some unitary u in $\mathcal{M}$ by 8.9.3. Moreover, since $\alpha_t \,|\, p\mathcal{M}p$ is derivable we may choose u commuting with p such that $\alpha_s(up) = up$ for all s in G by 8.6.6. But then $\alpha_s(u) = u$ by 8.9.1 since $c(p) = 1$, so that u is a fixed-point.

8.9.5. PROPOSITION. *Let $(\mathcal{M}, G, \alpha)$ be a W^*-dynamical a-system and assume that $Sp(\alpha)/\Gamma(\alpha)$ is compact. If $t \in G$ such that $Sp(\alpha_t)$ is not the whole unit circle, then $\alpha_{nt} = \mathrm{Ad}\, u$ for some n in $\mathbf{N}$ and some unitary fixed-point u in $\mathcal{M}$.*

Proof. By 8.1.14 the set

$$E = \{(t, \sigma) \,|\, \sigma \in \Gamma_B(\alpha)\}$$

is not dense in $\boldsymbol{T}$. Moreover, by 8.8.5 E is the union of cyclic groups. It follows that

$$E \subset \{\exp(2\pi i k n^{-1}) \,|\, 1 \leqslant k \leqslant n\}$$

for a suitable n in $\mathbf{N}$. Consequently nt annihilates $\Gamma_B(\alpha)$, whence 8.9.4 applies.

8.9.6. COROLLARY. *If α is an automorphism of a von Neumann algebra such that $Sp(\alpha)$ is not the whole unit circle then α^n is an inner automorphism for some n.*

8.9.7. THEOREM. *Let (A, G, α) be a C^*-dynamical system, and denote by G_0 the annihilator of $\Gamma_B(\alpha)$ in G. If $\alpha_t = \mathrm{Ad}\, u$ for some unitary u in the fixed-point algebra of G in $M(A)$, then $t \in G_0$. The converse holds if A is simple and $Sp(\alpha)/\Gamma(\alpha)$ is compact (in $\Gamma/\Gamma(\alpha)$).*

Proof. If $\alpha_t = \mathrm{Ad}\, u$ there exists by a standard maximality argument for each $\varepsilon > 0$ a set $\{f_i\}$ of continuous positive functions on $Sp(u)$ and a corresponding set of spectral values $\{\lambda_i\}$ such that $|\lambda - \lambda_i| \leqslant \frac{1}{2}\varepsilon$ for each λ in $Sp(u) \cap \mathrm{supp}.f_i$ and such that the hereditary C^*-subalgebras B_i of A generated by $f_i(u)Af_i(u)$ are contained in pairwise orthogonal closed ideals of A. Let B be the direct sum of the B_i's. Then $B \in \mathscr{H}^\alpha_B(A)$ by the maximality of the family $\{B_i\}$ and if $x \in B$ we have an expansion $x = \Sigma x_i$ with centrally orthogonal elements x_i, whence

$$\begin{aligned}\|\alpha_t(x) - x\| &= \|ux - xu\| = \mathrm{Sup}\|ux_i - x_iu\| \\ &\leqslant \mathrm{Sup}\|(u - \lambda_i)x_i\| + \|x_i(u - \lambda_i)\| \leqslant \varepsilon\|x\|.\end{aligned}$$

If $\sigma \in \Gamma_B(\alpha)$ then $(t, \sigma) \in Sp(\alpha_t \,|\, B)$ by 8.1.14, which implies that $|1 - (t, \sigma)| \leqslant \varepsilon$. Since ε is arbitrary $(t, \sigma) = 1$, i.e. $t \in G_0$.

Choose a neighbourhood Ω of θ such that $|1-(t,\Omega)|<\sqrt{3}$. By 8.8.7 there is a B in $\mathscr{H}^{\alpha}_{B}(A)$ such that

$$\mathrm{Sp}(\alpha\,|\,B)\subset\Gamma_B(\alpha)+\Omega.$$

Note that $\Gamma_B(\alpha)=\Gamma(\alpha)$ since A is now simple. It follows from 8.1.14 that $\nu(\iota-\alpha_t\,|\,B)<\sqrt{3}$ whence $\alpha_t\,|\,B$ is derivable by 8.7.7. Since A is simple so is B (by 4.11.10). Thus by 8.6.10 we have $\alpha_t\,|\,B=\mathrm{Ad}\,w$ for some unitary w in $M(B)$. Moreover, by 8.6.10 we may assume that $\alpha_s(w)=w$ for all s in G. Identifying B'' with $pA''p$ for some open projection p in A'' (cf. 3.11.10) we have $c(p)=1$ (otherwise $c(p)A''\cap A$ would be a non-trivial ideal of A). Applying 8.9.1 there is a unitary u in A'' such that $\alpha_t=\mathrm{Ad}\,u$ and $up=pu=w$. Since u is unique we have $\alpha''_s(u)=u$ for all s in G, i.e. u is a fixed-point. It is immediate to verify that the set

$$I=\{x\in A\,|\,ux\in A\}$$

is a closed ideal in A. Since $w\in M(B)$, $B\subset I$ whence $I=A$ since A was simple. Consequently $u\in M(A)$, as desired.

8.9.8. COROLLARY. *Let (A,G,α) be a C^*-dynamical a-system such that A is simple and $\mathrm{Sp}(\alpha)/\Gamma(\alpha)$ is compact. If $t\in G$ and $\mathrm{Sp}(\alpha_t)$ is not the whole unit circle then $\alpha_{nt}=\mathrm{Ad}\,u$ for some n in $\boldsymbol{N}$ and some unitary fixed-point u in $M(A)$.*

Proof. Exactly as in 8.9.5.

8.9.9. COROLLARY. *If α is an automorphism of a simple C^*-algebra such that $\mathrm{Sp}(\alpha)$ is not the whole unit circle then α^n is inner in $M(A)$ for some n.*

8.9.10. COROLLARY. *An automorphism α of a simple C^*-algebra A is inner in $M(A)$ if and only if $\Gamma(\alpha)=0$.*

8.9.11. Notes and remarks. Proposition 8.9.2 is due to Kaplansky, see [144]. The results in 8.9.3–8.9.6 were all established by Borchers in [24]. Theorem 8.9.7 and its corollaries are due to Olesen, see [182].

8.10. Connes spectrum and compact groups

8.10.1. In this section G will denote a compact abelian group and Γ its discrete dual group. Thus the elements of Γ (i.e. the characters on G) belong to $K^1(G)$. Consequently the theory of spectral subspaces reduces to questions of eigenspaces (cf. 8.1.8).

If (A,G,α) is a C^*-dynamical a-system and $\tau\in\Gamma$ we put

$$A_\tau=M^\alpha(\{\tau\})=\{x\in A\,|\,\alpha_t(x)=(t,\tau)x,\forall t\in G\}.$$

In particular, A_θ denotes the fixed-point algebra of G in A. Note from 8.1.9 that $\tau \in \mathrm{Sp}(\alpha)$ if and only if $A_\tau \neq 0$. Moreover, by 8.1.4 A is the norm closure of ΣA_τ.

For W^*-dynamical systems we employ the same notations and obtain analogous results.

8.10.2. LEMMA. *Let (A, G, α) be a C^*-dynamical system in which G is compact. For each closed normal subgroup H of G let A^H denote the C^*-subalgebra of fixed-points of H in A; and consider the C^*-dynamical system $(A^H, G/H, \dot{\alpha})$ obtained naturally from (A, G, α). Let $\mathscr{H}^\alpha(A)$ and $\mathscr{H}^{\dot{\alpha}}(A^H)$ be as in 8.8.2. Then the map $B \to B \cap A^H$ is a bijection of $\mathscr{H}^\alpha(A)$ onto $\mathscr{H}^{\dot{\alpha}}(A^H)$.*

Proof. The projection π of A onto A^H given by

$$\pi(x) = \int_H \alpha_t(x)\, \mathrm{d}_H(t), \qquad x \in A,$$

is faithful. If therefore $B \neq 0$ then $B \cap A^H \neq 0$. Thus the map is well-defined. Moreover, if $B_1 \cap A^H = B_2 \cap A^H$ take an approximate unit $\{u_\lambda\}$ for B_1. We may assume that $\{u_\lambda\} \subset A^H$, replacing it otherwise with $\{\pi(u_\lambda)\}$. But then $\{u_\lambda\} \subset B_2$ and for each x in B_1 we have

$$x = \mathrm{Lim}\, u_\lambda x u_\lambda \in B_2,$$

whence $B_1 = B_2$. Thus the map is injective.

Take now C in $\mathscr{H}^{\dot{\alpha}}(A^H)$. The smallest hereditary cone M of A_+ containing C_+ is the set of elements x in A_+ such that $x \leqslant y$ for some y in C_+. The closure of M is the positive part of a hereditary C^*-subalgebra B of A; and clearly $B \cap A^H = C$. From the construction we see that M is G-invariant, whence $B \in \mathscr{H}^\alpha(A)$. This shows that the map is surjective, and completes the proof.

8.10.3. PROPOSITION. *Let (A, G, α) be a C^*-dynamical a-system with G compact. For each closed subgroup H of G consider the C^*-dynamical a-system $(A^H, G/H, \dot{\alpha})$ described in 8.10.2. Identifying $(G/H)\hat{}$ with the annihilator $H^\perp$ of H in Γ we have*

$$\mathrm{Sp}(\dot{\alpha}) = \mathrm{Sp}(\alpha) \cap H^\perp; \qquad \Gamma(\dot{\alpha}) = \Gamma(\alpha) \cap H^\perp.$$

Proof. Take B in $\mathscr{H}^\alpha(A)$ and assume that $\sigma \in \mathrm{Sp}(\alpha \mid B) \cap H^\perp$. There is then a non-zero x in B such that $\alpha_t(x) = (t, \sigma)x$ for all t in G (cf. 8.10.1). In particular $\alpha_t(x) = x$ whenever $t \in H$, whence $x \in A^H$. Consequently $\sigma \in \mathrm{Sp}(\dot{\alpha} \mid B \cap A^H)$. The converse is obvious, so that

$$\mathrm{Sp}(\alpha \mid B) \cap H^\perp = \mathrm{Sp}(\dot{\alpha} \mid B \cap A^H).$$

Taking $B = A$ we obtain the first half of the proposition; and the second follows from 8.10.2.

8.10.4. THEOREM. *Let (A, G, α) be a C^*-dynamical a-system with G compact and assume that A is G-prime (cf. 7.9.4). Then the fixed-point algebra A_θ is prime if and only if* $\mathrm{Sp}(\alpha) = \Gamma(\alpha)$.

Proof. If A_θ is not prime then $xA_\theta y = 0$ for two non-zero elements x and y in A_θ. We may assume that x and y are positive replacing them otherwise with x^*x and yy^*. Since A is G-prime, $xay \neq 0$ for some a in A; and as the eigenspaces A_τ, $\tau \in \mathrm{Sp}(\alpha)$, span a dense set in A we may assume that $a \in A_\tau$ for some τ in $\mathrm{Sp}(\alpha)$, $\tau \neq \theta$. Using the facts that $A_\sigma A_\tau \subset A_{\sigma+\tau}$ and $A_\tau^* = A_{-\tau}$ we see that $xay \in A_\tau$ and that $z \in A_\theta$, where $z = (xay)^*(xay)$.

Let B be the hereditary C^*-subalgebra of A generated by z, i.e. $B = L^* \cap L$ where L is the closure of Az. Since z is a fixed-point, $B \in \mathscr{H}^\alpha(A)$. Since B is generated by z the sequence $\{(n^{-1} + z)^{-1}z\}$ is an approximate unit for B, and it follows that if $b \in B\backslash\{0\}$ then $zb \neq 0$ and $bz \neq 0$.

Take now b in $B \cap A_\tau$. Then $b(xay)^* \in A_\theta$, whence

$$zbz = (xay)^*xa(yb(xay)^*x)ay = 0,$$

because $yA_\theta x = (xA_\theta y)^* = 0$. It follows that $b = 0$ and thus $B \cap A_\tau = 0$. Consequently $\tau \notin \mathrm{Sp}(\alpha | B)$, whence $\tau \notin \Gamma(\alpha)$. Thus $\Gamma(\alpha) \neq \mathrm{Sp}(\alpha)$.

For the converse assume that A_θ is prime and take τ in $\mathrm{Sp}(\alpha)$. There is then a non-zero a in A_τ. If $B \in \mathscr{H}^\alpha(A)$ there is a non-zero b in $B \cap A_\theta$. We claim that $bxa \neq 0$ for some x in A_θ. Otherwise $bA_\theta aa^* = 0$, a contradiction since $aa^* \in A_\theta$ and A_θ is prime. Put $z = bxa$. Then $z \in A_\tau$, whence $zyb \neq 0$ for some y in A_θ, reasoning as before. Consequently

$$0 \neq bxayb \in B \cap A_\tau,$$

whence $\tau \in \mathrm{Sp}(\alpha | B)$. This holds for all B in $\mathscr{H}^\alpha(A)$ and we conclude that $\tau \in \Gamma(\alpha)$. Thus $\mathrm{Sp}(\alpha) = \Gamma(\alpha)$, completing the proof.

8.10.5. COROLLARY. *If $(\mathscr{M}, G, \alpha)$ is a W^*-dynamical a-system where G is compact and $\mathscr{M}$ is a factor then* $\mathrm{Sp}(\alpha) = \Gamma(\alpha)$ *if and only if $\mathscr{M}_\theta$ is a factor.*

Proof. If x, y are positive elements then $x\mathscr{M}y = 0$ if and only if $c(x)c(y) = 0$ (cf. 2.6). Thus a von Neumann algebra is prime if and only if it is a factor. Moreover, if $\mathscr{M}$ is a factor then any σ-weakly dense C^*-subalgebra of $\mathscr{M}$ must be prime. The result is therefore immediate from 8.8.10 and 8.10.4. (Note that $\mathscr{M}_\theta \subset \mathscr{M}^c$ so that $(\mathscr{M}^c)_\theta = \mathscr{M}_\theta$).

8.10.6. PROPOSITION. *Let (A, G, α) be a C^*-dynamical a-system with G compact and A G-prime. Assume that* $\mathrm{Sp}(\alpha)/\Gamma(\alpha)$ *is finite. There exists then B in $\mathscr{H}^\alpha(A)$ such that $B \cap A_\theta$ is prime.*

Proof. Since $\mathrm{Sp}(\alpha)/\Gamma(\alpha)$ is finite there is by 8.8.7 a B in $\mathscr{H}^\alpha(A)$ such that $\mathrm{Sp}(\alpha | B) = \Gamma(\alpha)$. (Note that $\Gamma_B(\alpha) = \Gamma(\alpha)$ and $\mathscr{H}_B^\alpha(A) = \mathscr{H}^\alpha(A)$ since A is G-

prime). Since A is G-prime, so is B; and as $\Gamma(\alpha \mid B) = \Gamma(\alpha) = \mathrm{Sp}(\alpha \mid B)$ we conclude from 8.10.4 that $B \cap A_\theta$ is prime.

8.10.7. Consider a C^*-dynamical a-system (A, G, α) with G compact and let $M(A)$ be the multiplier algebra of A. We shall examine the relative commutant of A_θ in $M(A)$, i.e. the set

$$A_\theta^{rc} = A'_\theta \cap M(A).$$

Note that A_θ^{rc} is also the relative commutant of $M(A)_\theta$ in $M(A)$; for if $x \in A_\theta^{rc}$ and $y \in M(A)_\theta$ take an approximate unit $\{u_\lambda\}$ for A contained in A_θ. Then $yu_\lambda \in A_\theta$ and $yu_\lambda \to y$ σ-weakly in A'', whence

$$xy - yx = \mathrm{Lim}\, xyu_\lambda - yu_\lambda x = 0.$$

8.10.8. LEMMA. *Let (A, G, α) be a C^*-dynamical a-system with G compact and assume that the centre of $M(A)_\theta$ is the scalar s. Then $\mathrm{Sp}(\alpha \mid A_\theta^{rc})$ is a subgroup of Γ and for each τ in $\mathrm{Sp}(\alpha \mid A_\theta^{rc})$ there is a unitary u in $A_\theta^{rc} \cap M(A)_\tau$ such that*

$$A_\theta^{rc} \cap M(A)_{n\tau} = \boldsymbol{C}u^n, \qquad n \in \boldsymbol{Z}.$$

If H denotes the stabilizer of u in G then the C^*-algebra $C^*(u)$ generated by u is the centre of the fixed-point algebra $M(A)^H$ of H in $M(A)$.

Proof. If $\tau \in \mathrm{Sp}(\alpha \mid A_\theta^{rc})$ take non-zero elements x and y in $A_\theta^{rc} \cap M(A)_\tau$. Then x^*y and yx^* belong to the centre of $M(A)_\theta$ (cf. 8.10.7), which is trivial by assumption. Taking $x = y$ we see that x is a scalar multiple of a unitary u in $A_\theta^{rc} \cap M(A)_\tau$. Taking $x = u$ we see that $u^*y \in \boldsymbol{C}1$, and consequently $y \in \boldsymbol{C}u$, i.e.

$$A_\theta^{rc} \cap M(A)_\tau = \boldsymbol{C}u. \tag{$*$}$$

Since a product of unitaries is never zero we see from $(*)$ that $\mathrm{Sp}(\alpha \mid A_\theta^{rc})$ is a group. Note that $u^n \in M(A)_{n\tau}$ for every n in $\boldsymbol{Z}$, whence

$$A_\theta^{rc} \cap M(A)_{n\tau} = \boldsymbol{C}u^n.$$

Let H be the stabilizer of u in G. Since $\alpha_t(u) = (t, \tau)u$ for every t in G we see that H can also be defined as the annihilator of the cyclic group $\boldsymbol{Z}\tau$ in Γ.

If $x \in M(A)^H \cap M(A)_{n\tau}$ for some n in $\boldsymbol{Z}$ then

$$x = xu^{-n}u^n \in M(A)_\theta u^n,$$

so that u commutes with x. Considering the system $(M(A)^H, G/H, \dot{\alpha})$ we see from 8.10.3 that $\mathrm{Sp}(\dot{\alpha}) \subset \boldsymbol{Z}\tau$. Thus the eigenspaces $M(A)^H \cap M(A)_{n\tau}$, $n \in \boldsymbol{Z}$, have dense span in $M(A)^H$, whence u commutes with $M(A)^H$. Since $u \in M(A)_\tau$ and $M(A)_\tau \subset M(A)^H$ it follows that $C^*(u)$ is contained in the centre of $M(A)^H$. On the other hand, $M(A)_\theta \subset M(A)^H$, whence $(M(A)^H)' \subset (M(A)_\theta)'$, so that the centre of $M(A)^H$ is contained in $A_\theta^{rc} \cap M(A)^H$. As $\mathrm{Sp}(\dot{\alpha}) \subset \boldsymbol{Z}\tau$ and

$$(A_\theta^{rc} \cap M(A)^H)_{n\tau} = \boldsymbol{C}u^n$$

for every n in $\boldsymbol{Z}$ it follows that the centre of $M(A)^H$ equals $C^*(u)$.

8.10.9. COROLLARY. *If (A, G, α) is a C^*-dynamical system with G compact and Γ cyclic, and if the centre of $M(A)_\theta$ is the scalars then A_θ^{rc} is commutative.*

Proof. Since $\mathrm{Sp}(\alpha \mid A_\theta^{rc})$ is a subgroup of a cyclic group it is itself cyclic so that we may take the τ in 8.10.8 as a generator for $\mathrm{Sp}(\alpha \mid A_\theta^{rc})$. It follows from 8.10.8 that $A_\theta^{rc} = C^*(u)$.

8.10.10. THEOREM. *Let (A, G, α) be a C^*-dynamical system where G is either the circle group $\boldsymbol{T}$ or cyclic of prime order, and assume that A is G-prime. If $\mathrm{Sp}(\alpha) = \Gamma(\alpha)$ then $A_\theta^{rc} = \boldsymbol{C}1$ and no $\alpha_t \neq \iota$ is inner in $M(A)$.*

Proof. From 8.10.4 we know that A_θ is prime, and thus the centre of $M(A)_\theta$ is the scalars, since $M(A)_\theta = M(A_\theta)$ and the multiplier algebra of a prime algebra is prime (A is essential in $M(A)$ and 4.1.11 applies). Applying 8.10.8 we obtain a unitary u in $M(A)$ and a closed subgroup H of G such that

$$A_\theta^{rc} = C^*(u) = M(A)^H \cap (M(A)^H)'.$$

If $H = \{\theta\}$ then $M(A)^H = M(A)$ which is prime. Thus $A_\theta^{rc} = \boldsymbol{C}1$ (and moreover $G = \{0\}$, since G is the stabilizer for u $(= 1)$). If $H = G$ then $M(A)^H = M(A)_\theta$ which has trivial centre. Thus again $A_\theta^{rc} = \boldsymbol{C}1$. For a cyclic group of prime order these are the only possibilities. To finish the proof we must therefore exclude the case where $G = \boldsymbol{T}$ and where H is a finite non-zero subgroup of $\boldsymbol{T}$.

Applying 8.10.6 to the system $(A, H, \alpha \mid H)$ we obtain a non-zero hereditary C^*-subalgebra B of A which is H-invariant, such that $B \cap A^H$ is prime. Since A_θ^{rc} is the centre of $M(A)^H$ the set

$$I = \{x \in A_\theta^{rc} \mid x(B \cap A^H) = 0\}$$

is a closed ideal of A_θ^{rc}. Moreover, the quotient A_θ^{rc}/I is isomorphic to a set of central multipliers of $B \cap A^H$. Since $B \cap A^H$ is prime so is $M(B \cap A^H)$, whence $A_\theta^{rc}/I \subset \boldsymbol{C}1$. It follows that $u - \lambda 1 \in I$ for some scalar λ with $|\lambda| = 1$. Assuming, as we may that $\lambda = 1$ this means that $ub = b$ for every b in $B \cap A^H$. Take a non-zero b in $(B \cap A^H)_+$. Then

$$b\alpha_t(b) = b\alpha_t(ub) = (t, \tau)b\,\alpha_t(b).$$

If $t \notin H$ then $(t, \tau) \neq 1$, whence $b\alpha_t(b) = 0$. However, $\alpha_t(b) \to b$ as $t \to 0$ in $\boldsymbol{T}$, which leads to a contradiction. Consequently $A^{rc\theta} = \boldsymbol{C}1$ also when $G = \boldsymbol{T}$.

If $\alpha_t = \mathrm{Ad}\, w$ for some w in $M(A)$ then $w \in A_\theta^{rc}$. In our case this implies that $\alpha_t = \iota$.

8.10.11. COROLLARY. *Let $(\mathcal{M}, G, \alpha)$ be a W^*-dynamical system where G is either the circle group of cyclic of prime order, and assume that $\mathcal{M}$ is a factor. If* $\mathrm{Sp}(\alpha) = \Gamma(\alpha)$ *then* $(\mathcal{M}_\theta)' \cap \mathcal{M} = \boldsymbol{C}1$ *and no* $\alpha_t \neq \iota$ *is inner.*

Proof. Consider the C^*-dynamical system $(\mathcal{M}^c, G, \alpha \mid \mathcal{M}^c)$. If $\mathcal{M}$ is a factor then $\mathcal{M}^c$ is prime. Thus by 8.8.10 and 8.10.10 the relative commutant of $(\mathcal{M}^c)_\theta$ $(=\mathcal{M}_\theta)$ in $\mathcal{M}^c$ is trivial. If therefore $x \in (\mathcal{M}_\theta)' \cap \mathcal{M}$ then $\alpha_f(x) \in \boldsymbol{C}1$ for every f in $L^1(G)$, whence $x \in \boldsymbol{C}1$. Thus $(\mathcal{M}_\theta)' \cap \mathcal{M} = \boldsymbol{C}1$ and as before this implies that no $\alpha_t \neq \iota$ is inner.

8.10.12. THEOREM. *Let (A, G, α) be a C^*-dynamical a-system with G compact and assume that A is simple. If* $\mathrm{Sp}(\alpha)$ *is finite, in particular if G is finite, the following conditions are equivalent:*

(i) $\mathrm{Sp}(\alpha) = \Gamma(\alpha)$;

(ii) *A_θ is simple;*

(iii) *A_θ is prime;*

(iv) *The centre of $M(A)_\theta$ is the scalars;*

(v) *No $\alpha_t \neq \iota$ is implemented by a unitary in $M(A)_\theta$.*

Proof. By 8.10.4, (i)$\Leftrightarrow$(iii) and trivially (ii)$\Rightarrow$(iii)$\Rightarrow$(iv)$\Rightarrow$(v). Moreover, (v)$\Rightarrow$(i) since $\Gamma(\alpha) \neq \mathrm{Sp}(\alpha)$ implies the existence of a t in G that annihilates $\Gamma(\alpha)$ but not $\mathrm{Sp}(\alpha)$. Thus $\alpha_t \neq \iota$ and by 8.9.7 $\alpha_t = \mathrm{Ad}\, u$ for some unitary u in $M(A)_\theta$.

(i)$\Rightarrow$(ii). Suppose I_0 is a non-trivial closed ideal of A_θ. If x, y are elements in some A_τ, $\tau \in \mathrm{Sp}(\alpha)$ then $y^* A_\theta x \subset A_\theta$. Simple algebraic manipulations show that the set of linear combinations y^*ax, with x, y in A_τ and a in I_0 form an ideal in A_θ. We denote by I_τ its closure and note that it is a non-zero ideal in A_θ, since $I_0 x = 0$ for some non-zero x in A_τ would imply the existence of an ideal orthogonal to I_0 in A_θ. However, this is impossible since A_θ is prime.

Let $I = \cap I_\tau$, $\tau \in \mathrm{Sp}(\alpha)$. Since A_θ is prime and $\mathrm{Sp}(\alpha)$ is finite we see that $I \neq 0$. We claim that $x^* I x \subset I$ for every x in A_τ, $\tau \in \mathrm{Sp}(\alpha)$. To see this it suffices to remark that by the construction of the I_τ's we have $x^* I_\sigma x \subset I_{\tau+\sigma}$ for every x in A_τ.

Let p denote the strong limit in A'' of an approximate unit $\{u_\tau\}$ for I. If $x \in A_\tau$ then $x^* u_\tau x \leqslant \|x\|^2 p$, since p is a unit for I; whence in the limit $x^* p x \leqslant \|x\|^2 p$. Let u be a unitary in $\tilde{A}$. Then we have a decomposition $u = \Sigma x_\tau$, $\tau \in \mathrm{Sp}(\alpha)$, where $x_\tau \in A_\tau$ for $\tau \neq \theta$ and $x_\theta \in A_\theta + \boldsymbol{C}1$. By successive applications of the inequality $(a+b)^* p (a+b) \leqslant 2(a^* p a + b^* p b)$ we obtain

$$u^* p u \leqslant 2^n \left(\sum \|x_\tau\|^2\right) p,$$

where n is the cardinality of $\mathrm{Sp}(\alpha)$. This implies that $u^* p u$ is orthogonal to $1 - p$; whence $u^* p u \leqslant p$. Since this holds for every unitary u in $\tilde{A}$ we conclude that p is an open central projection in A''. By 3.11.10 there is a closed ideal J in A such that $J = A''p \cap A$. We have $J \neq 0$ since $I \subset J$. But also $J \neq A$, because $I \neq A_\theta$ so that $\phi(p) = 0$ for some normal state of A'' that extends a state of A_θ which annihilates I. Since A was simple we have reached a contradiction. Consequently A_θ is simple.

8.10.13. PROPOSITION. *Let (A, G, α) be a C^*-dynamical a-system with G compact and assume that* $\mathrm{Sp}(\alpha)$ *is finite and A is simple. Then $A_\theta^{rc} = \mathbf{C}1$ if and only if every $\alpha_t \neq \iota$ is an outer automorphism of $M(A)$.*

Proof. As we saw in the proof of 8.10.10, if $\alpha_t = \mathrm{Ad}\, w$, $w \in M(A)$ then $w \in A_\theta^{rc}$; so the condition $A_\theta^{rc} = \mathbf{C}1$ is sufficient. Assume therefore that $A_\theta^{rc} \neq \mathbf{C}1$. Since $A_\theta^{rc} \cap M(A)_\theta = \mathbf{C}1$ if every $\alpha_t \neq \iota$ is outer (by 8.10.12) this implies that there is a non-zero element τ in $\mathrm{Sp}(\alpha \,|\, A_\theta^{rc})$. By 8.10.8 there is a unitary u in $A_\theta^{rc} \cap M(A)_\tau$ such that $C^*(u)$ is the centre of $M(A)^H$ where H is the stabilizer of u in G. Applying 8.10.12 to the system $(A, H, \alpha \,|\, H)$ we see that $\Gamma(\alpha \,|\, H) \neq \mathrm{Sp}(\alpha \,|\, H)$. By 8.9.7 there is an $\alpha_t \neq \iota$, $t \in H$, such that $\alpha_t = \mathrm{Ad}\, w$ for some unitary w in $C^*(u)$, contrary to our assumption.

8.10.14. Let α be a periodic automorphism of a prime C^*-algebra A and denote by n the period of α. There is then a smallest number m $(1 \leqslant m \leqslant n)$ such that $\alpha^m = \mathrm{Ad}\, w$ for some w in $M(A)$; and m divides n. Since $\alpha^m = \mathrm{Ad}\, \alpha(w)$ as well, we see that $\alpha(w) = \gamma w$, where $\gamma \in \mathbf{T}$. Clearly $\alpha^m(w) = w$ so that $\gamma^m = 1$. Let k denote the order of γ so that k divides m. We say that the product km is the *minimal period* of α.

8.10.15. LEMMA. *The minimal period of an automorphism of a prime C^*-algebra is the smallest number q for which $\alpha^q = \mathrm{Ad}\, u$ with u in (the centre of) the fixed-point algebra of α in $M(A)$.*

Proof. Assume that $\alpha^q = \mathrm{Ad}\, u$ with $\alpha(u) = u$. Since m is the smallest exponent for which α^m is inner in $M(A)$ we have $q = rm$. But then

$$\mathrm{Ad}\, u = \alpha^{mr} = \mathrm{Ad}\, w^r,$$

whence $r \geqslant k$, so that $q \geqslant km$. On the other hand, $\alpha^{km} = \mathrm{Ad}\, w^k$, and $\alpha(w^k) = \gamma^k w^k = w^k$, completing the proof.

8.10.16. THEOREM. *If α is a periodic automorphism of a simple C^*-algebra A (respectively a factor $\mathcal{M}$) the following conditions are equivalent:*

(i) *The minimal period of α is equal to the period;*

(ii) $\mathrm{Sp}(\alpha) = \Gamma(\alpha)$;

(iii) *If $\alpha^q = \mathrm{Ad}\, w$ for some w in $M(A)$* (resp. $\mathcal{M}$) *and $q <$ period α then $\alpha(w) \neq w$.*

If moreover the period of α is the product of distinct primes these conditions are again equivalent to:

(iv) *No* $\alpha^q, q \leqslant$ *period* α *is inner in* $M(A)$ (resp. $\mathcal{M}$).

Proof. (i) $\Rightarrow$ (iii) is 8.10.15. (iii) $\Rightarrow$ (ii) and (ii) $\Rightarrow$ (i) follow from 8.9.7 (resp. 8.9.4), and (iv) $\Rightarrow$ (ii) is evident.

Now assume (i) (and (ii), (iii)) and let n be the period of α. From 8.10.14 we have $\alpha^m = \operatorname{Ad} w$ and $\alpha(w) = \gamma w$ with $\gamma^k = 1$. Moreover, m divides n and k divides m. If $m < n$ then $1 < m$ and $1 < k$ (if $m = 1$ then $k = 1$ whence $n = km = 1$ by (i); and if $k = 1$ then $m = n$ by (iii)). Thus by (i)

$$n = km = k^2(m/k).$$

Since n contains no squares, $k = 1$; whence $m = n$ as desired.

8.10.17. Notes and remarks. This section is taken from the paper [188] by Olesen, Størmer and the author. Theorem 8.10.4 was found independently by Kishimoto and Takai [150]. The inspiration to the paper was the von Neumann algebra results in 8.10.5, 8.10.11 and 8.10.16 obtained by Connes [38] and [41]. In [41] there is an example of an automorphism α of the Fermion algebra such that $\alpha^4 = \iota$, $\alpha^2 = \operatorname{Ad} w$ for some unitary w with $\alpha(w) = -w$. This shows that the implication (ii) $\Rightarrow$ (iv) in 8.10.16 is not true in general. It also shows that 8.10.10 is not necessarily true if the cyclic group G does not have prime order.

8.11. Connes spectrum and crossed products

8.11.1. We return to the case of a general locally compact (separable) abelian group of automorphisms, and consider a C^*-dynamical a-system (A, G, α). The algebra A^G of fixed-points of G in A (or $M(A)$) will usually be of little interest if G is not compact, since we no longer have a projection from A to A^G. Instead of the pair (A, A^G) we shall instead consider the pair $(G \times_\alpha A, A)$. Note that A is contained in the fixed-point algebra of Γ in $M(G \times_\alpha A)$ (cf. 7.8.2 (i)) and that the operator I defined in 7.8.4 is a densely defined map from $G \times_\alpha A$ into A by 7.8.7.

When G is compact the function $p: G \to M(A)$, given by $p(t) = 1$ for all t, is a projection in $M(G \times_\alpha A)$ (cf. 7.6.2), and a straightforward computation shows that $p(G \times_\alpha A)p = A^G p$ (regarding A as a subalgebra of $M(G \times_\alpha A)$). This observation may be used to deduce some of the results from the previous section (viz. 8.10.4, 8.10.5, 8.10.12) from the corresponding results in this section (8.11.10, 8.11.15, 8.11.12).

8.11.2. LEMMA. *Let* (A, G, α) *and* (A, G, β) *be* C^**-dynamical systems. There is then a* C^**-dynamical system* $(A \otimes M_2, G, \gamma)$ *such that*

$$\gamma_t\begin{pmatrix} x & 0 \\ 0 & y \end{pmatrix} = \begin{pmatrix} \alpha_t(x) & 0 \\ 0 & \beta_t(y) \end{pmatrix}, \qquad x, y \in A, \qquad t \in G,$$

if and only if there is a function $t \to u_t$ from G into the unitary group of $M(A)$ satisfying the conditions:

(i) $u_{st} = u_s\alpha_s(u_t)$;

(ii) $\beta_t = (\mathrm{Ad}\, u_t)\alpha_t$;

(iii) $t \to u_t x$ *is continuous for each x in A.*

Proof. If a system $(A \otimes M_2, G, \gamma)$ exists as described above we extend γ to $M(A \otimes M_2)$ $(= M(A) \otimes M_2)$ and put

$$\gamma_t\begin{pmatrix} 0 & 0 \\ 1 & 0 \end{pmatrix} = \begin{pmatrix} a & c \\ u_t & b \end{pmatrix} = m \in M(A) \otimes M_2.$$

Since γ_t is an automorphism and $\alpha_t(1) = \beta_t(1) = 1$ we know that

$$m^*m = \begin{pmatrix} 0 & 0 \\ 0 & 1 \end{pmatrix} \quad \text{and} \quad mm^* = \begin{pmatrix} 1 & 0 \\ 0 & 0 \end{pmatrix}.$$

This implies that $a = b = c = 0$ and that u_t is unitary in $M(A)$. Moreover,

$$\begin{pmatrix} 0 & 0 \\ u_{st} & 0 \end{pmatrix} = \gamma_{st}\begin{pmatrix} 0 & 0 \\ 1 & 0 \end{pmatrix} = \gamma_s\begin{pmatrix} 0 & 0 \\ 1 & 0 \end{pmatrix}\begin{pmatrix} u_t & 0 \\ 0 & 0 \end{pmatrix},$$

whence $u_{st} = u_s\alpha_s(u_t)$. Further,

$$\begin{pmatrix} 0 & 0 \\ 0 & \beta_t(x) \end{pmatrix} = \gamma_t\begin{pmatrix} 0 & 0 \\ 1 & 0 \end{pmatrix}\begin{pmatrix} x & 0 \\ 0 & 0 \end{pmatrix}\begin{pmatrix} 0 & 1 \\ 0 & 0 \end{pmatrix},$$

whence $\beta_t(x) = u_t\alpha_t(x)u_t^*$ so that $\beta_t = (\mathrm{Ad}\, u_t)\circ\alpha_t$. Finally, the function

$$t \to \gamma_t\begin{pmatrix} 0 & 0 \\ 1 & 0 \end{pmatrix}\begin{pmatrix} x & 0 \\ 0 & 0 \end{pmatrix} = \begin{pmatrix} 0 & 0 \\ u_t\alpha_t(x) & 0 \end{pmatrix}$$

is continuous for every x in A, so that also $t \to u_t x$ is continuous.

Conversely, if $t \to u_t$ is a function satisfying the three conditions above, define

$$\gamma_t\begin{pmatrix} x_{11} & x_{12} \\ x_{21} & x_{22} \end{pmatrix} = \begin{pmatrix} \alpha_t(x_{11}) & \alpha_t(x_{12})u_t^* \\ u_t\alpha_t(x_{21}) & \beta_t(x_{22}) \end{pmatrix}.$$

Elementary computations show that γ_t is an automorphism (by (ii)); that $t \to \gamma_t$ is a representation (by (i)); and that each function $t \to \gamma_t(m)$, $m \in A \otimes M_2$, is continuous (by (iii)). Thus $(A \otimes M_2, G, \gamma)$ is a C^*-dynamical system as desired.

8.11.3. We say that two systems (A, G, α) and (A, G, β) are *exterior equivalent* if they satisfy the conditions in 8.11.2. Straightforward computations with the equations (i) and (ii) in 8.11.2 show that this is indeed an equivalence relation.

A function satisfying condition (i) is called a unitary *cocycle* (more precisely a one-cocycle). Note that if a unitary cocycle $\{u_t \mid t \in G\}$ satisfies condition (iii) in 8.11.2 then we can *define* a new system (A, G, β) from the given system (A, G, α) by $\beta_t = (\mathrm{Ad}\, u_t)\alpha_t$.

We say that two W^*-dynamical systems $(\mathcal{M}, G, \alpha)$ and $(\mathcal{M}, G, \beta)$ are *exterior equivalent* if there exists a σ-weakly continuous function $t \to u_t$ from G into the unitary group of $\mathcal{M}$ satisfying conditions (i) and (ii) of 8.11.2. It is easy to show that this is equivalent with the existence of a W^*-dynamical system $(\mathcal{M} \otimes \mathbf{M}_2, G, \gamma)$ such that

$$\gamma_t\begin{pmatrix} x & 0 \\ 0 & y \end{pmatrix} = \begin{pmatrix} \alpha_t(x) & 0 \\ 0 & \beta_t(y) \end{pmatrix}, \qquad x, y \in \mathcal{M}, \qquad t \in G.$$

8.11.4. LEMMA. *Let (A, G, α) (resp. $(\mathcal{M}, G, \alpha)$) be a C^*dynamical (resp., W^*-dynamical) system and suppose that p and q are G-invariant equivalent projections in $M(A)$ (resp. $\mathcal{M}$). Then*

$$\Gamma(\alpha \mid pAp) = \Gamma(\alpha \mid qAq) \ (\textit{resp. } \Gamma(\alpha \mid p\mathcal{M}p) = \Gamma(q\mathcal{M}q)).$$

Proof. By hypothesis there is a partial isometry v in $M(A)$ such that $v^*v = p$ and $vv^* = q$. Suppose that $\sigma \in \Gamma(\alpha \mid pAp)$ and take B in $\mathscr{H}^\alpha(qAq)$ (notations as in 8.8.2). For each neighbourhood Ω of σ we must show that $R^\alpha(\Omega) \cap B \neq \{0\}$.

Choose neighbourhoods Ω_0 and Ω_1 of θ and σ, respectively, such that $\Omega_0 + \Omega_1 \subset \Omega$. Then find a covering of Γ with open sets Ω_i, $i \in I$, such that $\Omega_i - \Omega_i \subset \Omega_0$ for every i. If b is a non-zero element in B then $bv \neq 0$. Consequently $\alpha_f(bv) \neq 0$ for some f in $K^1(G)$ with $\mathrm{supp.} f \subset \Omega_i$ for some i. Otherwise

$$bv \in R^{\alpha'}(\Omega_i)^\perp = M^\alpha(\Gamma \backslash \Omega_i)$$

for all i, whence $bv = 0$ since $\cup\, \Omega_i = \Gamma$. Set $x = \alpha_f(bv)$ and let C be the hereditary C^*-subalgebra of A generated by the orbit $\{\alpha_t(x^*x) \mid t \in G\}$. Then $C \in \mathscr{H}^\alpha(pAp)$ and since $\sigma \in \Gamma(\alpha \mid pAp)$ there is a non-zero y in $C \cap R^\alpha(\Omega_1)$. From the construction of C we see that

$$z = \alpha_t(x) y\, \alpha_s(x^*) \neq 0$$

for some s and t in G. But $z \in B$ since $xAx^* \subset B$ and B is G-invariant. Finally, by 8.3.3

$$z \in R^\alpha(\Omega_i) R^\alpha(\Omega_1) R^\alpha(\Omega_i)^* \subset R^\alpha(\Omega_i + \Omega_1 - \Omega_i) \subset R^\alpha(\Omega).$$

Thus $R^\alpha(\Omega) \cap B \neq \{0\}$ and since Ω and B were arbitrary it follows that

N

$\sigma \in \Gamma(\alpha | qAq)$. Consequently $\Gamma(\alpha | pAp) \subset \Gamma(\alpha | qAq)$ and a symmetric argument yields the converse inclusion, and thus equality.

The proof for $(\mathcal{M}, G, \alpha)$ is analogous.

8.11.5. PROPOSITION. *If (A, G, α) and (A, G, β) (resp. $(\mathcal{M}, G, \alpha)$ and $(\mathcal{M}, G, \beta)$) are exterior equivalent C^*-dynamical (resp. W^*-dynamical) a-systems then $\Gamma(\alpha) = \Gamma(\beta)$.*

Proof. Consider the system $(A \otimes M_2, G, \gamma)$ obtained from 8.11.1 and let $p = \begin{pmatrix} 1 & 0 \\ 0 & 0 \end{pmatrix}$ and $q = \begin{pmatrix} 0 & 0 \\ 0 & 1 \end{pmatrix}$. Then p and q are G-invariant equivalent projections in $M(A) \otimes M_2$, whence by 8.11.4

$$\Gamma(\alpha) = \Gamma(\gamma | p(A \otimes M_2)p) = \Gamma(\gamma | q(A \otimes M_2)q) = \Gamma(\beta).$$

The W^*-dynamical situation is analogous.

8.11.6. PROPOSITION. *Let (A, G, α) (resp. $(\mathcal{M}, G, \alpha)$) be a C^*-dynamical (resp. W^*-dynamical) a-system and let $(\Gamma \times_{\hat{\alpha}} G \times_{\alpha} A, G, \hat{\hat{\alpha}})$ (resp. $(\Gamma \times_{\hat{\alpha}} G \times_{\alpha} \mathcal{M}, G, \hat{\hat{\alpha}})$) denote its double dual system (cf. 7.9.3 and 7.10.6). Then $\Gamma(\alpha) = \Gamma(\hat{\hat{\alpha}})$.*

Proof. We claim that the two systems

$$(A \otimes C(L^2(G)), G, \alpha \otimes \iota) \qquad \text{and} \qquad (A \otimes C(L^2(G)), G, \alpha \otimes \text{Ad}\,\tilde{\lambda})$$

are exterior equivalent. To see this define $u_t = 1 \otimes \tilde{\lambda}_t$ and note that u is a unitary representation of G into the algebra of fixed-points (under $\alpha \otimes \iota$) of G in $M(A) \otimes C(L^2(G))$). In particular, u satisfies condition (i) and (ii) of 8.11.2, and clearly also (iii). Thus $\Gamma(\alpha \otimes \iota) = \Gamma(\alpha \otimes \text{Ad}\,\tilde{\lambda})$ by 8.11.5. From 7.9.3 we know that $\Gamma(\alpha \otimes \text{Ad}\,\tilde{\lambda}) = \Gamma(\hat{\hat{\alpha}})$ and it only remains to show that $\Gamma(\alpha) = \Gamma(\alpha \otimes \iota)$. Choosing a basis for $L^2(G)$ we obtain a matrix representation for $C(L^2(G))$ and thus for each point n on the diagonal we have a covariant embedding ι_n of A as a hereditary C^*-subalgebra of $A \otimes C(L^2(G))$. From the definition of the Connes spectrum this implies that $\Gamma(\alpha \otimes \iota) \subset \Gamma(\alpha)$. On the other hand, if B is a non-zero G-invariant hereditary C^*-subalgebra of $A \otimes C(L^2(G))$ then $B \cap \iota_n(A) \neq \{0\}$ for some diagonal point n whence $\text{Sp}(\alpha \otimes \iota | B) \supset \text{Sp}(\alpha | C)$, where C denotes the hereditary C^*-subalgebra of A such that $\iota_n(C) = B \cap \iota_n(A)$. Consequently $\Gamma(\alpha \otimes \iota) \supset \Gamma(\alpha)$, whence $\Gamma(\alpha \otimes \iota) = \Gamma(\alpha)$ and the proof is complete.

The W^*-dynamical situation is analogous.

8.11.7. LEMMA. *Let (A, G, α) be a C^*-dynamical a-system and $(G \times_{\alpha} A, \Gamma, \hat{\alpha})$ its dual system (cf. 7.8.3). An element t in G belongs to the Connes spectrum $G(\hat{\alpha})$ of the dual system if and only if $I \cap \alpha_t(I)$ is non-zero for every non-zero closed ideal I of A.*

Proof. The arguments depend heavily on the results in 7.8. Thus we will regard $G \times_\alpha A$ as a G-product (7.8.2) and identify A with the set of elements in $M(G \times_\alpha A)$ that satisfy Landstad's conditions (7.8.2).

If $I \cap \alpha_t(I) = \{0\}$ for some non-zero closed ideal I of A choose (by spectral theory) non-zero positive elements x and y in I such that $yx = y$. There is then a compact neighbourhood E of 0 such that $\|\alpha_s(x) - x\| < 1$ for every s in E. Consequently

$$z_s = \sum_{n=0}^{\infty} (\alpha_s(x) - x)^n \in A,$$

and

$$\alpha_s(y) = \alpha_s(y)(1 - (\alpha_s(x) - x))z_s = \alpha_s(y)xz_s \in I;$$

whence

$$(*) \qquad yA\alpha_{t-s}(y) = \alpha_{-s}(\alpha_s(y)A\alpha_t(y)) \subset \alpha_{-s}(IA\alpha_t(I)) = \{0\}$$

for every s in E.

Let B denote the hereditary C^*-subalgebra of $G \times_\alpha A$ generated by $y(G \times_\alpha A)y$, and note that $B \in \mathscr{H}^{\hat{\alpha}}(G \times_\alpha A)$ since y is a fixed-point (under Γ) in $M(G \times_\alpha A)$. Take now any g in $L^1(\Gamma)$ with $\operatorname{supp}\hat{g} \subset t - E$ and let b be an $\hat{\alpha}$-integrable element in B of the form

$$b = y\lambda_f^* a\lambda_f y, \qquad f \in L^1(G) \cap L^2(G), \qquad a \in G \times_\alpha A.$$

Then

$$\begin{aligned}
\hat{\alpha}_g(b) &= \iint \hat{\alpha}_\sigma(y\lambda_f^* a\lambda_f y)\overline{(s, \sigma)}\hat{g}(s)\, ds\, d\sigma \\
&= \iint y\hat{\alpha}_\sigma(\lambda_f^* a\lambda_f\lambda_{-s})\lambda_s y\hat{g}(s)\, d\sigma\, ds \\
&= \int yI(\lambda_f^* a\lambda_f\lambda_{-s})\alpha_s(y)\lambda_s\hat{g}(s)\, ds = 0,
\end{aligned}$$

since $I(\lambda_f^* a\lambda_f\lambda_{-s}) \in A$ by 7.8.7 and $yA\alpha_s(y) = \{0\}$ for all s in the support of $\hat{g}$ by $(*)$. Thus $\hat{\alpha}_g(b) = 0$ for a dense set of elements b in B; i.e. $\hat{\alpha}_g \mid B = 0$. It follows that $t \notin \operatorname{Sp}(\hat{\alpha} \mid B)$, whence $t \notin G(\hat{\alpha})$.

Conversely, if $t \notin G(\hat{\alpha})$ there is a neighbourhood E of t and a B in $\mathscr{H}^{\hat{\alpha}}(G \times_\alpha A)$ such that $\hat{\alpha}_g \mid B = 0$ whenever $g \in K^1(\Gamma)$ and $\operatorname{supp}\hat{g} \subset E$. Choose a smaller neighbourhood E_1 and a neighbourhood E_0 of 0 such that $E_1 + E_0 - E_0 \subset E$. Then for every pair of functions f_1, f_2 in $K(G)$ with supports in E_0 and every y in B we have, for each g in $K^1(\Gamma)$,

$$\begin{aligned}
\hat{\alpha}_g(\lambda_{f_1}^* y\lambda_{f_2}) &= \iiint \hat{\alpha}_\sigma(\lambda_r y\lambda_s)g(\sigma)\overline{f_1}(-r)f_2(s)\, ds\, dr\, d\sigma \\
&= \iiint \lambda_r\hat{\alpha}_\sigma(y)\lambda_s(r + s, \sigma)g(\sigma)\overline{f_1}(-r)f_2(s)\, ds\, dr\, d\sigma \\
&= \iiint \alpha_r(\hat{\alpha}_\sigma(y))\lambda_s(s, \sigma)g(\sigma)\overline{f_1}(-r)f_2(s - r)\, ds\, dr\, d\sigma \\
&= \iint \alpha_r(\hat{\alpha}_{sg}(y))\lambda_s\overline{f_1}(-r)f_2(s - r)\, ds\, dr.
\end{aligned}$$

If now $\text{supp.}\hat{g} \subset E_1$, then, whenever $-r \in \text{supp.} f_1$ and $s - r \in \text{supp.} f_2$, we have

$$\text{supp.}(sg) = \text{supp.}\hat{g} - s \subset E_1 + E_0 - E_0 \subset E,$$

whence $\hat{\alpha}_g(\lambda_{f_1}^* y \lambda_{f_2}) = 0$.

Let L denote the Γ-invariant, closed, left ideal of $G \underset{\alpha}{\times} A$ such that $L^* \cap L = B$ (cf. 1.5.2). Fix a non-zero function f in $K(G)$ with support in E_0 and let L_0 be the closure of the linear span of the set $\cup L\lambda_{\sigma f}$, $\sigma \in \Gamma$. Then L_0 is a Γ-invariant, closed, left ideal in $G \underset{\alpha}{\times} A$, whence $B_0 = L_0^* \cap L_0$ belongs to $\mathscr{H}^{\hat{\alpha}}(G \underset{\alpha}{\times} A)$. A dense set of elements in B have the form

$$y = \sum \lambda_{\sigma_i f}^* \, y_i^* \, y_j \, \lambda_{\sigma_j f}$$

for some finite sets $\{\sigma_i\}$, $\{\sigma_j\}$ in Γ and $\{y_i\}$, $\{y_j\}$ in L. For each such element we have $\hat{\alpha}_g(y) = 0$ whenever $g \in K^1(\Gamma)$ and $\text{supp.}\hat{g} \subset E_1$, by the computations above. Consequently $t \notin \text{Sp}(\hat{\alpha} \mid B_0)$.

Choose a non-zero, positive, $\hat{\alpha}$-integrable element y_0 in B_0 (which is possible, since B_0 contains a dense set of integrable elements, of the form $\lambda_f^* y \lambda_f$, $y \in B$; cf. 7.8.6), and put $x_0 = I(y_0)$. Then $x_0 \in A_+ \setminus \{0\}$ by 7.8.7; and if $\{g_i\}$ is an increasing net of positive functions in $K(\Gamma)$ converging pointwise to 1, we have an increasing net $\{x_i\}$ in B_0, where $x_i = \hat{\alpha}_{g_i}(y_0)$ and $x_i \nearrow x_0$. Realizing $G \underset{\alpha}{\times} A$ as operators on some Hilbert space $L^2(G, H)$ (cf. 7.7.1 and 7.7.7) we let $\mathscr{M}_0$ denote the strong closure of B_0. Since we are working in a covariant representation of $(G \underset{\alpha}{\times} A, \Gamma, \hat{\alpha})$, we obtain a W^*-dynamical system $(\mathscr{M}_0, \Gamma, \beta)$, where β is a σ-weakly continuous extension of $\hat{\alpha}$. Note that $\text{Sp}(\hat{\alpha} \mid B_0) = \text{Sp}(\beta \mid \mathscr{M}_0)$ by 8.8.9. In particular, $t \notin \text{Sp}(\beta \mid \mathscr{M}_0)$.

For any x in A, the element $x_0 x \lambda_t x_0$ belongs to $\mathscr{M}_0$, being the strong limit of the net $\{x_i x \lambda_t x_i\}$ in B_0. However, for each σ in Γ we have

$$\beta_\sigma(x_0 x \lambda_t x_0) = \hat{\alpha}_\sigma(x_0 x \lambda_t x_0) = (t, \sigma) x_0 x \lambda_t x_0$$

by (**) and (i) in 7.8.2. Since $t \notin \text{Sp}(\beta \mid \mathscr{M}_0)$ this implies that $x_0 x \lambda_t x_0 = 0$, i.e.

$$x_0 x \alpha_t(x_0) = (x_0 x \lambda_t x_0) \lambda_{-t} = 0.$$

This holds for any x in A and we have thus found a non-zero, closed ideal I of A (viz. the ideal generated by x_0), such that $I \cap \alpha_t(I) = \{0\}$.

8.11.8. PROPOSITION. *Let (A, G, α) be a C^*-dynamical a-system and $(G \underset{\alpha}{\times} A, \Gamma, \hat{\alpha})$ its dual system. An element σ in Γ belongs to the Connes spectrum $\Gamma(\alpha)$ if and only if $J \cap \hat{\alpha}_\sigma(J) \neq \{0\}$ for every non-zero closed ideal J of $G \underset{\alpha}{\times} A$.*

Proof. By 8.11.7 the condition is necessary and sufficient for σ to belong to the Connes spectrum $\Gamma(\hat{\hat{\alpha}})$ corresponding to the double dual system. However, $\Gamma(\hat{\hat{\alpha}}) = \Gamma(\alpha)$ by 8.11.6.

8.11.9. COROLLARY. *Let (A, G, α) be a C*-dynamical a-system and denote by C the centre of $M(G \times_\alpha A)$ (cf. 4.4.8). If $\sigma \in \Gamma(\alpha)$ then $\hat{\alpha}_\sigma | C$ is trivial.*

Proof. We have $C = C(\hat{C})$, where $\hat{C}$ is a compact Hausdorff space and by transposition we obtain for every σ in Γ a homeomorphism $\hat{\sigma}$ of $\hat{C}$ such that

$$\hat{\alpha}_\sigma(z)(t) = z(\hat{\sigma}(t)), \qquad z \in C, \qquad t \in \hat{C}.$$

If $\hat{\alpha}_\sigma | C$ is non-trivial then $\hat{\sigma}$ is non-trivial, whence $\hat{\sigma}(\Omega) \cap \Omega = \varnothing$ for some non-empty open subset Ω of $\hat{C}$. Take a non-zero element z in C with support in Ω and note that $\hat{\alpha}_\sigma(z)z = 0$. Thus with I as the closure of $z(G \times_\alpha A)$ we have a non-zero closed ideal in $G \times_\alpha A$ such that $\hat{\alpha}_\sigma(I) \cap I = \{0\}$. By 8.11.8 $\sigma \notin \Gamma(\alpha)$, as desired.

8.11.10. THEOREM. *Let (A, G, α) be a C*-dynamical a-system. The following two conditions are equivalent:*

(i) *$G \times_\alpha A$ is prime;*

(iia) *A is G-prime and* (iib) $\Gamma(\alpha) = \Gamma$.

Proof. (i) $\Rightarrow$ (iia) follows from 7.9.7 and (i) $\Rightarrow$ (iib) from 8.11.8.

(ii) $\Rightarrow$ (i). If $G \times_\alpha A$ is not prime there are two non-zero, orthogonal, closed ideals J_1 and J_2. Take σ in Γ and assume that $I = \hat{\alpha}_\sigma(J_1) \cap J_2 \neq \{0\}$. By (iib) and 8.11.8 we have

$$\{0\} \neq \hat{\alpha}_{-\sigma}(I) \cap I \subset J_1 \cap J_2,$$

a contradiction. Consequently $\hat{\alpha}_\sigma(J_1) \cap J_2 = \{0\}$ for all σ in Γ. Let J_0 denote the closed ideal generated by $\cup \hat{\alpha}_\sigma(J_1)$. Then $J_0 \cap J_2 = \{0\}$ so that J_0 is a non-zero Γ-invariant ideal in $G \times_\alpha A$. However, by (iia) and 7.9.7, $G \times_\alpha A$ is Γ-prime and we have reached a contradiction. Consequently $G \times_\alpha A$ is prime, as desired.

8.11.11. LEMMA. *Let (A, G, α) be a C*-dynamical a-system where G is compact. If A is G-simple and prime then it is actually simple.*

Proof. Assume that I is a non-trivial closed ideal in A, and choose (by spectral theory) non-zero positive elements x, y in I such that $yx = y$. As in the proof of 8.11.7 this implies that $\alpha_s(y) \in I$ for all s in some non-empty neighbourhood E of 0.

Since G is compact it has a finite covering by sets of the form $E - t_k, t_k \in G$, $1 \leqslant k \leqslant n$. Applying the fact that A is prime successively we find elements $a_1, \dots, a_{n-1}$ in A such that

$$y_0 = \alpha_{t_1}(y)a_1\alpha_{t_2}(y)a_2 \cdots \alpha_{t_{n-1}}(y)a_{n-1}\alpha_{t_n}(y) \neq 0.$$

If now $t \in G$ then $t \in E - t_k$ for some k, whence $\alpha_t(y_0) \in I$ since it contains a factor $\alpha_{t+t_k}(y)$ in I. Consequently the ideal I_0 generated by the orbit $\alpha_G(y_0)$ is non-trivial ($y_0 \in I_0 \subset I$) and G-invariant, in contradiction with A being G-simple.

8.11.12. THEOREM. *Let (A, G, α) be a C^*-dynamical a-system. If G is discrete the following conditions are equivalent:*

(i) *$G \underset{\alpha}{\times} A$ is simple;*

(iia) *A is G-simple and* (iib) $\Gamma(\alpha) = \Gamma$.

Proof. (i) $\Rightarrow$ (iia) follows from 7.9.6 and (i) $\Rightarrow$ (iib) from 8.11.10.

(ii) $\Rightarrow$ (i). By 8.11.10 $G \underset{\alpha}{\times} A$ is prime and by 7.9.6 it is Γ-simple. Since G is discrete, Γ is compact whence $G \underset{\alpha}{\times} A$ is simple by 8.11.11.

8.11.13. THEOREM. *Let $(\mathcal{M}, G, \alpha)$ be a W^*-dynamical a-system and $(G \underset{\alpha}{\times} \mathcal{M}, \Gamma, \hat{\alpha})$ its dual system. If $\mathcal{Z}$ denotes the centre of $G \underset{\alpha}{\times} \mathcal{M}$ then $\Gamma(\alpha)$ is equal to the kernel of the homomorphism $\sigma \to \hat{\alpha}_\sigma | \mathcal{Z}$ from Γ to* $\mathrm{Aut}(\mathcal{Z})$.

Proof. Put $\mathcal{N} = G \underset{\alpha}{\times} \mathcal{M}$ and with $\beta = \hat{\alpha} | \mathcal{N}^c$ consider the C^*-dynamical system $(\mathcal{N}^c, \Gamma, \beta)$. Note that $\Gamma \underset{\beta}{\times} \mathcal{N}^c$ is σ-weakly dense in $\Gamma \underset{\hat{\alpha}}{\times} \mathcal{N}$ and contained in $(\Gamma \underset{\alpha}{\times} \mathcal{N})^c$ (the last 'c' refers to the action of $\hat{\hat{\alpha}}$). Moreover, $\hat{\hat{\alpha}} | \Gamma \underset{\beta}{\times} \mathcal{N}^c = \hat{\beta}$. By 8.8.9 and 8.11.6 we have $\Gamma(\alpha) = \Gamma(\hat{\hat{\alpha}}) \subset \Gamma(\hat{\beta})$. If therefore $\sigma \in \Gamma(\alpha)$ then β_σ is trivial on the centre of $\mathcal{N}^c$ by (the dual version of) 8.11.9. However, the centre of $\mathcal{N}^c$ is $\mathcal{Z}^c$ which is σ-weakly dense in $\mathcal{Z}$. Consequently $\hat{\alpha}_\sigma | \mathcal{Z}$ is trivial.

Conversely, if $\sigma \notin \Gamma(\alpha)$ put $\gamma = \alpha | \mathcal{M}^c$ and note that $\Gamma(\alpha) = \Gamma(\gamma)$ by 8.8.10. By 8.11.8 there is a non-zero closed ideal J in $G \underset{\gamma}{\times} \mathcal{M}^c$ such that $\hat{\gamma}_\sigma(J) \cap J = \{0\}$. The σ-weak closure of J has the form $p(G \underset{\alpha}{\times} \mathcal{M})$ for some non-zero projection p in $\mathcal{Z}$, and since $\hat{\alpha} | G \underset{\gamma}{\times} \mathcal{M}^c = \hat{\gamma}$ we have $\hat{\alpha}_\sigma(p)p = 0$; whence $\hat{\alpha}_\sigma | \mathcal{Z}$ is non-trivial.

8.11.14. LEMMA. *Let $(\mathcal{M}, G, \alpha)$ be a W^*-dynamical a-system and $(G \underset{\alpha}{\times} \mathcal{M}, \Gamma, \hat{\alpha})$ its dual system. Then G is ergodic on the centre of $\mathcal{M}$ if and only if Γ is ergodic on the centre of $G \underset{\alpha}{\times} \mathcal{M}$.*

Proof. Put $\beta = \alpha | \mathcal{M}^c$ and note that $G \underset{\beta}{\times} \mathcal{M}^c$ is σ-weakly dense in $G \underset{\alpha}{\times} \mathcal{M}$ and that $\hat{\alpha} | G \underset{\beta}{\times} \mathcal{M}^c = \hat{\beta}$. If therefore $\hat{\alpha}$ is ergodic on the centre of $G \underset{\alpha}{\times} \mathcal{M}$ then $G \underset{\beta}{\times} \mathcal{M}^c$ is Γ-prime. By 7.9.7 this implies that $\mathcal{M}^c$ is G-prime and consequently there are no non-trivial central fixed-points under G in $\mathcal{M}$.
The converse statement follows by duality, cf. 7.10.6.

8.11.15. THEOREM. *Let $(\mathcal{M}, G, \alpha)$ be a W^*-dynamical a-system. The following two conditions are equivalent:*

(i) *$G \underset{\alpha}{\times} \mathcal{M}$ is a factor;*

(iia) *G is ergodic on the centre of $\mathcal{M}$ and* (iib) $\Gamma(\alpha) = \Gamma$.

Proof. (i) ⇒ (iia) follows from 8.11.14 and (i) ⇒ (iib) from 8.11.13.

(ii) ⇒ (i). From 8.11.14 and 8.11.13 we see that Γ must act ergodicly but also trivially on the centre of $G \times_\alpha \mathcal{M}$. This is only possible if $G \times_\alpha \mathcal{M}$ is a factor.

8.11.16. Notes and remarks. This section is based on the paper [186] by Olesen and the author. Exterior equivalence was defined for W^*-dynamical systems by Connes [38] where the von Neumann algebra versions of 8.11.4–8.11.6 were also obtained. Proposition 8.11.8 was conjectured by Takai inspired by 8.11.13, which was proved by Connes and Takesaki [45] together with 8.11.15.

8.12. The KMS condition

8.12.1. Consider a C^*-dynamical system $(A, \mathbf{R}, \alpha)$. We say that an element x in A is *analytic* for α if the function $t \to \alpha_t(x)$ has an extension, necessarily unique, to an analytic (entire) vector function $\zeta \to \alpha_\zeta(x)$, $\zeta \in \mathbf{C}$ (see A4, Appendix). If $x \in A$ put

$$x_n = \pi^{-1/2} n^{1/2} \int \alpha_t(x) \exp(-nt^2)\, dt.$$

Then x_n is analytic for α; indeed,

$$\alpha_\zeta(x_n) = \pi^{-1/2} n^{1/2} \int \alpha_t(x) \exp(-n(t-\zeta)^2)\, dt.$$

Since $x_n \to x$ as $n \to \infty$ we see that the set A^a of analytic elements is dense in A. Using for example (iv) of A4 we see that the product of analytic elements is again analytic (with $\alpha_\zeta(xy) = \alpha_\zeta(x)\alpha_\zeta(y)$), and it follows that A^a is a dense $*$-subalgebra of A.

If $(\mathcal{M}, \mathbf{R}, \alpha)$ is a W^*-dynamical system we show in the same manner that the set $\mathcal{M}^a$ of analytic elements for α is a σ-weakly $*$-subalgebra of $\mathcal{M}$ (contained in $\mathcal{M}^c$).

8.12.2. Given a C^*-dynamical system $(A, \mathbf{R}, \alpha)$ we say that a state ϕ of A satisfies the *Kubo–Martin–Schwinger* (KMS) *condition at* β $(0 < \beta < \infty)$ if for any x in A^a and y in A we have

$$\phi(y\alpha_{\zeta + i\beta}(x)) = \phi(\alpha_\zeta(x)y), \qquad \zeta \in \mathbf{C}.$$

We extend this definition to cover also the limiting cases $\beta = 0$ and $\beta = \infty$ as follows: If $\beta = 0$ we say that ϕ satisfies the KMS condition at zero if ϕ is an α-invariant trace (ϕ is said to be a *chaotic state*). If $\beta = \infty$ we say that ϕ satisfies the KMS condition at ∞ if for each x in A^a and y in A the analytic function $f(\zeta) = \phi(y\alpha_\zeta(x))$ satisfies

$$|f(\zeta)| \leqslant \|x\| \, \|y\| \qquad \text{whenever } \operatorname{Im}\zeta \geqslant 0.$$

(ϕ is said to be a *ground state*).

8.12.3. PROPOSITION. *Let $(A, \boldsymbol{R}, \alpha)$ be a C^*-dynamical system and fix β $(0 < \beta \leqslant \infty)$. The following conditions on a state ϕ of A are equivalent:*

(i) *ϕ is a β-KMS state;*

(ii) *ϕ satisfies the condition in 8.12.2 for just a dense set of elements x in A^a;*

(iii) *for all x, y in A there is a bounded continuous function f on the strip*

$$\Omega_\beta = \{\zeta \in \boldsymbol{C} \mid 0 \leqslant \operatorname{Im} \zeta \leqslant \beta\}$$

such that f is holomorphic in the interior of Ω_β and if $\beta < \infty$,

$$f(t) = \phi(y\alpha_t(x)), \qquad f(t + i\beta) = \phi(\alpha_t(x)y), \qquad t \in \boldsymbol{R};$$

if $\beta = \infty$,

$$f(t) = \phi(y\alpha_t(x)), \qquad t \in \boldsymbol{R}, \qquad \|f\| \leqslant \|x\| \, \|y\|.$$

Proof. Clearly (iii) $\Rightarrow$ (i) $\Rightarrow$ (ii). To prove that (ii) $\Rightarrow$ (iii) assume that the conditions in 8.12.2 are satisfied for all x in M, where $M \subset A^a$ and M is dense in A, and take x, y in A.

If $\beta < \infty$ let $\{x_n\}$ be a sequence in M converging to x. We then obtain a sequence of analytic functions $\{f_n\}$, where

$$f_n(\zeta) = \phi(y\alpha_\zeta(x_n)), \qquad f_n(\zeta + i\beta) = \phi(\alpha_\zeta(x_n(y).$$

Each f_n is bounded on Ω_β; in fact if $\zeta = s + it$ then

$$\|f_n(\zeta)\| \leqslant \|y\| \, \|\alpha_{s+it}(x_n)\| = \|y\| \, \|\alpha_{it}(x_n)\|,$$

which remains bounded since $0 \leqslant t \leqslant \beta$. It follows from the Phragmen–Lindelöf theorem (see e.g. Theorem 12.8 of [218]) that

$$|f_n(\zeta) - f_m(\zeta)| \leqslant \operatorname{Sup}\{|f_n(\zeta) - f_m(\zeta)| \mid \zeta \in \partial\Omega_\beta\}$$
$$= \operatorname{Sup}\{|\phi(y\alpha_t(x_n - x_m))| \vee |\phi(\alpha_t(x_n - x_m)y)|\} \leqslant \|y\| \, \|x_n - x_m\|.$$

Consequently $\{f_n\}$ is uniformly convergent to a function f in $C^b(\Omega_\beta)$ which is holomorphic in the interior of Ω_β. On the boundary we have

$$f(t) = \phi(y\alpha_t(x)), \qquad f(t + i\beta) = \phi(\alpha_t(x)y), \qquad t \in \boldsymbol{R}.$$

If $\beta = \infty$ we define $f_n(\zeta) = \phi(y\alpha_\zeta(x_n))$, and the KMS condition at infinity gives

$$|f_n(\zeta) - f_m(\zeta)| \leqslant \|x_n - x_m\| \, \|y\|,$$

whenever $\operatorname{Im} \zeta \geqslant 0$. It follows that $\{f_n\}$ is uniformly convergent to a function f in $C^b(\Omega_\infty)$ which is holomorphic in the interior. Moreover, $f(t) = (y\alpha_t(x))$ for all t and $\|f\| \leqslant \|x\| \, \|y\|$.

8.12.4. PROPOSITION. *If $(A, \boldsymbol{R}, \alpha)$ is a C^*-dynamical system and ϕ is a state of A satisfying the KMS condition at β $(0 \leqslant \beta \leqslant \infty)$ then ϕ is α-invariant.*

Proof. Take an approximate unit $\{u_\lambda\}$ for A. If $0 < \beta < \infty$ we have for each x in A^a that

$$\phi(u_\lambda \alpha_{\zeta + i\beta}(x)) = \phi(\alpha_\zeta(x) u_\lambda).$$

In the limit we obtain $\phi(\alpha_{\zeta+i\beta}(x)) = \phi(\alpha_\zeta(x))$. Thus the analytic function $f : \zeta \to \phi(\alpha_\zeta(x))$ is bounded on the strip Ω_β and is periodic with period $i\beta$. We conclude that f is bounded on $\boldsymbol{C}$ and therefore constant by Liouville's theorem. Since A^a is dense in A it follows that ϕ is α-invariant.

If $\beta = 0$ the α-invariance is included in the definition. If $\beta = \infty$ we use $\{u_\lambda\}$ as above and obtain for each x in A^a an analytic function $f : \zeta \to \phi(\alpha_\zeta(x))$ such that $|f(\zeta)| \leqslant \|x\|$ whenever $\operatorname{Im} \zeta \geqslant 0$. Since $\phi = \phi^*$ we have $\overline{\phi(\alpha_\zeta(x))} = \phi(\alpha_{\bar\zeta}(x^*))$. Thus for $\operatorname{Im} \zeta \leqslant 0$ we have

$$|f(\zeta)| \leqslant |\overline{\phi(\alpha_{\bar\zeta}(x^*))}| \leqslant \|x\|.$$

It follows that f is bounded, hence constant and thus again ϕ is α-invariant.

8.12.5. THEOREM. *Let $(A, \boldsymbol{R}, \alpha)$ be a C^*dynamical system. The following conditions on a state ϕ of A are equivalent:*

(i) *ϕ is α-invariant, and if $(\pi_\phi, u^\phi, H_\phi, \xi_\phi)$ is the cyclic covariant representation associated with ϕ (7.4.12) then $\operatorname{Sp}(u^\phi) \subset \hat{\boldsymbol{R}}_+$;*

(ii) *there is a positive (not necessarily bounded) operator h on H_ϕ with $h\xi_\phi = 0$ such that*

$$\exp(ith)\pi_\phi(x)\exp(-ith) = \pi_\phi(\alpha_t(x)), \qquad \forall t \in \boldsymbol{R}, \qquad x \in A;$$

(iii) *$\phi(y \cdot) \in M^{\alpha'}(\boldsymbol{R}_+)$ for every y in $\tilde{A}$;*

(iv) *ϕ satisfies the KMS condition at ∞, i.e. ϕ is a ground state.*

When these conditions are satisfied we have $u_t^\phi = \exp(ith)$ for all t and $u_{\boldsymbol{R}}^\phi \subset \pi_\phi(A)''$ (i.e. h is affiliated with $\pi_\phi(A)''$).

Proof. (i) ⇒ (ii). By Stone's theorem (7.1.7) there is a unique self-adjoint operator h on H_ϕ such that $u_t^\phi = \exp(ith)$ for every t in $\boldsymbol{R}$. Moreover, $\operatorname{Sp}(h) = \operatorname{Sp}(u^\phi)$ so that $h \geqslant 0$. For each η in $\mathscr{D}(h)$ we have

$$\begin{aligned}(\xi_\phi | h\eta) &= \operatorname{Lim}(it)^{-1}(\xi_\phi | (1\text{-}\exp(ith))\eta) \\ &= \operatorname{Lim}(it)^{-1}(\xi_\phi | (1 - u_t^\phi)\eta) = 0,\end{aligned}$$

since $u_t^\phi \xi_\phi = \xi_\phi$ for all t; whence $\xi_\phi \in \mathscr{D}(h)$ with $h\xi_\phi = 0$.

(ii) ⇒ (i). Define $u_t = \exp(ith)$ and note that u is a unitary representation of $\boldsymbol{R}$ with $\operatorname{Sp}(u) \subset \hat{\boldsymbol{R}}_+$. Since $h\xi_\phi = 0$ we have

$$\pi_\phi(\alpha_t(x))\xi_\phi = u_t\pi_\phi(x)u_{-t}\xi_\phi = u_t\pi_\phi(x)\xi_\phi$$

for every x in A. It follows immediately that ϕ is α-invariant and that $u_t = u_t^\phi$ for all t.

(ii) $\Rightarrow$ (iii). Take x, y in $\tilde{A}$ and f in $K^1(G)$. Then

$$\int \phi(y^*\alpha_t(x))f(t)\,dt = \int (\exp(ith)\pi_\phi(x)\exp(-ith)\xi_\phi \mid \pi_\phi(y)\xi_\phi)f(t)\,dt$$
$$= \int (\exp(ith)\xi_x \mid \xi_y)f(t)\,dt = (\hat{f}(h)\xi_x \mid \xi_y).$$

Since $h \geq 0$ we see that $\hat{f}(h) = 0$ if $\text{supp}.\hat{f} \subset -\hat{R}_+$. Consequently $\phi(y\,\cdot)$ annihilates $R^\alpha(\hat{R}\backslash\hat{R}_+)$, i.e. $\phi(y\,\cdot) \in M^{\alpha'}(R_+)$ for every y in $\tilde{A}$.

(iii) $\Rightarrow$ (i). Taking $y = 1$ we see that $\phi \in M^{\alpha'}(\hat{R}_+)$. But $\phi = \phi^*$ whence $\phi \in M^{\alpha'}(-\hat{R}_+)$ (cf. 8.3.3). Thus $\phi \in M^{\alpha'}(\{0\})$ by 8.1.4 (iv), i.e. ϕ is α-invariant by 8.1.8. In the covariant representation associated with ϕ we have $u_t^\phi = \exp(ith)$ for some self-adjoint operator h on H_ϕ. As above we get

$$\int \phi(y^*\alpha_t(x))f(t)\,dt = (\hat{f}(h)\xi_x \mid \xi_y)$$

for all x, y in A, and by assumption the integral is zero whenever $\text{supp}.\hat{f} \subset \hat{R}\backslash\hat{R}_+$. Consequently $h \geq 0$; i.e. $\text{Sp}(u^\phi) \subset R_+$ as desired.

(ii) $\Rightarrow$ (iv). If $h \geq 0$ we can define for x, y in A a function f in $C^b(\Omega_\infty)$, holomorphic in the interior, by

$$f(\zeta) = (\exp(i\zeta h)\xi_x \mid \xi_y).$$

Since $\exp(-sh) \leq 1$ for $s \geq 0$ it follows that $\|f\| \leq \|x\|\,\|y\|$. Furthermore, $f(t) = \phi(y^*\alpha_t(x))$ for all real t so that ϕ satisfies the KMS condition at infinity.

(iv) $\Rightarrow$ (i). We know from 8.12.4 that ϕ is α-invariant so that we may consider the covariant representation $(\pi_\phi, u^\phi, H_\phi, \xi_\phi)$ and write $u_t = \exp(ith)$ for some self-adjoint operator h on H_ϕ. If x is analytic for α then ξ_x is analytic for $\exp(ith)$ and we have an analytic function

$$f: \zeta \to (\exp(i\zeta h)\xi_x \mid \xi_x) = \phi(x^*\alpha_\zeta(x)).$$

By assumption $f(\zeta) \leq \|x\|^2$ if $\text{Im}\,\zeta \geq 0$. In particular

$$((\exp(-h))^s\xi_x \mid \xi_x) \leq \|x\|^2$$

for all $s \geq 0$. It follows that $\exp(-h) \leq 1$ i.e. $h \geq 0$, whence $\text{Sp}(u^\phi) \subset R_+$.

Clearly the operator h in (ii) is the infinitesimal generator for u^ϕ. Moreover, $u^\phi \subset \pi_\phi(A)''$ by 8.4.13 so that h is affiliated with $\pi_\phi(A)''$.

8.12.6. Let (A, R, α) be a C^*-dynamical system. We say that α is *approximately inner* if there is a net $\{h_\lambda\}$ in A_{sa} such that for each x in A^a we have

$$(*)\qquad \text{Lim}\,\|\alpha_\zeta(x) - \exp(i\zeta h_\lambda)x\exp(-i\zeta h_\lambda)\| = 0$$

uniformly in ζ on compact subsets of $\boldsymbol{C}$. This condition implies but is slightly stronger than the demand that for each x in A

$$(**) \qquad \operatorname{Lim} \|\alpha_t(x) - \exp(ith_\lambda)x\exp(-ith_\lambda)\| = 0$$

uniformly in t on compact subsets of $\boldsymbol{R}$.

Note that if α is uniformly continuous then $A^a = A$ since we can define $\alpha_\zeta(x) = \exp(\zeta\delta)(x)$, where δ is the *-derivation of A that generates α. This fact is used implicitly in the next result.

8.12.7. PROPOSITION. *Let $(A, \boldsymbol{R}, \alpha)$ be a C^*-dynamical system. Then α is approximately inner if there is a net $\{\alpha^\lambda_{\boldsymbol{R}}\}$ of uniformly continuous one-parameter groups of automorphisms such that $\|\alpha_\zeta(x) - \alpha^\lambda_\zeta(x)\| \to 0$ for each x in A^a, uniformly in ζ on compact subsets of $\boldsymbol{C}$.*

Proof. The condition is clearly necessary. Assume for a moment that α is uniformly continuous, and write $\alpha_t = \operatorname{Ad}(\exp(ith))$ where h is the minimal positive generator (cf. 8.5.5). From 8.6.12 we obtain an increasing net $\{h_\lambda\}$ in $\tilde{A}_+$ such that $h_\lambda \nearrow h$ and such that

$$\|hh_\lambda - h_\lambda h\| \to 0 \qquad \text{and} \qquad \|(h - h_\lambda)x - x(h - h_\lambda)\| \to 0$$

for every x in A (The restriction in 8.6.12 that A be separable is used only to obtain a sequence $\{h_n\}$ instead of a net $\{h_\lambda\}$). It follows from 8.6.13 that

$$\|\alpha_\zeta(x) - \exp(i\zeta h_\lambda)x\exp(-i\zeta h_\lambda)\| \to 0$$

for every x in A, uniformly in ζ on compact subsets of $\boldsymbol{C}$, so that α is approximately inner. If now α is a limit of uniformly continuous representations, as described in the proposition, it follows from the triangle inequality that α is approximately inner.

8.12.8. Approximately inner representations arise in many contexts. Suppose for example that $(A, \boldsymbol{R}, \alpha)$ is a C^*-dynamical system and that there is a net (under inclusion) of $\alpha_{\boldsymbol{R}}$-invariant C^*-subalgebras $\{A_\lambda\}$, such that $\cup A_\lambda$ is dense in A and $\alpha \mid A_\lambda$ is uniformly continuous for every λ. Then it follows readily from 8.12.7 that α is approximately inner.

Of particular interest is the case where A is the closure of an increasing net $\{A_\lambda\}$ of finite-dimensional C^*-subalgebras. We say then that A is *approximately finite-dimensional* (or that A is an *AF-algebra*). Note that every Glimm algebra is approximately finite-dimensional (cf. 6.4). If α is a continuous representation of $\boldsymbol{R}$ into $\operatorname{Aut}(A)$, where A is approximately finite-dimensional, we see from the above that α is approximately inner if it leaves each A_λ invariant (or if we can choose the net $\{A_\lambda\}$ to be $\alpha_{\boldsymbol{R}}$-invariant).

8.12.9. PROPOSITION. *Let $(A, \mathbf{R}, \alpha)$ be a C^*-dynamical system and assume that α is approximately inner and $1 \in A$. Then A has a ground state (i.e. a state satisfying the KMS condition at $\beta = \infty$).*

Proof. Let $\{h_\lambda\}$ be a net in A_{sa} satisfying the conditions in 8.12.6 and put $\alpha_t^\lambda = \mathrm{Ad}(\exp(ith_\lambda))$. Since the addition of a multiple of 1 to h_λ will not change α^λ we may assume that $h_\lambda \geqslant 0$ and that $0 \in \mathrm{Sp}(h_\lambda)$ for every λ. By 3.1.6 (applied to $C^*(1, h_\lambda)$) there is a net $\{\phi_\lambda\}$ of states of A such that $\phi_\lambda(h_\lambda) = 0$ for each λ. Since the state space is compact $(1 \in A)$ we may assume, passing if necessary to a subnet, that $\{\phi_\lambda\}$ is weak* convergent to a state ϕ of A.

If x, y belong to A and $\zeta \in \mathbf{C}$ with $\mathrm{Im}\,\zeta \geqslant 0$ then

$$|\phi_\lambda(y\alpha_\zeta^\lambda(x))| = |\phi_\lambda(y\exp(i\zeta h_\lambda)x\exp(-i\zeta h_\lambda))|$$
$$= |\phi_\lambda(y\exp(i\zeta h_\lambda)x)| \leqslant \|x\|\,\|y\|\,\|\exp(i\zeta h_\lambda)\| \leqslant \|x\|\,\|y\|,$$

since $\phi_\lambda(ah_\lambda) = 0$ for every a in A. Thus ϕ_λ is a ground state for α^λ. Moreover, if $x \in A^a$ then

$$|\phi(y\alpha_\zeta(x))| \leqslant |(\phi - \phi_\lambda)(y\alpha_\zeta(x))| + |\phi_\lambda(y(\alpha_\zeta(x) - \alpha_\zeta^\lambda(x)))| + |\phi_\lambda(y\alpha_\zeta^\lambda(x))|$$
$$\leqslant |(\phi - \phi_\lambda)(y\alpha_\zeta(x))| + \|y\|\,\|\alpha_\zeta(x) - \alpha_\zeta^\lambda(x)\| + \|x\|\,\|y\|$$

whence in the limit

$$|\phi(y\alpha_\zeta(x))| \leqslant \|x\|\,\|y\|,$$

so that ϕ is a ground state for A.

8.12.10. THEOREM. *Let $(A, \mathbf{R}, \alpha)$ be a C^*-dynamical system and assume that α is approximately inner and $1 \in A$. If A has a state satisfying the KMS condition at some $\beta_0 \neq \infty$, there is a KMS state for every β $(0 \leqslant \beta \leqslant \infty)$.*

Proof. Suppose that ϕ satisfies the KMS condition at β_0 $(\beta_0 \neq \infty)$. Let $\{h_\lambda\}$ be a net in A_{sa} satisfying the conditions in 8.12.6 and put $\alpha_t^\lambda = \mathrm{Ad}(\exp(ith_\lambda))$. Fix $\beta \geqslant 0$ and put $k_\lambda = \exp(\frac{1}{2}(\beta_0 - \beta)h_\lambda))$. Then consider the net $\{\psi_\lambda\}$ of states of A given by

$$\psi_\lambda(x) = \phi(k_\lambda^2)^{-1}\phi(k_\lambda x k_\lambda), \qquad x \in A.$$

Since the state space is compact $(1 \in A)$ we may assume that $\{\psi_\lambda\}$ is weak* convergent to a state ψ of A.

Take x in A^a and y in A. Using the KMS condition at β_0 and the fact that $k_\lambda^{-1}ak_\lambda = \alpha_{i(\beta_0-\beta)/2}^\lambda(a)$ for every a in A we get

$$\phi(k_\lambda\alpha_{i(\beta_0-\beta)/2}^\lambda(x)yk_\lambda) = \phi(xk_\lambda yk_\lambda)$$
$$\phi(k_\lambda yk_\lambda\alpha_{i\beta_0}(x)) = \phi(k_\lambda y\alpha_{i(\beta-\beta_0)/2}(\alpha_{i\beta_0}(x))k_\lambda).$$

Dividing by $\phi(k_\lambda^2)$ this gives

$$\psi_\lambda(\alpha^\lambda_{i(\beta_0-\beta)/2}(x)y) = \psi_\lambda(y\alpha^\lambda_{i(\beta-\beta_0)/2}(\alpha_{i\beta_0}(x))).$$

Passing to the limit we obtain

$$\psi(\alpha_{i(\beta_0-\beta)/2}(x)y) = \psi(y\alpha_{i(\beta+\beta_0)/2}(x)),$$

and replacing x with $\alpha_{i(\beta-\beta_0)/2}(x)$ we finally have

$$\psi(xy) = \psi(y\alpha_{i\beta}(x)),$$

so that ψ satisfies the KMS condition at β.

8.12.11. We shall study some groups of automorphisms on the Fermion algebra $\boldsymbol{F}$ defined in 6.4. Recall from 6.4.1 that we may identify each matrix algebra $\boldsymbol{M}_{2^n}$, $n \in \boldsymbol{N}$, with a C^*-subalgebra of $\boldsymbol{F}$ in such a way that $\boldsymbol{M}_{2^{n+1}} = \boldsymbol{M}_{2^n} \otimes \boldsymbol{M}_2$; and $\cup \boldsymbol{M}_{2^n}$ is dense in $\boldsymbol{F}$.

Fix a sequence $\{\mu_n\}$ in $]0, \infty[$ and for each n define a unitary representation u^n of $\boldsymbol{R}$ into $\boldsymbol{M}_{2^n}$ by

$$u^n_t = \bigotimes_{k=1}^{n} \begin{pmatrix} 1 & 0 \\ 0 & \mu_k^{it} \end{pmatrix}.$$

If $x \in \boldsymbol{M}_{2^m}$ with $m \leqslant n$ we define $\alpha_t(x) = u^n_t x u^n_{-t}$. Note that $\alpha_t(x)$ does not depend on n (as long as $m \leqslant n$) so that we obtain a continuous one-parameter group of automorphisms on $\cup \boldsymbol{M}_{2^m}$. Since this set is dense in $\boldsymbol{F}$ the automorphisms extend and we obtain a C^*-dynamical system $(\boldsymbol{F}, \boldsymbol{R}, \alpha)$.

Since each $\boldsymbol{M}_{2^n}$ is α-invariant it follows from 8.12.7 that α is approximately inner (cf. 8.12.8). As $1 \in \boldsymbol{F}$ and $\boldsymbol{F}$ has a unique tracial state τ (cf. 6.4.3) it follows from 8.12.10 that there are KMS states of $\boldsymbol{F}$ for every β $(0 \leqslant \beta \leqslant \infty)$. However, in this case it is possible to give concrete examples of KMS states, and these are useful for later examples.

Fix $\beta > 0$ and put $\lambda_n = (1 + \mu_n^\beta)^{-1}$, so that $0 < \lambda_n < 1$, and define

$$h_n = \bigotimes_{k=1}^{n} \begin{pmatrix} 2(1-\lambda_k) & 0 \\ 0 & 2\lambda_k \end{pmatrix} \in \boldsymbol{M}_{2^n}.$$

With τ as the tracial state on $\boldsymbol{F}$ we define

$$\phi(x) = \tau(h_n x), \qquad x \in \boldsymbol{M}_{2^m}, \qquad m \leqslant n.$$

Then ϕ is a state on $\boldsymbol{M}_{2^m}$, since $\tau(h_n) = 1$, and as $\phi(x)$ does not depend on n (as long as $m \leqslant n$) we obtain a state on $\cup \boldsymbol{M}_{2^m}$. By continuity we can extend this to a state ϕ of $\boldsymbol{F}$, and we see from (***) in 6.5.3 that ϕ is a product state of $\boldsymbol{F}$. Put $\varepsilon_n = \Pi_{k=1}^n (2(1-\lambda_k))$. Then a simple calculation shows that

$$\varepsilon_n^{it} u^n_{\beta t} = h_n^{-it}, \qquad t \in \boldsymbol{R}.$$

It follows that $\alpha_{\beta t}(x) = h_n^{-it} x h_n^{it}$ whenever $x \in \boldsymbol{M}_{2^m}$ and $m \leqslant n$. To show that the state $\tau(h_n \cdot)$ satisfies the KMS condition at 1 with respect to the group

$t \to \mathrm{Ad}(h_n^{-it})$ on M_{2^n} is straightforward (cf. 8.14.13), and it follows by continuity that ϕ satisfies the KMS condition at 1 with respect to the group $t \to \alpha_{t\beta}$ on F; i.e. ϕ is a KMS state at β for the group $t \to \alpha_t$ on F.

8.12.12. The most important case of the above obtains by taking $\mu_n = e$ for all n. The ensuing C^*-dynamical system appears in the applications to quantum statistical mechanics where the automorphisms α_t are known as *gauge transformations*. Since $\alpha_{2\pi} = \iota$ we are in fact dealing with a C^*-dynamical system (F, T, α) (identifying T with $R/2\pi Z$. We have $\lambda_n = \lambda = (1 + e^{\beta})^{-1}$ so that $0 < \lambda < \frac{1}{2}$. The product state arising from λ is thus seen to be the state ϕ_λ defined in 6.5.10. We summarize our observations in the following proposition.

8.12.13. PROPOSITION. *Let (F, T, α) denote the C^*-dynamical system of gauge transformations on the Fermion algebra, and for each λ in $]0, \frac{1}{2}[$ let ϕ_λ denote the permutation-invariant product state defined in 6.5.10. Then ϕ_λ satisfies the KMS condition at $\beta = \log(\lambda^{-1} - 1)$ for the system (F, T, α).*

8.12.14. Notes and remarks. The KMS condition was initially proposed in [152] and [166] as a boundary condition determining the solution of an infinite set of differential equations; conditions fulfilled by the Green functions describing equilibrium states. Later it was realized by Haag, Hugenholtz and Winnink [110] that the KMS condition can be substituted for the 'Gibbs Ansatz' as the first principle of equilibrium statistical mechanics; with the advantage that it is valid for infinite systems, while the Gibbs Ansatz can only be stated for finite systems ('in a box'), and requires the procedure of 'passing to the thermodynamic limit' to become applicable to infinite systems. To exemplify this, let A be a finite-dimensional matrix algebra with canonical trace Tr and consider the unitary group $\{\exp(ith) \mid t \in R\}$ and the state $\phi(x) = \mathrm{Tr}(ax)(\mathrm{Tr}(a))^{-1}$, $x \in A$. Elementary calculations show that ϕ satisfies the KMS condition at some β with respect to the automorphism group $\alpha_t(x) = \exp(ith)x\exp(-ith)$, $x \in A$, if and only if $a = \exp(-\beta h)$ (Gibbs Ansatz). If A is an inductive limit of finite-dimensional subalgebras $\{A_n\}$ we may have the condition above satisfied for each n, even though the limit state cannot be expressed with the aid of a trace and the limit automorphism group is outer. An example of this phenomenon is given in 8.12.11. The survey article [145] by Kastler contains a wealth of information on KMS states and their rôle in quantum statistical mechanics.

Theorem 8.12.5 is due to Borchers, see [19] and [24]. The notion (*) of approximately inner automorphism groups employed in 8.12.6 is not the usual one. Powers and Sakai introduced the terminology in [212] with the condition (**) in 8.12.6 and proved 8.12.9 and the version of 8.12.10 where $\beta_0 = 0$. Later Jørgensen in [122] established 8.12.10 with a condition on α which is only a little stronger than (**) and probably much weaker than (*). It was proved in [187] that a C^*-dynamical system $(A, \boldsymbol{R}, \alpha)$ for which there is an isometry v in A (i.e. $v^*v = 1$) such that $\alpha_t(v) = \exp(ist)v$, $s \neq 0$, cannot be approximately inner. Systems of such type appear in [158] and [46]. Taking $\beta_0 = \infty$ or $\beta_0 = \log n$, $n \in \boldsymbol{N}$, there is a system $(A, \boldsymbol{R}, \alpha)$ with $1 \in A$ such that α admits only one β-KMS state, namely for $\beta = \beta_0$. These systems are thus highly 'outer', see [158] and [187].

8.13. The Tomita–Takesaki theory

8.13.1. In this section we prove that to each faithful normal state ϕ of a von Neumann algebra $\mathcal{M}$ there is a unitary representation u of $\boldsymbol{R}$ on H_ϕ such that $u_t \pi_\phi(\mathcal{M}) u_{-t} = \pi_\phi(\mathcal{M})$ for all t in $\boldsymbol{R}$. Furthermore we show in section 8.14 that ϕ satisfies the KMS condition at $\beta = 1$ for the automorphism group associated with u. Whereas in C^*-algebra theory one assumes the existence of a one-parameter group of automorphisms and then search for KMS states (cf. 8.12.9 and 8.12.10) the situation is quite the reverse for von Neumann algebras: when a faithful state is given an automorphism group appears.

It will be clear from the construction that if the faithful normal state ϕ is a trace (and only then) the unitary representation u is trivial. On the other hand the KMS condition furnishes a link between the values $\phi(xy)$ and $\phi(yx)$ for all x, y in $\mathcal{M}$. The philosophy behind the Tomita–Takesaki theory is that with the extra information about ϕ contained in the representation u one can deal with ϕ as if it was a trace.

An analogous construction exists for weights. It is instructive to realize that if $\mathcal{M}$ is the group von Neumann algebra corresponding to a non-unimodular group G and if ϕ is the σ-weight ϕ_e defined in 7.2.7 then the unitary group associated with ϕ_e is given by

$$(u_t\xi)(s) = \Delta(s)^{it}\xi(s), \qquad \xi \in L^2(G), \qquad t \in \boldsymbol{R},$$

where Δ is the modular function on G that links left and right Haar measure (see 8.13.8).

8.13.2. Given a Hilbert space H we define

$$\langle \xi \mid \eta \rangle = \mathrm{Re}(\xi \mid \eta), \qquad \xi, \eta \in H.$$

Then $(H, \langle \cdot \mid \cdot \rangle)$ becomes a real Hilbert space which we shall denote by H_r. Note the conversion formula

$$(\xi \mid \eta) = \langle \xi \mid \eta \rangle - i\langle i\xi \mid \eta \rangle. \qquad (*)$$

Throughout this section we shall assume that there is a closed (real) subspace K of H_r satisfying the two conditions

$$K \cap \mathrm{i}K = \{0\} \quad \text{and} \quad (K + \mathrm{i}K)^\perp = \{0\}.$$

We let p and q denote the (real) orthogonal projections of H_r on K and $\mathrm{i}K$, respectively, and define

$$(**) \qquad a = p + q, \qquad jb = p - q,$$

where jb is the polar decomposition of $p - q$ in H_r.

8.13.3. LEMMA. *The operators p, q, a, b and j satisfy*:

(i) *a and b are complex linear and* $0 \leqslant a \leqslant 2$, $\quad 0 \leqslant b \leqslant 2$;

(ii) *a, $2 - a$ and b are injective and* $b = a^{1/2}(2 - a)^{1/2}$;

(iii) *j is a conjugate linear isometry with $j^2 = 1$ and $(j\xi \mid \eta) = \overline{(\xi \mid j\eta)}$ for all ξ, η in H*;

(iv) *b commutes with p, q, a and j*;

(v) $jp = (1 - q)j$, $\quad jq = (1 - p)j$, $\quad ja = (2 - a)j$.

Proof. (i) Simple calculations show that $\mathrm{i}p = q\mathrm{i}$. Hence $a = p + q$ is complex linear, while $p - q$ is conjugate linear. Since $b^2 = (p - q)^2$, it follows that b^2 and therefore also b is complex linear. Consequently j is conjugate linear since $jb = p - q$. In H_r the operators p and q are positive; thus a and b are positive and j is self-adjoint in H_r. Since a and b are linear it follows from the conversion formula $(*)$ in 8.13.2 that they are self-adjoint and therefore positive also in H. It is clear that $\|a\| \leqslant 2$ and $\|b\| \leqslant 2$ so that (i) is proved.

(ii) If $a\xi = 0$ we have

$$\|p\xi\|^2 + \|q\xi\|^2 = \langle p\xi \mid \xi\rangle + \langle q\xi \mid \xi\rangle = \langle a\xi \mid \xi\rangle = 0.$$

Thus $\xi \in K^\perp \cap (\mathrm{i}K)^\perp = (K + \mathrm{i}K)^\perp$; whence $\xi = 0$. It follows that a is injective, and applying the same argument to $1 - p$ and $1 - q$ it follows that $2 - a$ is injective. Since p and q are idempotents we have $b^2 = a(2 - a)$, whence also b is injective and (ii) is established.

(iii) Since $p - q$ is self-adjoint in H_r, j is self-adjoint in H_r, and as b is injective, j is an injective isometry. Thus $j^2 = 1$. Moreover, by the conversion formula $(*)$

$$\overline{(j\xi \mid \eta)} = \langle j\xi \mid \eta\rangle - \mathrm{i}\langle \mathrm{i}j\xi \mid \eta\rangle = \langle \xi \mid j\eta\rangle + \mathrm{i}\langle \mathrm{i}\xi \mid j\eta\rangle = \overline{(\xi \mid j\eta)}.$$

(iv) By algebra, $(p - q)^2$ commutes with both p and q. It follows that b commutes with p, q and a. Since $p - q$ is self-adjoint in H_r we see that b also commutes with j.

(v) We have

$$bjp = (p - q)p = (1 - q)(p - q) = (1 - q)bj = b(1 - q)j.$$

Since b is injective it follows that $jp = (1 - q)j$. Taking adjoints in H_r we obtain $jq = (1 - q)j$; and adding these two equations we finally get $ja = (2 - a)j$.

8.13.4. LEMMA. *Define* $\Delta = a^{-1}(2 - a)$. *Then* Δ *(the modular operator) is a (possibly unbounded) self-adjoint, positive, injective operator, and* $\Delta^{-1} = j\Delta j$. *Moreover,* $K + iK \subset \mathscr{D}(\Delta^{1/2})$ *and for all* ξ, η *in* K

$$j\Delta^{1/2}(\xi + i\eta) = \xi - i\eta.$$

Proof. Since $0 \leqslant a \leqslant 2$ and both a and $2 - a$ are injective it follows from the spectral theorem that Δ is self-adjoint, positive and injective. The equality $\Delta^{-1} = j\Delta j$ follows from 8.13.3 (v).

Take ξ, η in K. Then

$$(2 - p - q)\xi = (p - q)\xi \qquad \text{and} \qquad (2 - p - q)(i\eta) = -(p - q)(i\eta).$$

It follows that $(2 - a)(\xi + i\eta) = jb(\xi - i\eta)$. For each ψ in $\mathscr{D}(a^{-1})$ we therefore have

$$\begin{aligned}(\xi + i\eta \mid \Delta\psi) &= ((2 - a)(\xi + i\eta) \mid a^{-1}\psi)\\ &= (jb(\xi - i\eta) \mid a^{-1}\psi) = (j(\xi - i\eta) \mid \Delta^{1/2}\psi),\end{aligned}$$

using 8.13.3(ii) and the fact that $\mathscr{D}(\Delta^{1/2}) \supset \mathscr{D}(\Delta) \supset \mathscr{D}(a^{-1})$. In particular,

$$|(\xi + i\eta \mid \Delta^{1/2}(\Delta^{1/2}\psi))| \leqslant \|\xi - i\eta\| \, \|\Delta^{1/2}\psi\|.$$

Since $\Delta^{1/2}\mathscr{D}(a^{-1})$ is dense in H it follows that $\xi + i\eta \in \mathscr{D}(\Delta^{1/2})$ and that $\Delta^{1/2}(\xi + i\eta) = j(\xi - i\eta)$.

8.13.5. LEMMA. *The function* $t \to \Delta^{it}$ *is a unitary representation of* $\boldsymbol{R}$ *in* H *satisfying* $j\Delta^{it} = \Delta^{it}j$ *and* $\Delta^{it}K = K$ *for all* t.

Proof. Since Δ is positive and injective it follows from the spectral theorem that $t \to \Delta^{it}$ is a strongly continuous one-parameter unitary group in H with $\log(\Delta)$ as infinitesimal generator. From the definition of Δ (8.13.4) we see that

$$(***) \qquad \Delta^{it} = (2 - a)^{it}a^{-it}.$$

It follows from 8.13.3 (v) that $ja^{it} = (2 - a)^{-it}j$, where the minus sign in the second exponent stems from the conjugate linearity of j. From this and (***) it is immediate that $j\Delta^{it} = \Delta^{it}j$. Thus Δ^{it} commutes with both a, b and j, and therefore commutes with p and q. In particular,

$$\Delta^{it}K = \Delta^{it}pH_r = p\Delta^{it}H_r = pH_r = K.$$

8.13.6. If (u, H) is a unitary representation of $\boldsymbol{R}$ we say that a vector ξ in H is analytic for u, if the function $t \to u_t\xi$ has an extension (necessarily unique) to an analytic function $\zeta \to u_\zeta\xi$ from $\boldsymbol{C}$ to H (see A4, Appendix and 8.12.1). If

$\xi \in H$ we define

$$\xi_n = (n/\pi)^{1/2} \int \exp(-nt^2) u_t \xi \, dt.$$

Then ξ_n is analytic for u, with

$$u_\zeta \xi_n = (n/\pi)^{1/2} \int \exp(-n(t-\zeta)^2 u_t \, dt.$$

It follows as in 8.12.1 that the set of analytic vectors for u is a dense subspace of H. Moreover, if K is a closed real subspace of H_r such that $u_t K = K$ for all t, then since $\exp(-nt^2)$ is real we see that $\xi_n \in K$ whenever $\xi \in K$. Thus in this case the analytic vectors for u in K are dense in K.

8.13.7. Let $\mathcal{M}$ be a von Neumann algebra on a Hilbert space H and let ξ_0 be a vector which is cyclic and separating for $\mathcal{M}$ (hence also for $\mathcal{M}'$, cf. 2.8.2). Note that if ϕ is a faithful normal functional on $\mathcal{M}$ with cyclic representation $(\pi_\phi, H_\phi, \xi_\phi)$ then π_ϕ is an isomorphism so that we may identify $\mathcal{M}$ and $\pi_\phi(\mathcal{M})$; and ξ_ϕ is cyclic and separating for $\mathcal{M}$.

Let K be the closure of $\mathcal{M}_{sa}\xi_0$. Then we have the following proposition.

8.13.8. PROPOSITION. *The closed real subspace K defined above satisfies the conditions of 8.13.2, and the closed operator $j\Delta^{1/2}$ obtained from K is an extension of the densely defined map $x\xi_0 \to x^*\xi_0$, $x \in \mathcal{M}$.*

Proof. Since ξ_0 is cyclic $K + iK$ is dense in H. To prove that $K \cap iK = \{0\}$ we show that $\mathcal{M}'_{sa}\xi_0$ and $i\mathcal{M}_{sa}\xi_0$ are orthogonal in H_r. Indeed, if $x' \in \mathcal{M}'_{sa}$ and $x \in \mathcal{M}_{sa}$ then $(x'\xi_0 | ix\xi_0) = -(ix\xi_0 | x'\xi_0)$, whence $\langle x'\xi_0 | ix\xi_0 \rangle = 0$. It follows that $\mathcal{M}'_{sa}\xi_0 \subset (iK)^\perp$ and similarly $i\mathcal{M}'_{sa}\xi_0 \subset K^\perp$. Consequently,

$$(K \cap iK)^\perp = K^\perp + (iK)^\perp \supset \mathcal{M}'\xi_0,$$

and since $\mathcal{M}'\xi_0$ is dense in H, $K \cap iK = \{0\}$.

The second assertion is immediate from 8.13.4.

8.13.9. LEMMA. *Let $\mathcal{M}$, ξ_0 and K be as in 8.13.7. We have $q\xi_0 = 0$ and*

$$p\xi_0 = a\xi_0 = b\xi_0 = j\xi_0 = \Delta\xi_0 = \xi_0.$$

Moreover, for each x' in $\mathcal{M}'_{sa}$ there is an x in $\mathcal{M}_{sa}$ such that $jbx'\xi_0 = x\xi_0$.

Proof. We have $\xi_0 \in K$ and since $\mathcal{M}'_{sa}\xi_0 \subset (iK)^\perp$ we also have $\xi_0 \in (iK)^\perp$. It follows that $p\xi_0 = \xi_0$ and $q\xi_0 = 0$, whence $a\xi_0 = b\xi_0 = \xi_0$. From this it follows that $j\xi_0 = \xi_0$ and $\Delta\xi_0 = \xi_0$.

To prove the second assertion assume first that $0 \leqslant x' \leqslant 1$. Then the functional ψ in $\mathcal{M}_*$ given by $\psi(y) = (y\xi_0 | x'\xi_0)$ is positive and dominated by ϕ. By 5.3.2 (with $\lambda = \frac{1}{2}$) there is an x in $\mathcal{M}$, $0 \leqslant x \leqslant 1$, such that

$$(y\xi_0 | x'\xi_0) = \tfrac{1}{2}((xy + yx)\xi_0 | \xi_0)$$

for all y in $\mathcal{M}$. In particular, for y in $\mathcal{M}_{\text{sa}}$

$$\langle y\xi_0 \mid x'\xi_0\rangle = \langle y\xi_0 \mid x\xi_0\rangle,$$

from which we conclude that $x\xi_0 = p(x'\xi_0)$. Since $q(x'\xi_0) = 0$, as we saw, we have $jbx'\xi_0 = x\xi_0$, as desired.

8.13.10. LEMMA. *For each x' in $\mathcal{M}'$ and each complex λ with* $\operatorname{Re}\lambda > 0$ *there is an x in $\mathcal{M}$ such that*

$$bjx'jb = \lambda(2-a)xa + \bar{\lambda}ax(2-a).$$

Proof. By linearity we may assume that $x' \in \mathcal{M}'_{\text{sa}}$. If $x' = 1$ we can take $x = (2\operatorname{Re}\lambda)^{-1}1$, so by adding a scalar multiple of 1 we may assume that $0 \leqslant x'$ (and also that $\|x'\| \leqslant 1$). As in the proof of 8.13.9 we use 5.3.2 to find x in $\mathcal{M}_+$ such that

$$(y\xi_0 \mid x'\xi_0) = ((\lambda xy + \bar{\lambda}yx)\xi_0 \mid \xi_0)$$

for every y in $\mathcal{M}$. Substituting z^*y for y (where $y, z \in \mathcal{M}$) we obtain

$$(y\xi_0 \mid x'z\xi_0) = \lambda(y\xi_0 \mid zx\xi_0) + \bar{\lambda}(yx\xi_0 \mid z\xi_0).$$

Given arbitrary y', z' in $\mathcal{M}'_{\text{sa}}$ there are corresponding elements y, z in $\mathcal{M}_{\text{sa}}$ satisfying the condition in 8.13.9. Substituting $jby'\xi_0$ and $jbz'\xi_0$ for $y\xi_0$ and $z\xi_0$ and rearranging the terms using 8.13.3 (iii) we obtain

$$(bjx'jbz'\xi_0 \mid y'\xi_0) = \lambda(jby'\xi_0 \mid zx\xi_0) + \bar{\lambda}(yx\xi_0 \mid jbz'\xi_0).$$

Since $c\xi_0 = j\Delta^{1/2}c^*\xi_0$ for each c in $\mathcal{M}$ by 8.13.8 we can reduce the equation above further:

$$\begin{aligned}(bjx'jbz'\xi_0 \mid y'\xi_0) &= \lambda(jby'\xi_0 \mid j\Delta^{1/2}xz\xi_0) + \bar{\lambda}(j\Delta^{1/2}xy\xi_0 \mid jbz'\xi_0)\\ &= \lambda(\Delta^{1/2}xz\xi_0 \mid by'\xi_0) + \bar{\lambda}(bz'\xi_0 \mid \Delta^{1/2}xy\xi_0)\\ &= \lambda(xz\xi_0 \mid (2-a)y'\xi_0) + \bar{\lambda}((2-a)z'\xi_0 \mid xy\xi_0)\\ &= \lambda(xjbz'\xi_0 \mid (2-a)y'\xi_0) + \bar{\lambda}((2-a)z'\xi_0 \mid xjby'\xi_0).\end{aligned}$$

Finally, since $a - jb = 2q$ and $q\mathcal{M}'_{\text{sa}}\xi_0 = 0$ we obtain

$$\begin{aligned}(bjx'jbz'\xi_0 \mid y'\xi_0) &= \lambda(xaz'\xi_0 \mid (2-a)y'\xi_0) + \bar{\lambda}((2-a)z'\xi_0 \mid xay'\xi_0)\\ &= ((\lambda(2-a)xa + \bar{\lambda}ax(2-a))z'\xi_0 \mid y'\xi_0).\end{aligned}$$

Since y' and z' were arbitrary and $\mathcal{M}'_{\text{sa}}\xi_0$ is total in H we conclude that

$$bjx'jb = \lambda(2-a)xa + \bar{\lambda}ax(2-a).$$

8.13.11. LEMMA. *If $\lambda = \exp(i\theta/2)$, $|\theta| < \pi$ and f is an analytic function which is* *bounded on the strip* $\{\zeta\in \boldsymbol{C}\,|\,|\mathrm{Re}\,\zeta|\leqslant\frac{1}{2}\}$ *then*

$$f(0)=\tfrac{1}{2}\int \exp(-\theta t)(\cosh(\pi t))^{-1}(\lambda f(\mathrm{i}t+\tfrac{1}{2})+\bar{\lambda}f(\mathrm{i}t-\tfrac{1}{2}))\,\mathrm{d}t.$$

Proof. Define $g(\zeta)=\pi\exp(\mathrm{i}\theta\zeta)(\sin(\pi\zeta))^{-1}f(\zeta)$. Then g is meromorphic in the strip, with zero as the only pole. The pole is simple and the residue is $f(0)$. Since g tends rapidly to zero at infinity within the strip, because f is bounded and $|\theta|<\pi$, we can apply Cauchy's integral formula to obtain

$$f(0)=(2\pi\mathrm{i})^{-1}(\int g(\mathrm{i}t+\tfrac{1}{2})\mathrm{i}\,\mathrm{d}t-\int g(\mathrm{i}t-\tfrac{1}{2})\mathrm{i}\,\mathrm{d}t).$$

Now $\sin(\pi(\mathrm{i}t+\frac{1}{2}))=\cos(\pi\mathrm{i}t)=\cosh(\pi t)$, and $\sin(\pi(\mathrm{i}t-\frac{1}{2}))=-\cosh(\pi t)$; whereas $\exp(\mathrm{i}\theta(\mathrm{i}t+\frac{1}{2}))=\lambda\exp(-\theta t)$ and $\exp(\mathrm{i}\theta(\mathrm{i}t-\frac{1}{2}))=\bar{\lambda}\exp(-\theta t)$. Thus the desired formula follows by substitution.

8.13.12. LEMMA. *If* x', λ *and* x *are as in 8.13.10 then with* $\lambda=\exp(\mathrm{i}\theta/2)$, $|\theta|<\pi$, *we have*

$$x=\tfrac{1}{2}\int \Delta^{\mathrm{i}t}jx'j\Delta^{-\mathrm{i}t}\exp(-\theta t)(\cosh(\pi t))^{-1}\,\mathrm{d}t.$$

Proof. Let ξ and η be analytic vectors for $\{\Delta^{\mathrm{i}t}\}$ in K and define the analytic function

$$f(\zeta)=(bxb\Delta^{-\zeta}\xi\,|\,\Delta^{\bar{\zeta}}\eta).$$

Since f is bounded on every vertical strip, 8.13.11 can be applied. Moreover, using 8.13.4 and 8.13.3 (ii)

$$f(\mathrm{i}t+\tfrac{1}{2})=(bxb\Delta^{-\mathrm{i}t}\Delta^{-1/2}\xi\,|\,\Delta^{-\mathrm{i}t}\Delta^{1/2}\eta)=(\Delta^{\mathrm{i}t}(2-a)xa\Delta^{-\mathrm{i}t}\xi\,|\,\eta),$$

$$f(\mathrm{i}t-\tfrac{1}{2})=(bxb\Delta^{-\mathrm{i}t}\Delta^{1/2}\xi\,|\,\Delta^{-\mathrm{i}t}\Delta^{-1/2}\eta)=(\Delta^{\mathrm{i}t}ax(2-a)\Delta^{-\mathrm{i}t}\xi\,|\,\eta);$$

so that from 8.13.10 we have

$$\lambda f(\mathrm{i}t+\tfrac{1}{2})+\bar{\lambda}f(\mathrm{i}t-\tfrac{1}{2})=(\Delta^{\mathrm{i}t}bjx'jb\Delta^{-\mathrm{i}t}\xi\,|\,\eta).$$

Applying 8.13.11 we get

$$\begin{aligned}(bxb\xi\,|\,\eta)&=\tfrac{1}{2}\int \exp(-\theta t)(\cosh(\pi t))^{-1}(\Delta^{\mathrm{i}t}bjx'jb\Delta^{-\mathrm{i}t}\xi\,|\,\eta)\,\mathrm{d}t\\&=(\tfrac{1}{2}\int \Delta^{\mathrm{i}t}jx'j\Delta^{-\mathrm{i}t}\exp(-\theta t)(\cosh(\pi t))^{-1}\,\mathrm{d}t\,b\xi\,|\,b\eta),\end{aligned}$$

from which the formula follows since K is total in H and the range of b is dense.

8.13.13. LEMMA. *For each* t *in* $\boldsymbol{R}$ *and* x' *in* $\mathscr{M}'$ *we have* $\Delta^{\mathrm{i}t}jx'j\Delta^{-\mathrm{i}t}\in\mathscr{M}$.

Proof. Take y' in $\mathscr{M}'$ and ξ,η in H and define

$$g(t)=((\Delta^{\mathrm{i}t}jx'j\Delta^{-\mathrm{i}t}y'-y'\Delta^{\mathrm{i}t}jx'j\Delta^{-\mathrm{i}t})\xi\,|\,\eta).$$

It follows from 8.13.12 that for every θ with $|\theta|<\pi$ we have

$$\int g(t)\exp(-\theta t)(\cosh(\pi t))^{-1}\,\mathrm{d}t=0.$$

Now the function f defined for $|\mathrm{Re}\,\zeta| < \pi$ by

$$f(\zeta) = \int g(t)\exp(-\zeta t)(\cosh(\pi t))^{-1}\,\mathrm{d}t$$

is holomorphic. Since it vanishes for real ζ it vanishes everywhere. In particular it vanishes for $\zeta = is$, $s \in \boldsymbol{R}$, i.e.

$$\int g(t)(\cosh(\pi t))^{-1}\exp(-ist)\,\mathrm{d}t = 0.$$

From the uniqueness of the Fourier transform it follows that $g = 0$. Consequently $\Delta^{it}jx'j\Delta^{-it} \in \mathcal{M}'' = \mathcal{M}$.

8.13.14. THEOREM. *Let ξ_0 be a cyclic and separating vector for a von Neumann algebra $\mathcal{M}$ on a Hilbert space H. There is then a positive (unbounded) injective operator Δ on H called the modular operator and a conjugate linear isometry j with $j^2 = 1$, such that $j\mathcal{M}j = \mathcal{M}'$ and $\Delta^{it}\mathcal{M}\Delta^{-it} = \mathcal{M}$ for every t in $\boldsymbol{R}$. Moreover, $j\xi_0 = \xi_0$ and $\mathcal{M}\xi_0 \subset \mathcal{D}(\Delta^{1/2})$ with*

$$j\Delta^{1/2}x\xi_0 = x^*\xi_0, \qquad x \in \mathcal{M}.$$

Proof. We let K be the closure of $\mathcal{M}_{sa}\xi_0$ and see from 8.13.8 that it satisfies the conditions of 8.13.2. Thus from the preceding we obtain a positive injective operator Δ and a conjugate linear isometry j with $j^2 = 1$. Moreover, $j\xi_0 = \xi_0$ (8.13.9) and $\mathcal{M}\xi_0 \subset \mathcal{D}(\Delta^{1/2})$ with $j\Delta^{1/2}x\xi_0 = x^*\xi_0$, for each x in $\mathcal{M}$ (8.13.8). It remains to show that $j\mathcal{M}j = \mathcal{M}'$ and that $\Delta^{it}\mathcal{M}\Delta^{-it} = \mathcal{M}$.

From 8.13.13 with $t = 0$ we have $j\mathcal{M}'j \subset \mathcal{M}$. To obtain the converse inclusion take x, y in $\mathcal{M}_{sa}$. Since $j\xi_0 = \xi_0$ we have

$$(yjxj\xi_0 \,|\, \xi_0) = (\xi_0 \,|\, xjyj\xi_0).$$

However, this equation is linear in both x and y and is therefore valid for all x, y in $\mathcal{M}$. Thus we can take x, y in $\mathcal{M}_{sa}$ and y' in $\mathcal{M}'$ and replace y by $yjy'j$ (which belongs to $\mathcal{M}$ by 8.13.13) to obtain

$$(y(jy'j(jxj)\xi_0 \,|\, \xi_0) = (\xi_0 \,|\, xj(yjy'j)j\xi_0).$$

Using this and 8.13.3 (iii) we get

$$(xjyj\xi_0 \,|\, y'\xi_0) = (jyjx\xi_0 \,|\, y'\xi_0).$$

Since $\mathcal{M}'\xi_0$ is dense in H this implies that $xjyj\xi_0 = jyjx\xi_0$. However, this equation is linear in x and is therefore valid for all x in $\mathcal{M}$. So replace x with xz, x, z in $\mathcal{M}$ to obtain

$$jyjxz\xi_0 = xzjyj\xi_0 = xjyjz\xi_0,$$

whence $jyjx = xjyj$ since $\mathcal{M}\xi_0$ is dense, and $jyj \in \mathcal{M}'$. Consequently,

$$\mathcal{M} = j^2\mathcal{M}j^2 \subset j\mathcal{M}'j.$$

From $j\mathcal{M}'j = \mathcal{M}$ it follows immediately from 8.13.13 that $\Delta^{it}\mathcal{M}\Delta^{-it} = \mathcal{M}$ for all t in $\boldsymbol{R}$.

8.13.15. Notes and remarks. Theorem 8.13.14 was proved by Tomita [267]. The first version of Tomita's theory was published by Takesaki [259], and contained a host of additional results. Since then an avalanche of papers have covered the subject. The proof of 8.13.14 employed here is taken from Rieffel and VanDaele [216], but we have also benefited from unpublished notes by Haagerup.

8.14. The modular group

8.14.1. Given a unitary representation (u, H) of $\boldsymbol{R}$ and an invariant closed real subspace K of H, we say that u satisfies the *modular condition* with respect to K if for any two vectors ξ, η in K there is a bounded continuous function f defined on the strip

$$\Omega_{-1} = \{\zeta \in \boldsymbol{C} \mid -1 \leqslant \operatorname{Im}\zeta \leqslant 0\}$$

such that f is holomorphic in the interior of Ω_{-1} and satisfies the boundary conditions

$$f(t) = (u_t\xi \mid \eta), \qquad f(t-i) = (\eta \mid u_t\xi), \qquad t \in \boldsymbol{R}.$$

As in the case of automorphism groups (8.12.2) the modular condition has an equivalent formulation in terms of analytic vectors in K.

8.14.2. PROPOSITION. *Let K be a subspace satisfying the conditions in 8.13.2 and let Δ be the modular operator defined in 8.13.4. Then the representation $t \to \Delta^{it}$ is the unique unitary representation of $\boldsymbol{R}$ that satisfies the modular condition with respect to K.*

Proof. Note that $\Delta^{it}K = K$ by 8.13.5. Take ξ, η in K. Since $K \subset \mathscr{D}(\Delta^s)$ for $0 \leqslant s \leqslant \frac{1}{2}$ by 8.13.4 we can define a bounded continuous function f on Ω_{-1} by

$$f(t - is) = (\Delta^{it}\Delta^{s/2}\xi \mid \Delta^{s/2}\eta), \qquad t - is \in \Omega_{-1}.$$

The function $\zeta \to \Delta^{i\zeta}\xi$ is holomorphic for $-\frac{1}{2} < \operatorname{Im}\zeta < 0$ by the spectral theorem (with derivative $i\Delta^{i\zeta}\log\Delta\xi$). Since we can write any ζ_0 in the interior of Ω_{-1} in the form $\zeta_0 = \zeta - ir$, where $-\frac{1}{2} < \operatorname{Im}\zeta < 0$ and $0 < r < \frac{1}{2}$, whence $f(\zeta_0) = (\Delta^{i\zeta}\xi \mid \Delta^r\eta)$; it follows that f is holomorphic in the interior of Ω_{-1}. For $\zeta_0 = t - i$ we obtain by 8.13.4, 8.13.5 and 8.13.3 (iii)

$$f(t-i) = (\Delta^{it}\Delta^{1/2}\xi \mid \Delta^{1/2}\eta) = (\Delta^{it}j\xi \mid j\eta) = (j\Delta^{it}\xi \mid j\eta) = (\eta \mid \Delta^{it}\xi).$$

Suppose now that (u, H) was another unitary representation of $\boldsymbol{R}$ satisfying the modular condition with respect to K. If ξ_1 is an analytic vector for u in K the analytic function $\zeta \to u_\zeta \xi_1$ must satisfy

$$(u_{t-i}\xi_1 \mid \eta) = (\eta \mid u_t \xi_1)$$

for every η in K. Taking η_1 to be analytic for the group $t \to \Delta^{it}$ we must similarly have

$$(\xi \mid \Delta^{it+1}\eta_1) = (\Delta^{it}\eta_1 \mid \xi)$$

for every ξ in K.

Define an analytic function g by

$$g(\zeta) = (u_\zeta \xi_1 \mid \Delta^{i(\bar{\zeta}-i)}\eta_1).$$

Since $\Delta^{it}K = K$ we have

$$g(t-i) = (u_{t-i}\xi_1 \mid \Delta^{it}\eta_1) = (\Delta^{it}\eta_1 \mid u_t \xi_1).$$

Similarly, since $u_t K = K$ we get

$$g(t) = (u_t \xi_1 \mid \Delta^{it+1}\eta_1) = (\Delta^{it}\eta_1 \mid u_t \xi_1).$$

It follows that $g(\zeta - i) = g(\zeta)$ for all ζ, since it is true for all ζ in $\boldsymbol{R}$. The functions $\zeta \to u_\zeta \xi_1$ and $\zeta \to \Delta^{i\zeta}\eta_1$ are bounded on horizontal strips, in particular g is bounded on Ω_{-1}. The periodicity of g implies that g is bounded on $\boldsymbol{C}$ and therefore constant, by Liouville's theorem. Consequently

$$(\Delta^{it}\eta_1 \mid u_t \xi_1) = g(t) = g(0) = (\eta_1 \mid \xi_1);$$

and since this is valid for a dense set of vectors in K and $K + iK$ is dense in H we conclude that $u_{-t}\Delta^{it} = 1$ for all t, as desired.

8.14.3. Proposition. *Let $(A, \boldsymbol{R}, \alpha)$ be a C^*-dynamical system and suppose that a state ϕ of A satisfies the KMS condition at $\beta = 1$. Consider the covariant, cyclic representation $(\pi_\phi, u^\phi, H_\phi, \xi_\phi)$ associated with ϕ and let K be the closure of $\pi_\phi(A_{sa})\xi_\phi$. Then u^ϕ satisfies the modular condition with respect to K and K satisfies the conditions in 8.13.2.*

Proof. Since ϕ is α-invariant by 8.12.5 we obtain a cyclic covariant representation $(\pi_\phi, u^\phi, H_\phi, \xi_\phi)$ (cf. 7.4.12). From the definition of u^ϕ we see that

$$u_t^\phi \pi_\phi(x)\xi_\phi = \pi_\phi(\alpha_t(x))\xi_\phi$$

for each x in A_{sa}, whence $u_t^\phi K = K$ for all t.

Take ξ and η in K and choose sequences $\{x_n\}$ and $\{y_n\}$ in A_{sa} such that $\pi_\phi(x_n)\xi_\phi \to \xi$ and $\pi_\phi(y_n)\xi_\phi \to \eta$. By assumption there is for each n a bounded continuous function f_n on the strip

$$\Omega_1 = \{\zeta \in \boldsymbol{C} \mid 0 \leqslant \operatorname{Im} \zeta \leqslant 1\}$$

such that f_n is holomorphic in the interior of Ω_1 and satisfies the boundary conditions

$$f_n(t) = \phi(y_n \alpha_t(x_n)) = (u_t^\phi \pi_\phi(x_n)\xi_\phi \mid \pi_\phi(y_n)\xi_\phi);$$

$$f_n(t+\mathrm{i}) = \phi(\alpha_t(x_n)y_n) = (\pi_\phi(y_n)\xi_\phi \mid u_t^\phi \pi_\phi(x_n)\xi_\phi).$$

Since the sequence $\{f_n\}$ is uniformly bounded and uniformly convergent on the boundary of Ω_1 it follows from the Phragmen–Lindelöf theorem (cf. the proof of 8.12.2) that $\{f_n\}$ converges to a function f which is holomorphic in the interior of Ω_1 and satisfies the boundary conditions

$$f(t) = (u_t^\phi \xi \mid \eta); \qquad f(t+\mathrm{i}) = (\eta \mid u_t^\phi \xi).$$

Setting $g(\bar{\zeta}) = \bar{f}(\zeta)$ it follows immediately that u^ϕ satisfies the modular condition with respect to K.

It is clear that $(K + \mathrm{i}K)^\perp = \{0\}$ (since ξ_ϕ is cyclic) so we need only prove the first condition in 8.13.2. To do so take ξ in $K \cap \mathrm{i}K$ and η in K. Since u^ϕ satisfies the modular condition there are bounded continuous functions f_1 and f_2 on the strip

$$\Omega_{-1} = \{\zeta \in \boldsymbol{C} \mid -1 \leqslant \operatorname{Im} \zeta \leqslant 0\}$$

such that these functions are holomorphic in the interior of Ω_{-1} and satisfy the boundary conditions

$$f_1(t) = (u_t^\phi \xi \mid \eta); \qquad f_1(t-\mathrm{i}) = (\eta \mid u_t^\phi \xi);$$

$$f_2(t) = (u_t^\phi \mathrm{i}\xi \mid \eta); \qquad f_2(t-\mathrm{i}) = (\eta \mid u_t \mathrm{i}\xi).$$

We have $\mathrm{i}f_1(t) = f_2(t)$ and $-\mathrm{i}f_1(t-\mathrm{i}) = f_2(t-\mathrm{i})$, which implies that $\mathrm{i}f_1(\zeta) = f_2(\zeta)$ and $-\mathrm{i}f_1(\zeta) = f_2(\zeta)$ for all ζ in Ω_{-1}. It follows that $f_1 = f_2 = 0$, and since this holds for all η in K we conclude that $\zeta = 0$ because K is total in H_ϕ.

8.14.4. Corollary. *If $(A, \boldsymbol{R}, \alpha)$ is a C^*-dynamical system and ϕ is a state of A satisfying the KMS condition at some $\beta < \infty$, then $L_\phi = \ker \pi_\phi$ and the vector ξ_ϕ is cyclic and separating for $\pi_\phi(A)''$.*

Proof. If $\beta = 0$ the result follows from 5.3.3. If $\beta > 0$ define $\alpha'_t = \alpha_{\beta t}$. Then ϕ satisfies the KMS-condition at 1 for the system $(A, \boldsymbol{R}, \alpha')$ so that 8.14.3 can be applied. Thus if $x \in \pi_\phi(A)''$ such that $x\xi_\phi = 0$, then assuming as we may that $x \geqslant 0$ we observe that for every y in $\pi_\phi(A)''_{\mathrm{sa}}$ we have $(xy + yx)\xi_\phi \in K$ and $(xy - yx)\xi_\phi \in \mathrm{i}K$. Since $K \cap \mathrm{i}K = \{0\}$ this implies that $xy\xi_\phi = 0$ for all y, whence $x = 0$. Thus ξ_ϕ is separating and of course also cyclic for $\pi_\phi(A)''$. If $x \in L_\phi$, i.e. $\phi(x^*x) = 0$, then $\pi_\phi(x)\xi_\phi = 0$, whence $\pi_\phi(x) = 0$ so that $x \in \ker \pi_\phi$. Consequently $L_\phi = \ker \pi_\phi$.

8.14.5. THEOREM. *To each faithful normal state ϕ of a von Neumann algebra $\mathcal{M}$ there is a unique W^*-dynamical system $(\mathcal{M}, \boldsymbol{R}, \sigma)$ such that ϕ satisfies the KMS condition at $\beta = 1$. We say that $\{\sigma_t | t \in \boldsymbol{R}\}$ is the modular group associated with ϕ.*

Proof. Given $\mathcal{M}$ and ϕ consider the normal cyclic representation $(\pi_\phi, H_\phi, \xi_\phi)$ of $\mathcal{M}$ associated with ϕ. Since ϕ is faithful so is π_ϕ, and we may identify $\mathcal{M}$ with its image $\pi_\phi(\mathcal{M})$ in $B(H_\phi)$. Moreover, ξ_ϕ is separating so that 8.13.14 applies. Consequently we can define

$$\sigma_t(x) = \Delta^{it} x \Delta^{-it}, \qquad x \in \mathcal{M}, \qquad t \in \boldsymbol{R},$$

and obtain a W^*-dynamical system $(\mathcal{M}, \boldsymbol{R}, \sigma)$. By 8.14.2 the representation $t \to \Delta^{it}$ satisfies the modular condition with respect to the subspace $K = (\mathcal{M}_{sa}\xi_\phi)^-$. If therefore $x, y \in \mathcal{M}_{sa}$ there is a bounded continuous function f on Ω_{-1}, holomorphic in the interior of Ω_{-1} and satisfying the boundary conditions

$$f(t) = (\Delta^{it} x \xi_\phi | y \xi_\phi) = (\Delta^{it} x \Delta^{-it} \xi_\phi | y \xi_\phi) = \phi(y \sigma_t(x)),$$

$$f(t - i) = (y \xi_\phi | \Delta^{it} x \xi_\phi) = (y \xi_\phi | \Delta^{it} x \Delta^{-it} \xi_\phi) = \phi(\sigma_t(x) y).$$

Taking $g(\zeta) = \bar{f}(\bar{\zeta})$ we obtain a function holomorphic in the interior of the strip

$$\Omega_1 = \{\zeta \in \boldsymbol{C} | 0 \leqslant \operatorname{Im} \zeta \leqslant 1\}$$

and satisfying the correct boundary conditions. Thus ϕ is a KMS state at $\beta = 1$ for the system $(\mathcal{M}, \boldsymbol{R}, \sigma)$.

Conversely, if ϕ is a KMS-state at $\beta = 1$ for some W^*-dynamical system $(\mathcal{M}, \boldsymbol{R}, \alpha)$ then applying 8.14.3 to the C^*-dynamical system $(\mathcal{M}^c, \boldsymbol{R}, \alpha | \mathcal{M}^c)$ we see from 8.14.2 that $u_t^\phi = \Delta^{it}$ for all t, whence

$$\alpha_t(x) = u_t^\phi x u_{-t}^\phi = \Delta^{it} x \Delta^{-it} = \sigma_t(x)$$

for all x in $\mathcal{M}$ and t in $\boldsymbol{R}$.

8.14.6. LEMMA. *Let ϕ be a faithful normal state of a von Neumann algebra $\mathcal{M}$. Let $\sigma_{\boldsymbol{R}}$ denote the modular group associated with ϕ and let $\mathcal{M}^\sigma$ denote the fixed-point algebra of $\sigma_{\boldsymbol{R}}$ in $\mathcal{M}$. Then $x \in \mathcal{M}^\sigma$ if and only if $\phi(xy - yx) = 0$ for all y in $\mathcal{M}$.*

Proof. Given x, y in $\mathcal{M}$ there is a function f in $C^b(\Omega_1)$, holomorphic in the interior of Ω_1, such that

$$f(t) = \phi(y \sigma_t(x)), \qquad f(t + i) = \phi(\sigma_t(x) y).$$

If $x \in \mathcal{M}^\sigma$ then f is constant on $\boldsymbol{R}$, hence on Ω_1, so that

$$\phi(yx) = f(0) = f(i) = \phi(xy).$$

Conversely, if $\phi(xy - yx) = 0$ for all y then the same is true for $\sigma_t(x)$, since ϕ is σ_R-invariant. Thus the KMS-function satisfies $f(t) = f(t + i)$ for all t. But then f has an analytic extension to $\boldsymbol{C}$ which is periodic (with period i) and consequently bounded. By Liouville's theorem f is constant. Fix s and put $y = (\sigma_s(x) - x)^*$. Then

$$0 = f(s) - f(0) = \phi((\sigma_s(x) - x)^*(\sigma_s(x) - x)),$$

whence $\sigma_s(x) = x$. Since s is arbitrary, $x \in \mathcal{M}^\sigma$.

8.14.7. PROPOSITION. *Let ϕ be a faithful normal state of a von Neumann algebra $\mathcal{M}$ and denote by σ_R its modular group. If $\psi \in \mathcal{M}_*$, $0 \leqslant \psi \leqslant \phi$, and ψ is σ_R-invariant there is a unique h in $\mathcal{M}_{sa}$ such that $\psi = \phi(h\,\cdot)$. Moreover, $0 \leqslant h \leqslant 1$ and $h \in \mathcal{M}^\sigma$.*

Proof. By 5.3.2 there is a unique element h in $\mathcal{M}$ with $0 \leqslant h \leqslant 1$ such that $\psi = \frac{1}{2}(\phi(h\,\cdot) + \phi(\cdot\, h))$. Since ψ and ϕ are both σ_R-invariant, and h is unique, it follows from 8.14.6 that $h \in \mathcal{M}^\sigma$, whence $\psi = \phi(h\,\cdot)$ $(= \phi(\cdot\, h))$.

8.14.8. PROPOSITION. *Let ϕ and ψ be faithful normal states on a von Neumann algebra $\mathcal{M}$ and assume that they have the same modular group σ_R. There is then a unique positive injective operator h affiliated with $\mathcal{M} \cap \mathcal{M}'$ such that $\psi = \phi(h\,\cdot)$ (and $\phi = \psi(h^{-1}\,\cdot)$).*

Proof. Assume first that $\psi \leqslant \lambda\phi$ for some $\lambda > 0$. By 8.14.7 we have $\psi = \phi(h\,\cdot)$ where $h \in \mathcal{M}^\sigma$. Take a unitary u in $\mathcal{M}$ and an arbitrary x in $\mathcal{M}$. Applying the KMS condition for ϕ to the elements u^* and hux we obtain a function f in $C^b(\Omega_1)$, holomorphic in the interior of Ω_1, such that

$$f(t) = \phi(u^*\sigma_t(hux)), \qquad f(t + i) = \phi(\sigma_t(hux)u^*).$$

Applying the KMS condition for ψ to the elements u^* and ux we obtain a similar function g, where

$$g(t) = \psi(u^*\sigma_t(ux)), \qquad g(t + i) = \psi(\sigma_t(ux)u^*).$$

Since $h \in \mathcal{M}^\sigma$ we observe that $f(t + i) = g(t + i)$, whence $f = g$. In particular,

$$\phi(u^*hux) = f(0) = g(0) = \phi(hx).$$

The uniqueness of h (8.14.7) implies that $u^*hu = h$ for all u, whence $h \in \mathcal{M} \cap \mathcal{M}'$.

In the general case note first that σ_R is also the modular group for $\phi + \psi$. Applying the first part of the proof we obtain positive operators h_1 and h_2 in $\mathcal{M} \cap \mathcal{M}'$ such that

$$\psi = (\phi + \psi)(h_1\,\cdot), \qquad \phi = (\phi + \psi)(h_2\,\cdot).$$

Since ϕ and ψ are faithful, h_1 and h_2 are injective (and $h_1 + h_2 = 1$). Thus with $h = h_1 h_2^{-1}$ we have a positive injective operator affiliated with $\mathcal{M} \cap \mathcal{M}'$ such that $\psi = \phi(h\,\cdot)$.

8.14.9. LEMMA. *If $\sigma_{\boldsymbol{R}}$ is the modular group associated with a faithful normal state ϕ on a von Neumann algebra $\mathcal{M}$ and if α is an automorphism of $\mathcal{M}$ then $\{\alpha^{-1} \circ \sigma_t \circ \alpha \mid t \in \boldsymbol{R}\}$ is the modular group associated with $\phi \circ \alpha$.*

Proof. Take x, y in $\mathcal{M}$. Applying the KMS condition to the elements $\alpha(x), \alpha(y)$, we obtain a function f in $C^b(\Omega_1)$, holomorphic in the interior of Ω_1, such that

$$f(t) = \phi(\alpha(y)\sigma_t(\alpha(x))) = \phi \circ \alpha(y(\alpha^{-1} \circ \sigma_t \circ \alpha(x)));$$

$$f(t + \mathrm{i}) = \phi(\sigma_t(\alpha(x))\alpha(y)) = \phi \circ \alpha((\alpha^{-1} \circ \sigma_t \circ \alpha(x))y).$$

The result is now immediate from 8.14.5.

8.14.10. THEOREM. *Let ϕ and ψ be faithful normal states on a von Neumann algebra $\mathcal{M}$ and let $\sigma_{\boldsymbol{R}}^{\phi}$ and $\sigma_{\boldsymbol{R}}^{\psi}$ denote their modular groups, respectively. The following conditions are equivalent:*

(i) *ϕ is $\sigma_{\boldsymbol{R}}^{\psi}$-invariant;*

(i′) *ψ is $\sigma_{\boldsymbol{R}}^{\phi}$-invariant;*

(ii) *$\sigma_{\boldsymbol{R}}^{\psi}$ and $\sigma_{\boldsymbol{R}}^{\phi}$ commute;*

(iii) *there exists a unique positive injective operator h affiliated with $\mathcal{M}^{\sigma^{\phi}} \cap \mathcal{M}^{\sigma^{\psi}}$ such that $\psi = \phi(h\,\cdot)$;*

(iii′) *there exists a unique positive injective operator h' affiliated with $\mathcal{M}^{\sigma^{\phi}} \cap \mathcal{M}^{\sigma^{\psi}}$ such that $\phi = \psi(h'\,\cdot)$.*

Proof. (i)$\Leftrightarrow$(ii). By 8.14.9 the modular group for $\phi \circ \sigma_s^{\psi}$ is $\{\sigma_{-s}^{\psi} \circ \sigma_t^{\phi} \circ \sigma_s^{\psi} \mid t \in \boldsymbol{R}\}$. If therefore ϕ is $\sigma_{\boldsymbol{R}}^{\psi}$-invariant we have $\sigma_{-s}^{\psi} \circ \sigma_t^{\phi} \circ \sigma_s^{\psi} = \sigma_t^{\phi}$ by 8.14.5, which shows that $\sigma_{\boldsymbol{R}}^{\phi}$ and $\sigma_{\boldsymbol{R}}^{\psi}$ commute.

Conversely, if the modular groups commute we see from the above that $\phi \circ \sigma_s^{\psi} = \phi(h_s\,\cdot)$ for some positive operator h_s affiliated with $\mathcal{M} \cap \mathcal{M}'$ by 8.14.8. The uniqueness of h_s implies that $h_{s+t} = h_s h_t$ and thus $h_{ns} = h_s^n$ for all n in $\boldsymbol{Z}$. If p is a spectral projection of h_s corresponding to an interval disjoint from $\{1\}$ then $\{\phi(h_s^n p) \mid n \in \boldsymbol{Z}\}$ is unbounded unless $h_s p = 0$. However, $\phi(h_s^n p) = \phi(\sigma_{ns}^{\psi}(p)) \leq 1$, so $h_s p = 0$. Consequently $h_s = 1$ and ϕ is $\sigma_{\boldsymbol{R}}^{\psi}$-invariant.

(i′)$\Leftrightarrow$(ii) is analogous, and (iii)$\Rightarrow$(i′), (iii′)$\Rightarrow$(i) are immediate.

(i)$\Rightarrow$(iii). Consider the state $\rho = \frac{1}{2}(\phi + \psi)$ and denote by $\sigma_{\boldsymbol{R}}^{\rho}$ its modular group. Now ρ is $\sigma_{\boldsymbol{R}}^{\psi}$-invariant, and since (i)$\Leftrightarrow$(i′) it follows that ψ is $\sigma_{\boldsymbol{R}}^{\rho}$-invariant. As $\psi \leq 2\rho$ we see from 8.14.7 that $\psi = \rho(k\,\cdot)$ where $0 \leq k \leq 2$ and k is $\sigma_{\boldsymbol{R}}^{\rho}$-invariant. However, k is unique and both ψ and ρ are $\sigma_{\boldsymbol{R}}^{\psi}$-invariant and $\sigma_{\boldsymbol{R}}^{\phi}$-invariant. Consequently the same is true for k. Note that

$\phi = \rho((2-k)\cdot)$ and that both k and $2-k$ are injective, since ϕ and ψ are both faithful. Thus $h = k(2-k)^{-1}$ is a positive injective operator affiliated with $\mathcal{M}^{\sigma^\phi} \cap \mathcal{M}^{\sigma^\psi}$ such that $\psi = \phi(h\cdot)$. Clearly h is unique. Since $\phi = \psi(h^{-1}\cdot)$ we have also shown that (i) $\Rightarrow$ (iii') so the proof is complete.

8.14.11. PROPOSITION. *Let ϕ and ψ be faithful normal states on a von Neumann algebra $\mathcal{M}$, and denote by $\sigma_{\mathbf{R}}^\phi$ and $\sigma_{\mathbf{R}}^\psi$ the associated modular groups. Then the W^*-dynamical systems $(\mathcal{M}, \mathbf{R}, \sigma^\phi)$ and $(\mathcal{M}, \mathbf{R}, \sigma^\psi)$ are exterior equivalent.*

Proof. Define a faithful normal state ρ of $\mathcal{M} \otimes M_2$ by

$$\rho\left(\begin{pmatrix} x_{11} & x_{12} \\ x_{21} & x_{22} \end{pmatrix}\right) = \tfrac{1}{2}(\phi(x_{11}) + \psi(x_{22})).$$

If $\sigma_{\mathbf{R}}^\rho$ denotes the modular group associated with ρ we see from the uniqueness part of 8.14.5 that

$$\sigma_t^\rho\left(\begin{pmatrix} x & 0 \\ 0 & y \end{pmatrix}\right) = \begin{pmatrix} \sigma_t^\phi(x) & 0 \\ 0 & \sigma_t^\psi(y) \end{pmatrix}, \qquad x, y \in \mathcal{M}, \qquad t \in \mathbf{R}.$$

It follows from 8.11.3 that the two systems are exterior equivalent; the cocycle u linking σ^ψ with σ^ϕ being determined by

$$\sigma_t^\rho\left(\begin{pmatrix} 0 & 0 \\ 1 & 0 \end{pmatrix}\right) = \begin{pmatrix} 0 & 0 \\ u_t & 0 \end{pmatrix}.$$

8.14.12. PROPOSITION. *Let the assumptions be as in 8.14.11. The cocycle u linking σ^ψ with σ^ϕ is a group representation if and only if ϕ and ψ satisfy the conditions in 8.14.10. Moreover, in that case $u_t = h^{-it}$ where h is the Radon–Nikodym derivative of ψ with respect to ϕ, i.e. $\psi = \phi(h\cdot)$.*

Proof. We have $\sigma_t^\psi(x) = u_t\sigma_t^\phi(x)u_t^*$, $x \in \mathcal{M}$, and $u_{s+t} = u_s\sigma_s^\phi(u_t)$ (cf. 8.11.3). If ϕ and ψ satisfy the conditions in 8.14.10 then ϕ is invariant under both $\sigma_{\mathbf{R}}^\phi$ and $\sigma_{\mathbf{R}}^\psi$, whence

$$\phi(x) = \phi(\sigma_t^\psi(\sigma_{-t}^\phi(x))) = \phi(u_t x u_t^*)$$

for all x in $\mathcal{M}$. It follows that $\phi(u_t x) = \phi(u_t x u_t u_t^*) = \phi(x u_t)$, whence $u_t \in \mathcal{M}^{\sigma^\phi}$ by 8.14.6. Consequently u is a group representation.

Conversely, if u is a group representation then from the cocycle equation we see that each u_t is $\sigma_{\mathbf{R}}^\phi$-invariant. Thus

$$\sigma_s^\phi(\sigma_t^\psi(x)) = \sigma_s^\phi(u_t\sigma_t^\phi(x)u_{-t}) = u_t\sigma_{s+t}^\phi(x)u_{-t} = \sigma_t^\psi(\sigma_s^\phi(x))$$

for all x in $\mathcal{M}$, so that $\sigma_{\mathbf{R}}^\phi$ and $\sigma_{\mathbf{R}}^\psi$ commute.

Assume now that ϕ and ψ satisfy the conditions in 8.14.10. We can then write $\psi = \phi(h\cdot)$ and the function $t \to h^{-it}$ is a σ-weakly continuous unitary representation of $\mathbf{R}$ into $\mathcal{M}^{\sigma^\phi}$. Define a W^*-dynamical system $(\mathcal{M} \otimes M_2, \mathbf{R}, \sigma)$

by

$$\sigma_t\left(\begin{pmatrix} x_{11} & x_{12} \\ x_{21} & x_{22} \end{pmatrix}\right) = \begin{pmatrix} \sigma_t^\phi(x_{11}) & \sigma_t^\phi(x_{12})h^{it} \\ h^{-it}\sigma_t^\phi(x_{21}) & h^{-it}\sigma_t^\phi(x_{22})h^{it} \end{pmatrix}.$$

Let ρ be as in the proof of 8.14.12 and for $x = (x_{ij})$ and $y = (y_{ij})$ in $\mathcal{M} \otimes M_2$ consider the two functions

$$2\rho(y\sigma_t(x)) = \phi(y_{11}\sigma_t^\phi(x_{11}) + y_{12}h^{-it}\sigma_t^\phi(x_{21})) + \psi(y_{21}\sigma_t^\phi(x_{12})h^{it} + y_{22}h^{-it}\sigma_t^\phi(x_{22})h^{it});$$

$$2\rho(\sigma_t(x)y) = \phi(\sigma_t^\phi(x_{11})y_{11} + \sigma_t^\phi(x_{12})h^{it}y_{21}) + (h^{-it}\sigma_t^\phi(x_{21})y_{12} + h^{-it}\sigma_t^\phi(x_{22})h^{it}y_{22}).$$

We will show that these functions are the boundary values of a function f in $\bar{C}^b(\Omega_1)$, which is holomorphic in the interior of Ω_1. If suffices, by the Phragmen–Lindelöf theorem to prove this for all x in a σ-weakly dense *-algebra of $\mathcal{M} \otimes M_2$. So we may assume that all x_{ij} are analytic for σ_R^ϕ and belong to $p_n\mathcal{M}p_n$, where p_n is the spectral projection of h corresponding to the interval $[n^{-1}, n]$. In that case we see that $2\rho(y\sigma_t(x)) = \Sigma_{k=1}^4 f_k(t)$, where each f_k has an analytic extension to $\boldsymbol{C}$.

Using the KMS condition we immediately get $f_1(t + \mathrm{i}) = \phi(\sigma_t^\phi(x_{11})y_{11})$. Moreover, since $\psi = \phi(h\,\cdot)$ we have

$$f_2(t + \mathrm{i}) = \phi(y_{12}h^{-it}h\sigma_{t+\mathrm{i}}^\phi(x_{21})) = \phi(y_{12}\sigma_{t+\mathrm{i}}^\phi(hh^{-it}x_{21})) = \phi(\sigma_t^\phi(hh^{-it}x_{21})y_{12}) = \psi(h^{-it}\sigma_t^\phi(x_{21})y_{12}).$$

In the same manner we obtain

$$f_3(t + \mathrm{i}) = \psi(y_{21}\sigma_{t+\mathrm{i}}^\phi(x_{12})h^{it}h^{-1}) = \phi(y_{21}\sigma_{t+\mathrm{i}}^\phi(x_{12}h^{it})) = \phi(\sigma_t^\phi(x_{12})h^{it}y_{21}).$$

Finally

$$f_4(t + \mathrm{i}) = \psi(y_{22}h^{-it}h\sigma_{t+\mathrm{i}}^\phi(x_{22})h^{-it}h^{-1}) = \phi(y_{22}\sigma_{t+\mathrm{i}}^\phi(hh^{-it}x_{22}h^{it})) = \phi(\sigma_t^\phi(hh^{-it}x_{22}h^{it})y_{22}) = \psi(h^{-it}\sigma_t^\phi(x_{22})h^{it}y_{22}).$$

Now observe that $\Sigma_{k=1}^4 f_k(t + \mathrm{i}) = 2\rho(\sigma_t(x)y)$. Thus by 8.14.5 σ_R is the modular group associated with ρ, whence $h^{-it} = u_t$, as desired.

8.14.13. PROPOSITION. *The modular group associated with a faithful normal state ϕ of a von Neumann algebra $\mathcal{M}$ is inner if and only if $\mathcal{M}$ is semi-finite.*

Proof. If $\mathcal{M}$ is semifinite let τ be a faithful normal semi-finite trace on $\mathcal{M}$. By 5.3.11 there is a unique positive operator h affiliated with $\mathcal{M}$ such that $\phi = \tau(h\,\cdot)$. Since ϕ is faithful, h is injective. Let p_n be the spectral projection of h corresponding to the interval $[n^{-1}, n]$. Then τ is bounded on $p_n\mathcal{M}p_n$ and, being a trace, its modular group is trivial (cf. 8.14.5). It follows from 8.14.12 that the modular group of ϕ on $p_n \mathcal{M} p_n$ is given by $\sigma_t^\phi(x) = h^{-it}xh^{it}$; and since this holds for every n, and $p_n \nearrow 1$ as $n \to \infty$, we conclude that the modular group is inner.

Conversely, if $t \to u_t$ is a σ-weakly continuous representation of $\boldsymbol{R}$ into $\mathcal{M}$ such that $\sigma_t^\phi(x) = u_t x u_{-t}$ for all x in $\mathcal{M}$ and t in $\boldsymbol{R}$, then we have $u_t = h^{it}$, $t \in \boldsymbol{R}$, for some positive injective operator h affiliated with $\mathcal{M}^{\sigma^\phi}$. Let p_n be the spectral projection of h corresponding to the interval $[n^{-1}, n]$, and define $\tau_n = \phi(hp_n \cdot)$ on $p_n \mathcal{M} p_n$. By 8.14.12 the modular group of τ_n is trivial which implies that τ_n is a faithful normal trace on $p_n \mathcal{M} p_n$. Thus $p_n \mathcal{M} p_n$ is finite, and since $p_n \nearrow 1$ as $n \to \infty$ we conclude that $\mathcal{M}$ is semi-finite (with $\phi(h \cdot)$ as a trace).

8.14.14. COROLLARY. *Let $(A, \boldsymbol{R}, \alpha)$ be a C*-dynamical system and suppose that ϕ and ψ are KMS states of A corresponding to different non-zero, finite values β_1 and β_2. If $\pi_\phi(A)''$ and $\pi_\psi(A)''$ are von Neumann algebras of type III the representations (π_ϕ, H_ϕ) and (π_ψ, H_ψ) are disjoint.*

Proof. If the representations are not disjoint there exists non-zero central projections p and q in $\pi_\phi(A)''$ and $\pi_\psi(A)''$, respectively, and an isomorphism $\Phi: p\pi_\phi(A)'' \to q\pi_\psi(A)''$. Since central elements are fixed under the modular group by 8.14.6 we obtain a W^*-dynamical system $(\mathcal{M}, \boldsymbol{R}, \alpha)$ (say with $\mathcal{M} = q\pi_\psi(A)''$) and faithful positive normal functionals $\tilde{\phi}$ and $\tilde{\psi}$ with modular groups given by $t \to \alpha_{\beta_1 t}$ and $t \to \alpha_{\beta_2 t}$, respectively. It follows from 8.14.10 that $\tilde{\psi} = \tilde{\phi}(h \cdot)$ for some positive, injective operator affiliated with $\mathcal{M}^{\alpha(\boldsymbol{R})}$, whence by 8.14.12

$$\alpha_{\beta_2 t}(x) = h^{-it}\alpha_{\beta_1 t}(x)h^{it}, \qquad x \in \mathcal{M}, \qquad t \in \boldsymbol{R},$$

so that

$$\alpha_{(\beta_2 - \beta_1)t}(x) = h^{-it}xh^{it}, \qquad x \in \mathcal{M}, \qquad t \in \boldsymbol{R}.$$

Since $\beta_2 \neq \beta_1$ we see from 8.14.13 that $\mathcal{M}$ is semi-finite, contrary to our assumption. Thus the representations are disjoint.

8.14.15. Notes and remarks. The results 8.14.2–8.14.10 are due to Takesaki, cf. [259]. The extension of 8.14.10 to weights was studied in detail in [210]. The simple, but ingenious 2 × 2-matrix argument giving 8.14.11 and 8.14.12 is due to Connes, see [38]. Propositions 8.14.13 and 8.14.14 were proved by Takesaki [259] and [260]. As noted by Takesaki, 8.14.13 shows that every factor $\mathcal{M}$ of type III on a separable Hilbert space has outer automorphisms. For if we take a dynamical system $(\mathcal{M}, \boldsymbol{R}, \sigma)$ and assume that each σ_t is an inner automorphism, then by a result of Kadison [130] the group is inner. A simple proof of Kadison's result is presented in [113]. Further generalizations of the implementation procedures are found in [137], [168] and [114].

8.15. A classification of type III algebras

8.15.1. Let $\mathcal{M}$ be a von Neumann algebra which is σ-finite (cf. 3.8.3). There is then a faithful normal state ϕ of $\mathcal{M}$ and thus by 8.14.5 we obtain a W^*-dynamical system $(\mathcal{M}, \boldsymbol{R}, \sigma)$, where $\sigma_{\boldsymbol{R}}$ is the modular group associated with ϕ. Any other modular group associated with a faithful normal state of $\mathcal{M}$ will produce an exterior equivalent system by 8.14.11. The Connes spectrum $\Gamma(\sigma)$ of σ is therefore independent of ϕ and σ, and is an algebraic invariant for $\mathcal{M}$ by 8.11.5. We denote it by $\Gamma(\mathcal{M})$ and recall from 8.8.4 (and 8.8.10) that it is a closed subgroup of $\boldsymbol{R}$.

If p is a non-trivial central projection in $\mathcal{M}$ then by definition of the Connes spectrum we have

$$\Gamma(\mathcal{M}) = \Gamma(p\mathcal{M}) \cap \Gamma((1-p)\mathcal{M}).$$

For this reason we shall be almost exclusively interested in the case where $\mathcal{M}$ is a factor.

8.15.2. PROPOSITION. *Let $\mathcal{M}$ be a σ-finite factor. If $\Gamma(\mathcal{M}) \neq \{0\}$ the annihilator of $\Gamma(\mathcal{M})$ (under the duality $(s,t) = \exp(ist)$) consists of the real numbers t such that for every modular group $\sigma_{\boldsymbol{R}}$ associated with a faithful normal state of $\mathcal{M}$ there is a unitary u in the centre of $\mathcal{M}^\sigma$ with $\sigma_t = \operatorname{Ad} u$.*

Proof. This follows directly from 8.9.4 since every non-trivial quotient group of $\boldsymbol{R}$ is compact.

8.15.3. PROPOSITION. *If $\mathcal{M}$ is semi-finite then $\Gamma(\mathcal{M}) = \{0\}$.*

Proof. Combine 8.14.13 and 8.9.4.

8.15.4. Let ϕ be a normal state of a von Neumann algebra $\mathcal{M}$. There is then a smallest projection p in $\mathcal{M}$ for which $\phi(p) = 1$ (p is the *support* of ϕ) and thus ϕ is faithful on $p\mathcal{M}p$. We can therefore define the modular group $\sigma^\phi_{\boldsymbol{R}}$ and the modular operator Δ_ϕ associated with $\phi | p\mathcal{M}p$. Note that $\sigma^\phi_{\boldsymbol{R}}$ is an automorphism group of $p\mathcal{M}p$ and that Δ_ϕ is an operator on the Hilbert space $\pi_\phi(p)H_\phi$. Define

$$S(\mathcal{M}) = \cap \operatorname{Sp}(\Delta_\phi),$$

where ϕ ranges over the set of all normal states of $\mathcal{M}$.

8.15.5. PROPOSITION. *Let $\mathcal{M}$ be a von Neumann algebra. If $\mathcal{M}$ is of type III then $0 \in S(\mathcal{M})$. Otherwise $S(\mathcal{M}) = \{1\}$.*

Proof. If $0 \notin S(\mathcal{M})$ choose a normal state ϕ with support p and modular objects $\sigma^\phi_{\boldsymbol{R}}$ and Δ_ϕ, such that $0 \notin \operatorname{Sp}(\Delta_\phi)$. Since $j\Delta_\phi j = \Delta_\phi^{-1}$ by 8.13.4 it follows that $\operatorname{Sp}(\Delta_\phi) = \operatorname{Sp}(\Delta_\phi^{-1})$ (cf. 8.15.6). In the case at hand this implies that Δ_ϕ is bounded. But then σ_R^ϕ is uniformly continuous on $p\mathcal{M}p$ and therefore inner by 8.5.3 so that $p\mathcal{M}p$ is semi-finite by 8.15.3. Since $p \neq 0$, $\mathcal{M}$ is not of type III.

Conversely, if $\mathcal{M}$ is not of type III it contains a non-zero finite projection p, and we may assume that p is the support of a normal tracial state ϕ. Since then $\Delta_\phi = 1$ (regarded as an operator in pH) it follows that $S(\mathcal{M}) = \{1\}$.

8.15.6. LEMMA. *Let ξ_0 be a cyclic and separating vector for a von Neumann algebra $\mathcal{M}$, and denote by Δ and σ_R the modular operator and the modular group associated with ξ_0, respectively. Then for every real s we have $s \in \mathrm{Sp}(\sigma)$ if and only if $e^s \in \mathrm{Sp}(\Delta)$.*

Proof. Take f in $L^1(\boldsymbol{R})$ and x in $\mathcal{M}$. Then

$$f(\log(\Delta))x\xi_0 = \int \exp(it\log(\Delta))x\xi_0 f(t)\,dt$$
$$= \int \Delta^{it}x\xi_0 f(t)\,dt = \int \sigma_t(x)\xi_0 f(t)\,dt = \sigma_f(x)\xi_0.$$

It follows that $\sigma_f(x) = 0$ for all x in $\mathcal{M}$ if and only if $f(\log(\Delta)) = 0$. By 8.1.9 we know that $s \in \mathrm{Sp}(\sigma)$ if and only if $\sigma_f \neq 0$ for all f in $L^1(\boldsymbol{R})$ such that

$$f(\log(e^s)) = f(s) \neq 0,$$

from which the result is immediate.

8.15.7. PROPOSITION. *If $\mathcal{M}$ is a σ-finite von Neumann algebra and $s \in \boldsymbol{R}$ then*

$$s \in \Gamma(\mathcal{M}) \Leftrightarrow e^s \in S(\mathcal{M}).$$

Proof. If ϕ is a normal state of $\mathcal{M}$ with support p and modular objects σ_R^ϕ and Δ_ϕ, choose a faithful normal state ψ of $(1-p)\mathcal{M}(1-p)$ and consider the faithful state $\rho = \frac{1}{2}(\phi + \psi)$ on $\mathcal{M}$. If σ_R denotes the modular group associated with ρ it follows from the unicity of σ_R^ϕ on $p\mathcal{M}p$ that

$$\sigma_t(x) = \sigma_t^\phi, \qquad x \in p\mathcal{M}p, \qquad t \in \boldsymbol{R}.$$

Moreover, $p \in \mathcal{M}^\sigma$ by 8.14.6. Thus

$$\Gamma(\mathcal{M}) = \Gamma(\sigma) \subset \Gamma(\sigma \mid p\mathcal{M}p) = \Gamma(\sigma^\phi).$$

It follows from 8.15.6 that if $s \in \Gamma(\mathcal{M})$ then $e^s \in \mathrm{Sp}(\Delta_\phi)$; and since ϕ is arbitrary this implies that $e^s \in S(\mathcal{M})$.

Conversely, take a faithful normal state ϕ of $\mathcal{M}$ with modular group σ_R. For each non-zero projection p in $\mathcal{M}^\sigma$ we see, again from 8.14.5, that $\sigma_R \mid p\mathcal{M}p$ is the modular group associated with the faithful normal state $\phi(p)^{-1}\phi$ of $p\mathcal{M}p$. Thus if $e^s \in S(\mathcal{M})$ it follows from 8.15.6 that $s \in \mathrm{Sp}(\sigma \mid p\mathcal{M}p)$; and since p is arbitrary this implies that $s \in \Gamma(\sigma)$ $(= \Gamma(\mathcal{M}))$.

8.15.8. LEMMA. *Let ξ_0 be a cyclic and separating vector for a von Neumann algebra $\mathcal{M}$ and let Δ be the modular operator associated with ξ_0. Then $\mathrm{Sp}(\Delta)$*

consists of those numbers $\lambda \geqslant 0$ such that for every $\varepsilon > 0$ there are elements x in $\mathcal{M}$ and y in $\mathcal{M}'$ with

$$\|(\lambda^{1/2}x - y^*)\xi_0\| + \|(x^* - \lambda^{1/2}y)\xi_0\| < \varepsilon\|x\xi_0\|.$$

Proof. If $\lambda \in \mathrm{Sp}(\Delta)$ there is for each $\varepsilon > 0$ an x in $\mathcal{M}$ such that

$$\|x\xi_0\| = 1, \qquad \|\Delta^{1/2}x\xi_0 - \lambda^{1/2}x\xi_0\| < \tfrac{1}{2}\varepsilon,$$

since $\mathcal{M}\xi_0$ is dense in $\mathcal{D}(\Delta^{1/2})$, equipped with the graph norm. Let j be the conjugation associated with ξ_0 (cf. 8.13.4) and put $y = jxj$. Then $y \in \mathcal{M}'$ and using 8.13.8 we get

$$\|(\lambda^{1/2}x - y^*)\xi_0\| + \|(x^* - \lambda^{1/2}y)\xi_0\| = \|(\lambda^{1/2}x - jx^*)\xi_0\| + \|(x^* - \lambda^{1/2}jx)\xi_0\|$$
$$\|(\lambda^{1/2} - \Delta^{1/2})x\xi_0\| + \|j(\Delta^{1/2} - \lambda^{1/2})x\xi_0\| < \varepsilon.$$

Conversely, if we can find x and y satisfying the conditions above, then by computation

$$\begin{aligned}(\Delta - \lambda)(\Delta^{1/2} + 1)^{-1}x\xi_0 &= (\Delta^{1/2} - \lambda\Delta^{-1/2})(1 + \Delta^{-1/2})^{-1}x\xi_0 \\ &= \Delta^{1/2}(1 + \Delta^{-1/2})^{-1}x\xi_0 - \lambda^{1/2}\Delta^{-1/2}(1 + \Delta^{-1/2})^{-1} \\ &\quad \times (\lambda^{1/2}x\xi_0 - y^*\xi_0) - \lambda^{1/2}\Delta^{-1/2}(1 + \Delta^{-1/2})^{-1}y^*\xi_0 \\ &= (1 + \Delta^{-1/2})^{-1}(\Delta^{1/2}x\xi_0 - \lambda^{1/2}\Delta^{-1/2}y^*\xi_0) \\ &\quad - \lambda^{1/2}\Delta^{-1/2}(1 + \Delta^{-1/2})^{-1}(\lambda^{1/2}x\xi_0 - y^*\xi_0).\end{aligned}$$

Since $\|(1 + \Delta^{-1/2})^{-1}\| \leqslant 1$ and $\|\Delta^{-1/2}(1 + \Delta^{-1/2})^{-1}\| \leqslant 1$ it follows that

$$\begin{aligned}&\|(\Delta - \lambda)(\Delta^{1/2} + 1)^{-1}x\xi_0\| \\ &\quad \leqslant \|\Delta^{1/2}x\xi_0 - \lambda^{1/2}\Delta^{-1/2}y^*\xi_0\| + \lambda^{1/2}\|\lambda^{1/2}x\xi_0 - y^*\xi_0\| \\ &\quad = \|j\Delta^{1/2}x\xi_0 - \lambda^{1/2}j\Delta^{-1/2}y^*\xi_0\| + \lambda^{1/2}\|\lambda^{1/2}x\xi_0 - y^*\xi_0\| \\ &\quad = \|(x^* - \lambda^{1/2}y)\xi_0\| + \lambda^{1/2}\|(\lambda^{1/2}x - y^*)\xi_0\| \\ &\quad < (1 + \lambda^{1/2})\varepsilon\|x\xi_0\|,\end{aligned}$$

since $j\Delta^{1/2}x\xi_0 = x^*\xi_0$ by 8.13.8 and

$$j\Delta^{-1/2}y^*\xi_0 = (j\Delta^{-1/2}j)(jy^*j)\xi_0 = \Delta^{1/2}jy^*j\xi_0 = j(jyj)\xi_0 = y\xi_0$$

by 8.13.4. Since ε is arbitrary we see that $0 \in \mathrm{Sp}((\Delta - \lambda)(\Delta^{1/2} + 1)^{-1})$, whence $\lambda \in \mathrm{Sp}(\Delta)$ by spectral theory.

8.15.9. THEOREM. *Let $\mathcal{M}$ be a σ-finite von Neumann algebra on a Hilbert space H. Then $\Gamma(\mathcal{M})$ consists of those real numbers s such that for each unit vector ξ in H and every $\varepsilon > 0$ there are elements x in $\mathcal{M}$ and y in $\mathcal{M}'$ with $\|x\xi\| > 1$ and*

$$\|x\xi - y\xi\| < \varepsilon, \qquad \|x^*\xi - e^s y^*\xi\| < \varepsilon.$$

Proof. Assume that $s \in \Gamma(\mathcal{M})$ and take ξ in H and $\varepsilon > 0$. Put $p = [\mathcal{M}'\xi]$ and $p' = [\mathcal{M}\xi]$. Then ξ is cyclic and separating for the von Neumann algebra $\mathcal{N} = p\mathcal{M}pp'$ on the Hilbert space $pp'H$. Moreover, $\mathcal{N}' = p'\mathcal{M}'p'p$. Since $c(p') = [\mathcal{M}\mathcal{M}'\xi] \geq p$ (in $B(H)$) it follows from 2.6.7 that $\mathcal{N}$ is isomorphic to $p\mathcal{M}p$. Therefore

$$s \in \Gamma(\mathcal{M}) \subset \Gamma(p\mathcal{M}p) = \Gamma(\mathcal{N}),$$

so that $e^s \in S(\mathcal{N})$ by 8.15.7. By 8.15.8 there are elements xp' and yp in $\mathcal{N}$ and $\mathcal{N}'$, respectively (i.e. $x \in p\mathcal{M}p$ and $y \in p'\mathcal{M}'p'$) such that $\|x\xi\| = \|xp'\xi\| > 1$ and

$$\|x\xi - y\xi\| = \|xp'\xi - yp\xi\| < \varepsilon;$$
$$\|x^*\xi - e^s y^*\xi\| = \|x^*p'\xi - e^s y^* p\xi\| < \varepsilon,$$

as desired.

Conversely, if s satisfies the conditions choose a maximal family $\{\xi_n\}$ of unit vectors in H such that with $p_n = [\mathcal{M}'\xi_n]$ we have $p_n p_m = 0$ if $n \neq m$. Then $\Sigma p_n = 1$ and the family is countable since $\mathcal{M}$ is σ-finite. Define

$$\phi(x) = \Sigma 2^{-n}(x\xi_n \mid \xi_n), \qquad x \in \mathcal{M}.$$

Then ϕ is a faithful normal state of $\mathcal{M}$ (for if $x\xi_n = 0$ then $xp_n = 0$) and we denote by σ_R its associated modular group. Note that $\{p_n\} \subset \mathcal{M}^\sigma$ by construction of ϕ (cf. 8.14.6). Take a non-zero projection p_0 in $\mathcal{M}^\sigma$. Since $\Sigma p_n = 1$ we have $p_0 p_n \neq 0$ for some n. From the polar decomposition of $p_0 p_n$ we obtain a non-zero partial isometry v in $\mathcal{M}^\sigma$ such that

$$q_0 = vv^* \leq p_0, \qquad q_n = v^*v \leq p_n.$$

Since $v \in \mathcal{M}^\sigma$ it provides a covariant isomorphism of $q_0\mathcal{M}q_0$ onto $q_n\mathcal{M}q_n$, whence

$$\mathrm{Sp}(\sigma \mid q_0\mathcal{M}q_0) = \mathrm{Sp}(\sigma \mid q_n\mathcal{M}q_n).$$

By 8.14.5 the modular group associated with the faithful normal state $\psi = \phi(q_n)^{-1}\phi$ on $q_n\mathcal{M}q_n$ is $\sigma \mid q_n\mathcal{M}q_n$. Moreover,

$$\psi(x) = \phi(q_n)^{-1}\phi(q_n x q_n) = \|q_n\xi_n\|^{-2}(xq_n\xi_n \mid q_n\xi_n).$$

Put $\xi_0 = \|q_n\xi_n\|^{-1}q_n\xi_n$ and let $p = [\mathcal{M}'\xi_0]$, $p' = [\mathcal{M}\xi_0]$. By assumption there are elements x in $\mathcal{M}$ and y in $\mathcal{M}'$ such that $\|x\xi_0\| > 1$ and

$$\|(x - y)\xi_0\| < \varepsilon, \qquad \|(x^* - e^s y^*)\xi_0\| < \varepsilon.$$

Put $x_1 = pxpp'$ and $y_1 = p'yp'p$. Then

$$x_1\xi_0 = px\xi_0 = p(x - y)\xi_0 + y\xi_0 = (p - 1)(x - y)\xi_0 + x\xi_0.$$

Consequently $\|x_1\xi_0\| > 1 - 2\varepsilon$, and moreover

$$\|(x_1 - y_1)\xi_0\| = \|pp'(x - y)\xi_0\| < \varepsilon;$$
$$\|x_1^* - e^s y_1^* \xi_0\| = \|pp'(x^* - e^s y^*)\xi_0\| < \varepsilon.$$

Working in the von Neumann algebra $\mathcal{N} = p\mathcal{M}pp'$ we see from 8.15.8 that $e^s \in \mathrm{Sp}(\Delta_\psi)$, whence $s \in \mathrm{Sp}(\sigma \mid q_n \mathcal{M} q_n)$ by 8.15.6. It follows that

$$s \in \mathrm{Sp}(\sigma \mid q_0 \mathcal{M} q_0) \subset \mathrm{Sp}(\sigma \mid p_0 \mathcal{M} p_0),$$

and since p_0 is arbitrary we conclude that $s \in \Gamma(\mathcal{M})$.

8.15.10. COROLLARY. $\Gamma(\mathcal{M}) = \Gamma(\mathcal{M}')$.

8.15.11. Let $\mathcal{M}$ be a σ-finite factor and consider $\Gamma(\mathcal{M})$. By 8.15.3 we may as well assume that $\mathcal{M}$ is of type III. There are three cases:

(i) $\Gamma(\mathcal{M}) = \{0\}$;

(ii) $\Gamma(\mathcal{M}) = \{n \log \lambda \mid n \in \boldsymbol{Z}\}$, where $0 < \lambda < 1$;

(iii) $\Gamma(\mathcal{M}) = \boldsymbol{R}$.

It is customary (but slightly bewildering at first) to refer to (i) as the case $\lambda = 0$ and to (iii) as the case $\lambda = 1$. Note though that the subgroups $\boldsymbol{Z} \log \lambda$ of $\boldsymbol{R}$ increase in size as $\lambda \to 1$. Thus each factor $\mathcal{M}$ of type III has been assigned a number λ in $[0, 1]$. We say that $\mathcal{M}$ is *of type* III_λ.

8.15.12. LEMMA. *Fix λ in $]0, \frac{1}{2}[$ and let $(\boldsymbol{F}, \boldsymbol{T}, \alpha)$ and ϕ_λ be as in 8.12.13. For each t in $\boldsymbol{R}$ let σ_t denote the extension of $\alpha_{\beta t}$ from $\boldsymbol{F}$ to $\mathcal{M}_\lambda = \pi_\lambda(\boldsymbol{F})''$, where $\beta = \log(\lambda^{-1} - 1)$. Then we have a W*-dynamical system $(\mathcal{M}_\lambda, \boldsymbol{R}, \sigma)$ where $\sigma_{\boldsymbol{R}}$ is the modular group associated with ϕ_λ on $\mathcal{M}_\lambda$. Moreover, $\sigma_{\boldsymbol{R}}$ is compact and the fixed-point algebra $(\mathcal{M}_\lambda)^\sigma$ is a factor (of type II_1).*

Proof. Since ϕ_λ is a KMS state at β for the system $(\boldsymbol{F}, \boldsymbol{T}, \alpha)$ it follows from 8.14.4 and 8.14.5 that ϕ_λ extends to a faithful normal state on $\mathcal{M}_\lambda$ and that $\sigma_{\boldsymbol{R}}$ is the modular group associated with ϕ_λ. If $\beta t = 2\pi$ we have $\sigma_t = \iota$ so that $\sigma_{\boldsymbol{R}}$ is compact.

Take now a projection p in the centre of $(\mathcal{M}_\lambda)^\sigma$ and put $\psi = \phi_\lambda(p\,\cdot)$. Let $\{u_t \mid t \in \Pi\}$ denote the unitary group corresponding to the finite permutations of the tensor factors, described in 6.5.11. Then $u_t \in (\mathcal{M}_\lambda)^\sigma$ for every t in Π, because $\phi_\lambda(u_t \cdot u_t^*) = \phi_\lambda$ by 6.5.12. Since p belongs to the centre of $(\mathcal{M}_\lambda)^\sigma$ we see that also $\psi_\lambda(u_t \cdot u_t^*) = \psi$ for all t. But then $\psi = \phi_\lambda(p)\phi_\lambda$ by 6.5.14. In particular

$$0 = \psi(1 - p) = \phi_\lambda(p)\phi_\lambda(1 - p),$$

so that either $p = 0$ or $p = 1$. It follows that $(\mathcal{M}_\lambda)^\sigma$ is a factor, as desired.

8.15.13. PROPOSITION. *The factor $\mathcal{M}_\lambda$ arising from a permutation-invariant product state ϕ_λ of the Fermion algebra (cf. 6.5.15) is of type $III_{\lambda'}$, where $\lambda' = \lambda(1-\lambda)^{-1}$ and $0 < \lambda < \frac{1}{2}$.*

Proof. Let σ_R denote the modular group associated with ϕ_λ on $\mathcal{M}_\lambda$. In the proof of 8.15.12 we showed that $\sigma_t = \iota$ when

$$t = 2\pi\beta^{-1} = 2\pi(\log(\lambda^{-1} - 1))^{-1} = 2\pi(-\log\lambda')^{-1}$$

thus

$$\mathrm{Sp}(\sigma) \subset \log(\lambda^{-1} - 1)\boldsymbol{Z} = (\log\lambda')\boldsymbol{Z}.$$

On the other hand, regarding the matrix $x = \begin{pmatrix} 0 & 0 \\ 1 & 0 \end{pmatrix}$ as an element of F we see from 8.12.11 that

$$\sigma_t(x) = \alpha_{\beta t}(x) = u^{1}_{\beta t} x u^{1}_{-\beta t} = \mu^{i\beta t} x = (\lambda')^{-it} x$$

so that $\log\lambda' \in \mathrm{Sp}(\sigma)$. Since $(\mathcal{M}_\lambda)^\sigma$ is a factor and σ_R is compact by 8.15.12, it follows from 8.10.5 that $\mathrm{Sp}(\sigma) = \Gamma(\sigma)$. Consequently

$$\Gamma(\mathcal{M}_\lambda) = \Gamma(\sigma) = (\log\lambda')\boldsymbol{Z}$$

which proves that $\mathcal{M}_\lambda$ is of type $\mathrm{III}_{\lambda'}$.

8.15.14. Notes and remarks. This section is taken from Connes' thesis [38]. It is a common joke to refer to Dixmier's two books as 'the old and the new testament' (inspired maybe by the sober black binding of the copies in Størmer's library). There are people who, in the same spirit, would liken [38] to the Koran; indeed some know it by heart. Together with the Tomita–Takesaki theory it has revolutionized von Neumann algebra theory. Without any attempts of completeness we cite the following papers for major achievements arising from the Tomita–Takesaki–Connes theories: [262], [263], [46], [45], [42], and [43].

8.15.15. Notes and remarks. It is time to end this book. However, the author feels that a few words should be said about the very important, but in this treatise completely absent, subject of C^*-tensor products.

If A and B are C^*-algebras there are in general many C^*-norms on the algebraic tensor product $A \otimes B$ which satisfy the cross norm property $\|x \otimes y\| = \|x\|\,\|y\|$ for all x in A and y in B. Turumaru [270] gave the first example of a C^*-cross norm: realizing A and B as operators on Hilbert spaces H and K let $A \otimes_{\min} B$ denote the norm closure of $A \otimes B$ in its natural embedding as a subalgebra of $B(H \otimes K)$. Takesaki [256] showed that this cross norm is the smallest among all C^*-cross norms (so that any C^*-completion of $A \otimes B$ admits $A \otimes_{\min} B$ as a quotient). In 1965 Guichardet showed the existence of a largest C^*-cross norm, corresponding to a C^*-tensor product $A \otimes_{\max} B$, see [107].

Following Lance we say that a C^*-algebra A is *nuclear* if

$$A \otimes_{\min} B = A \otimes_{\max} B$$

for all C^*-algebras B; i.e. if all C^*-cross norms coincide. Takesaki showed in [256] that all C^*-algebras of type I are nuclear and that an inductive limit of nuclear C^*-algebras is again nuclear. But the reduced group C^*-algebra $C_r^*(F_2)$, F_2 the free group on two generators, is not nuclear. Lance [156] generalized this considerably by showing that for any discrete group G, $C_r^*(G)$ is nuclear if and only if G is amenable. Guichardet had previously noted that $C^*(G)$ is nuclear for any locally compact amenable group.

For every n and every linear map ϕ between partially ordered Banach spaces A and B there is a canonical extension to a linear map $\bar{\phi}: A \otimes \boldsymbol{M}_n \to B \otimes \boldsymbol{M}_n$ given by $\bar{\phi}(x \otimes e_{ij}) = \phi(x) \otimes e_{ij}$. We say that ϕ is completely positive if $\bar{\phi} \geqslant 0$ for all n. Stinespring [241] showed that a positive linear map $\phi: A \to \boldsymbol{B}(H)$, A a C^*-algebra, is completely positive if and only if there is a representation (π, K) of A and a partial isometry $v: K \to H$ such that $\phi(x) = v\pi(x)v^*$ for all x in A. Any positive map between C^*-algebras A and B is completely positive if either A or B is commutative; but already $\boldsymbol{M}_2$ admits a positive map onto itself which is not completely positive. The memoir by Evans and Lewis [87] contains a wealth of information about (groups and semi-groups of) completely positive maps and their applications to quantum statistical mechanics.

Completely positive maps play a distinguished rôle in the theory of nuclear C^*-algebras. Unfortunately a rather complicated set of definitions is necessary before the conclusive result can be stated.

A C^*-algebra A is called *injective* if for any two C^*-algebras B and C with $B \subset C$ and any completely positive contraction $\phi: B \to A$ there is a completely positive contraction $\tilde{\phi}: C \to A$ extending ϕ. Using a result of Arveson it is shown in [78] that A is injective if and only if for each C^*-algebra B containing A there is a completely positive contraction $\phi: B \to A$ which is a left inverse for the inclusion map $\iota: A \to B$.

We say that a C^*-algebra A has the *completely positive approximation property* if the identity map on A^* can be approximated in the topology of simple norm (or weak*) convergence by completely positive contractions of finite rank. A von Neumann algebra $\mathcal{M}$ is *semi-discrete* if the identity map on $\mathcal{M}$ can be approximated in the topology of simple σ-weak convergence by normal completely positive contradictions of finite rank. As shown in [78] this is equivalent to the condition that the natural $*$-homomorphism $\eta: \mathcal{M} \otimes \mathcal{M}' \to \boldsymbol{B}(H)$ (where $\mathcal{M} \subset \boldsymbol{B}(H)$) given by $\eta(x \otimes y) = xy$ extends to a morphism of the C^*-algebra $\mathcal{M} \otimes_{\min} \mathcal{M}'$.

Connes showed in [43] that for a factor $\mathcal{M}$ on a separable Hilbert space the following conditions are equivalent:

(i) $\mathscr{M}$ is hyperfinite (cf. 6.4.8);

(ii) $\mathscr{M}$ is injective;

(iii) $\mathscr{M}$ is semi-discrete.

With this information at hand the papers [30], [31], [78] by Choi, Effros and Lance give the following equivalent conditions on a C^*-algebra A:

(i) A is nuclear;

(ii) A has the completely positive approximation property;

(iii) A'' is semi-discrete;

(iv) A'' is injective.

References

[1] J. F. Aarnes and R. V. Kadison, Pure states and approximate identities. *Proc. Amer. Math. Soc.* **21** (1969), 749–752.
[2] C. A. Akemann, G. A. Elliott, G. K. Pedersen and J. Tomiyama, Derivations and multipliers of C^*-algebras. *Amer. J. Math.* **98** (1976), 679–708.
[3] C. A. Akemann and P. Ostrand, On a tensor product C^*-algebra associated with the free group on two generators. *J. Math. Soc. Japan* **27** (1975), 589–599.
[4] C. A. Akemann and G. K. Pedersen, Complications of semicontinuity in C^*-algebra theory. *Duke Math. J.* **40** (1973), 785–795.
[5] C. A. Akemann and G. K. Pedersen, Ideal perturbations of elements in C^*-algebras, *Math. Scand.* **41** (1977), 117–139.
[6] C. A. Akemann and G. K. Pedersen, Central sequences and inner derivations of separable C^*-algebras. *Amer. J. Math.*, to appear.
[7] C. A. Akemann, G. K. Pedersen and J. Tomiyama, Multipliers of C^*-algebras. *J. Functional Anal.* **13** (1973), 277–301.
[8] W. Ambrose, Structure theorems for a special class of Banach algebras. *Trans. Amer. Math. Soc.* **57** (1945), 364–386.
[9] J. Anderson and J. Bunce, A-type II_∞ factor representation of the Calkin algebra. *Amer. J. Math.*
[10] H. Araki, On the algebra of all local observables. *Progr. Theoret. Phys.* **32** (1964), 956–965.
[11] H. Araki and E. J. Woods, A classification of factors. *Publ. RIMS Kyoto Univ.* **4** (1968), 51–130.
[12] W. B. Arveson, Subalgebras of C^*-algebras. *Acta Math.* **123** (1969), 141–224.
[13] W. B. Arveson, On groups of automorphisms of operator algebras. *J. Functional Anal.* **15** (1974), 217–243.
[14] W. B. Arveson, Notes on extensions of C^*-algebras. *Duke Math. J.* **44** (1977), 329–355.
[15] W. B. Arveson, 'An invitation to C^*-algebra'. Graduate Texts in Mathematics. Springer-Verlag, New York, Heidelberg, Berlin, 1976.
[16] J. Bendat and S. Sherman, Monotone and convex operator functions. *Trans. Amer. Math. Soc.* **79** (1955), 58–71.
[17] S. Berberian, Baer *-rings *in* 'Ergebnisse der Mathematik'. Springer-Verlag, Berlin, Heidelberg, New York, 1972.
[18] H.-J. Borchers, On the structure of the algebra of field operators. *Nuovo Cimento* **24** (1962), 214–236.

[19] H.-J. Borchers, Energy and momentum as observables in quantum field theory. *Comm. Math. Phys.* **2** (1966), 49–54.

[20] H.-J. Borchers, On groups of automorphisms with semi-bounded spectrum. *In* 'Systèmes à une nombre infini de degrés de liberté.' Colloques Gif-sur-Yvette, Mai 1969 (CNRS).

[21] H.-J. Borchers, On the implementability of automorphism groups. *Comm. Math. Phys.* **14** (1969), 305–314.

[22] H.-J. Borchers, Über C^*-algebren mit lokalkompakten Symmetriegruppen. *Nachr. Göttinger Akad.* 1973 No. 1.

[23] H.-J. Borchers, Über Ableitungen von C^*-Algebren. *Nachr. Göttinger Akad.* 1973, No. 2.

[24] H.-J. Borchers, Characterization of inner *-automorphisms of W^*-algebras. *Publ. RIMS Kyoto Univ.* **10** (1974), 11–49.

[25] O. Bratteli, Inductive limits of finite dimensional C^*-algebras. *Trans. Amer. Math. Soc.* **171** (1972), 195–234.

[26] O. Bratteli and D. W. Robinson, 'Algebraic methods and mathematical physics'. In preparation.

[27] L. G. Brown, Stable isomorphism of hereditary subalgebras of C^*-algebras. *Pacific J. Math.* **71** (1977), 335–348.

[28] L. G. Brown, P. Green and M. A. Rieffel, Stable isomorphism and strong Morita equivalence of C^*-algebras. *Pacific J. Math.* **71** (1977), 349–363.

[29] R. C. Busby, Double centralizers and extensions of C^*-algebras. *Trans. Amer. Math. Soc.* **132** (1968), 79–99.

[30] M.-D. Choi and E. G. Effros, Separable nuclear C^*-algebras and injectivity. *Duke Math. J.* **43** (1976), 309–322.

[31] M.-D. Choi and E. G. Effros, Nuclear C^*-algebras and injectivity: the general case. *Indiana Univ. Math. J.*

[32] M.-D. Choi and E. G. Effros, The completely positive lifting problem for C^*-Algebras. *Ann. of Math.* **104** (1976), 585–609.

[33] E. Christensen, Non-commutative integration for monotone sequentially closed C^*-algebras. *Math. Scand.* **31** (1972), 171–190.

[34] F. Combes, Poids sur une C^*-algèbre. *J. Math. Pures Appl.* **47** (1968), 57–100.

[35] F. Combes, Sur les faces d'une C^*-algèbre. *Bull. Sci. Math.* **93** (1969), 37–62.

[36] F. Combes, Quelques propriétés des C^*-algèbres. *Bull. Sci. Math.* **94** (1970), 165–192.

[37] F. Combes, Poids associe à une algèbre hilbertienne a gauche. *Compositio Math.* **23** (1971), 49–77.

[38] A. Connes, Une classification des facteurs de type III. *Ann. Sci. Ecole Norm. Sup., Paris* (4) **6** (1973), 133–252.

[39] A. Connes, On hyperfinite factors of type III_0 and Krieger's factors. *J. Functional Anal.* **18** (1975), 318–327.

[40] A. Connes, Almost periodic states and factors of type III_0. *J. Functional Anal.* **16** (1974), 415–445.

[41] A. Connes, Periodic automorphisms of the hyperfinite factor of type II_1. *Acta Sci. Math. (Szeged)* **39** (1977), 39–66.

[42] A. Connes, A factor not anti-isomorphic to itself. *Ann. of Math.* **101** (1975), 536–554.

[43] A. Connes, Classification of injective factors. *Ann. of Math.* **104** (1976), 73–116.

[44] A. Connes, On the cohomology of operator algebras. *J. Functional Anal.*

[45] A. Connes and E. Størmer, Entropy for automorphisms of II_1 von Neumann algebras. *Acta Math.* **134** (1975), 289–306.
[46] A. Connes and M. Takesaki, The flow of weights on factors of type III. *Tôhoku Math. J.* **29** (1977), 473–575.
[47] J. Cuntz, Simple C^*-algebras generated by isometries. *Comm. Math. Phys.* **57** (1977), 173–185.
[48] J. Cuntz and G. K. Pedersen, Equivalence and traces on C^*-algebras. *J. Functional Analysis*, to appear.
[49] J. Cuntz and G. K. Pedersen, Equivalence and KMS states on periodic C^*-dynamical systems. *J. Functional Analysis*, to appear.
[50] E. B. Davies, On the Borel structure of C^*-algebras. *Comm. Math. Phys.* **8** (1968), 147–163.
[51] E. B. Davies, The Structure of Σ^*-algebras. *Quart. J. Math. Oxford.* (2) **20** (1969), 351–366.
[52] E. B. Davies, Decomposition of traces on separable C^*-algebras. *Quart. J. Math. Oxford*, (2) **20** (1969), 97–111.
[53] J. Dixmier, Les anneaux d'opérateurs de classe finie. *Ann. Sci. Ecole Norm. Sup., Paris* (3) **66** (1949), 209–261.
[54] J. Dixmier, Les fonctionnelles linéaires sur l'ensemble des opérateurs bornés d'un espace de Hilbert. *Ann. of Math.* **51** (1950), 387–408.
[55] J. Dixmier, Sur certaines espaces considérés par M. H. Stone. *Summa Brasil. Math.* **2** (1951), 151–182.
[56] J. Dixmier, Algèbres quasi-unitaires. *Comment. Math. Helv.* **26** (1952), 275–322.
[57] J. Dixmier, Formes lineaires sur un anneau d'opérateurs. *Bull. Soc. Math. France* **81** (1953), 9–39.
[58] J. Dixmier, Sur les anneaux d'opérateurs dans les espaces hilbertiens. *C. R. Acad. Sci., Paris* **238** (1954), 439–441.
[59] J. Dixmier, 'Les algèbres d'opérateurs dans l'espace hilbertien.' 2(eme) édition. Gauthier–Villars, Paris, 1969.
[60] J. Dixmier, Sur les C^*-algèbres. *Bull. Soc. Math. France* **88** (1960), 95–112.
[61] J. Dixmier, Sur les structures boréliennes du spectre d'une C^*-algèbre. *Publ. Inst. Hautes Etudes Sci.* no. 6 (1960), 297–303.
[62] J. Dixmier, Points séparés dans le spectre d'une C^*-algèbre. *Acta. Sci. Math.* (Szeged) **22** (1961), 115–128.
[63] J. Dixmier, Dual et quasi-dual d'une algèbre de Banach involutive. *Trans. Amer. Math. Soc.* **104** (1962), 278–283.
[64] J. Dixmier, Traces sur les C^*-algèbres. *Ann. Inst. Fourier* (*Grenoble*) **13** (1963), 219–262.
[65] J. Dixmier, 'Les C^*-algèbres et leurs représentations'. Gauthier–Villars, Paris, 1964.
[66] J. Dixmier, On some C^*-algebras considered by Glimm *J. Functional. Anal.* **1** (1967), 182–203.
[67] J. Dixmier, Idéal center of a C^*-algebra. *Duke Math. J.* **35** (1968), 375–382.
[68] J. Dixmier and A. Douady, Champs continus d'espaces hilbertiens et de C^*-algèbres. *Bull. Soc. Math. France* **91** (1963), 227–284.
[69] S. Doplicher, R. V. Kadison, D. Kastler and D. W. Robinson, Asymptotically abelian systems. *Comm. Math. Phys.* **6** (1967), 101–120.
[70] S. Doplicher, D. Kastler and D. W. Robinson, Covariance algebras in field theory and statistical mechanics. *Comm. Math. Phys.* **3** (1966), 1–28.
[71] S. Doplicher, D. Kastler and E. Størmer, Invariant states and asymptotic abelianness. *J. Functional Anal.* **3** (1969), 419–434.

[72] N. Dunford and J. T. Schwartz, 'Linear operators,' part I. Interscience, New York, 1958.

[73] H. A. Dye, The Radon–Nikodym theorem for finite rings of operators. *Trans. Amer. Math. Soc.* **72** (1952), 243–280.

[74] E. G. Effros, Order ideals in a C^*-algebra and its dual. *Duke Math. J.* **30** (1963), 391–412.

[75] E. G. Effros, Transformation groups and C^*-algebras. *Ann. of Math.* **81** (1965), 38–55.

[76] E. G. Effros, The canonical measures for a separable C^*-algebra. *Amer. J. Math.* **92** (1970), 56–60.

[77] E. G. Effros and F. Hahn, 'Locally compact transformation groups and C^*-algebras'. *Mem. Amer. Math. Soc.* **75**, 1967.

[78] E. G. Effros and E. C. Lance, Tensor products of operator algebras. *Advances in Math.* **25** (1977), 1–34.

[79] G. A. Elliott, Derivations of matroid C^*-algebras. *Invent. Math.* **9** (1970), 253–269.

[80] G. A. Elliott, Derivations of matroid C^*-algebras, II. *Ann. of Math.* **100** (1974), 407–422.

[81] G. A. Elliott, On the classification of inductive limits of sequences of semi-simple finite-dimensional algebras. *J. Algebra* **38** (1976), 29–44.

[82] G. A. Elliott, On approximately finite-dimensional von Neumann algebras. *Math. Scand.* **39** (1976), 91–101.

[83] G. A. Elliott, The Mackey–Borel structure on the spectrum of an approximately finite-dimensional separable C^*-algebra. *Trans. Amer. Math. Soc.* 233 (1977), 59–68.

[84] G. A. Elliott, Some C^*-algebras with outer derivations, III, *Ann. of Math.* **106** (1977), 121–143.

[85] G. A. Elliott and D. Olesen, A simple proof of the Dauns–Hofmann theorem. *Math. Scand.* **34** (1974), 231–234.

[86] J. Ernest, A decomposition theory for unitary representations of locally compact groups. *Trans. Amer. Math. Soc.* **104** (1962), 252–277.

[87] D. E. Evans and J. T. Lewis, 'Dilations of irreversible evolutions in algebraic quantum theory'. *Comm. Dublin Inst. Adv. Stud.* Series A, No. 24, 1977.

[88] P. Eymard, 'Moyennes Invariantes et Représentations Unitaires.' Lecture Notes in Math., no. 300, Springer-Verlag, Berlin, Heidelberg, New York, 1972.

[89] J. M. G. Fell, The dual spaces of C^*-algebras. *Trans. Amer. Math. Soc.* **94** (1960), 365–403.

[90] J. M. G. Fell, C^*-algebras with smooth dual. *Illinois J. Math.* **4** (1960), 221–230.

[91] J. M. G. Fell, The structure of algebras of operator fields. *Acta Math.* **106** (1961), 233–280.

[92] I. M. Gelfand and M. A. Naimark, On the embedding of normed rings into the ring of operators in Hilbert space. *Mat. Sb.* **12** (1943), 197–213.

[93] I. M. Gelfand and D. Raikov, Irreducible unitary representations of locally bicompact groups. *Mat. Sb.* **13** (1943), 301–316.

[94] R. M. Gillette and D. C. Taylor, A characterization of the Pedersen ideal of $C_0(T, B_0(H))$ and a counterexample. *Proc. Amer. Math. Soc.* **68** (1978), 59–63.

[95] J. Glimm, On a certain class of operator algebras. *Trans. Amer. Math. Soc.* **95** (1960), 318–340.

[96] J. Glimm, A Stone–Weierstrass theorem for C^*-algebras. *Ann. of Math.* **72** (1960), 216–244.

[97] J. Glimm, Type I C^*-algebras. *Ann. of Math.* **73** (1961), 572–612.
[98] J. Glimm, Families of induced representations. *Pacific J. Math.* **12** (1962), 885–911.
[99] J. Glimm and R. V. Kadison, Unitary operators in C^*-algebras. *Pacific J. Math.* **10** (1960), 547–556.
[100] R. Godement, Théorèmes taubériens et théorie spectrale. *Ann. Sci. Ecole Norm. Sup.* (3), **63** (1947), 119–138.
[101] R. Godement, Les fonctions de types positif et la théorie de groupes. *Trans. Amer. Math. Soc.* **63** (1948), 1–84.
[102] R. Godement, Théorie des caracteres, II. *Ann. of Math.* **59** (1954), 63–85.
[103] F. P. Greenleaf, 'Invariant means on topological groups'. Mathematical Studies No. 16, van Nostrand–Reinhold, New York, 1969.
[104] A. Grothendieck, Un résultat sur le dual d'une C^*-algèbre. *J. Math. Pures Appl.* **36** (1957), 97–108.
[105] A. Guichardet, Sur un problème posé par G. W. Mackey, *C. R. Acad. Sci., Paris* **250** (1960), 962–963.
[106] A. Guichardet, Caractères des algèbres de Banach involutives. *Ann. Inst. Fourier (Grenoble)* **13** (1962), 1–81.
[107] A. Guichardet, 'Tensor products of C^*-algebras'. Arhus University Lecture Notes, Series 12 (1969).
[108] A. Guichardet, 'Systèmes dynamiques non commutatifs'. *Soc. Math. France*, Astérisque no. 13–14, 1974.
[109] A. Guichardet and D. Kastler, Désintégration des états quasi-invariants des C^* algèbres. *J. Math. Pures Appl.* **49** (1970), 349–380.
[110] R. Haag, N. Hugenholtz and M. Winnink, On the equilibrium states in quantum statistical mechanics. *Comm. Math. Phys.* **5** (1967), 215–236.
[111] U. Haagerup, Normal weights on W^*-algebras. *J. Functional Anal.* **19** (1975), 302–317.
[112] H. Halpern, Quasi-equivalence classes of normal representations for a separable C^*-algebra. *Trans. Amer. Math. Soc.* **203** (1975), 129–140.
[113] F. Hansen, Inner one-parameter groups acting on a factor. *Math. Scand.* **41** (1977), 113–116.
[114] F. Hansen and D. Olesen, Perturbations of centre-fixing dynamical systems. *Math. Scand.* **41** (1977), 295–307.
[115] E. Hewitt and K. A. Ross, 'Abstract Harmonic Analysis, I'. Springer-Verlag, Berlin, Heidelberg, New York, 1963.
[116] E. Hewitt and K. A. Ross, 'Abstract Harmonic Analysis, II'. Springer–Verlag, Berlin, Heidelberg, New York, 1970.
[117] N. Hugenholtz, On the factor type of equilibrium states in quantum statistical mechanics. *Comm. Math. Phys.* **6** (1967), 189–193.
[118] A. Hulanicki, Means and Følner conditions on locally compact groups. *Studia Math.* **27** (1966), 87–104.
[119] K. Jacobs, Neuere Methoden und Ergebnisse der Ergoden-theorie. *In* 'Ergebnisse der Mathematik' 29, Springer-Verlag, Berlin, Heidelberg, New York, 1960.
[120] N. Jacobson, A topology for the set of primitive ideals in an arbitrary ring. *Proc. Nat. Acad. Sci. USA* **31** (1945), 333–338.
[121] B. E. Johnson, An introduction to the theory of centralizers. *Proc. London Math. Soc.* **14** (1964), 299–320.
[122] P. E. T. Jørgensen, Trace states and KMS states for approximately inner one-parameter groups of *-automorphisms. *Comm. Math. Phys.* **53** (1977), 177–182.

[123] R. V. Kadison, A representation theory for commutative topological algebras. *Mem. Amer. Math. Soc.* **7** (1951).

[124] R. V. Kadison, Isometries of operator algebras. *Ann. of Math.* **54** (1951), 325–338.

[125] R. V. Kadison, A generalized Schwarz inequality and algebraic invariants for operator algebras. *Ann. of Math.* **56** (1952), 494–503.

[126] R. V. Kadison, On the additivity of the trace in finite factors. *Proc. Nat. Acad. Sci. USA* **41** (1955), 385–387.

[127] R. V. Kadison, Operator algebras with a faithful weakly-closed representation. *Ann. of Math.* **64** (1956), 175–181.

[128] R. V. Kadison, Irreducible operator algebras. *Proc. Nat. Acad. Sci. USA* **43** (1957), 273–276.

[129] R. V. Kadison, Unitary invariants for representations of operator algebras. *Ann. of Math.* **66** (1957), 304–379.

[130] R. V. Kadison, Transformation of states in operator theory and dynamics. *Topology* **3**, suppl. 2 (1965), 177–198.

[131] R. V. Kadison, Derivations of operator algebras. *Ann. of Math.* **83** (1966), 280–293.

[132] R. V. Kadison, Strong continuity of operator functions. *Pacific J. Math.* **26** (1968), 121–129.

[133] R. V. Kadison, A note on derivations of operator algebras. *Bull. London Math. Soc.* **7** (1975), 41–44.

[134] R. V. Kadison, E. C. Lance and J. R. Ringrose, Derivations and automorphisms of operator algebras, II. *J. Functional Anal.* **1** (1967), 204–221.

[135] R. V. Kadison and G. K. Pedersen, Equivalence in operator algebras. *Math. Scand.* **27** (1970), 205–222.

[136] R. V. Kadison and J. R. Ringrose, Derivations and automorphisms of operator algebras. *Comm. Math. Phys.* **4** (1967), 32–63.

[137] R. R. Kallman, Groups of inner automorphisms of von Neumann algebras. *J. Functional Anal.* **7** (1971), 43–60.

[138] I. Kaplansky, Normed algebras. *Duke Math. J.* **16** (1949), 399–418.

[139] I. Kaplansky, A theorem on rings of operators. *Pacific J. Math.* **1** (1951), 227–232.

[140] I. Kaplansky, Projections in Banach algebras. *Ann. of Math.* **53** (1951), 235–249.

[141] I. Kaplansky, The structure of certain operator algebras. *Trans. Amer. Math. Soc.* **70** (1951), 219–255.

[142] I. Kaplansky, Group algebras in the large. *Tôhoku Math. J.* **3** (1951), 249–256.

[143] I. Kaplansky, Algebras of type I. *Ann. of Math.* **56** (1952), 460–472.

[144] I. Kaplansky, Modules over operator algebras. *Amer. J. Math.* **75** (1953), 839–853.

[145] D. Kastler, Equilibrium states of matter and operator algebras. *Symposia Math.* **20** (1976), 49–107.

[146] D. Kastler and D. W. Robinson, Invariant states in statistical mechanics. *Comm. Math. Phys.* **3** (1966), 151–180.

[147] E. T. Kehlet, On the monotone sequential closure of a C^*-algebra. *Math. Scand.* **25** (1969), 59–70.

[148] J. L. Kelley and R. L. Vaught, The positive cone in Banach algebras. *Trans. Amer. Math. Soc.* **74** (1953), 44–55.

[149] A. Kishimoto, Dissipations and derivations. *Comm. Math. Phys.* **47** (1976), 25–32.

[150] A. Kishimoto and H. Takai, On the invariant $\Gamma(\alpha)$ in C^*-dynamical systems. *Tôhoku Math. J.* **30** (1978), 83–94.

[151] I. Kovács and J. Szücs, Ergodic type theorems in von Neumann algebras. *Acta Sci. Math.* (Szeged) **27** (1966), 233–246.

[152] R. Kubo, Statistical–mechanical theory of irreversible processes. I—General theory and simple applications to magnetic and conduction problems. *J. Phys. Soc. Japan* **12** (1967), 570–588.

[153] E. C. Lance, Automorphisms of certain operator algebras. *Amer. J. Math.* **91** (1967), 160–174.

[154] E. C. Lance, Inner automorphisms of UHF algebras. *J. London Math. Soc.* **43** (1968), 681–688.

[155] E. C. Lance, 'Notes on the Glimm–Sakai theorem'. Lecture notes (mimeographed), Newcastle-upon-Tyne, 1971.

[156] E. C. Lance, On nuclear C^*-algebras. *J. Functional Anal.* **12** (1973), 157–176.

[157] E. C. Lance, Direct integrals of left Hilbert algebras. *Math. Ann.* **216** (1975), 11–28.

[158] E. C. Lance and A. Niknam, Unbounded derivations of group C^*-algebras. *Proc. Amer. Math. Soc.* **61** (1976), 310–314.

[159] M. B. Landstad, Duality theory of covariant systems. *Trans. Amer. Math. Soc.*, to appear.

[160] K. B. Laursen and A. M. Sinclair, Lifting matrix units in C^*-algebras, II. *Math. Scand.* **37** (1975), 167–172.

[161] A. J. Lazar and D. C. Taylor, 'Multipliers of Pedersen's ideal'. *Mem. Amer. Math. Soc.* **169** (1976).

[162] L. H. Loomis, 'An introduction to abstract harmonic analysis'. Van Nostrand, Princeton, 1953.

[163] K. Löwner, Über monotone Matrixfunktionen. *Math. Z.* **38** (1934), 177–216.

[164] G. W. Mackey, Borel structure in groups and their duals. *Trans. Amer. Math. Soc.* **85** (1957), 134–165.

[165] G. W. Mackey, Point realizations of transformation groups. *Illinois J. Math.* **6** (1962), 327–335.

[166] P. C. Martin and J. Schwinger, Theory of many particles systems, I. *Phys. Rev.* **115** (1959), 1342–1344.

[167] J. Moffat, Connected topological groups acting on von Neumann algebras. *J. London Math. Soc.* **9** (1975), 411–417.

[168] C. C. Moore, Group extensions and cohomology for locally compact groups, III. *Trans. Amer. Math. Soc.* **221**, (1976), 1–33.

[169] F. J. Murray and J. von Neumann, On rings of operators. *Ann. of Math.* **37** (1936), 116–229.

[170] F. J. Murray and J. von Neumann, On rings of operators, II. *Trans. Amer. Math. Soc.* **41** (1937), 208–248.

[171] F. J. Murray and J. von Neumann, On rings of operators, IV. *Ann. of Math.* **44** (1943), 716–808.

[172] J. von Neumann, Zur algemeinen Theorie des Masses. *Fund. Math.* **13** (1929), 73–116.

[173] J. von Neumann, Zur Algebra der Funktionaloperationen und Theorie der normalen Operatoren. *Math. Ann.* **102** (1929), 370–427.

[174] J. von Neumann, Über adjungierte Funktionaloperatoren. *Ann. of Math.* **33** (1932), 294–310.

[175] J. von Neumann, On a certain topology for rings of operators. *Ann. of Math.* **37** (1936), 111–115.

[176] J. von Neumann, On rings of operators, III. *Ann. of Math.* **41** (1940), 94–161.

[177] J. von Neumann, On rings of operators: Reduction theory. *Ann. of Math.* **50** (1949), 401–485.
[178] O. A. Nielsen, Borel sets of von Neumann algebras. *Amer. J. Math.* **95** (1973), 145–164.
[179] T. Ogasawara, A theorem on operator algebras. *J. Sci. Hiroshima Univ.* **18** (1955), 307–309.
[180] D. Olesen, Derivations of AW^*-algebras are inner. *Pacific J. Math.* **53** (1974), 555–561.
[181] D. Olesen, On norm-continuity and compactness of spectrum. *Math. Scand.* **35** (1974), 223–236.
[182] D. Olesen, Inner automorphisms of simple C^*-algebras. *Comm. Math. Phys.* **44** (1975), 175–190.
[183] D. Olesen, On spectral subspaces and their applications to automorphism groups. *Symposia Math.* **20** (1976), 253–296.
[184] D. Olesen and G. K. Pedersen, Derivations of C^*-algebras have semi-continuous generators. *Pacific J. Math.* **53** (1974), 563–572.
[185] D. Olesen and G. K. Pedersen, Groups of automorphisms with spectrum condition and the lifting problem. *Comm. Math. Phys.* **51** (1976), 85–95.
[186] D. Olesen and G. K. Pedersen, Applications of the Connes spectrum to C^*-dynamical systems. *J. Functional Anal.*, **30** (1978), 179–197.
[187] D. Olesen and G. K. Pedersen, Some C^*-dynamical systems with a single KMS state. *Math. Scand.* **42** (1978), 111–118.
[188] D. Olesen, G. K. Pedersen and E. Størmer, Compact abelian groups of automorphisms of simple C^*-algebras. *Inventiones Math.* **39** (1977), 55–64.
[189] T. W. Palmer, Characterizations of C^*-algebras. *Bull. Amer. Math. Soc.* **74** (1968), 538–540.
[190] G. K. Pedersen, Measure theory for C^*-algebras. *Math. Scand.* **19** (1966), 131–145.
[191] G. K. Pedersen, Measure theory for C^*-algebras II. *Math. Scand.* **22** (1968), 63–74.
[192] G. K. Pedersen, A decomposition theorem for C^*-algebras. *Math. Scand.* **22** (1968), 266–268.
[193] G. K. Pedersen, Measure theory for C^*-algebras III. *Math. Scand.* **25** (1969), 71–93.
[194] G. K. Pedersen, Measure theory for C^*-algebras IV. *Math. Scand.* **25** (1969), 121–127.
[195] G. K. Pedersen, On weak and monotone σ-closures of C^*-algebras. *Comm. Math. Phys.* **11** (1969), 221–226.
[196] G. K. Pedersen, The 'Up-Down' problem for operator algebras. *Proc. Nat. Acad. Sci. USA*, **68** (1971), 1896–1897.
[197] G. K. Pedersen, Monotone closures in operator algebras. *Amer. J. Math.* **94** (1972), 955–962.
[198] G. K. Pedersen, Applications of weak* semicontinuity in C^*-algebra theory. *Duke Math. J.* **39** (1972), 431–450.
[199] G. K. Pedersen, Operator algebras with weakly closed abelian subalgebras. *Bull. London Math. Soc.* **4** (1972), 171–175.
[200] G. K. Pedersen, C^*-integrals, an approach to non-commutative measure theory. Thesis, Copenhagen, 1972.
[201] G. K. Pedersen, Borel structure in operator algebras. *Danske Vid. Selsk. Mat. Fys. Medd.* **39**, 5 (1974).

[202] G. K. Pedersen, The trace in semi-finite von Neumann algebras. *Math. Scand.* **37** (1975), 142–144.
[203] G. K. Pedersen, A non-commutative version of Souslin's theorem. *Bull. London Math. Soc.* **8** (1976), 87–90.
[204] G. K. Pedersen, Lifting derivations from quotients of separable C^*-algebras. *Proc. Nat. Acad. Sci.-USA* **73** (1976), 1414–1415.
[205] G. K. Pedersen, Lifting groups of automorphisms. *Symposia Math.* **20** (1976), 161–167.
[206] G. K. Pedersen, Isomorphisms of UHF algebras. *J. Functional Anal.* **30**, (1978), 1–16.
[207] G. K. Pedersen, Groupes localement compacts d'automorphismes d'une C^*-algèbre et conditions spectrales. Preprint. Unpublished.
[208] G. K. Pedersen and N. H. Petersen, Ideals in a C^*-algebra. *Math. Scand.* **27** (1970), 193–204.
[209] G. K. Pedersen and E. Størmer, Automorphisms and equivalence in von Neumann algebras, II. *Indiana Univ. Math. J.* **23** (1973), 121–129.
[210] G. K. Pedersen and M. Takesaki, The Radon–Nikodym theorem for von Neumann algebras. *Acta Math.* **130** (1973), 53–87.
[211] R. T. Powers, Representations of uniformly hyperfinite algebras and their associated von Neumann rings. *Ann. of Math.* **86** (1967), 138–171.
[212] R. T. Powers and S. Sakai, Existence of ground states and KMS states for approximately inner dynamics. *Comm. Math. Phys.* **39** (1975), 273–288.
[213] L. Pukanszky, Some examples of factors. *Publ. Math. Debrecen* **4** (1956), 135–156.
[214] C. Rickart, Banach algebras with an adjoint operation. *Ann. of Math.* **47** (1946), 528–550.
[215] M. A. Rieffel, Morita equivalence for C^*-algebras and W^*-algebras. *J. Pure Appl. Algebra* **5** (1974), 51–96.
[216] M. Rieffel and A. van Daele, A bounded operator approach to Tomita–Takesaki theory. *Pacific J. Math.* **69** (1977), 187–221.
[217] W. Rudin, 'Fourier analysis on groups'. Interscience, New York, 1962.
[218] W. Rudin, 'Real and complex analysis'. McGraw-Hill, New York, 1966.
[219] D. Ruelle, States of physical systems. *Comm. Math. Phys.* **3** (1966), 133–150.
[220] K. Saitô, Non-commutative extension of Lusin's theorem. *Tôhoku Math. J.* **19** (1967), 332–340.
[221] S. Sakai, A characterization of W^*-algebras. *Pacific J. Math.* **6** (1956), 763–773.
[222] S. Sakai, On topological properties of W^*-algebras. *Proc. Japan Acad.* **33** (1957), 439–444.
[223] S. Sakai, On linear functionals of W^*-algebras. *Proc. Japan Acad.* **34** (1958), 571–574.
[224] S. Sakai, On a conjecture of Kaplansky. *Tôhoku Math. J.* **12** (1960), 31–33.
[225] S. Sakai, A. Radon–Nikodym theorem in W^*-algebras. *Bull. Amer. Math. Soc.* **71** (1965), 149–151.
[226] S. Sakai, On the central decomposition for positive functionals on C^*-algebras. *Trans. Amer. Math. Soc.* **118** (1965), 406–419.
[227] S. Sakai, Derivations of W^*-algebras. *Ann. of Math.* **83** (1966), 287–293.
[228] S. Sakai, On a characterization of type I C^*-algebras. *Bull. Amer. Math. Soc.* **72** (1966), 508–512.
[229] S. Sakai, On type I C^*-algebras. *Proc. Amer. Math. Soc.* **18** (1967), 861–863.
[230] S. Sakai, Derivations of simple C^*-algebras. *J. Functional Anal.* **2** (1968), 202–206.

[231] S. Sakai, C^*-algebras and W^*-algebras. 'Ergebnisse der Mathematik' 60, Springer-Verlag, Berlin, Heidelberg, New York, 1971.

[232] S. Sakai, Derivations of simple C^*-algebras, II. *Bull. Soc. Math. France* **99** (1971), 259–263.

[233] S. Sakai, 'The theory of unbounded derivations in C^*-algebras.' To appear.

[234] J. Schwartz, Type II factors in a central decomposition. *Comm. Pure Appl. Math.* **16** (1963), 247–252.

[235] I. E. Segal, Irreducible representations of operator algebras. *Bull. Amer. Math. Soc.* **53** (1947), 73–88.

[236] I. E. Segal, Two-sided ideals in operator algebras. *Ann. of Math.* **50** (1949), 856–865.

[237] I. E. Segal, Equivalence of measure spaces. *Amer. J. Math.* **73** (1951), 275–313.

[238] I. E. Segal, 'Decomposition of operator algebras, I.' *Mem. Amer. Math. Soc.* **9** (1951), 1–67.

[239] I. E. Segal, 'Decomposition of operator algebras, II.' *Mem. Amer. Math. Soc.* **9** (1951), 1–66.

[240] I. E. Segal, A non-commutative extension of abstract integration. *Ann. of Math.* **57** (1953), 401–457.

[241] W. F. Stinespring, Positive functions on C^*-algebras. *Proc. Amer. Math. Soc.* **6** (1955), 211–216.

[242] E. Størmer, Large groups of automorphisms. *Comm. Math. Phys.* **5** (1967), 1–22.

[243] E. Størmer, Two-sided ideals in C^*-algebras. *Bull. Amer. Math. Soc.* **73** (1967), 254–257.

[244] E. Størmer, Types of von Neumann algebras associated with extremal invariant states. *Comm. Math. Phys.* **6** (1967), 194–204.

[245] E. Størmer, States and invariant maps of operator algebras. *J. Functional Anal.* **5** (1970), 44–65.

[246] E. Størmer, Asymptotically abelian systems. 'Cargèse Lectures in Physics 4', pp. 195–213. Gordon and Breach, New York, London, Paris (1970).

[247] E. Størmer, Automorphisms and invariant states of operator algebras. *Acta Math.* **127** (1971), 1–9.

[248] E. Størmer, Invariant states of von Neumann algebras. *Math. Scand.* **30** (1972), 253–256.

[249] E. Størmer, Automorphisms and equivalence in von Neumann algebras. *Pacific J. Math.* **44** (1973), 371–383.

[250] E. Størmer, Spectra of states, and asymptotically abelian C^*-algebras. *Comm. Math. Phys.* **28** (1972), 279–294.

[251] E. Størmer, Spectra of ergodic transformations. *J. Functional Anal.* **15** (1974), 202–215.

[252] E. Størmer, Some aspects of ergodic theory in operator algebras. *Proc. Int. Congr. Vancouver, 1974*, pp. 111–114.

[253] C. E. Sutherland, The direct integral theory of weights and the Plancherel formula. Thesis, UCLA, 1973.

[254] H. Takai, On a duality for crossed products of C^*-algebras. *J. Functional Anal.* **19** (1975), 25–39.

[255] M. Takesaki, On the conjugate space of an operator algebra. *Tôhoku Math. J.* **10** (1958), 194–203.

[256] M. Takesaki, On the cross-norm of the direct product of C^*-algebras. *Tôhoku Math. J.* **16** (1964), 111–122.

[257] M. Takesaki, Covariant representations of C^*-algebras and their locally compact automorphism groups. *Acta Math.* **119** (1967), 273–303.

[258] M. Takesaki, A liminal crossed product of a uniformly hyperfinite C^*-algebra by a compact abelian automorphism group. *J. Functional Anal.* **7** (1971), 140–146.

[259] M. Takesaki, 'Tomita's theory of modular Hilbert algebras and its applications'. Lecture Notes in Math. No. 128, Springer–Verlag, Berlin, Heidelberg, New York, 1970.

[260] M. Takesaki, Disjointness of the KMS-states of different temperatures. *Comm. Math. Phys.* **7** (1969), 33–41.

[261] M. Takesaki, The theory of operator algebras. Lecture Notes, UCLA, 1971–72.

[262] M. Takesaki, The structure of a von Neumann algebra with a homogeneous periodic state. *Acta Math.* **131** (1973), 79–121.

[263] M. Takesaki, Duality for crossed products and the structure of von Neumann algebras of type III. *Acta Math.* **131** (1973), 249–310.

[264] E. Thoma, Über unitäre Darstellungen abzählbarer, diskreter Gruppen. *Math. Ann.* **153** (1964), 111–138.

[265] M. Tomita, Spectral theory of operator algebras, I. *Math. J. Okayama Univ.* **9** (1959), 63–98.

[266] M. Tomita, Spectral theory of operator algebras, II. *Math. J. Okayama Univ.* **10** (1960), 19–60.

[267] M. Tomita, Standard forms of von Neumann algebras. The Vth functional analysis symposium of the *Math. Soc. of Japan*, Sendai, 1967.

[268] J. Tomiyama, On the projection of norm one in W^*-algebras. *Proc. Japan Acad.* **33** (1957), 608–612.

[269] J. Tomiyama, Tensor products and projections of norm one in von Neumann algebras. Lecture Notes (mimeographed), University of Copenhagen, 1970.

[270] T. Turumaru, On the direct product of operator algebras, I. *Tôhoku Math. J.* **4** (1952), 242–251.

[271] T. Turumaru, Crossed products of operator algebras. *Tôhoku Math. J.* **10** (1958), 355–365.

[272] A. vanDaele, A Radon–Nikodym theorem for weights on von Neumann algebras. *Pacific J. Math.* **61** (1975), 527–542.

[273] A. vanDaele, 'Continuous crossed products and type III von Neumann Algebras'. London Math. Soc. Lecture Notes Series 31. Cambridge University Press (1978).

[274] W. Wils, Désintégration centrale des formes positives sur les C^*-algèbres. *C. R. Acad. Sci. Paris* **267** (1968), 810–812.

[275] W. Wils, The ideal center of partially ordered vector spaces. *Acta Math.* **127** (1971), 41–47.

[276] J. D. Maitland Wright, Wild AW^*-factors and Kaplansky–Rickart algebras. *J. London Math. Soc.* (2), **13** (1976), 83–89.

[277] G. Zeller-Meyer, Produits croisés d'une C^*-algèbre par un groupe d'automorphismes. *J. Math. Pures Appl.* **47** (1968), 101–239.

[278] J. Anderson, A C^*-algebra A for which Ext(A) is not a group. Preprint.

[279] B. Blackadar, Infinite tensor products of C^*-algebras. *Pacific J. Math.* **72** (1977), 313–334.

[280] O. Bratteli, Crossed products of UHF algebras by product type actions. *Duke Math. J.* To appear.

[281] L. G. Brown, R. Douglas and P. Fillmore, Extensions of C^*-algebras and K-homology. *Ann. of Math.* **105** (1977), 265–324.

[282] M.-D. Choi, A simple C^*-algebra generated by two finite-order unitaries. Preprint.

[283] M.-D. Choi and E. G. Effros, Injectivity and operator spaces. *J. Functional Anal.* **24** (1977), 73–115.

[284] M.-D. Choi and E. G. Effros, Lifting problems and the cohomology of C^*-algebras. *Canad. J. Math.* **29** (1977), 1092–1111.

[285] M.-D. Choi and E. G. Effros, Nuclear C^*-algebras and the approximation property. *Amer. J. Math.* **100** (1978), 61–79.

[286] E. Christensen, Generators of semigroups of completely positive maps. *Comm. Math. Phys.* **62** (1978), 167–171.

[287] A. Connes, Outer conjugacy classes of automorphisms of factors. *Ann. Sci. École Norm. Sup.* (4) **8** (1975), 383–420.

[288] E. G. Effros and J. Rosenberg, C^*-algebras with approximately inner flip. *Pacific J. Math.* **77** (1978), 417–443.

[289] E. G. Effros and C.-L. Shen, Approximately finite C^*-algebras and continued fractions. Preprint.

[290] E. G. Effros and C.-L. Shen, The geometry of finite rank dimension functions. Preprint.

[291] E. G. Effros and E. Størmer, Positive projections and Jordan structure in operator algebras. Preprint.

[292] Th. Fack et O. Maréchal, Sur la classification des symétries des C^*-algèbres UHF. *Canad. J. Math.* To appear.

[293] Th. Fack and O. Maréchal, Sur la classification des automorphismes périodiques des C^*-algèbres UHF. Preprint.

[294] E. C. Gootman and J. Rosenberg, The structure of crossed product C^*-algebras: A proof of the generalized Effros–Hahn conjecture. Preprint.

[295] P. Green, C^*-algebras of transformation groups with smooth orbit space. *Pacific J. Math.* **72** (1977), 71–97.

[296] P. Green, The local structure of twisted covariance algebras. *Acta Math.* **140** (1978), 191–250.

[297] P. Green, Morita equivalence of C^*-algebras. Preprint.

[298] U. Haagerup, An example of a non-nuclear C^*-algebra which has the metric approximation property. *Inventiones Math.* **50** (1979), 279–293.

[299] A. Kishimoto, On the fixed-point algebra of a UHF algebra under a periodic automorphism of product type. *Publ. RIMS Kyoto Univ.* **13** (1977), 777–791.

[300] A. Kishimoto and H. Takai, Some remarks on C^*-dynamical systems with a compact abelian group. *Publ. RIMS Kyoto Univ.* **14** (1978), 383–387.

[301] J. Kraus, Compact abelian groups of automorphisms of von Neumann algebras. Thesis, Berkeley, 1977.

[302] D. Olesen, A classification of ideals in crossed products. Preprint.

[303] D. Olesen, A note on free action and duality in C^*-algebra theory. Preprint.

[304] W. L. Paschke and N. Salinas, Matrix algebras over O_n. *Michigan J. Math.* To appear.

[305] W. L. Paschke and N. Salinas, C^*-algebras associated with free products of groups. Preprint.

[306] G. K. Pedersen, Remarks on the Connes spectrum for C^*-dynamical systems. Preprint.

[307] R. T. Powers, Simplicity of the C^*-algebra associated with the free group on two generators. *Duke Math. J.* **42** (1975), 151–156.

[308] M. A. Rieffel, On the uniqueness of the Heisenberg commutation relations. *Duke Math. J.* **39** (1972), 745–752.

[309] M. A. Rieffel, Induced representations of C^*-algebras. *Advances in Math.* **13** (1974), 176–257.

[310] M. A. Rieffel, The type of group measure space von Neumann algebras. *Mh. Math.* **85** (1978), 149–162.

[311] J. R. Ringrose, Derivations of quotients of von Neumann algebras. *Proc. London Math. Soc.* (3) **36** (1978), 1–26.

[312] J. Rosenberg, Amenability of crossed products of C^*-algebras. *Comm. Math. Phys.* **57** (1977), 187–191.

[313] J. L. Sauvageot, Ideaux primitifs de certains produits croisés. *Math. Ann.* **231** (1977), 61–76.

[314] C.-L. Shen, On the classification of ordered groups associated with the approximately finite-dimensional C^*-algebras. Preprint.

[315] D. Voiculescu, A non-commutative Weyl–von Neumann theorem. *Rev. Roum. Math. Pures et Appl.* **21** (1976), 97–113.

Appendix

A1. Let A be a complex algebra with unit. If x, y are elements of A then

$$\mathrm{Sp}(xy)\backslash\{0\} = \mathrm{Sp}(yx)\backslash\{0\}.$$

For if $\lambda \notin \mathrm{Sp}(xy) \cup \{0\}$ then

$$z(\lambda - xy) = (\lambda - xy)z = 1$$

for some z in A. An elementary computation yields

$$(1 + yzx)(\lambda - yx) = (\lambda - yx)(1 + yzx) = \lambda,$$

which shows that $\lambda^{-1}(\lambda + yzx)$ is the inverse of $\lambda - yx$, whence $\lambda \notin \mathrm{Sp}(yx)$.

A2. Krein–Smulian theorem. A convex set in the dual space X^* of a Banach space X is weak* closed if its intersection with every positive multiple of the closed unit ball in X^* is weak* closed. (See V.5.7 of [72].)

A3. Integration of vector functions. Let f be a bounded function from a separable, locally compact Hausdorff space T into a Banach space X. Suppose first that X is the dual space of a Banach space X_* and that f is weak* continuous. For each bounded Radon measure μ on T there is then a unique element x in X such that

$$x(\phi) = \int f(t)(\phi)\, d\mu(t)$$

for each ϕ in X_*. We say that x is the *weak* integral* of f (with respect to μ) and denote it by $\int f\, d\mu$.

In the case where X is a general Banach space we assume that f is norm continuous. From the argument above we see that the weak integral $\int f\, d\mu$ exists as an element in X^{**}. To show that it actually belongs to X take $\varepsilon > 0$. Since μ is a Radon measure there is a compact set C in T such that $|\mu|(T\backslash C) < \varepsilon$. Since f is continuous there is a finite set $\{x_n\}$ in X and corresponding functions $\{f_n\}$ in $C_0(T)_+$ (giving a partition of unity on C) such that

$$\|f(t) - \sum f_n(t)x_n\| < \varepsilon(|\mu|(C))^{-1}$$

for all t in C. With $\gamma_n = \int f_n \, d\mu$ it follows that

$$\|\int f \, d\mu - \sum \gamma_n x_n\| < 3\|f\|\varepsilon,$$

assuming, as we may, that $\Sigma f_n(t) \leqslant 1$ for all t and that $\|x_n\| \leqslant \|f\|$ for all n. Since $\Sigma \gamma_n x_n \in X$ and ε is arbitrary it follows that $\int f \, d\mu \in X$.

A4. Holomorphic vector functions. Let f be a function from an open set Ω in $\boldsymbol{C}$ into a Banach space X. The following conditions are equivalent:

(i) f is differentiable in norm on Ω;

(ii) f is weak* differentiable on Ω;

(iii) f is norm continuous and for each open set D in Ω with sufficiently nice boundary δD we have for each ζ_0 in D

$$f(\zeta_0) = (2\pi i)^{-1} \oint_{\delta D} (\zeta - \zeta_0)^{-1} f(\zeta) \, d\zeta.$$

(iv) For each ζ_0 in Ω there is a sequence $\{x_n\}$ in X such that

$$f(\zeta) = \sum_{n=0}^{\infty} (\zeta - \zeta_0)^n x_n \qquad \text{(norm convergence)}$$

for all ζ in a neighbourhood of ζ_0.

(i) $\Rightarrow$ (ii) is trivial. To see that (ii) $\Rightarrow$ (iii) take ζ_0 in Ω and let $\{\varepsilon_n\}$ be a sequence in $\boldsymbol{C}$ tending to zero. For each ϕ in X^* the function $\zeta \to \phi(f(\zeta))$ is differentiable by assumption and thus the sequence

$$\{\varepsilon_n^{-1} \phi(f(\zeta_0 + \varepsilon_n) - f(\zeta_0))\}$$

is bounded. From the uniform boundedness principle it follows that the sequence $\{\|\varepsilon_n^{-1}(f(\zeta_0 + \varepsilon_n) - f(\zeta_0))\|\}$ is bounded. In particular, $f(\zeta_0 + \varepsilon_n) \to f(\zeta_0)$ which proves that f is norm continuous. From A3 we conclude that the line integral

$$(2\pi i)^{-1} \oint_{\delta D} (\zeta - \zeta_0)^{-1} f(\zeta) \, d\zeta$$

defines an element in X. Using the Cauchy integral formula for the scalar-valued holomorphic functions $\zeta \to \phi(f(\zeta)), \phi \in X^*$, we see that this element is equal to $f(\zeta_0)$.

(iii) $\Rightarrow$ (iv). For each ζ_0 in Ω let D be a closed disc in Ω with centre ζ_0 and define

$$x_n = (2\pi i)^{-1} \oint_{\delta D} (\eta - \zeta_0)^{-n-1} f(\eta) \, d\eta.$$

The exact same computation as in the scalar-valued case yields

$$\left\| f(\zeta) - \sum_{k=0}^{n} (\zeta - \zeta_0)^k x_k \right\| \leqslant (2\pi)^{-1} \oint_{\delta D} |(\zeta - \zeta_0)(\eta - \zeta_0)^{-1}|^{n+1} \, d\eta$$

for every ζ in the interior of D. Consequently the series $\Sigma(\zeta - \zeta_0)^n x_n$ is norm convergent to $f(\zeta)$. Having the series expansion for f we immediately see that f is infinitely often differentiable in norm with $f^{(n)}(\zeta_0) = n! x_n$. In particular (iv) $\Rightarrow$ (i).

Index